ATOMIC AND LASER SPECTROSCOPY

ALAN CORNEY

ATOMIC AND LASER SPECTROSCOPY

CLARENDON PRESS · OXFORD

1977

Oxford University Press, Walton Street, Oxford OX2 6DP

OXFORD LONDON GLASGOW NEW YORK
TORONTO MELBOURNE WELLINGTON CAPE TOWN
IBADAN NAIROBI DAR ES SALAAM LUSAKA ADDIS ABABA
KUALA LUMPUR SINGAPORE JAKARTA HONG KONG TOKYO
DELHI BOMBAY CALCUTTA MADRAS KARACHI

ISBN 0 19 851138 8

First published 1977
Reprinted with corrections 1978

Printed in Great Britain
by Thomson Litho Ltd
East Kilbride

Preface

It is now over forty years since *Resonance radiation and excited atoms* by A.C.G. Mitchell and M.W. Zemansky, (Cambridge University Press, 1934), first appeared. Since then there have been many advances, and in teaching for the Final Honour School of Physics at Oxford I have often felt the need of an up-to-date account of the progress that has been made in the field of optical physics, particularly during the last quarter of a century. This volume is an attempt to fill that need. The first five chapters of the book prepare the foundations of atomic physics, electromagnetism, and quantum mechanics which are necessary for an understanding of the interaction of electromagnetic radiation with free atoms. The application of these concepts to processes involving the spontaneous emission of radiation is then developed in Chapters 6, 7, and 8 while stimulated transitions and the properties of gas and tunable dye lasers form the subject matter of Chapters 9 to 14. The last four chapters are concerned with the physics and applications of resonance fluorescence, optical double-resonance, optical pumping and atomic beam magnetic resonance experiments. It proved remarkably difficult, once the manuscript had been completed, to find a title which accurately described the content of the book. *Atomic and laser spectroscopy* is intended to indicate the wide range of the material treated within, but it is slightly misleading in the sense that I have been concerned not so much with an explanation of the gross spectra of atoms but rather with demonstrating how the techniques of atomic spectroscopy and lasers have been applied to a wide range of problems in atomic and molecular physics. In a sense therefore this text is complementary to those well-known books *Atomic spectra* by H.G. Kuhn (2nd edn Longmans 1970) and *Elementary atomic structure* by G.K. Woodgate (McGraw-Hill 1970).

This volume is an extended version of a set of lectures on atomic physics given to third year undergraduates at the

University of Oxford. By this stage these undergraduates have already completed introductory courses in electromagnetism, atomic physics, and quantum mechanics, courses which I regard as prerequisite for a full understanding of this book. This text should then provide an excellent basis for an advanced undergraduate course in atomic physics. However, the later sections of many chapters are more suitable for work at the graduate level and for this reason the book should also prove valuable to those engaged in research in the fields of atomic physics, lasers, astrophysics, and physical chemistry. Since this book is intended mainly for experimentalists, I have not attempted to develop the theory of the interaction of atoms and radiation in a rigorous manner but have made use of classical or semi-classical calculations whenever possible. This approach has the advantage of simplicity and moreover it often yields great insight into the physics of the problems under discussion. For those students who wish to progress to a more rigorous development of the theory I strongly recommend *The quantum theory of light* by R. Loudon, (Clarendon Press, Oxford 1973).

Throughout the book the equations are given in SI units although readers who prefer to work in the c.g.s system can make the transposition in most cases by removing factors of $(1/4\pi\varepsilon_0)$ or $(\mu_0/4\pi)$ from the expressions given. The problems at the ends of the chapters form an important part of the book since they are designed to develop a feel for the order of magnitude of the quantities involved and to illustrate and extend the topics discussed in the main text. The bibliography at the end of each chapter includes only those papers which I consider most important for the particular topics under discussion. However, these references, together with the suggestions for further reading, should enable the student to progress easily to the current literature in most fields.

In spite of the length of the book some important topics had to be omitted; these include photodissociation, photoionization, and free-free transitions. Perhaps more important still it was not possible to include a discussion of the non-resonant scattering of light, the refractive index of gases, and the non-linear susceptibility of atomic vapours. Similarly, although tensor operators are briefly mentioned, it was felt that a complete discussion of radiation processes in general and resonance experiments in particular using irreducible tensor operators was beyond the level of this book.

Over the years my understanding of atomic and laser spectroscopy has benefited considerably from discussions with my colleagues and students in the Clarendon Laboratory. I am especially grateful to Professor G.W. Series who first aroused my interest in atomic physics and under whose expert and enthusiastic guidance I performed my first research. I wish to thank Dr C.E. Webb for advice on laser population mechanisms and for his constructive criticism of Chapters 11-13, and Dr G.K. Woodgate who read the complete manuscript. Many of the improvements in the final draft are due to his invaluable assistance. I also want to thank Professor J.N. Dodd and the staff members of the Department of Physics, University of Otago, New Zealand for providing the hospitality and seclusion which enabled me to complete a substantial part of this book. Of my many antipodean friends I am particularly indebted to Drs C.G. Carrington and D.M. Warrington who read the manuscript in an incomplete state and made many helpful suggestions.

Finally I thank my wife not only for her patience and encouragement but also for computer programming and curve plotting, for tracing the line drawings of most of the figures, and for a true labour of love in typing the whole manuscript.

A.C.

Clarendon Laboratory, Oxford.
July 1976.

Contents

cont.....

cont.....

1
Introduction

The purpose of this book is to present a unified account of the exciting developments in the field of quantum electronics which have occurred during the last two decades. Following introductory chapters on electromagnetism and quantum mechanics, the basic processes involving the interaction of radiation and atoms are discussed in some detail. We believe that a thorough understanding of this material is essential for later chapters.

However, in order that the reader shall not lose sight of our ultimate objective, we present here a brief historical introduction showing how our understanding of light and matter has developed since the end of the nineteenth century. It is hoped that this will allow the student to set the material which follows in the proper context and will give him or her an appreciation of the beauty and fascination of the field of optical physics.

1.1. Planck's radiation law

By the end of the nineteenth century the development of the wave theory of light, commenced by Young and Fresnel, seemed to have reached its culmination in the brilliant theoretical work of Maxwell. In another branch of physics the theory of heat leading on to the kinetic theory of gases and to statistical mechanics as developed by Clausius, Boltzmann, Maxwell, and Gibbs also seemed to be complete in most of its essential points. This encouraged Michelson to write in 1899:

'The more important fundamental laws and facts of physical science have all been discovered, and these are so firmly established that the possibility of their ever being supplanted in consequence of new discoveries is exceedingly remote.'

There was, however, an unresolved discrepancy in the theory of black-body radiation. In fact the solution of this apparently minor problem by Planck (1901) was to be the starting point of the quantum theory and of our present understanding of the structure of atoms and molecules.

Following the work of Kirchhoff on the connection between emission and absorption coefficients, it had been proved that the radiation inside a totally enclosed cavity maintained at a uniform temperature was a function of the temperature alone and was identical with the radiation which would be emitted by a perfectly black body at the same temperature. The spectral distribution of the radiation had been investigated experimentally and it was found that the intensity increased slowly with decreasing frequency until it reached a peak, after which it decreased very rapidly (Fig.1.1). However, all attempts to derive an equation giving the intensity as a function of frequency had failed. The most convincing approach made by Rayleigh and Jeans on the basis of classical thermodynamics gave the result

$$\rho(\omega) = \frac{\omega^2}{\pi^2 c^3} kT , \qquad (1.1)$$

where $\rho(\omega)\, d\omega$ represents the energy of radiation per unit volume in the angular frequency range between ω and $\omega+d\omega$. The term kT in equation (1.1) comes from the equipartition of energy applied to a linear oscillator which is assumed to act as the emitter or absorber of radiation of frequency $\omega/2\pi$. We see from equation (1.1) and Fig. 1.1 that the intensity predicted by the Rayleigh-Jeans law increases indefinitely at higher frequencies, the well-known ultraviolet catastrophe, in complete contradiction with the experimental results.

Planck realized that the difficulty lay in the principle of equipartition which could only be circumvented by a complete departure from classical mechanics. He postulated that an oscillator of frequency $\omega/2\pi$, instead of being able to assume all possible energy values, could exist only in one of a set of equally spaced energy levels having the values 0, $\hbar\omega$, $2\hbar\omega, \ldots$, $m\hbar\omega$, $\ldots$. m is an integer and $\hbar = h/2\pi$, where h is a constant now known as Planck's constant.

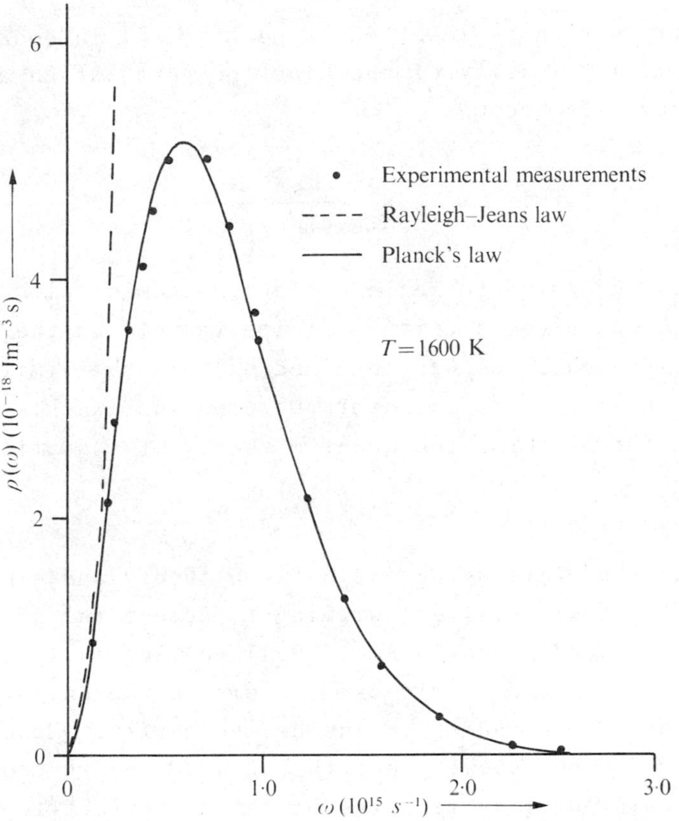

Fig.1.1. Energy density of black-body radiation as a function of the angular frequency ω

The unit of energy $\hbar\omega$, which cannot be further subdivided, Planck called a *quantum*. If we now assume that the probability of finding an oscillator with the energy $m\hbar\omega$ is given by the Boltzmann factor $\exp(-m\hbar\omega/kT)$, then the average energy of an oscillator is given by

$$\bar{\varepsilon} = \frac{\sum_{m=0}^{\infty} m\hbar\omega \exp(-m\hbar\omega/kT)}{\sum_{m=0}^{\infty} \exp(-m\hbar\omega/kT)}$$

$$= \frac{\hbar\omega}{\exp(\hbar\omega/kT)-1} \quad . \tag{1.2}$$

When this expression is inserted in place of kT in equation (1.1), the energy density of the black-body radiation at the frequency $\omega/2\pi$ becomes

$$\rho(\omega) = \frac{\omega^2}{\pi^2 c^3} \frac{\hbar\omega}{\exp(\hbar\omega/kT)-1} . \tag{1.3}$$

This expression proved to be in perfect agreement with experiment when $\hbar$ was given a certain finite value, and the classical Rayleigh-Jeans law was obtained only in the limit as $\hbar \to 0$. Thus the turn of the century ushered in Planck's law and with it the birth of the quantum theory of radiation.

1.2. *The photoelectric effect*

In spite of this agreement between theory and experiment, not all physicists were willing to accept the quantum hypothesis. Indeed Planck himself believed initially that it applied only to the atomic oscillators in the cavity walls and not to the radiation inside the cavity. However, Einstein (1905) soon showed that the quantum theory could explain certain puzzling features of the photoelectric effect in a very simple way.

The emission of electrons from the surface of a metal which is illuminated by ultraviolet light was discovered by Hertz in 1887. Later experiments by Lenard showed that when the intensity of the incident light was changed, without altering its spectral distribution, the kinetic energy of the ejected electrons remained constant and only the number emitted varied. We now know that at very low intensities the rate of emission can become so small that only a few electrons per second are ejected. In this limit, it appears as if all the energy in the light wave falling on the surface would have to be concentrated on a single electron in order to give it the observed kinetic energy. This seems to be at variance with classical electrodynamics, for the energy in a light wave is usually assumed to be distributed uniformly across the wavefront.

To resolve this difficulty, Einstein made the bold hypothesis that the energy in the radiation field actually existed as discrete quanta, now called *photons*, each having an energy of $\hbar\omega$ and that in interactions between light and matter this energy is essentially localized at one electron. In the photoelectric effect an electron at the surface of the metal gains an energy $\hbar\omega$ by the absorption of a photon and the maximum kinetic energy of the ejected electron is observed to be

$$\frac{1}{2}\,mv^2_{max} = \hbar\omega - e\phi \tag{1.4}$$

where ϕ is the work function of the surface measured in electron volts. According to this theory the rate of emission of photoelectrons will be proportional to the flux of quanta incident upon the surface, in agreement with Lenard's experiments. Moreover Einstein's theory predicts that the maximum kinetic energy of the photoelectrons should be a linear function of the frequency of the incident light. This was later confirmed experimentally by Millikan (1916), and led to a value of $\hbar$ which agreed to within 0.5 per cent with the value derived by fitting equation (1.3) to the black-body spectrum. With this clarification of the quantum aspects of the interaction of light and matter, the stage was then set for further advances in the field of atomic spectroscopy.

1.3. *Early atomic spectroscopy*

The spectroscopy of gases had been studied since the middle of the nineteenth century, first with prism spectroscopes and later with spectrographs using diffraction gratings. Atoms excited in gas discharges were found to emit spectra consisting of many sharp lines of definite frequency, in contrast to the continuous spectra emitted by black-body sources or by solids at high temperatures. Although most atoms had very complicated spectra, that of hydrogen was relatively simple and in 1885 Balmer discovered an empirical formula which fitted the wavelengths of the visible lines to surprisingly high accuracy. For many purposes it is more convenient to work with the wavenumbers, $\tilde{\nu}$, which are related to the wavelengths *in vacuo*, λ_{vac}, by $\tilde{\nu}=1/\lambda_{vac}$. The Balmer formula for hydrogen then becomes

$$\tilde{\nu} = R_H\left(\frac{1}{2^2} - \frac{1}{n^2}\right) \qquad (1.5)$$

where n is an integer taking the values 3, 4, 5, ... and R_H is a constant having the value 109 677·58 cm^{-1}. Further work by Rydberg in the 1890s showed that the lines of the alkali metals could be classified into a number of spectral series, each of which was described by a similar formula

$$\tilde{\nu} = \tilde{\nu}_\infty - \frac{R}{(n-d)^2} \qquad (1.6)$$

where n is again an integer and d is a constant. The Rydberg constant, R, was shown to have the same value, to within 0·05 per cent, for all elements and for the first time indicated a common link between the spectra of different chemical elements.

In both the Balmer and the Rydberg formulae the wavenumber of a spectral line is given by the difference of two quantities. In 1908 Ritz showed experimentally that in any spectrum it was possible to set up tables of quantities called terms, having dimensions of cm^{-1}, such that the wavenumbers of the observed spectral lines could be written as the difference of two terms. This is known as the Ritz combination principle. In hydrogen, new spectral series were predicted with lines given by

$$\tilde{\nu} = R_H\left(\frac{1}{m^2} - \frac{1}{n^2}\right)$$

where m and n are integers and m is fixed for a given series. When, in the same year, Paschen discovered one of these new series in the infrared region, the principle received striking confirmation.

These spectral lines at precisely related frequencies seemed to be in violent disagreement with the picture of the atom proposed by Rutherford (1911) to explain experiments on the scattering of α-particles. In this model, described more clearly by Bohr (1913), the atom was assumed to consist of a massive positively-charged nucleus, radius $\approx 10^{-12}$ cm, which was surrounded by a cloud of negatively-charged electrons moving in orbits of radius $\approx 10^{-8}$ cm. But according to classical mechanics, an electron moving in the Coulomb field of the nucleus would radiate electromagnetic energy at the frequency of its orbital motion. As the electron lost energy it would slowly spiral into the nucleus, emitting radiation of continuously increasing frequency. Obviously this did not happen in real atoms for they are observed to be stable in their normal state and to radiate only certain discrete frequencies when excited. The first major step forward in the solution of this problem was taken by Bohr (1913).

1.4. The postulates of Bohr's theory of atomic structure

In order to develop a quantitative theory of the hydrogen atom, Bohr put forward three basic postulates.

Postulate 1. A bound atomic or molecular system can exist only in certain discrete energy levels denoted by the values $E_1, E_2, \ldots, E_i, \ldots$. These energies are determined by superimposing certain quantum conditions (Postulate 3) on to classical mechanics.

Postulate 2. In the absence of any interaction with radiation, the atomic system will continue to exist permanently in one of the allowed energy levels which are therefore called stationary states. The emission or absorption of radiation is connected with a process in which the atom passes from one stationary state to another. In emission, the energy lost by the atomic system appears as energy in the radiation field and the frequency of the light emitted is given by

$$\hbar\omega_{ji} = E_j - E_i \tag{1.7}$$

where $\hbar$ is Planck's constant divided by 2π. The energy in the radiation field appears in the form of a photon and, following Einstein's hypothesis, has many of the characteristics of a material particle. In absorption, a photon is removed from the radiation field and the atom makes a transition from a lower to a higher energy level.

These two postulates were confirmed in the following year by the experiments of Franck and Hertz (1914) on the critical potentials of atoms. They found that when atoms were bombarded with a monoenergetic beam of electrons, only elastic collisions occurred at low energies. As the accelerating voltage is increased, a point is reached at which the electrons have sufficient kinetic energy to raise the atoms to the lowest excited energy level. Inelastic collisions occur and the atoms start to emit light in the transition connecting the lowest excited state to the ground level. As the electron energy is increased again, further excitation thresholds are reached and the spectrum of the atom appears line by line. Finally the bombarding electrons have sufficient energy to release an electron from the atom, so ionizing the gas. These experiments provide a direct, although rather imprecise, measurement of the energy levels and ionization potentials of atoms.

These two postulates also explain the existence of spectral terms and the Ritz combination principle. We see from equation (1.7) that the wavenumbers of spectral lines are given by differences of quantities of the form E_i/hc. These must therefore be the spectral terms. Thus the work of the spectroscopists in deducing the terms from the observed spectra immediately provides one with a table of atomic energy levels. Following this important clarification and the later development of selection rules for radiative transitions, the classification and analysis of atomic and molecular spectra proceeded rapidly. Energy-level tables are now available for the majority of atoms, ions, and simple molecules, although the analyses of the complex spectra of the rare earth metals and of highly-ionized atoms are still often incomplete.

Postulate 3. For the hydrogen atom Bohr assumed that the electron moved in circular orbits about the nucleus under the influence of the Coulomb attraction. The allowed circular orbits were then determined simply by the requirement that the angular momentum of the electron should be an integral multiple of $\hbar$. This restricts the energies of the hydrogen atom, in states of angular momentum $n\hbar$, to the quantized values

$$E_n = -\frac{1}{(4\pi\varepsilon_0)^2}\frac{\mu e^4}{2\hbar^2}\frac{1}{n^2}, \tag{1.8}$$

where ε_0 is the permittivity of free space and μ is the reduced mass of the electron, given by $\mu = m/(1+m/M)$ where m and M are the masses of the electron and proton respectively. Not only do the predicted energy levels have the $1/n^2$ dependence, in agreement with the formulae for the wavenumbers of the lines of the Lyman, Balmer, and Paschen series in hydrogen, but the empirical Rydberg constant is, for the first time, given in terms of atomic constants:

$$R_H = \frac{\mu e^4}{(4\pi\varepsilon_0)^2\, 4\pi\hbar^3 c} \tag{1.9}$$

When this expression is evaluated, it agrees with the spectroscopic value within the combined errors of the various constants involved.

1.5. *Development of quantum mechanics*

This tremendous accomplishment of Bohr stimulated other theoreticians, and soon Sommerfeld (1916) had generalized Bohr's quantum conditions and applied the results to elliptical orbits in the hydrogen atom. By including relativistic effects he was able to interpret many of the details of the hydrogen fine structure. The introduction of an additional quantum number describing the allowed orientations of the orbits in space also gave a partial explanation of the observed Zeeman effect. However, one of the most important theoretical developments was Bohr's own correspondence principle. This states that in the limit of large quantum numbers, the results of quantum theory must be in agreement with the physical laws deduced from a combination of classical mechanics and electromagnetism. When this principle was combined with the relations connecting the transition probabilities for spontaneous and stimulated transitions derived by Einstein (1917), many of the results of the modern theory of radiation were obtained.

However, the old quantum theory was incomplete. It gave no definite basis for the calculation of the relative or absolute intensities of spectral lines. It also failed to give satisfactory results when attempts were made to calculate the energy levels of atoms containing more than one electron. Even in an apparently simple atom like helium, numerous attempts to calculate the energy of the ground state failed to reach agreement with the experimentally determined value.

The starting point of a new quantum theory of atoms was the hypothesis put forward by de Broglie (1924) that a particle moving with linear momentum p has associated with it a wave whose wavelength is given by

$$\lambda = h/p. \tag{1.10}$$

This idea was improved and extended by Schrödinger (1926) and led to the development of the equation which now bears his name. Almost simultaneously Heisenberg (1925) developed a theory of matrix mechanics which was soon shown to be mathematically equivalent to Schrödinger's wave mechanics. The development of the quantum theory of atomic structure and a review of some of the important results are continued in Chapter 3. We therefore now turn our attention to the development of the quantum theory of radiation processes.

1.6. *Interaction of atoms and radiation 1926-39*

In the period between 1926 and 1939 the development of our understanding of radiation processes parallels the development of the theory of atomic and molecular structure. Many of the results which had been derived by the use of the correspondence principle and the old quantum theory were re-derived in a more rigorous and satisfactory manner. The quantum-mechanical expression for the refractive index of a gas or vapour is an example of this type of progress. In hydrogenic systems the rates of radiative transitions were calculated and the theoretical lifetimes of the different excited levels were derived. In other atoms only approximate estimates of the transition probabilities were possible, but the lifetime measurements made by canal rays and fluorescence from atomic beams were not sufficiently accurate to demand more refined calculations.

Thus by 1939, it appeared once again that the understanding of atoms and light was almost complete, although it is true that in many cases satisfactory agreement between theory and experiment was only reached with the use of the more sophisticated techniques developed after the end of the second world war. A relativistically-invariant theory of one-electron atoms which included the effects of electron spin in a consistent manner had been developed by Dirac. This appeared to be the final solution to the problem of the hydrogen atom. However, it should be remembered that a few experimentalists were not entirely happy and questioned whether the Dirac theory of the fine structure was completely correct. There was therefore a need to develop new experimental techniques which would increase the accuracy of fine and hyperfine-structure measurements.

1.7. *Optical physics since 1945*

The work on radar and radio communications carried out during the second world war proved to be of immense assistance to one of the new spectroscopic techniques. This was the magnetic resonance method first applied by Rabi *et al.* (1939) to atomic beams, and extended after the war to liquids and solids by Bloch, Purcell, and their colleagues (Bloch *et al.* 1946; Purcell *et al.* 1946). The first major shock to atomic physics came, however, when Lamb and Retherford (1947) applied this new technique to metastable hydrogen atoms and showed that the $2^2S_{\frac{1}{2}}$ and $2^2P_{\frac{1}{2}}$ levels of atomic hydrogen were not degenerate as predicted by the Dirac theory. This experiment stimulated the development of the theory of quantum electrodynamics and led to the precise calculation of radiative corrections to atomic energy levels. Further applications of the magnetic resonance method were opened up by the invention of the optical double resonance and optical pumping techniques by Kastler and his pupils in Paris (Brossel and Kastler (1949), Kastler (1950)). The extraordinary precision of these experiments has enabled a detailed study of the interactions of atoms with light and radio waves to be made, and has revealed subtle new effects which were previously hidden or unsuspected.

More recently the invention of microwave masers and of lasers operating at optical frequencies has transformed the field of optical physics in a remarkable way. In solid state physics, Franken's demonstration of the frequency doubling of ruby laser light immediately opened up a new field of non-linear optics. In atomic and molecular physics the applications of laser radiation were initially less rapid and spectacular. However, following the development of narrow-bandwidth tunable dye lasers in 1970, the interest in this field of research has experienced an explosive growth which as yet shows no signs of slowing down.

1.8. The present situation (1975)

We now believe that quantum mechanics, extended where necessary by quantum electrodynamics, is able to explain satisfactorily all features of the energy-level structure of atoms and molecules. Precise and detailed calculations of the energy levels of many atoms have now been made, and there is generally excellent agreement with the empirical data obtained by spectroscopists. Similarly, we now believe that we have a good understanding of the processes involving the interaction of atoms with light, at least for experiments involving incoherent light or coherent radiation of low intensity. However, the effects of coherent, high intensity, radio- and optical-frequency radiation on atoms and molecules are at present only partially understood. Thus a great deal of experimental and theoretical work on the nonlinear optics and spectroscopy of gases is currently in progress.

It is therefore safe only to say that this book is an attempt to present the state of our knowledge of the interaction of atoms and radiation as it exists now. In certain areas, theory and experiment dove-tail so neatly together that these processes must be considered to be well understood. In other areas, disagreement exists or work is still in progress and we should certainly be prepared for further surprises and unexpected developments in the future.

Problems

1.1. By considering the electromagnetic field confined within a large cubical box having perfectly reflecting sides, show that the number of modes of the radiation field per unit volume in the angular frequency range between ω and $\omega+d\omega$ is $\omega^2 d\omega/\pi^2 c^3$.

1.2. (a) From Planck's expression, equation (1.3), show that the angular frequency at which the energy density of black-body radiation is a maximum at a fixed temperature is proportional to the temperature. This is Wien's displacement law.

(b) By integrating equation (1.3) over all frequencies, show that the total energy density of black-body radiation is proportional to T^4 and that the constant of proportionality is given by

$$\sigma' = \frac{4\sigma}{c} = \frac{\pi^2 k^4}{15c^3\hbar^3} = 7{\cdot}565 \times 10^{-16}\ \mathrm{J\ m^{-3}\ K^{-4}}$$

This is the Stefan-Boltzmann law. $\int_0^\infty \frac{x^3 dx}{e^x - 1} = \frac{\pi^4}{15}$

1.3. A classical model of a one-electron atom consists of a positive charge of amount +e uniformly distributed throughout the volume of a sphere, radius R, together with a point electron of charge -e which is free to move within the sphere. Show that the electron will oscillate about its equilibrium position with simple harmonic motion and find the frequency of oscillation, $\omega_0/2\pi$.

If the atomic radius R is equal to the radius of the first Bohr orbit, $a_0 = 4\pi\varepsilon_0\hbar^2/me^2$, show that ω_0 is twice the Rydberg angular frequency,

$$\omega_R = \frac{1}{2}\left(\frac{e^2}{4\pi\varepsilon_0 m a_0^3}\right)^{1/2}$$

1.4. In air, Ångström determined the wavelengths of the spectral lines of the visible series of atomic hydrogen as

H_α	6562·10 Å
H_β	4860·74 Å
H_γ	4340·10 Å
H_δ	4101·20 Å

Show that these lines are accurately described by the Balmer formula and from them calculate a value for the Rydberg constant, R_H. The refractive index of air in the visible region of the spectrum may be taken as 1·00028 at standard temperature and pressure.

By replacing the factor $1/2^2$ in equation (1.5) by $1/1^2$ and $1/3^2$ in turn, calculate the wavelengths *in vacuo* of the first four lines of the Lyman and Paschen series respectively.

1.5. Using the Bohr formula, equation (1.8), and the modern value of the Rydberg constant,

$R_\infty = 109\ 737\ cm^{-1} \equiv 13{\cdot}6$ eV, construct energy-level diagrams for atomic hydrogen and singly-ionized helium to the same scale.

1.6. Show that as $n \to \infty$, the frequency of radiation emitted by a hydrogen atom in the transition from a level of principal quantum number n to one of quantum number n-1 tends to the frequency of the orbital motion of the electron as deduced from classical mechanics.

This is an illustration of Bohr's Correspondence Principle.

References

Bloch, F., Hansen, W.W., and Packard, M. (1946). *Phys.Rev.* 69, 127.

Bohr, N. (1913). *Phil. Mag.* 26, 476.

Brossel, J., and Kastler, A. (1949). *C.R. hebd.Séanc.Acad. Sci., Paris* 229, 1213.

de Broglie, L. (1924). *Phil. Mag.* 47, 446.

Einstein, A. (1905). *Annln.Phys.* 17, 132.

____________ (1917). *Phys. Z.* 18, 121.

Franck, J., and Hertz, G. (1914). *Verh.dt.phys.Ges.* 16, 457 and 512.

Heisenberg, W. (1925). *Z.Phys.* 33, 879.

Kastler, A. (1950). *J.Phys., Paris* 11, 255.

Lamb, W.E., and Retherford, R.C. (1947). *Phys.Rev.* 72, 241.

Millikan, R.A. (1916). *Phys.Rev.* 7, 355.

Planck, M. (1901). *Annln.Phys.* 4, 553.

Purcell, E.M., Torrey, H.C., and Pound, R.V. (1946). *Phys. Rev.* 69, 37.

Rabi, I., Millman, S., Kusch, P., and Zacharias, J.R. (1939), *Phys.Rev.* 55, 526.

Rutherford, E. (1911), *Phil.Mag.* 21, 669.

Schrödinger, E. (1926), *Annln.Phys.* 79, 361, 489, 734.

Sommerfeld, A. (1916), *Annln.Phys.* 51, 1.

General references and further reading

ter Haar, D. (1967). *The old quantum theory*, Pergamon Press, Oxford.

Contains translations and commentaries on several of the important articles listed above.

Slater, J.C. (1960). *Quantum theory of atomic structure*, Vol. 1. McGraw-Hill, New York.

Chapters 1 and 2 give an account of the historical development of modern physics and the references suggested on p.47 are particularly useful.

A comprehensive monograph on this subject has been written by

Whittaker, E.T. (1954). *A history of theories of aether and electricity*, (*The modern theories, 1900-1926*). Philosophical Library, New York.

2

Review of classical electrodynamics

This chapter gives a concise account of those sections of classical electromagnetism which we shall require later. We commence by introducing Maxwell's equations for time-dependent electric and magnetic fields and from them derive the electromagnetic wave equation. The properties and polarization of plane electromagnetic waves are examined and expressions for the energy flux and energy density obtained. We then introduce the scalar and vector potentials and discuss the radiation of electromagnetic waves by time-varying distributions of current and charge. This enables explicit expressions for the fields produced by oscillating electric and magnetic dipoles to be obtained. We derive expressions for the rate at which these dipoles radiate energy and angular momentum which will later be fundamental to our treatment of the spontaneous emission of radiation by excited atoms. Finally, we consider the radiation fields of electric quadrupole charge distributions and show how the angular distribution and rate of radiation differ from those of dipole sources.

2.1. Maxwell's equations

We now recognize that many apparently unrelated physical phenomena such as γ-rays, X-rays, visible light, thermal radiation, and radio waves are all forms of electromagnetic radiation, having frequencies ranging from 10^{20} Hz for γ-rays to 10^4 Hz for radio waves. The idea that electromagnetic waves could be propagated through space was first suggested by Faraday and was later confirmed by the brilliant theoretical work of Maxwell published in 1864. Maxwell started from the basic laws of electricity and magnetism for macroscopic phenomena, which, when stated in their differential form in the rationalized M.K.S. system are as follows:

(a) Gauss's theorem applied to Coulomb's Law:

$$\operatorname{div} \underline{D} = \rho \tag{2.1}$$

(b) the absence of free magnetic poles:

$$\operatorname{div} \underline{B} = 0 \tag{2.2}$$

(c) Faraday's and Lenz's law of electromagnetic induction

$$\operatorname{curl} \underline{E} = -\frac{\partial \underline{B}}{\partial t} \tag{2.3}$$

(d) Ampere's law:

$$\operatorname{curl} \underline{H} = \underline{J}'. \tag{2.4}$$

In these equations $\underline{D}$ is the electric displacement and $\underline{B}$ the magnetic induction, while $\underline{E}$ and $\underline{H}$ are the electric and magnetic fields. The electric charge density is given by ρ.

Maxwell observed that the equations are inconsistent in this present form if $\underline{J}'$ is identified with the ordinary current density $\underline{J}$. This is easily proved by taking the divergence of both sides of equation (2.4) and recalling that div curl of any vector is identically zero. This leads to the equation $\operatorname{div} \underline{J}' = 0$ and if $\underline{J}' = \underline{J}$, this would conflict with the equation of continuity

$$\operatorname{div} \underline{J} + \frac{\partial \rho}{\partial t} = 0, \tag{2.5}$$

at least in time-dependent situations. Maxwell realized that the continuity equation could be converted into a vanishing divergence by substituting for ρ from equation (2.1), giving

$$\operatorname{div} \underline{J} + \frac{\partial}{\partial t} (\operatorname{div} \underline{D}) = \operatorname{div}(\underline{J} + \frac{\partial \underline{D}}{\partial t}) = 0. \tag{2.6}$$

Hence if $\underline{J}'$ is defined by

$$\underline{J}' = \underline{J} + \frac{\partial \underline{D}}{\partial t} \tag{2.7}$$

the inconsistency in the equations for time-dependent situations is removed. The additional term in equation (2.7), $\frac{\partial \underline{D}}{\partial t}$, is called the displacement current. Since $\underline{D} = \underline{P} + \varepsilon_0\underline{E}$, the displacement current can be regarded as arising from the changing polarization of the medium and the vacuum under the effect of time-varying electric fields.

The four equations:

$$\text{div}\ \underline{D} = \nabla.\underline{D} = \rho \tag{2.8}$$

$$\text{div}\ \underline{B} = \nabla.\underline{B} = 0 \tag{2.9}$$

$$\text{curl}\ \underline{E} = \nabla\wedge\underline{E} = -\frac{\partial \underline{B}}{\partial t} \tag{2.10}$$

$$\text{curl}\ \underline{H} = \nabla\wedge\underline{H} = \underline{J} + \frac{\partial \underline{D}}{\partial t} \tag{2.11}$$

are known as Maxwell's equations. When combined with Newton's second law of motion and the Lorentz expression for the force on a particle of charge q moving with velocity $\underline{v}$,

$$\underline{F} = q(\underline{E} + \underline{v}\wedge\underline{B}),$$

they give a complete description of electromagnetic fields and the non-relativistic motion of charged particles.

In macroscopic media Maxwell's equations must be supplemented by the constitutive relations which enable the average effect of a large number of atoms to be taken into account without requiring detailed knowledge of the response of individual atoms to the effects of electric and magnetic fields. For an isotropic permeable conducting dielectric these relations are of the form:

$$\underline{D} = \varepsilon\varepsilon_0\underline{E} = \varepsilon_0(1 + \chi_e)\underline{E} \tag{2.12}$$

$$\underline{B} = \mu\mu_0\underline{H} = \mu_0(1 + \chi_m)\underline{H} \tag{2.13}$$

$$\underline{J} = \sigma\underline{E}. \tag{2.14}$$

In the M.K.S. system, ε_0 is called the permittivity of free space and is experimentally determined to have the value

$8{\cdot}85 \times 10^{-12}$ C^2 N^{-1} m^{-2} while μ_0 is the permeability of free space and is defined as having the exact value $4\pi \times 10^{-7}$ H m^{-1}. The properties of the medium are given either by the dielectric constant, ε, and the magnetic permeability, μ, or by the electric and magnetic susceptibilities represented by χ_e and χ_m respectively. The third relation, equation (2.14), is just a generalization of Ohm's law and σ is the specific electrical conductivity of the medium. In the following sections we assume that the susceptibilities χ_e and χ_m are independent of the strengths of the fields $\underline{E}$ and $\underline{H}$. We therefore ignore any non-linear effects such as optical second- and third-harmonic generation, optical sum and difference frequency generation, etc., which have recently been observed using focused beams of high-intensity laser radiation.

2.2. *The electromagnetic wave equations*

For the case of an uncharged infinite, isotropic, non-conducting medium, substitution of the constitutive relations into Maxwell's equations yields

$$\text{div}\ \underline{E} = 0 \tag{2.15}$$

$$\text{div}\ \underline{H} = 0 \tag{2.16}$$

$$\text{curl}\ \underline{E} = -\mu\mu_0 \frac{\partial H}{\partial t} \tag{2.17}$$

$$\text{curl}\ \underline{H} = \varepsilon\varepsilon_0 \frac{\partial E}{\partial t}\ . \tag{2.18}$$

By taking the curl of equations (2.17) and (2.18) we can transform these first-order differential equations which involve both $\underline{E}$ and $\underline{H}$ into second-order equations for either $\underline{E}$ or $\underline{H}$ alone. We make use of the vector identity

$$\text{curl curl} \equiv \text{grad div} - \nabla^2$$

and note that the first term on the right will be zero in both cases because of equations (2.15) and (2.16). We have therefore

$$\text{curl curl } \underline{E} = -\nabla^2 \underline{E} = -\mu\mu_0 \frac{\partial}{\partial t} (\text{curl } \underline{H})$$

and

$$\text{curl curl } \underline{H} = -\nabla^2 \underline{H} = \varepsilon\varepsilon_0 \frac{\partial}{\partial t} (\text{curl } \underline{E}).$$

Finally by using equations (2.17) and (2.18) again, we obtain

$$\left.\begin{aligned} \nabla^2 \underline{E} &= \mu\mu_0\varepsilon\varepsilon_0 \frac{\partial^2 \underline{E}}{\partial t^2} \\ \nabla^2 \underline{H} &= \mu\mu_0\varepsilon\varepsilon_0 \frac{\partial^2 \underline{H}}{\partial t^2} . \end{aligned}\right\} \quad (2.19)$$

In order to present these equations in a more familiar form, we may write the electric field as $\underline{E} = E_x \hat{\underline{\imath}} + E_y \hat{\underline{\jmath}} + E_z \hat{\underline{k}}$. We see then that each cartesian component of the fields $\underline{E}$ and $\underline{H}$ satisfies the equation

$$\nabla^2 W = \frac{\partial^2 W}{\partial x^2} + \frac{\partial^2 W}{\partial y^2} + \frac{\partial^2 W}{\partial z^2} = \frac{1}{v^2} \frac{\partial^2 W}{\partial t^2} , \quad (2.20)$$

which is the equation governing wave motion in three dimensions. Maxwell's equations therefore lead directly to the prediction of wave-like solutions for $\underline{E}$ and $\underline{H}$ which propagate with the phase velocity, v, given by

$$v = (\mu\mu_0\varepsilon\varepsilon_0)^{-1/2} . \quad (2.21)$$

It is important to note that equations (2.17) and (2.18) predict that every electric wave will be accompanied by an associated magnetic wave, and vice versa. The waves are therefore more properly termed electromagnetic waves. In 1888, following an experimental search for the magnetic effects of Maxwell's displacement current, Hertz discovered waves which had exactly this electromagnetic character and possessed all the other properties which can be predicted from Maxwell's equations. Our present sophisticated radio and telecommunication systems have all developed from these first primitive experiments.

In free space, where $\mu = \varepsilon = 1$, these electromagnetic waves travel with a velocity given by

$$c = (\mu_0\varepsilon_0)^{-1/2} = 299\ 784 \pm 10\ \text{km s}^{-1}, \tag{2.22}$$

where the most accurate value of ε_0 has been used to evaluate equation (2.22). Within the experimental error, this value is identical with the velocity of light

$$c = 299\ 792\ 457{\cdot}4 \pm 1{\cdot}1\ \text{m s}^{-1},$$

where we have quoted the result of the most accurate of a series of recent measurements of this important fundamental quantity. The inescapable conclusion is that visible light is another example of electromagnetic radiation.

In material media the phase velocity is given by

$$v = c/(\mu\varepsilon)^{1/2} = c/n \tag{2.23}$$

where $n = (\mu\varepsilon)^{1/2}$ is the refractive index of the medium. In general, the dielectric constant is frequency dependent, $\varepsilon(\omega)$, and waves of different frequencies propagate with different phase velocities. Consequently a wavepacket made up of waves of different frequencies will change its shape as it travels through the medium. In addition, the centre of the packet will travel at the group velocity $v_g \neq v$ and the medium is said to be dispersive. However, since the Fourier integral theorem enables any disturbance to be represented by a superposition of monochromatic waves, there is no loss of generality if we consider from now on only sinusoidal oscillations at the angular frequency ω.

2.3. *Plane wave solutions*

We can show by direct substitution that the scalar wave equation (2.20) has solutions of the form (Problem 2.1)

$$W = W_0 \exp\{i(\underline{k}.\underline{r} - \omega t)\} \tag{2.24}$$

where the angular frequency ω and the magnitude k of the wavevector $\underline{k}$ are related to the phase velocity by

$$v = \omega/k. \tag{2.25}$$

In terms of the wavelength λ we have $k = 2\pi/\lambda$. At any instant t, the amplitude of the wave is constant over the surface

$$\underline{k}.\underline{r} = k_x x + k_y y + k_z z = \text{constant}. \qquad (2.26)$$

This is the equation of a plane normal to the vector $\underline{k}$ and the solution (2.24) therefore represents a plane wavefront travelling with velocity v in the direction $\underline{k}$ (Problem 2.2).

However, the electric and magnetic fields are vector quantities and we therefore assume that the plane-wave solutions of equations (2.19) are of the form:

$$\underline{E}(\underline{r},t) = \underline{\varepsilon}_1 \mathcal{E}_0 \exp\{i(\underline{k}.\underline{r} - \omega t)\} \qquad (2.27)$$

$$\underline{H}(\underline{r},t) = \underline{\varepsilon}_2 \mathcal{H}_0 \exp\{i(\underline{k}.\underline{r} - \omega t)\} \qquad (2.28)$$

where $\underline{\varepsilon}_1$, $\underline{\varepsilon}_2$ are unit vectors which we assume initially to be in arbitrary directions and $\mathcal{E}_0$, $\mathcal{H}_0$ are complex amplitudes. We adopt the convention that the physical fields are obtained by taking the real part of the complex quantities. This notation is very useful since all operations which are linear in the fields can be performed directly on the complex expressions of equations (2.27) and (2.28) and the real part is taken at the end of the calculation. Unfortunately expressions involving products of the fields cannot be treated in this simple way, as we explain in section 2.5 below.

By substituting equations (2.27) and (2.28) into the equations div $\underline{E} = 0$ and div $\underline{H} = 0$ we find that

$$\underline{\varepsilon}_1.\underline{k} = 0, \quad \underline{\varepsilon}_2.\underline{k} = 0 . \qquad (2.29)$$

This means that both $\underline{E}$ and $\underline{H}$ are perpendicular to the direction of propagation $\underline{k}$, and these solutions to Maxwell's equations therefore describe transverse waves. Important relations connecting the magnitudes and directions of $\underline{E}$ and $\underline{H}$ are obtained by the application of equation (2.17), giving

$$\{(\underline{k}\wedge\underline{\varepsilon}_1)\, \mathcal{E}_0 - \underline{\varepsilon}_2\, \mu\mu_0\omega\, \mathcal{H}_0\} = 0 . \qquad (2.30)$$

Since the vector part of equation (2.30) is satisfied if

$$\underline{\varepsilon}_2 = (\underline{k}\wedge\underline{\varepsilon}_1)/k \tag{2.31}$$

we find that ($\underline{\varepsilon}_1$, $\underline{\varepsilon}_2$, $\underline{k}$) form a right-handed orthogonal set of vectors, as illustrated in Fig.2.1. The scalar part of equation (2.30) shows that the fields $\underline{E}$ and $\underline{H}$ have the same phase and that the amplitudes have a constant ratio, Z, given by

$$Z = \mathcal{E}_0/\mathcal{H}_0 = \mu\mu_0\omega/k = \mu\mu_0 v = \left(\frac{\mu\mu_0}{\varepsilon\varepsilon_0}\right)^{1/2}. \tag{2.32}$$

This ratio has the dimensions of resistance and is called the intrinsic impedance of the medium. For free space the intrinsic impedance, Z_0, is given by

$$Z_0 = \mu_0 c = (\mu_0/\varepsilon_0)^{\frac{1}{2}} = 1/\varepsilon_0 c \approx 376{\cdot}7\Omega.$$

Thus once one of the electromagnetic fields is specified, the direction and magnitude of the accompanying field is uniquely determined by equations (2.31) and (2.32).

2.4. *Linear and circular polarizations*

The electromagnetic plane wave given by equations (2.27) and (2.28) has its electric vector pointing in the constant direction $\underline{\varepsilon}_1$ and is said to be linearly polarized with polarization vector $\underline{\varepsilon}_1$. To describe a general state of polarization for the transverse electromagnetic wave, we need to specify another wave which is linearly polarized in a direction orthogonal to the first. We choose the two linearly independent solutions

$$\left.\begin{aligned} \underline{E}_1 &= \underline{\varepsilon}_1\,\mathcal{E}_1 \exp\{i(\underline{k}.\underline{r} - \omega t)\} \\ \underline{E}_2 &= \underline{\varepsilon}_2\,\mathcal{E}_2 \exp\{i(\underline{k}.\underline{r} - \omega t)\} \end{aligned}\right\} \tag{2.33}$$

with the magnetic fields given by

$$\underline{H}_j = Z^{-1}\left(\frac{\underline{k}\wedge\underline{E}_j}{k}\right), \; j = 1,2.$$

The most general solution for a plane-wave propagating in the direction $\underline{k}$ is given by the linear combination

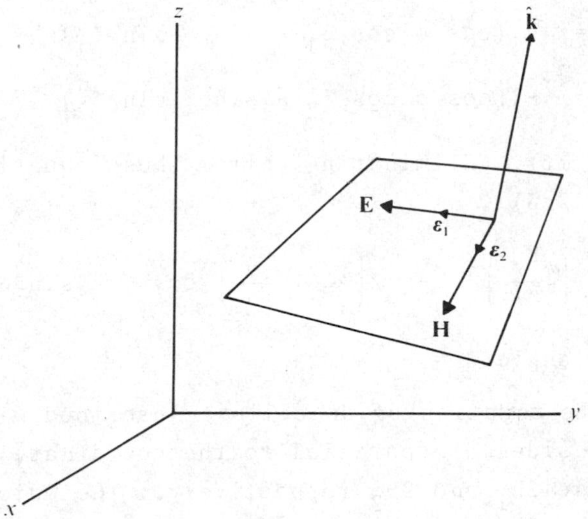

Fig.2.1. The electric and magnetic fields of a plane wave are perpendicular to the direction of propagation $\hat{\underline{k}}$, and $(\underline{E},\underline{H},\hat{\underline{k}})$ form an orthogonal, right-handed set of vectors.

$$\underline{E}(\underline{r},t) = (\underline{\varepsilon}_1 \mathcal{E}_1 + \underline{\varepsilon}_2 \mathcal{E}_2) \exp\{i(\underline{k}.\underline{r} - \omega t)\}. \qquad (2.34)$$

When $\mathcal{E}_1$ and $\mathcal{E}_2$ have the same phase, equation (2.34) represents a linearly-polarized wave with its polarization vector making an angle $\theta = \tan^{-1}(\mathcal{E}_2/\mathcal{E}_1)$ with the direction $\underline{\varepsilon}_1$ and having magnitude $E = (|\mathcal{E}_1|^2 + |\mathcal{E}_2|^2)^{1/2}$.

When $\mathcal{E}_1$ and $\mathcal{E}_2$ have different phases, the tip of the electric vector traces out an ellipse in any fixed plane perpendicular to $\underline{k}$ and the wave is said to be elliptically polarized. This general case is most easily discussed if we choose a coordinate system in which the z-axis coincides with the direction of propagation $\underline{k}$ and $\underline{\varepsilon}_1$, $\underline{\varepsilon}_2$ are in the directions 0x, 0y respectively. If the complex amplitudes are given by $\mathcal{E}_1 = E_1\exp(i\delta_1)$, $\mathcal{E}_2 = E_2\exp(i\delta_2)$, then the components of the electric field, obtained by taking the real part of equation (2.34) are

$$\left.\begin{aligned} E_x &= E_1(\cos\phi\cos\delta_1 - \sin\phi\sin\delta_1) \\ E_y &= E_2(\cos\phi\cos\delta_2 - \sin\phi\sin\delta_2) \end{aligned}\right\} \quad (2.35)$$

where $\phi = (kz - \omega t)$. Eliminating ϕ from these equations, we obtain (problem 2.3)

$$\left(\frac{E_x}{E_1}\right)^2 + \left(\frac{E_y}{E_2}\right)^2 - 2\left(\frac{E_x}{E_1}\frac{E_y}{E_2}\right)\cos\delta = \sin^2\delta \quad (2.36)$$

where the phase angle $\delta = \delta_2 - \delta_1$.

This is the equation of an ellipse inscribed within a rectangle whose sides are parallel to the coordinate axes and whose lengths are $2E_1$ and $2E_2$ respectively. The major axis of the ellipse makes an angle ψ with the direction 0x given by

$$\tan 2\psi = \frac{2E_1E_2\cos\delta}{E_1^2 - E_2^2} \quad (2.37)$$

which is illustrated in Fig.2.2(a).

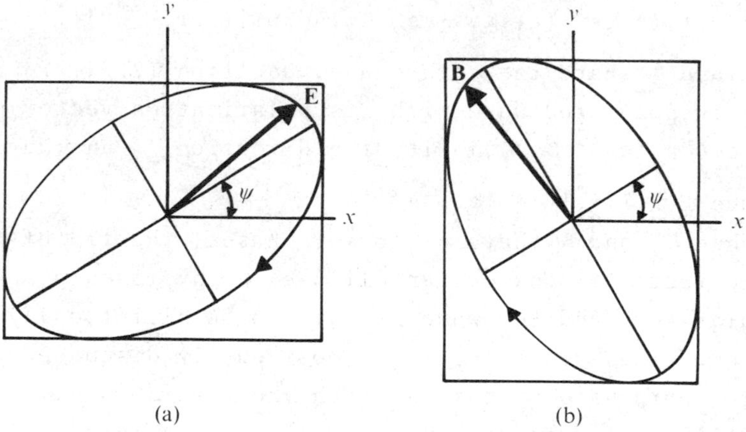

Fig.2.2. Fields of a right-handed elliptically-polarized electromagnetic wave. The direction of propagation is towards the reader. (a) Electric field; (b) Magnetic induction.

We can distinguish two cases of elliptical polarization depending on the sense in which the tip of the electric vector traces out the ellipse. The convention is that if an observer, facing into the approaching wave, sees the ellipse described in a clockwise sense, the polarization is said to be right-handed. If we consider the values of E_x and E_y given by equation (2.35) at two time instants separated by a quarter of a period, we see that the ellipse is described in a clockwise sense if $\sin \delta < 0$. If $\sin \delta > 0$ the ellipse is traced out anticlockwise and the polarization is said to be left-handed. From equations (2.33) we see that the magnetic field $\underline{H}$ and the ellipse it traces out are obtained by rotating the vector $\underline{E}$ and its ellipse by $\pi/2$ about the direction of $\underline{k}$ in such a sense that $\underline{E}\wedge\underline{H}$ points in the direction $\underline{k}$ as shown in Fig.2.2(b).

Two cases of elliptical polarization are of particular interest. These arise when the amplitudes are equal, $E_1 = E_2 = E_0$, and the phase difference $\delta = \pm \pi/2$. At a fixed point in space the electric vector then traces out a circle at the angluar frequency ω and the field is given by

$$\underline{E}(\underline{r},t) = E_0(\underline{\varepsilon}_1 \pm i\underline{\varepsilon}_2) \exp\{i(\underline{k}.\underline{r} - \omega t)\}. \qquad (2.38)$$

In this equation the positive sign corresponds to anticlockwise rotation for an observer facing into the oncoming wave, as we can verify by taking the real part of equation (2.38) to obtain the actual electric field components. When combined with the time dependence $\exp(-i\omega t)$, the vectors $(\underline{\varepsilon}_1 \pm i\underline{\varepsilon}_2)$ correspond to left and right circularly-polarized waves respectively.

The two circularly-polarized waves form an alternative set of independent field polarizations which is especially useful in many problems concerning the interaction of light with atoms subjected to external magnetic fields. If we define the basis of spherical orthogonal unit vectors by the relations

$$\underline{\varepsilon}_{\pm 1} = \mp \frac{1}{\sqrt{2}} (\underline{\varepsilon}_1 \pm i\, \underline{\varepsilon}_2)$$
$$\underline{\varepsilon}_0 = \underline{\varepsilon}_3, \tag{2.39}$$

then any electromagnetic field can be expanded in the form

$$\underline{E}(\underline{r},t) = \sum_{q=-1}^{1} (-1)^q\, \mathcal{E}_q\, \underline{\varepsilon}_{-q} \exp\{i(\underline{k}.\underline{r}-\omega t)\} \tag{2.40}$$

where again the quantities $\mathcal{E}_q$ are the complex amplitudes (Problem 2.4).

2.5. *The energy density and the Poynting vector*

By applying the principle of conservation of energy to a finite volume V bounded by the surface S, we can derive an expression for the energy flux associated with time-dependent electromagnetic fields. Since the rate at which an external electromagnetic field does work on a charge q is $q\underline{v}.\underline{E}$, where $\underline{v}$ is the velocity of the particle, the total rate at which work is done by the fields on a continuous distribution of charges contained within the volume V is given by

$$\int \underline{E}.\underline{J}\, dV$$

Substituting for $\underline{J}$ from equation (2.11) we obtain

$$\int \underline{E}.\underline{J}\, dV = \int (\underline{E}.\text{curl}\, \underline{H} - \underline{E}.\frac{\partial \underline{D}}{\partial t})\, dV. \tag{2.41}$$

We now make use of the vector identity

$$\text{div}(\underline{E}\wedge\underline{H}) = \underline{H}.\text{curl}\, \underline{E} - \underline{E}.\text{curl}\, \underline{H}$$

and Faraday's law, equation (2.10), in the right-hand side of equation (2.41), giving

$$\int \underline{E}.\underline{J}\, dV = -\int \{\text{div}(\underline{E}\wedge\underline{H}) + \underline{E}.\frac{\partial \underline{D}}{\partial t} + \text{H}.\frac{\partial \underline{B}}{\partial t}\}\, dV. \tag{2.42}$$

If we assume that the dielectric constant ε and the magnetic permeability μ are independent of t, i.e. the fields $\underline{E}$ and $\underline{H}$ are not large enough to introduce non-linear effects, and use

Gauss's theorem to transform the first term on the right-hand side of equation (2.42) into a surface integral, we obtain

$$\int \underline{E}.\underline{J}\ dV = -\int (\underline{E}\wedge\underline{H}).dS - \frac{\partial}{\partial t}\int \frac{1}{2}(\underline{D}.\underline{E} + \underline{B}.\underline{H})\ dV. \quad (2.43)$$

The rate at which work is done by the field is now seen to equal the sum of two quite distinct terms. The first represents the rate at which energy flows into the closed volume across the bounding surface. The vector

$$\underline{N} = \underline{E}\wedge\underline{H} \quad (2.44)$$

is called the Poynting vector. In most cases it represents the energy flux crossing unit area of the boundary, although rigorously only the surface integral of $\underline{N}$ has a physical significance (Problem 2.5). The second term represents the decrease in the energy stored in the electromagnetic fields within the volume concerned. The quantity

$$U = \frac{1}{2}(\underline{D}.\underline{E} + \underline{B}.\underline{H}) \quad (2.45)$$

may be regarded as the energy density of the electromagnetic field.

The Poynting vector $\underline{N}$ represents the instantaneous rate at which energy flows through unit area, but in a sinusoidal wave $\underline{E}$ and $\underline{H}$ are rapidly oscillating functions of time and we are generally interested in the average value of $\underline{N}$ over a complete period. In general, the time average of the product of the real parts of two complex vectors $\underline{F}$ and $\underline{G}$, both of which vary as $\exp(-i\omega t)$, is given by (Problem 2.6).

$$\overline{(\mathrm{Re}\ \underline{F})\cdot(\mathrm{Re}\ \underline{G})} = \frac{1}{2}\mathrm{Re}\ (\underline{F}\cdot\underline{G}^*) = \frac{1}{2}\mathrm{Re}\ (\underline{F}^*.\underline{G}).$$

Using the properties of the plane-wave solutions discussed in section 2.3, we find that the time-averaged energy flux is given by the real part of the complex Poynting vector

$$\begin{aligned}\overline{\underline{N}} &= \frac{1}{2}\underline{E}\wedge\underline{H}^* \\ &= \frac{1}{2}\mathcal{E}_0\,\mathcal{H}_0^*\,\underline{\varepsilon}_3 = \frac{|\mathcal{E}_0|^2}{2Z}\,\underline{\varepsilon}_3\end{aligned} \quad (2.46)$$

where $\underline{\varepsilon}_3$ is a unit vector in the direction of the wavevector $\underline{k}$. Similarly the time-averaged energy density $\overline{U}$ is given by

$$\begin{aligned}\overline{U} &= \frac{1}{4}\,(\varepsilon\varepsilon_0\,\underline{E}.\underline{E}^* + \mu\mu_0\,\underline{H}.\underline{H}^*) \\ &= \frac{1}{2}\,\varepsilon\varepsilon_0\,|\mathcal{E}_0|^2.\end{aligned} \tag{2.47}$$

From equations (2.46) and (2.47) we see that

$$\overline{\underline{N}} = v\overline{U}\,\underline{\varepsilon}_3,$$

showing that the energy density associated with a plane wave is propagating with the same velocity as that of the fields.

The Poynting vector has a dual role for it can be shown that the electromagnetic radiation fields transport momentum as well as energy, and that the momentum density is given by $\overline{\underline{N}}/c^2$. This relationship is most easily derived by making use of the idea that radiation consists of photons of energy $\hbar\omega$ whose momentum *in vacuo* is $\hbar\omega/c$, which follows from the quantum theory of radiation (Problem 2.7).

2.6. *Vector and scalar potentials*

The solutions of Maxwell's equations in terms of $\underline{E}$ and $\underline{H}$ which we have discussed in sections 2.3-2.5 illustrate in a direct and simple way the properties of electromagnetic waves. However, we have avoided the much more difficult problem of how these fields are generated by time-varying distributions of current and charge. This problem can be simplified by formulating the differential equations in terms of the scalar potential Φ and the magnetic vector potential $\underline{A}$, rather than attempting the direct solution in terms of $\underline{E}$ and $\underline{H}$. The magnetic induction $\underline{B}$ is related to the vector potential $\underline{A}$ through the definition

$$\underline{B} = \text{curl}\,\underline{A}. \tag{2.48}$$

Since div curl = 0 this automatically satisfies one of Maxwell's relations, equation (2.9). Substituting equation (2.48) into equation (2.10) we obtain

$$\text{curl }(\underline{E} + \frac{\partial \underline{A}}{\partial t}) = 0. \tag{2.49}$$

Now curl grad operating on any scalar function Φ is identically zero, thus the solution of equation (2.49) is given by

$$\underline{E} = -\text{grad }\Phi - \frac{\partial \underline{A}}{\partial t}. \tag{2.50}$$

This reduces to the usual relation between $\underline{E}$ and the electrostatic potential for problems involving static fields.

The remaining two of Maxwell's equations, equations (2.8) and (2.11), can now be expressed in terms of the potentials:

$$\nabla^2\Phi + \frac{\partial}{\partial t}(\text{div }\underline{A}) = -\rho/\varepsilon\varepsilon_0 \tag{2.51}$$

$$\nabla^2\underline{A} - \frac{1}{v^2}\frac{\partial^2\underline{A}}{\partial t^2} - \text{grad}(\text{div }\underline{A} + \frac{1}{v^2}\frac{\partial\Phi}{\partial t}) = -\mu\mu_0\underline{J}. \tag{2.52}$$

These equations are coupled by the third term in equation (2.52) and would be difficult to solve as they stand. However, because of the way the potentials are defined, it can be shown that it is possible to impose the additional condition (Panofsky and Phillips 1962, Ch.13)

$$\text{div }\underline{A} + \frac{1}{v^2}\frac{\partial\Phi}{\partial t} = 0 \tag{2.53}$$

without affecting the values of $\underline{E}$ and $\underline{H}$ obtained. This relation, equation (2.53), is known as the Lorentz gauge condition and its use in equations (2.51) and (2.52) uncouples those equations, enabling them to be solved for Φ and $\underline{A}$ separately:

$$\nabla^2\Phi - \frac{1}{v^2}\frac{\partial^2\Phi}{\partial t^2} = -\rho/\varepsilon\varepsilon_0 \tag{2.54}$$

$$\nabla^2\underline{A} - \frac{1}{v^2}\frac{\partial^2\underline{A}}{\partial t^2} = -\mu\mu_0\underline{J}. \tag{2.55}$$

These are known as the inhomogeneous wave equations. The solutions of equations (2.54) and (2.55) consist of particular solutions involving integrals over the charge and current distributions together with complementary solutions obtained by setting $\underline{J} = \rho = 0$. The latter are obviously just the electromagnetic waves discussed previously, since the scalar and vector potentials then satisfy the same homogeneous wave equations as the fields $\underline{E}$ and $\underline{H}$.

For a system of static charges, equation (2.54) reduces to Poisson's equation, for which the solution is known to be of the form

$$\Phi(\underline{r}) = \frac{1}{4\pi\varepsilon\varepsilon_0}\int\frac{\rho(\underline{r}')\,dV'}{|\underline{r} - \underline{r}'|} .$$

The time-dependent equation (2.54) possesses a similar solution given by

$$\Phi(\underline{r},t) = \frac{1}{4\pi\varepsilon\varepsilon_0}\int\frac{[\rho(\underline{r}',t')]\,dV'}{|\underline{r} - \underline{r}'|} . \qquad (2.56)$$

The square brackets in equation (2.56) indicate that the integrand is to be evaluated at the retarded time $t' = t - |\underline{r} - \underline{r}'|/v$. This ensures that the potential at the space-time point $(\underline{r},t)$ is the result of disturbances which have travelled with the velocity of light, v, from the charge at the point $\underline{r}'$, which existed at some appropriate earlier time t'. The solution of equation (2.55) for the retarded vector potential has a similar form:

$$\underline{A}(\underline{r},t) = \frac{\mu\mu_0}{4\pi}\int\frac{[\underline{J}(\underline{r}',t')]\,dV'}{|\underline{r} - \underline{r}'|} \qquad (2.57)$$

Mathematically correct solutions also exist in which t,t' are related by $t' = t + |\underline{r} - \underline{r}'|/v$. These are known as the advanced potentials, but since they appear to contradict the causality relation they seem to have no physical significance. Nevertheless they have been incorporated into a number of attempts to solve certain theoretical difficulties which arise in electrodynamics, one of which is briefly mentioned in section 4.4.1 in connection with the spontaneous emission of radiation.

2.7. *Electric dipole radiation*

We now investigate solutions of equations (2.56) and (2.57) for simple systems of charges and currents which vary in time. Since we can make a Fourier analysis of the time dependence and then treat each monochromatic component separately, there is no loss of generality if we assume that the system of charges and currents oscillates sinusoidally at the angular frequency ω:

$$\left.\begin{aligned} \rho(\underline{r},t) &= \rho(\underline{r})\exp(-i\omega t) \\ \underline{J}(\underline{r},t) &= \underline{J}(\underline{r})\exp(-i\omega t). \end{aligned}\right\} \quad (2.58)$$

Again we must take the real part of these expressions if we wish to obtain actual physical quantities. We are generally interested in the radiation fields in free space outside the current and charge distributions. Thus setting $\mu = \varepsilon = 1$ and making the dependence on the retarded time t' explicit in equation (2.57), we have

$$\underline{A}(\underline{r},t) = \frac{\mu_0}{4\pi}\exp(-i\omega t)\int \frac{\underline{J}(\underline{r}')\exp(ik|\underline{r}-\underline{r}'|)}{|\underline{r}-\underline{r}'|}\,dV' \quad (2.59)$$

since $\omega/k = c$. If the source is confined to a volume whose dimensions are small compared with the distance to the point P at which the radiation fields are observed, as shown in Fig.2.3, then

$$|\underline{r}-\underline{r}'| \cong r - \hat{\underline{r}}.\underline{r}'$$

where $\hat{\underline{r}}$ is a unit vector along the direction from the origin to P. Substituting into equation (2.59) we have

$$\underline{A}(\underline{r},t) = \frac{\mu_0}{4\pi}\exp\{i(kr-\omega t)\}\int \frac{\underline{J}(\underline{r}')\exp(-ik\hat{\underline{r}}.\underline{r}')}{(r-\hat{\underline{r}}.\underline{r}')}\,dV' \quad (2.60)$$

In this book we shall be mainly concerned with radiation from excited electrons in atoms or molecules, so that the dimensions of the charge distributions are of the order

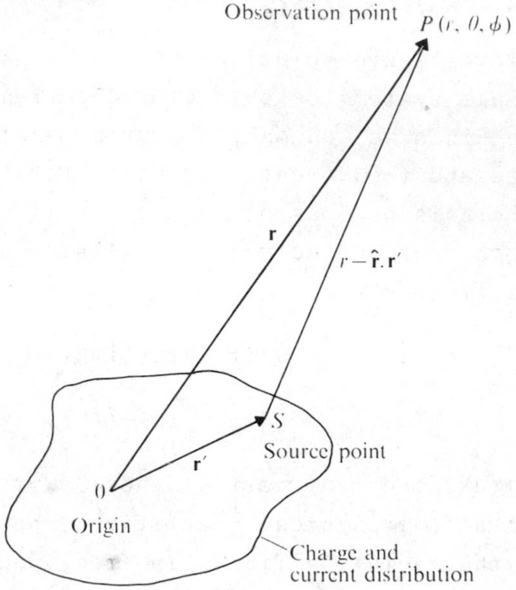

Fig.2.3. Coordinates of the source point S and a distant point of observation P referred to an origin O within the charge and current distribution.

$r' \approx 10^{-8}$cm. If we consider radiation in the centre of the visible region $\lambda = 5000$ Å, then $k \approx 1 \times 10^{5}\ \text{cm}^{-1}$. Consequently $kr' \approx 1 \times 10^{-3}$ and we can expand the exponential term in equation (2.60) as a power series:

$$\exp(-ik\hat{\underline{r}}.\underline{r}') = \{1 - ik\hat{\underline{r}}.\underline{r}' + \frac{i^2k^2}{2!}(\hat{\underline{r}}.\underline{r}')^2 - .....\} \quad (2.61)$$

For the present discussion we retain only the first term in this expansion and also set the denominator in equation (2.60) equal to r. This is known as the electric dipole approximation and we see that $\underline{A}(\underline{r},t)$ has the form

$$\underline{A}(\underline{r},t) = \frac{\mu_0}{4\pi}\frac{\exp\{i(kr-\omega t)\}}{r}\int \underline{J}(\underline{r}')\, dV' \quad (2.62)$$

We now transform the integral in equation (2.62) by using the expansion formula

$$\text{div}(x\underline{J}) = J_x + x \text{ div } \underline{J}$$

and the continuity equation, equation (2.5), applied to the sinusoidal currents and charges of equation (2.58):

$$\text{div } \underline{J} - i\omega \rho(\underline{r}) = 0. \tag{2.63}$$

Since the surface integral of $x\underline{J}$ vanishes for any surface outside the current distribution, we have

$$\underline{A}(\underline{r},t) = -\frac{\mu_0}{4\pi} i\omega \frac{\exp\{i(kr-\omega t)\}}{r} \underline{p} \tag{2.64}$$

where $\underline{p}$ is the electric dipole moment of the system of charges defined by

$$\underline{p} = \int \underline{r}' \rho(\underline{r}') \, dV'. \tag{2.65}$$

We see that the vector potential, $\underline{A}(\underline{r},t)$, has the space-time dependence characteristics of a spherical wave expanding from the origin.

The magnetic field produced by the oscillating electric dipole moment is given by equation (2.48)

$$\underline{H} = -\frac{i\omega}{4\pi} \text{curl}\left\{\frac{\underline{p}\exp\{i(kr-\omega t)\}}{r}\right\}.$$

Making use of the vector expansion

$$\text{curl}\{\underline{p}\, f(r)\} = f(r) \text{ curl } \underline{p} - \underline{p} \wedge \text{grad } f(r)$$

we have

$$\underline{H} = \frac{\omega k}{4\pi} \left(\frac{1}{r} + \frac{i}{kr^2}\right) (\hat{\underline{r}} \wedge \underline{p}) \exp\{i(kr-\omega t)\}. \tag{2.66}$$

Outside the source, the electric field is obtained by applying Maxwell's relation equation (2.18), taking into account the sinusoidal nature of the fields:

$$\underline{E} = \frac{i \operatorname{curl} \underline{H}}{\omega\varepsilon_0}$$

$$= \frac{1}{4\pi\varepsilon_0}\left[\frac{k^2}{r}(\hat{\underline{r}}\wedge\underline{p})\wedge\hat{\underline{r}} + \{3\hat{\underline{r}}(\hat{\underline{r}}.\underline{p}) - \underline{p}\}\left(\frac{1}{r^3} - \frac{ik}{r^2}\right)\right] \times \quad (2.67)$$

$$\times \exp\{i(kr-\omega t)\}.$$

We see that the magnetic field is always perpendicular to the radius vector $\hat{\underline{r}}$ but that in general the electric field has components both parallel and perpendicular to $\hat{\underline{r}}$. Further discussion of equations (2.66) and (2.67) is simplified if we consider the fields in three spatial regions defined in the following way:

(a) The near zone, $d \ll r \ll \lambda$. Here d is the size of the source distribution and we have already imposed the condition $d \ll \lambda$ in the expansion of equation (2.61). In this region the electric field is given by

$$\underline{E} = \frac{1}{4\pi\varepsilon_0}\left(\frac{3\hat{\underline{r}}(\hat{\underline{r}}.\underline{p}) - \underline{p}}{r^3}\right)\exp(-i\omega t). \quad (2.68)$$

Except for the time-dependent factor, this expression is identical to the field produced by a static electric dipole. This term is responsible for the interatomic forces between neutral atoms (van der Waals forces) as we discuss in Chapter 8. These forces are only appreciable when the atoms come close together, as in a collision, due to the $1/r^3$ dependence of the static field. The magnetic field in this region is smaller by a factor of kr and is therefore negligible.

(b) The intermediate zone, $d \ll r \approx \lambda$. In this zone the fields are rather complicated since there is still an appreciable radial component of $\underline{E}$. The situation is illustrated in Fig.2.4 for several successive instants of time. We see that in this region closed loops are formed in the streamlines of the electric field and that there is no well-defined wavelength which characterizes the field distribution.

(c) The wave zone, $d \ll \lambda \ll r$. In this zone the situation again becomes simplified since only those field components which decrease as 1/r need be considered. The radiation fields are given by

$$\underline{H} = \frac{k^2}{4\pi\varepsilon_0} (\hat{\underline{r}} \wedge \underline{p}) \frac{\exp\{i(kr-\omega t)\}}{r} \frac{1}{Z_0} \tag{2.69}$$

$$\underline{E} = \frac{k^2}{4\pi\varepsilon_0} (\hat{\underline{r}} \wedge \underline{p}) \wedge \hat{\underline{r}} \frac{\exp\{i(kr-\omega t)\}}{r} \tag{2.70}$$

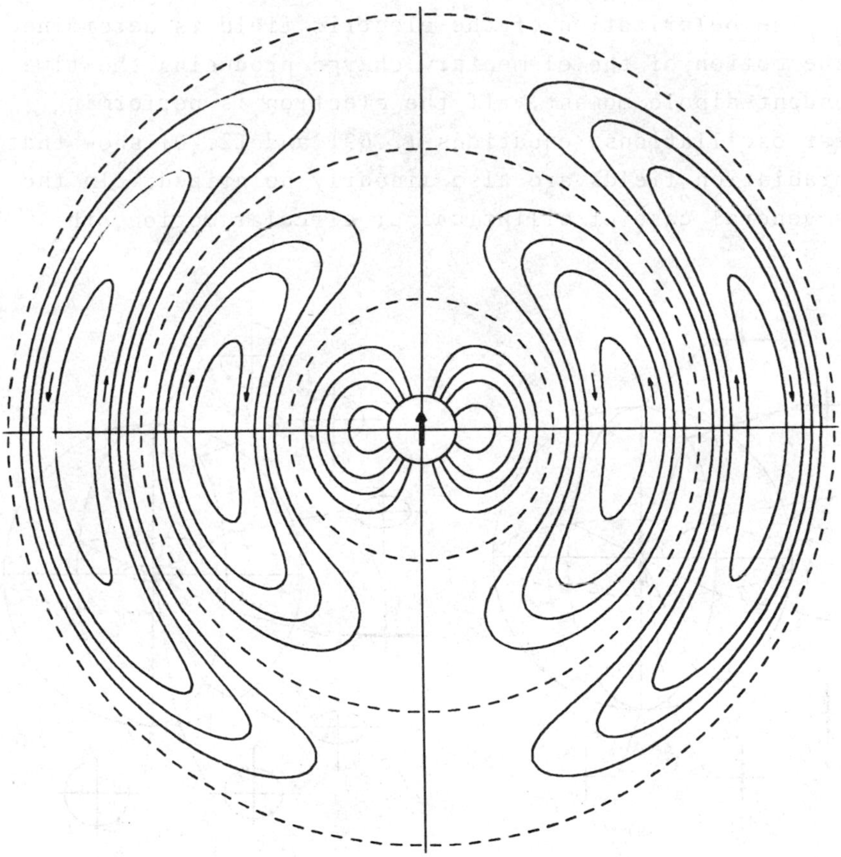

Fig.2.4. Streamlines of the electric field about a linear dipole oscillator of moment $\underline{p}=\underline{p}_0 \cos \omega t$.

where we have introduced the impedance of free space $Z_0 = 1/\varepsilon_0 c$. We see that $\underline{E}$ and $\underline{H}$ have the form of spherical waves expanding from the origin with a characteristic wavelength $\lambda = 2\pi/k$. The fields have all the properties characteristic of the wave fields which were discussed in section 2.3. In particular, $\underline{E}$ and $\underline{H}$ are perpendicular; they are transverse to the direction of propagation $\hat{\underline{r}}$; and they have amplitudes bearing the constant ratio $|\mathcal{E}_0/\mathcal{H}_0| = Z_0$ in the case of a linearly oscillating dipole. As $r\to\infty$ the spherical waves become closer and closer approximations to the plane waves discussed previously.

The polarization of the electric field is determined by the motion of the elementary charge producing the time-dependent dipole moment. If the electron is performing linear oscillations, equations (2.69) and (2.70) show that the radiation fields are also linearly polarized. In the more general case of elliptical or circular motion, the

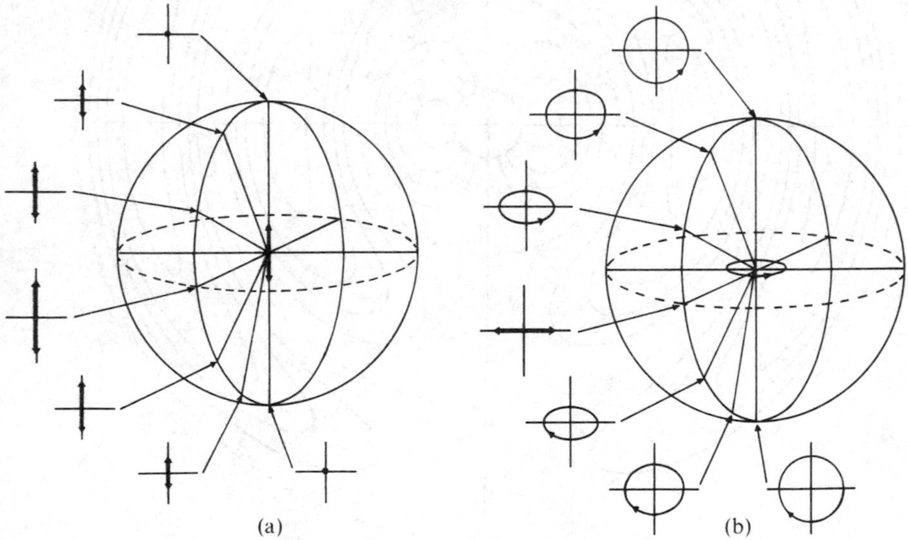

Fig.2.5. Polarization of the electric field seen by looking from different directions at a dipole oscillator executing (a) linear oscillations and (b) circular oscillations. The observer is at a large fixed distance and the amplitude of oscillation of the electron has been greatly exaggerated.

radiation is generally elliptically polarized and the shape of the ellipse depends on the direction of observation. This is illustrated in Fig.2.5 for the case of (a) a linear oscillation, and (b) a circular oscillation of an electron, and is discussed in more detail in section 5.2.2.

2.8. *Rate of radiation by an electric dipole oscillator*

The time-averaged intensity of the radiation at the point $P(r,\theta,\phi)$ is given by the Poynting vector $\underline{N} = \mathrm{Re}(\underline{E}\wedge\underline{H}^*)/2$ and the average power radiated into unit solid angle, $\frac{dP}{d\Omega}$, is therefore

$$\frac{dP}{d\Omega} = \frac{1}{2}\,\mathrm{Re}\,\{r^2\,\hat{\underline{r}}.(\underline{E}\wedge\underline{H}^*)\}\ .$$

Making use of equations (2.69) and (2.70) we see that the result is independent of the radius of the sphere r:

$$\frac{dP}{d\Omega} = \frac{1}{2Z_0}\left(\frac{k^2}{4\pi\varepsilon_0}\right)^2 |\hat{\underline{r}}\wedge(\hat{\underline{r}}.\underline{p})|^2\ . \tag{2.71}$$

If $\underline{p}$ is a linear oscillator, the angular distribution of the radiation is given by

$$\frac{dP}{d\Omega} = \frac{1}{2Z_0}\left(\frac{k^2}{4\pi\varepsilon_0}\right)^2 |\underline{p}|^2\ \sin^2\theta, \tag{2.72}$$

where θ is the angle between the direction of observation and that of $\underline{p}$. The $\sin^2\theta$ angular dependence, illustrated in Fig.2.6(a), is characteristic of dipole radiation leading to a maximum intensity in the plane $\theta = \pi/2$ and zero intensity in the directions $\theta = 0,\pi$.

For the case of circular oscillation in the x-y plane, the angular distribution can be shown to be (Problem 2.8)

$$\frac{dP}{d\Omega} = \frac{1}{4Z_0}\left(\frac{k^2}{4\pi\varepsilon_0}\right)^2 |\underline{p}|^2\ (1 + \cos^2\theta) \tag{2.73}$$

which is shown in Fig.2.6(b). In this case there is no direction in which the intensity is zero although the intensity in the direction $\theta = \pi/2$ is only half that along the polar axis.

The total power radiated by an oscillating electric dipole moment, P_{E1}, is obtained by integrating over the solid angle $d\Omega = \sin\theta \, d\theta \, d\phi$. In both cases we have

$$P_{E1} = \frac{k^4 c}{12\pi\varepsilon_0} |\underline{p}|^2 \quad . \tag{2.74}$$

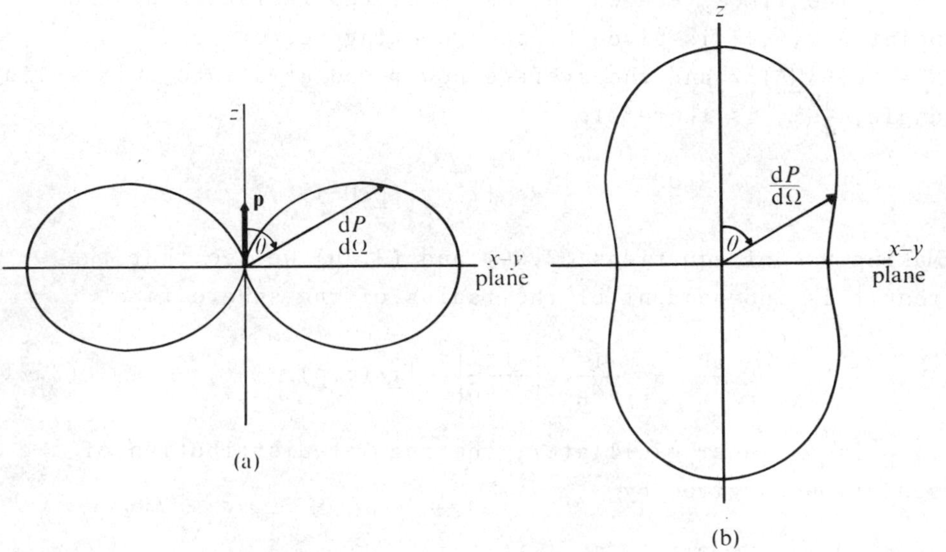

Fig.2.6. Angular distribution of electric dipole radiation for the case (a) of linear oscillation parallel to the z-axis, and (b) of circular motion in the x-y plane. The length of the radius vector is proportional to $\frac{dP}{d\Omega}$.

2.9. *Angular momentum of dipole radiation*

It has been shown experimentally that a beam of circularly-polarized light exerts a torque on any optical component, such as a quarter- or half-wave plate, which changes the state of polarization of the light. This corresponds to a transfer of angular momentum from the radiation fields to the material system. To calculate the rate at which angular momentum is radiated by an electric dipole source we consider the oscillating current distribution to be surrounded by a large, perfectly absorbing sphere.

Associated with the time-averaged flux, $\underline{\overline{N}}$, in the direction $\hat{\underline{k}}$ is an energy density U given by $U\hat{\underline{k}} = \underline{\overline{N}}/c$. The relativistic energy equation

$$U^2 = \underline{G}^2 c^2 + m_0^2 c^4$$

shows that there is also an associated momentum density $\underline{G} = \underline{\overline{N}}/c^2$ since the rest mass m_0 of a photon is zero. Thus the torque exerted on an element of area dS of the absorbing sphere is c dS $(\underline{r}\wedge\underline{\overline{N}}/c^2)$ and the angular momentum radiated per unit time is given by

$$\underline{L} = \frac{1}{c}\int (\underline{r}\wedge\underline{\overline{N}})\, dS\,. \tag{2.75}$$

Substituting for the time-averaged Poynting vector from equation (2.46) and expanding the triple vector product, we obtain $\underline{L}$ in terms of the fields:

$$\underline{L} = \frac{1}{2c}\int \{(\underline{r}.\underline{H}^*)\underline{E} - (\underline{r}.\underline{E})\underline{H}^*\}\, dS. \tag{2.76}$$

For electric dipole radiation the first term in equation (2.76) is identically zero since the magnetic field $\underline{H}^*$ is everywhere perpendicular to $\underline{r}$, equation (2.66). The electric field in the wavezone, equation (2.70), is also perpendicular to $\underline{r}$ and therefore in the second term of equation (2.76) we retain the intermediate field contribution to $\underline{E}$ which varies as $(1/r^2)$. We obtain

$$\underline{L} = \frac{ik}{4\pi\varepsilon_0 c}\int (\underline{r}.\underline{p})\underline{H}^*\, d\Omega. \tag{2.77}$$

Finally if we use the term in $(1/r)$ for the magnetic field , equation (2.69), we obtain a result which is independent of the radius r of the absorbing sphere:

$$\underline{L} = \frac{ik^3}{16\pi^2\varepsilon_0}\int (\hat{\underline{r}}.\underline{p})(\hat{\underline{r}}\wedge\underline{p}^*)\, d\Omega. \tag{2.78}$$

We observe that although the radiated energy is carried by the fields in the wave zone, the angular momentum is transported by a subtle combination of the intermediate zone $\underline{E}$ field and the wave zone $\underline{H}$ field.

An expression for the z-component of angular momentum, L_z, in terms of the components of the electric dipole moment (p_x, p_y, p_z) can be obtained by introducing the explicit angular dependence of the cartesian components of $\underline{r}$ and performing the angular integration, giving

$$L_z = \frac{ik^3}{12\pi\varepsilon_0}(p_x p_y^* - p_y p_x^*). \tag{2.79}$$

Similar expressions may be obtained for L_x, L_y by cyclic permutation. We find that a given component of $\underline{L}$ depends on the perpendicular components of $\underline{p}$ and moreover that these perpendicular components must be out of phase, since if p_x, p_y are real or have the same phase L_z, from equation (2.79), is zero. Thus a linear dipole radiates zero z-component of angular momentum.

The maximum angular momentum for a fixed value of $|\underline{p}|^2$ is radiated when $\underline{p}$ has two perpendicular components which are of equal magnitude and $\pi/2$ out of phase and a third component which is zero. Then if $p_y = \pm ip_x$, corresponding to left and right circularly-polarized radiation respectively, we find

$$L_z = \pm\frac{k^3|\underline{p}|^2}{12\pi\varepsilon_0}. \tag{2.80}$$

The significance of this result becomes evident if we take the ratio of maximum z-component of angular momentum radiated per unit time to the total power radiated, P_{E1}. From equations (2.74) and (2.80) we have

$$\frac{L_z}{P_{E1}} = \pm\frac{1}{\omega} = \pm\frac{\hbar}{\hbar\omega}.$$

By introducing the factor of $\hbar$ we are able to interpret this classically-derived relation in terms of the quantum theory. We see that each quantum $\hbar\omega$ of circularly-polarized light emitted by an oscillating dipole moment transports a z-component of angular momentum of $\pm\hbar$, depending on the sense of circular polarization. We shall meet these ideas again in section 5.2 where we discuss the selection rules and polarization of electric dipole radiation (Problem 2.9).

2.10. Magnetic dipole radiation

If, by reason of symmetry, the electric dipole moment of the system (2.65) is zero, we must investigate higher-order contributions to the vector potential. The next term in the expansion of equation (2.60) includes contributions from both the denominator and the exponential factor and has the form

$$\underline{A}(\underline{r},t) = \frac{\mu_0}{4\pi}\frac{\exp\{i(kr-\omega t)\}}{r}\left(\frac{1}{r} - ik\right)\int \underline{J}(\underline{r}')(\hat{\underline{r}}.\underline{r}')dV'. \quad (2.81)$$

It is convenient to separate the integrand into two parts, one symmetric and the other antisymmetric in $\underline{J}$ and $\underline{r}'$:

$$(\hat{\underline{r}}.\underline{r}')\underline{J} = \tfrac{1}{2}\{(\hat{\underline{r}}.\underline{r}')\underline{J} + (\hat{\underline{r}}.\underline{J})\underline{r}'\} + \tfrac{1}{2}\{(\underline{r}'\wedge\underline{J})\wedge\hat{\underline{r}}\}. \quad (2.82)$$

If we consider first the antisymmetric component, we see that it is related to the magnetic dipole moment $\underline{m}$ of the current distribution:

$$\underline{m} = \frac{1}{2}\int \underline{r}'\wedge\underline{J}(\underline{r}')dV'. \quad (2.83)$$

The symmetric component is related to the electric quadrupole moment of the charge distribution and will be considered in detail in the next section.

From equation (2.81) the oscillating magnetic dipole moment gives rise to the magnetic vector potential:

$$\underline{A}_{M1}(\underline{r}) = \frac{ik\mu_0}{4\pi}\left(\frac{1}{r} + \frac{i}{kr^2}\right)\hat{\underline{r}}\wedge\underline{m}\,\exp\{i(kr-\omega t)\}. \quad (2.84)$$

Comparing equations (2.66) and (2.84), we see that the vector potential for magnetic dipole radiation, $\underline{A}_{M1}(\underline{r})$, is proportional to the magnetic field of electric dipole radiation, $\underline{H}_{E1}(\underline{r})$, provided we replace $\underline{p}$ by $\underline{m}$:

$$\underline{A}_{M1}(\underline{r}) = \frac{i\mu_0}{\omega} \underline{H}_{E1}(\underline{r}). \tag{2.85}$$

It follows that the magnetic field for magnetic dipole radiation can be obtained directly from equation (2.67)

$$\underline{H}_{M1}(\underline{r}) = \frac{1}{4\pi} [\frac{k^2}{r}(\hat{\underline{r}}\wedge\underline{m})\wedge\hat{\underline{r}} + \{3\hat{\underline{r}}(\hat{\underline{r}}.\underline{m}) - \underline{m}\} \left(\frac{1}{r^3} - \frac{ik}{r^2}\right)]\times$$
$$\times \exp\{i(kr-\omega t)\}. \tag{2.86}$$

Similarly, from equations (2.50) and (2.85), we find that the electric field for the magnetic dipole source is $(-\mu_0)$ times the magnetic field of an electric dipole. Thus

$$\underline{E}_{M1}(\underline{r}) = -\frac{\mu_0}{4\pi} \omega k \left(\frac{1}{r} + \frac{i}{kr^2}\right)(\hat{\underline{r}}\wedge\underline{m})\exp\{i(kr-\omega t)\}. \tag{2.87}$$

All the remarks concerning the behaviour of the fields in the near and wave zone which we made in section 2.7 in connection with the electric dipole fields apply in the magnetic dipole case if we make the interchanges

$$\underline{E}_{E1} \to \underline{H}_{M1}/\varepsilon_0, \; \underline{H}_{E1} \to -\underline{E}_{M1}/\mu_0, \text{ and } \underline{p}\to\underline{m}.$$

This, of course, means that the plane of polarization is different in the two cases. For electric dipole radiation the electric vector lies in the plane defined by $\hat{\underline{r}}$ and $\underline{p}$, while for magnetic dipole radiation it is perpendicular to the plane containing $\hat{\underline{r}}$ and $\underline{m}$.

The angular distribution of the radiation is the same for the two kinds of dipole, but the total power radiated by an oscillating magnetic dipole is given by

$$P_{M1} = \frac{\mu_0 k^4 c}{12\pi} |\underline{m}|^2. \tag{2.88}$$

2.11. *Electric quadrupole radiation*

We now consider the first term in equation (2.82) which is symmetric in $\underline{J}$ and $\underline{r}'$. This can be transformed using the vector identity

$$\text{div}\{x'(\hat{\underline{r}}.\underline{r}')\underline{J}\} = (\hat{\underline{r}}.\underline{r}')J_x + x'(\hat{\underline{r}}.\underline{J}) + x'(\hat{\underline{r}}.\underline{r}')\text{div}\underline{J},$$

and the continuity equation in the form $\mathrm{div}\underline{J} = i\omega\rho(\underline{r})$. The corresponding vector potential equation (2.81) is given by

$$\underline{A}(\underline{r}) = -\frac{\mu_0 c}{8\pi} k^2 \frac{\exp\{i(kr-\omega t)\}}{r}\left(1 + \frac{i}{kr}\right) \int \underline{r}'(\hat{\underline{r}}.\underline{r}')\rho(\underline{r}')dV'. \quad (2.89)$$

The integral in equation (2.89) involves second moments of the charge distribution and we can therefore identify this contribution to the vector potential as originating from an oscillating electric quadrupole moment.

The general expressions for the fields produced by an electric quadrupole source are of rather complicated form. However, we are mainly interested in the fields in the wave zone since this will enable us to discuss the angular distribution and total power radiated. When using $\underline{B} = \mathrm{curl}\ \underline{A}$ we therefore retain only the term which varies as $1/r$. The magnetic field is found to have the following form:

$$\underline{H} = \frac{ik}{\mu_0} \hat{\underline{r}}\wedge\underline{A} \quad (2.90)$$

and the electric field in the wave zone is related to $\underline{H}$ in the usual way

$$\underline{E} = \frac{ikZ_0}{\mu_0} (\hat{\underline{r}}\wedge\underline{A})\wedge\hat{\underline{r}}.$$

The explicit expression for the magnetic field is obtained by combining equations (2.89) and (2.90), giving

$$\underline{H} = -\frac{ik^3 c}{8\pi} \frac{\exp\{i(kr-\omega t)\}}{r} \int (\hat{\underline{r}}\wedge\underline{r}')(\hat{\underline{r}}.\underline{r}')\rho(\underline{r}')dV'. \quad (2.91)$$

The integrand in equation (2.91) can be rewritten in a more convenient form as

$$\hat{\underline{r}}\wedge \int \underline{r}'(\hat{\underline{r}}.\underline{r}')\rho(\underline{r}')dV' = \frac{1}{3} \hat{\underline{r}}\wedge\underline{Q}(\hat{\underline{r}}). \quad (2.92)$$

In equation (2.92) $\underline{Q}(\hat{\underline{r}})$ is a vector whose components are defined by

$$Q_\alpha = \sum_\beta Q_{\alpha\beta}\, r_\beta\,. \quad (2.93)$$

In this equation r_α are the direction cosines of the unit vector $\hat{\underline{r}}$ and $Q_{\alpha\beta}$ are the components of the electric quadrupole moment tensor

$$Q_{\alpha\beta} = \int (3x_\alpha x_\beta - r^2 \delta_{\alpha\beta})\rho(\underline{r})dV \tag{2.94}$$

where $(x_\alpha, x_\beta, x_\gamma)$ are the cartesian components of $\underline{r}$.

With the use of these definitions, the magnetic field is given by

$$\underline{H} = -\frac{ik^3c}{24\pi} \cdot \frac{\exp\{i(kr-\omega t)\}}{r} \hat{\underline{r}}\wedge\underline{Q}(\hat{\underline{r}}). \tag{2.95}$$

The time-averaged power radiated per unit solid angle by the oscillating electric quadrupole moment is given by

$$\begin{aligned}\frac{dP}{d\Omega} &= \frac{1}{2} Z_0 |\underline{H}|^2 \\ &= \frac{k^6 c|\hat{\underline{r}}\wedge\underline{Q}(\hat{\underline{r}})|^2}{1152\pi^2\varepsilon_0}.\end{aligned} \tag{2.96}$$

Using the definition of $\underline{Q}(\hat{\underline{r}})$ (equation (2.93)), we can expand the angular dependence of equation (2.96) in the form

$$\begin{aligned}|\hat{\underline{r}}\wedge\underline{Q}(\hat{\underline{r}})|^2 &= \underline{Q}^*.\underline{Q} - |\hat{\underline{r}}.\underline{Q}|^2 \\ &= \sum_{\alpha,\beta,\gamma} Q^*_{\alpha\beta} Q_{\alpha\gamma} r_\beta r_\gamma - \sum_{\substack{\alpha,\beta\\ \gamma,\delta}} Q^*_{\alpha\beta} Q_{\gamma\delta} r_\alpha r_\beta r_\gamma r_\delta.\end{aligned} \tag{2.97}$$

It is apparent that the angular distribution of electric quadrupole radiation is generally a rather complicated function of θ,ϕ but a simple example will serve to illustrate the main features. We consider an oscillating spheroidal charge distribution. In this case the off-diagonal elements of the electric quadrupole moment tensor vanish because of the symmetry of the system. If the z-axis is taken as the axis of symmetry we have $Q_{11} = Q_{22}$, and since the tensor is defined in such a way that the sum of the elements on the leading diagonal is zero, we may write

$$Q_{33} = Q_0; \; Q_{11} = Q_{22} = -\frac{1}{2} Q_0.$$

Substituting these values into equation (2.97) and using explicit expressions for the cartesian components of $\hat{\underline{r}}$, we find that the angular distribution of the radiated power is given by

$$\frac{dP}{d\Omega} = \frac{ck^6 Q_0^2}{512\pi^2 \varepsilon_0} \sin^2\theta \cos^2\theta. \qquad (2.98)$$

This is a four-lobed pattern, as shown in Fig. 2.7, with zero intensity in the directions $\theta = 0, \pi/2, \pi$, and with maximum intensity in the directions $\theta = \pi/4, 3\pi/4$. It is quite different from the distributions for an oscillating dipole shown in Fig.2.6.

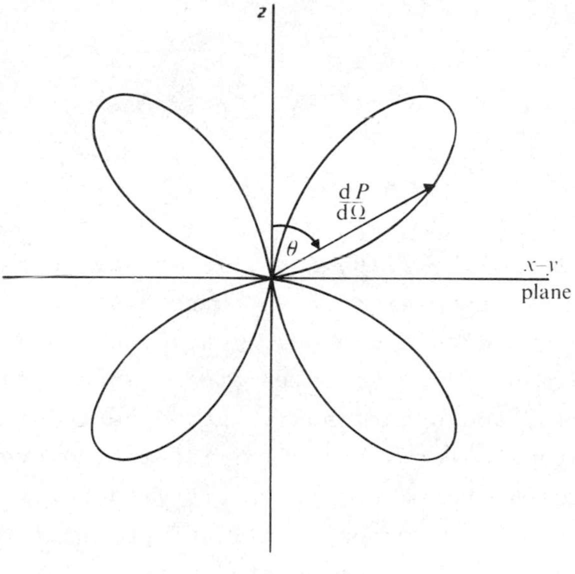

Fig.2.7. Angular distribution of quadrupole radiation from an oscillating spheroidal charge distribution.

We are also interested in the total power radiated since this will enable us to derive an expression for the transition probability for electric quadrupole radiation in section 7.2. From equations (2.96) and (2.97) we see that we require integrals over products of the cartesian components of $\hat{\underline{r}}$ expressed in terms of spherical polar coordinates. These are readily shown to be

$$\left.\begin{aligned}\int r_\beta r_\gamma \, d\Omega &= \frac{4\pi}{3}\,\delta_{\beta\gamma}\\ \int r_\alpha r_\beta r_\gamma r_\delta \, d\Omega &= \frac{4\pi}{15}\,(\delta_{\alpha\beta}\delta_{\gamma\delta} + \delta_{\alpha\gamma}\delta_{\beta\delta} + \delta_{\alpha\delta}\delta_{\beta\gamma}).\end{aligned}\right\} \tag{2.99}$$

Thus we find

$$\int |\hat{\underline{r}}\wedge\underline{Q}(\hat{\underline{r}})|^2 \, d\Omega = \frac{4\pi}{15}\left[3\sum_{\alpha,\beta} |Q_{\alpha\beta}|^2 - \sum_\alpha Q^*_{\alpha\alpha}\sum_\gamma Q_{\gamma\gamma}\right].$$

The second term in this expression is zero since the tensor is defined in such a way that its trace is zero. Thus the total power radiated by an electric quadrupole source is given by (Problem 2.10)

$$P_{E2} = \frac{ck^6}{1440\pi\varepsilon_0}\sum_{\alpha,\beta} |Q_{\alpha\beta}|^2. \tag{2.100}$$

2.12. *Multipole fields*

In sections 2.7, 2.10, and 2.11 we have expanded the radiation fields of an arbitrary charge distribution by the use of simple and direct methods. This technique is adequate for a discussion of excited atoms since only the electric dipole, and occasionally the magnetic dipole and electric quadrupole terms are significant. However, in nuclear physics higher-order terms are frequently required and it is clear that a more general and powerful technique should be used. In this method the charge distribution is expanded in terms of its multipole moments and the radiation fields are constructed from spherical waves which have well-defined values of the square of the total angular momentum $\underline{L}^2$ and of its z-component, L_z. This enables changes in the angular momentum and parity of excited states of nuclei in radiative γ-ray transitions to be discussed in a very elegant way. The same techniques are applicable to excited states of atoms and molecules, but unfortunately they are beyond the level of this book and the reader is referred to Rose (1955) for a complete discussion.

Problems

2.1. Show that equation (2.20) may be reduced to the form

$$\frac{\partial^2 W}{\partial \zeta^2} = \frac{1}{v^2}\frac{\partial^2 W}{\partial t^2}$$

where $\zeta = \underline{r}.\hat{\underline{k}}$ and $\hat{\underline{k}}$ is a unit vector in an arbitrary direction. Hence prove that the general solution of the wave equation is given by

$$W = W_1(p) + W_2(q)$$

where $p = \zeta - vt$ and $q = \zeta + vt$.

2.2. By considering the surfaces on which $W(\underline{r},t) = W_0\exp\{i(\underline{k}.\underline{r}-\omega t)\}$ is constant, prove that this represents a harmonic plane wave propagating in the direction $\underline{k}$ with a wavelength given by $\lambda = 2\pi/k = 2\pi v/\omega$.

2.3. Verify that equation (2.35) may be reduced to equation (2.36) by eliminating the common phase angle ϕ. By a rotation of the coordinate axes about the Oz direction or otherwise, show that the tip of the electric vector specified by equation (2.35) traces out an ellipse with semi-axes E_1 and E_2 and that the major axis of the ellipse is inclined at an angle ψ, given by equation (2.37), to the Ox direction.

2.4. Prove that the spherical unit vectors specified by equation (2.39) satisfy the orthonormality condition $\hat{\underline{\varepsilon}}_i.\hat{\underline{\varepsilon}}_j^* = \delta_{ij}$ where $\delta_{ij}=1$ if $i=j$ and $\delta_{ij}=0$ otherwise.

Show that the real parts of

$$\underline{E}(\underline{r},t) = E_0\, \hat{\underline{\varepsilon}}_{\pm 1} \exp\{i(\underline{k}.\underline{r}-\omega t)\}$$

represent left and right circularly-polarized waves respectively.

2.5. A long straight cylindrical conductor of radius a and resistivity ρ carries a steady current I. Find an expression for the Poynting vector at a radius $r < a$. Show that for a cylindrical shell in the conductor of inner and outer radii r_1 and r_2, the net flow of energy from the electromagnetic field into the shell is equal to the ohmic heating in the shell.

2.6. Prove that the time average of the product of the real parts of two complex vectors, both of which depend on time as $\exp(-i\omega t)$, is given by one-half of the real part of the product of one with the complex conjugate of the other.

2.7. The numerical value of the Poynting vector in a collimated, single-frequency laser beam is 0.13 W cm^{-2}. Calculate the amplitudes of the fields $\underline{E}$ and $\underline{H}$ and the pressure exerted on a completely absorbing surface normal to the beam.

2.8. A particle of charge $-e$ executes a circular path in the x-y plane given by $\underline{r}=r_0(\hat{\underline{i}}\cos\omega t + \hat{\underline{j}}\sin\omega t)$. Show, by using equation (2.71), that the angular distribution of the emitted radiation varies as $(1+\cos^2\theta)$.

2.9. Circularly-polarized waves of finite extent propagating in the z-direction are given approximately by

$$\underline{E}(\underline{r},t) \cong \left\{E_0(x,y)(\hat{\underline{i}}\pm i\hat{\underline{j}}) + \frac{i}{k}\left(\frac{\partial E_0}{\partial x} \pm i\frac{\partial E_0}{\partial y}\right)\hat{\underline{k}}\right\}\exp\{i(kz-\omega t)\}$$

where $E_0(x,y)$ is some arbitrary function of x and y. Show that the magnetic field is given approximately by

$$\underline{B} \cong \mp i(\varepsilon_0\mu_0)^{1/2}\,\underline{E}.$$

Calculate the time-averaged components of the angular momentum L_z parallel to the direction of propagation and show that the ratio of L_z to the electromagnetic energy density U is given by $L_z/U = \pm(1/\omega)$.

2.10 A linear quadrupole oscillator consists of a charge +2e which is stationary at the origin and two negative charges of -e at z_1 and z_2 given by

$$z_1 = -z_2 = a\cos(\omega t/2).$$

Compute the electric and magnetic fields at a large distance from the origin and find the rate at which energy is radiated by the system.

General references and further reading

Bleaney, B.I. and Bleaney, B. (1976). *Electricity and magnetism.* (3rd edition). Oxford University Press, London.

Born, M. and Wolf, E. (1970). *Principles of optics*, Pergamon Press, Oxford.

Jackson, J.D. (1962). *Classical electrodynamics*, Wiley, New York.

Panofsky, W.K.H. and Phillips, M. (1962). *Classical electricity and magnetism*, Addison-Wesley, Reading, Mass.

Rose, M.E. (1955). *Multipole fields*, Wiley, New York.

Stone, J.M. (1963). *Radiation and optics*, McGraw-Hill, New York.

3

Review of quantum mechanics

The main aim of this chapter is to obtain the form of the wavefunctions which describe atomic energy levels. These wavefunctions will be essential in our discussion of radiation processes in later chapters. We start by introducing Schrödinger's equation and summarize some of the operations which are fundamental to quantum mechanics. We then show how the angular part of Schrödinger's equation can be solved exactly in the case of spherically symmetric potentials. The orbital angular momentum operator is defined and the physical significance of the quantum numbers l,m is explained. The radial dependence of the wavefunctions is illustrated by considering hydrogenic systems. Next the concept of intrinsic spin is discussed and the effect of spin-orbit interaction in one-electron atoms is treated in some detail. We outline the extension of these results to atoms containing several electrons by considering the central-field approximation. Finally Russell-Saunders or L-S coupling and the fine structure of atomic energy levels is discussed since this introduces the spectroscopic notation which will be used frequently in later chapters.

3.1. *The Schrödinger wave equation*

In his doctoral dissertation de Broglie postulated that particles such as the electron, proton, etc. should also possess wave-like properties in exact analogy with the particle-like properties exhibited by electromagnetic waves in the quantum theory of radiation. For motion in one dimension he postulated that the momentum of the particle p and its kinetic energy E were related to the wavevector k and angular frequency ω of the guiding wave, Ψ, by the relations

$$p = \hbar k; \quad E = \hbar\omega. \tag{3.1}$$

In the case of a free particle of mass m moving along the x-axis, the energy and momentum satisfy the equation

$$E = p^2/2m.$$

The wavefunction $\Psi(x,t) = \exp\{i(kx-\omega t)\}$ describing the motion of the particle must therefore satisfy an equation involving a first-order derivative with respect to t and a second-order derivative with respect to x. One such possible equation is

$$i\hbar \frac{\partial \Psi}{\partial t} = -\frac{\hbar^2}{2m} \frac{\partial^2 \Psi}{\partial x^2} . \tag{3.2}$$

To extend this to three dimensions we use

$$\underline{p} = \hbar \underline{k}, \qquad k = |\underline{k}| = 2\pi/_\lambda$$

and the wave equation becomes

$$i\hbar \frac{\partial \Psi}{\partial t} = -\frac{\hbar^2}{2m} \nabla^2 \Psi, \tag{3.3}$$

with the complex solution

$$\Psi(r,t) = A \exp\{i(\underline{k}.\underline{r}-\omega t)\}.$$

This suggests that the energy and momentum of a particle can be represented by differential operators which act on the wavefunction $\Psi(r,t)$ where

$$E \to i\hbar \frac{\partial}{\partial t} \ ; \ \underline{p} \to -i\hbar \ \mathrm{grad}. \tag{3.4}$$

For a particle of mass m moving in a potential $V(\underline{r})$ the sum of the kinetic and potential energy of the system is constant. Schrödinger postulated, therefore, that equation (3.3) could be generalized to include this case and that the equation governing the motion of the particle is then

$$i\hbar \frac{\partial \Psi(\underline{r},t)}{\partial t} = \{-\frac{\hbar^2}{2m} \nabla^2 + V(\underline{r})\} \Psi(\underline{r},t). \tag{3.5}$$

The validity of Schrödinger's wave equation rests not on the plausibility arguments we have used to justify it but on the excellent agreement between theoretical predictions made from equation (3.5) and experimental results.

The usual interpretation of the wavefunction is that the probability $P(\underline{r},t)\,dv$ of finding the particle in the volume element dv centred on the point $\underline{r}$ at time t is given by

$$P(\underline{r},t)\,dv = |\Psi(\underline{r},t)|^2\,dv. \qquad (3.6)$$

The time-dependent Schrödinger equation has solutions of the form

$$\Psi(\underline{r},t) = \psi(\underline{r})\exp(-iEt/\hbar) \qquad (3.7)$$

where $\psi(\underline{r})$ is a solution of the time-independent equation

$$[-\frac{\hbar^2}{2m}\nabla^2 + V(\underline{r})]\,\psi(\underline{r}) = E\,\psi(\underline{r}). \qquad (3.8)$$

This equation may be written symbolically in the form

$$\mathcal{H}_0\,\psi = E\,\psi \qquad (3.9)$$

where

$$\mathcal{H}_0 = -\frac{\hbar^2}{2m}\nabla^2 + V(\underline{r})$$

is called the Hamiltonian of the system, and in a conservative system represents the sum of the kinetic and potential energies. Thus equation (3.9) is in the form of an eigenvalue equation in which the wavefunction $\psi(\underline{r})$ is the *eigenfunction* and E is the *eigenvalue* of the operator $\mathcal{H}_0$. The eigenvalue is in fact the measurable value of the physical observable represented by the operator, which in this case gives the total energy of the system. If $V(\underline{r})$ is finite, we require the wavefunction $\psi(\underline{r})$ to be continuous, finite, and single valued at every point in space. This ensures that a definite physical situation can be represented uniquely by a given wavefunction and that the probability density $P(\underline{r},t)$ is everywhere finite and continuous (Problem 3.1).

3.2. *Expectation values and matrix elements*

Before solving the time-independent Schrödinger equation, equation (3.9), explicitly we briefly review some of the algebraic properties of wavefunctions. All the information about a quantum-mechanical system is contained in the wavefunction $\Psi(\underline{r},t)$ which is usually subjected to the normalization condition

$$\int \Psi^*(\underline{r},t)\ \Psi(\underline{r},t)\ dv = 1 \qquad (3.10)$$

where the integration is over all space. To obtain this information we must make a series of measurements on the system. We assume that every observable such as position, momentum, energy, etc. can be represented by an operator which acts on the wavefunction. If Γ is such an operator, we define the *expectation value* of Γ as

$$\langle \Gamma \rangle = \langle \Psi | \Gamma | \Psi \rangle = \int \Psi^* \Gamma \Psi\ dv \qquad (3.11)$$

where Ψ^* is the complex conjugate of Ψ. The expectation value can be regarded as the average of the results of a large number of measurements made on independent systems all of which are described by the same wavefunction Ψ. If $\Psi = u_i$ is an eigenfunction of the operator Γ with the eigenvalue γ_i then

$$\Gamma u_i = \gamma_i u_i$$

and the expectation value of Γ is simply

$$\langle \Gamma \rangle = \int u_i^*\ \Gamma u_i\ dv = \gamma_i \int u_i^*\ u_i\ dv = \gamma_i \qquad (3.12)$$

since u_i is assumed to be normalized (Problem 3.2). In this case every measurement on the system gives the result γ_i with zero statistical spread in the values observed.

If Ψ is not an eigenfunction of Γ it can be expanded in terms of any complete set of normalized orthogonal eigenfunctions w_i:

$$\Psi = \sum_i a_i\ w_i \quad \text{where} \quad \int w_j^*\ w_i\ dv = \delta_{ji}\ .$$

The expectation value of Γ is now given by

$$\langle \Gamma \rangle = \sum_{i,j} a_j^* a_i \int w_j^* \Gamma w_i \, dv$$

$$= \sum_{i,j} a_j^* a_i \langle j|\Gamma|i \rangle.$$

Thus the operator Γ is now represented by a square matrix with the element in the jth row and i^{th} column being given by $\langle j|\Gamma|i \rangle$. The quantity $\langle j|\Gamma|i \rangle = \Gamma_{ji}$ is called the matrix element of the operator Γ. It depends not only on the form of the operator but also on the *basis set* w_i which was chosen for the *representation* of the operator. If the functions w_i are eigenfunctions of the operator Γ corresponding to the eigenvalue γ_i, then we have

$$\langle \Gamma \rangle = \sum_{i,j} a_j^* a_i \gamma_i \delta_{ij},$$

and the matrix of the operator is *diagonal* in this representation. This of course means that the basis set of eigenfunctions is a particularly simple and useful representation in which to work.

3.3. Solution of Schrödinger's equation for spherically-symmetric potentials

We shall be concerned mainly with the wavefunctions of electrons in atoms and in this case the predominant contribution to the potential comes from the Coulomb attraction of the nucleus. This potential is spherically symmetric and therefore $V(\underline{r})$ is a function of the radial coordinate r alone. This enables the Schrödinger equation to be separated into three differential equations which involve r, θ, and ϕ separately. If we consider the motion of a single electron of mass m about a nucleus of mass M we can separate off the centre-of-mass motion and consider only the relative motion of the electron. In spherical polar coordinates equation (3.8) becomes (Problem 3.3)

$$-\frac{\hbar^2}{2\mu}\left[\frac{1}{r^2}\frac{\partial}{\partial r}\left(r^2\frac{\partial}{\partial r}\right)+\frac{1}{r^2\sin\theta}\frac{\partial}{\partial\theta}\left(\sin\theta\frac{\partial}{\partial\theta}\right)+\frac{1}{r^2\sin^2\theta}\frac{\partial^2}{\partial\phi^2}\right]\psi+$$

$$+\ V(r)\ \psi = E\psi \qquad (3.13)$$

where $\mu = mM/(m+M)$ is the reduced mass of the electron. If we write the wavefunction in the form

$$\psi(r,\theta,\phi) = R(r)\ Y(\theta,\phi)$$

and substitute into equation (3.13), we obtain

$$\frac{1}{R}\frac{d}{dr}\left(r^2\frac{dR}{dr}\right)+\frac{2\mu r^2}{\hbar^2}\{E - V(r)\} = -\frac{1}{Y}\left(\frac{1}{\sin\theta}\frac{\partial}{\partial\theta}\left(\sin\theta\frac{\partial Y}{\partial\theta}\right)+\right.$$

$$\left.+\frac{1}{\sin^2\theta}\frac{\partial^2 Y}{\partial\phi^2}\right). \qquad (3.14)$$

Since the left-hand side of equation (3.14) depends only on r and the right-hand side on θ and ϕ, both sides must equal a constant which we call λ. Thus Schrödinger's equation separates into a radial equation

$$\frac{1}{r^2}\frac{d}{dr}\left(r^2\frac{dR}{dr}\right)+\left[\frac{2\mu}{\hbar^2}\{E - V(r)\}-\frac{\lambda}{r^2}\right]R = 0 \qquad (3.15)$$

and an angular equation

$$\frac{1}{\sin\theta}\frac{\partial}{\partial\theta}\left(\sin\theta\frac{\partial Y}{\partial\theta}\right)+\frac{1}{\sin^2\theta}\frac{\partial^2 Y}{\partial\phi^2}+\lambda Y = 0. \qquad (3.16)$$

It is apparent from equation (3.15) that the energy of the electron is determined mainly by the radial part of the wavefunction R(r), just as we would expect from classical mechanics. However, the detailed form of the radial wavefunction can only be determined exactly when the dependence of V(r) on r is specified, and then only in certain special cases. By contrast the angular dependence of the wavefunctions can be obtained relatively simply, even in the general case, by means of the substitution $Y(\theta,\phi) = \Theta(\theta)\Phi(\phi)$. Equation (3.16) then separates into:

$$\frac{d^2\Phi}{d\phi^2} + m^2\Phi = 0 \tag{3.17}$$

$$\frac{1}{\sin\theta}\frac{d}{d\theta}\left(\sin\theta\frac{d\Theta}{d\theta}\right) + \left(\lambda - \frac{m^2}{\sin^2\theta}\right)\Theta = 0, \tag{3.18}$$

where m^2 is the separation constant. The constant is chosen in this form since equation (3.17) then has solutions of the type

$$\Phi(\phi) = A\exp(im\phi) + B\exp(-im\phi)\,. \tag{3.19}$$

The physical requirement that $\Phi(\phi)$ shall be continuous and single-valued in the range $0 \le \phi \le 2\pi$ can only be satisfied if m is an integer

$$m = 0, \pm 1, \pm 2, \ldots\ldots\,. \tag{3.20}$$

Unfortunately the solutions of the θ-dependent equation are considerably more involved. By substituting $\xi = \cos\theta$ in equation (3.18) we obtain

$$\frac{d}{d\xi}\left\{(1-\xi^2)\frac{d\Theta}{d\xi}\right\} + \left\{\lambda - \frac{m^2}{1-\xi^2}\right\}\Theta = 0\,. \tag{3.21}$$

Again we require for physical reasons that the solutions of equation (3.21) should be single-valued, finite, and continuous in the range $-1 \le \xi \le 1$. This can only be achieved if the constant λ has the form

$$\lambda = l(l+1), \quad l = 0,1,2,\ldots. \tag{3.22}$$

and if $\qquad |m| \le l.$

The physically acceptable solutions of equation (3.21) are known as associated Legendre functions, $\Theta(\theta) = P_l^m(\cos\theta)$. These are polynomials in $\cos\theta$ whose form depends on the particular values of m and l.

The solutions of equation (3.16) are now seen to have the form

$$Y_l^m(\theta,\phi) = N_l^m\, P_l^m(\cos\theta)\exp(im\phi)$$

where N_l^m is a constant whose value is chosen so that the total angular wavefunction is normalized when integrated over the surface of the unit sphere:

$$\int_0^{2\pi} d\phi \int_0^{\pi} Y_l^{m*} \, Y_l^m \sin\theta \, d\theta = 1. \tag{3.23}$$

These functions are called the normalized spherical harmonics and, for $m \geq 0$, are given by:

$$Y_l^m(\theta,\phi) = (-)^m \left\{ \frac{(2l+1)}{4\pi} \frac{(l-m)!}{(l+m)!} \right\}^{\frac{1}{2}} P_l^m(\cos\theta)\exp(im\phi). \tag{3.24}$$

The explicit form of the spherical harmonics of low-order numbers (l,m) are given in Table 3.1. It may be verified by direct integration that they also obey the orthogonality relation (Problem 3.4)

$$\int_0^{2\pi} d\phi \int_0^{\pi} Y_l^{m*} \, Y_{l'}^{m'} \sin\theta \, d\theta = \delta_{mm'} \, \delta_{ll'} \, . \tag{3.25}$$

TABLE 3.1.

The normalized spherical harmonics $Y_l^m(\theta,\phi)$

$$Y_0^0 = \left(\frac{1}{4\pi}\right)^{1/2}$$

$$Y_1^0 = \left(\frac{3}{4\pi}\right)^{1/2} \cos\theta$$

$$Y_1^{\pm 1} = \mp\left(\frac{3}{8\pi}\right)^{1/2} \sin\theta \, \exp(\pm i\phi)$$

$$Y_2^0 = \left(\frac{5}{16\pi}\right)^{1/2} (3\cos^2\theta - 1)$$

$$Y_2^{\pm 1} = \mp\left(\frac{5}{8\pi}\right)^{1/2} \sin\theta \cos\theta \, \exp(\pm i\phi)$$

$$Y_2^{\pm 2} = \left(\frac{15}{32\pi}\right)^{1/2} \sin^2\theta \, \exp(\pm 2i\phi)$$

The spherical harmonics of negative m are given by

$$Y_l^{-m}(\theta,\phi) = (-)^m\, Y_l^{m*}(\theta,\phi). \qquad (3.26)$$

These solutions to the angular part of the one-electron Schrödinger equation are very important since they determine many of the atomic properties in which we are interested. These include the polarization of emitted light, selection rules for the emission and absorption of radiation, and Zeeman and Stark splittings. Thus although we can often only find approximate solutions for the radial wavefunctions, equation (3.15), we can usually describe the angular properties exactly (Problem 3.5). As an illustration of this, we consider the effect of inverting the coordinates of the electron through the origin,

$$(r,\theta,\phi) \rightarrow (r,\ \pi-\theta,\ \phi+\pi).$$

This is known as the parity operation. The effect on the wavefunction obviously depends only on the properties of the spherical harmonics. Since it can be shown (Problem 3.6) that

$$Y_l^m(\pi-\theta,\ \phi+\pi) = (-)^l\, Y_l^m(\theta,\phi), \qquad (3.27)$$

the wavefunction will either change sign or remain unchanged, depending on whether l is odd or even. The wavefunction is said to possess odd or even parity. The atomic Hamiltonian is unchanged or *invariant* under the parity operation provided there are no external perturbations such as electric or magnetic fields acting on the atom. Hence the parity of the wavefunction is a constant of the motion. This has the immediate consequence that in a state of well-defined parity the expectation value of the electric dipole moment of an atom, defined by

$$\langle -e\underline{r}\rangle = \int\psi^*(\underline{r})(-e\underline{r})\psi(\underline{r})\ dv,$$

is zero. This follows since if we calculate this expectation value in a coordinate system obtained by reflection through the origin, the sign of $\psi^*\psi$ does not change while that of $\underline{r}$ does. We would therefore have the non-physical situation of obtaining two different results for $\langle -e\underline{r} \rangle$ depending on our choice of coordinate system. The parity of atomic wavefunctions is also important in determining the selection rules for radiative transitions between excited states, as we shall see in sections 5.2.3 and 7.3.1.

3.4. *Orbital angular momentum*

In classical mechanics the angular momentum of a particle is given by

$$\underline{L}_{\text{classical}} = \underline{r} \wedge \underline{p} \tag{3.28}$$

where $\underline{r}$ is the position vector with respect to the origin and $\underline{p}$ is the linear momentum. The quantum-mechanical orbital angular momentum operator is obtained by substituting $\underline{p} = -i\hbar$ grad, giving

$$\underline{L}_{\text{quantum}} = \hbar \underline{l} = -i\hbar\, \underline{r} \wedge \text{grad}. \tag{3.29}$$

The factor $\hbar$ is introduced to make $\underline{l}$ a dimensionless operator with cartesian components

$$l_x = -i\left(y \frac{\partial}{\partial z} - z \frac{\partial}{\partial y}\right), \tag{3.30}$$

where l_y and l_z are obtained by cyclic permutation. Transforming the operator l_z into spherical polar coordinates gives

$$l_z = -i \frac{\partial}{\partial \phi}, \tag{3.31}$$

and from equation (3.24) we see that $Y_l^m(\theta,\phi)$ is an eigenfunction of the operator $\hbar l_z$ with the eigenvalue $m\hbar$:

$$\hbar l_z\, Y_l^m(\theta,\phi) = -i\hbar \frac{\partial}{\partial \phi} Y_l^m(\theta,\phi) = m\hbar\, Y_l^m(\theta,\phi). \tag{3.32}$$

We can also show (Problem 3.7) that

$$\underline{l}^2 = -\left[\frac{1}{\sin\theta}\frac{\partial}{\partial\theta}\left(\sin\theta\frac{\partial}{\partial\theta}\right) + \frac{1}{\sin^2\theta}\frac{\partial^2}{\partial\phi^2}\right], \tag{3.33}$$

which enables equation (3.16) to be written as an eigenvalue equation

$$\underline{l}^2\, Y_l^m(\theta,\phi) \;=\; l(l+1)\, Y_l^m(\theta,\phi). \tag{3.34}$$

We can now give the quantum numbers l,m a physical significance: $\hbar^2 l(l+1)$ is the expectation value of the square of the orbital angular momentum operator and $m\hbar$ is the expectation value of the z-component of this operator. This corresponds to the classical result that the angular momentum is a constant of the motion in a spherically-symmetric potential, V(r). However, unlike the classical case, only the z-component of $\underline{l}$ has well-defined values, since $Y_l^m(\theta,\phi)$ is not an eigenfunction of l_x, l_y. In addition equations (3.20) and (3.22) restrict l_z to a discrete set of integer values, $2l+1$ in number. This is known as space quantization and m is called the magnetic or spatial quantum number. We can represent $\underline{l}$ diagrammatically by a vector of length $\{l(l+1)\}^{\frac{1}{2}}$ whose projection on the z-axis can only have integral values between $+l$ and $-l$, as shown in Fig.3.1. By allowing this vector to precess at a constant angle about the z-axis, we can include the fact that the components l_x and l_y are not observable, i.e. they are not constants of the motion (Problem 3.8).

3.5. *Hydrogenic wavefunctions*

In hydrogenic systems such as H, He^+, Li^{++}, ... the central potential has the simple form $V(r) = -Ze^2/4\pi\varepsilon_0 r$ where Z is the nuclear charge. In this case the radial wave-equation, equation (3.15), can be solved exactly. We make the following substitutions:

$$a^2 \;=\; -\,\hbar^2/8\mu E \tag{3.35}$$

$$\rho \;=\; r/a \tag{3.36}$$

$$n \;=\; \frac{2\mu Z e^2 a}{4\pi\varepsilon_0 \hbar^2}\,. \tag{3.37}$$

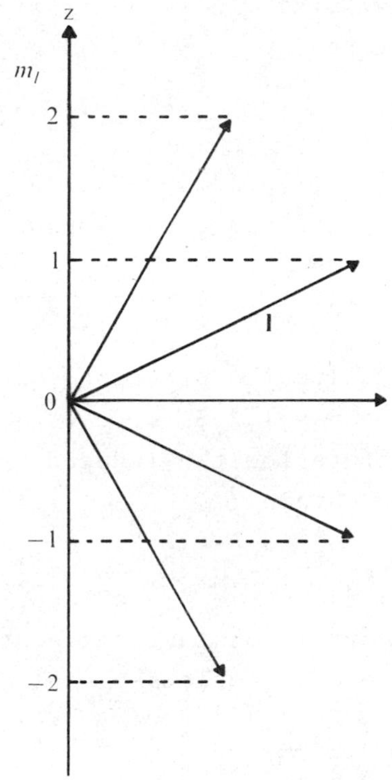

Fig.3.1. Space quantization of orbital angular momentum showing the $2l+1$ projections of the angular momentum vector $\underline{l}$ on the z-axis. The figure is drawn for the case $l=2$.

Recalling that $\lambda = l(l+1)$, we may transform equation (3.15) into

$$\frac{1}{\rho^2}\frac{d}{d\rho}\left(\rho^2\frac{dR}{d\rho}\right) + \left\{\frac{n}{\rho} - \frac{1}{4} - \frac{l(l+1)}{\rho^2}\right\}R = 0 . \quad (3.38)$$

Solutions of equation (3.38) which remain finite as $\rho\to\infty$ and also as $\rho\to 0$, and are therefore physically acceptable, have the form

$$R(\rho) = \exp(-\rho/2)\rho^l F(\rho) \quad (3.39)$$

where $F(\rho)$ is a polynomial in ρ. In order that the power series for $F(\rho)$ should terminate at the term $\rho^{n'}$, thus making $F(\rho)$ a finite polynomial, it is necessary that n should be a positive integer given by

$$n = n' + l + 1, \quad (n' = 0, 1, 2, \ldots). \tag{3.40}$$

From equations (3.35), (3.37), and(3.40) we see that the negative energy eigenvalues are given by

$$E_n = -\frac{Z^2 e^4 \mu}{2(4\pi\varepsilon_0)^2 \hbar^2 n^2}. \tag{3.41}$$

The integer n is called the principal quantum number and can have any value from n = 1,2,3, ...

If we substitute for the reduced mass $\mu = mM/(m+M)$ in equation (3.41), we obtain

$$E_n = -hc\, R_\infty \frac{Z^2}{n^2} \frac{1}{(1+m/M)} \tag{3.42}$$

where R_∞ is a universal constant called the Rydberg, corresponding to the case of a nucleus of infinite mass:

$$R_\infty = \frac{1}{hc}\left(\frac{me^4}{2(4\pi\varepsilon_0)^2\hbar^2}\right) = 109\ 737{\cdot}31\ \text{cm}^{-1}.$$

From equation (3.41) we see that states with the same value of n but different values of the quantum numbers l,m have the same energies. They are said to be *degenerate* and the total number of states with the same energy is given by

$$\sum_{l=0}^{n-1} (2l+1) = n^2.$$

The degeneracy with respect to the orbital quantum number l is a special feature of the energy-level schemes of hydrogenic systems brought about by the Coulomb potential $V(r) \propto 1/r$. It is not found in other effectively one-electron systems such as the alkalis, where the orbital states l of a given principal quantum number n usually differ considerably in energy. In these atoms the valence electron moves in a central potential in which the Coulomb potential of the nucleus is modified by the screening effect of the inner shells of electrons. The hydrogenic energy-level scheme is approached only at large values of n when the valence electron no longer penetrates this inner core. The accidental l-degeneracy in hydrogenic systems has the important consequence that the Stark splitting is a linear function of the strength of the applied electric field, whereas in other systems it varies quadratically with the field and is generally much smaller. The degeneracy with respect to the azimuthal quantum number m is a general feature of atomic energy levels. It can be lifted by the application of external electric or magnetic fields.

The shape of the radial part of the wavefunction is determined by the quantum numbers n and l. Therefore the total wavefunction has the form

$$\psi_{nlm}(r,\theta,\phi) = R_{nl}(r)\, Y_l^m(\theta,\phi). \tag{3.43}$$

Since the spherical harmonics $Y_l^m(\theta,\phi)$ have been normalized by integrating over the surface of the unit sphere, the complete wavefunction will be normalized over all space if

$$\int R^*_{nl}(r)\, R_{nl}(r)\, r^2\, dr = 1. \tag{3.44}$$

The explicit expressions for the first few normalized hydrogenic radial wavefunctions are given in Table 3.2 and plotted in Fig.3.2. The scale of the radial wavefunctions is determined mainly by the exponential term in equation (3.39) through the characteristic length

$$2a = \frac{n}{Z}\left\{\frac{(4\pi\varepsilon_0)\hbar^2}{\mu e^2}\right\} \cong \frac{na_0}{Z} \tag{3.45}$$

where a_0 is the radius of the first Bohr orbit for a nucleus of infinite mass:

$$a_0 = \left\{\frac{(4\pi\varepsilon_0)\hbar^2}{me^2}\right\} = 5{\cdot}29177 \times 10^{-11}\ \text{m}. \tag{3.46}$$

From the graphs of Fig. 3.2 it can be seen that larger values of n correspond to wavefunctions of greater radial extension in accord with the Bohr model (Problems 3.9 and 3.10).

The dependence of $R_{nl}(r)$ on the quantum number l which is apparent in equation (3.39) has the important consequence that states for which $l=0$ have a finite probability density at the origin,

$$|\psi_{n00}(0)|^2 = \frac{Z^3}{\pi a_0^3 n^3},$$

whereas for $l\neq 0$ this probability density is zero. The finite overlap with the nuclear charge distribution which occurs for states with $l=0$ leads to large magnetic hyperfine structures and to large nuclear volume effects in the isotope shifts of medium and heavy elements.

TABLE 3.2.

The normalized hydrogenic radial wavefunctions $R_{nl}(r)$

$$R_{10}(r) = \left(\frac{Z}{a_0}\right)^{3/2} 2\exp\left(-\frac{Zr}{a_0}\right)$$

$$R_{20}(r) = \left(\frac{Z}{2a_0}\right)^{3/2} 2\left(1-\frac{Zr}{2a_0}\right)\exp\left(-\frac{Zr}{2a_0}\right)$$

$$R_{21}(r) = \left(\frac{Z}{2a_0}\right)^{3/2} \frac{2}{\sqrt{3}}\left(\frac{Zr}{2a_0}\right)\exp\left(-\frac{Zr}{2a_0}\right)$$

$$R_{30}(r) = \left(\frac{Z}{3a_0}\right)^{3/2} 2\left\{1-\frac{2Zr}{3a_0}+\frac{2}{3}\left(\frac{Zr}{3a_0}\right)^2\right\}\exp\left(-\frac{Zr}{3a_0}\right)$$

$$R_{31}(r) = \left(\frac{Z}{3a_0}\right)^{3/2} \frac{4\sqrt{2}}{3}\left(\frac{Zr}{3a_0}\right)\left(1-\frac{Zr}{6a_0}\right)\exp\left(-\frac{Zr}{3a_0}\right)$$

$$R_{32}(r) = \left(\frac{Z}{3a_0}\right)^{3/2} \frac{2}{3}\left(\frac{2}{5}\right)^{1/2}\left(\frac{Zr}{3a_0}\right)^2\exp\left(-\frac{Zr}{3a_0}\right)$$

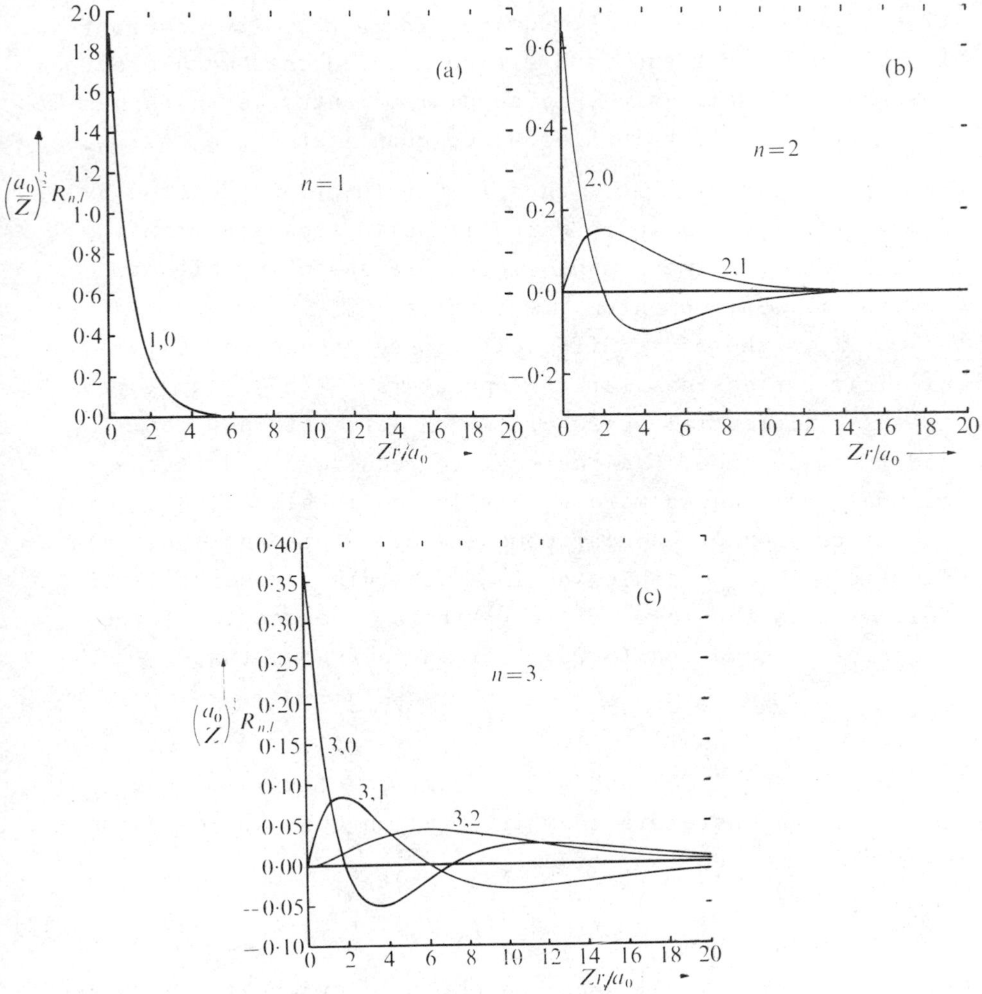

Fig.3.2. Radial wavefunctions for hydrogenic systems. Note the different ordinate scales.

3.6. *Spin angular momentum*

We have obtained a detailed description of the spatial wavefunction, $\psi_{nlm}(r,\theta,\phi)$, of one-electron atoms, equation (3.43). However, the Stern-Gerlach experiment, the anomalous Zeeman effect, and the fine structure of spectral lines provide ample evidence that the wavefunction is incomplete and that an additional degree of freedom or an additional quantum number is required for a complete description. In 1925 Uhlenbeck and Goudsmit postulated that each electron possesses an intrinsic spin angular momentum $\hbar\underline{s}$ which has only two possible values of space quantization, $m_s = \pm\frac{1}{2}$. The spin operator $\underline{s}$ acts on spin wavefunctions denoted by $\chi(m_s)$ in a special space which is quite separate from the ordinary three-dimensional coordinate space in which the orbital angular momentum $\hbar\underline{l}$ operates.

It is therefore difficult to see how we can obtain explicit expressions for the operators $\underline{s}^2$ and s_z analogous to the differential operators which were obtained for $\underline{l}^2$ and l_z in section 3.4. We seem to require a wider conceptual basis and a more generally applicable technique for the discussion of angular momentum operators and their eigenfunctions. One possible way in which this may be achieved follows from the form of the cartesian components of the operator $\underline{l}$, equation (3.30). It may be shown that

$$(l_x l_y - l_y l_x)f = i l_z f$$

where f is any differentiable function of x,y,z. This equation can therefore be written as an operator equation with the function f omitted:

$$l_x l_y - l_y l_x = [l_x, l_y] = i l_z. \tag{3.47}$$

Similar equations which can be obtained by cyclic permutation are conveniently summarized by

$$\underline{l} \wedge \underline{l} = i\underline{l} . \tag{3.48}$$

These are known as the commutation relations for orbital angular momentum.

By purely algebraic methods it can be shown that the whole quantum theory of angular momentum follows from these equations. In particular, if $\underline{j}$ is a Hermitian operator which is defined to obey the commutation relations

$$[j_x, j_y] = i j_z, \tag{3.49}$$

then there exist eigenfunctions ψ_{jm}, whose explicit form need not be specified, which are simultaneous eigenfunctions of $\underline{j}^2$ and j_z with eigenvalues given by

$$\begin{aligned} \underline{j}^2 \psi_{jm} &= j(j+1)\, \psi_{jm} \\ j_z \psi_{jm} &= m\, \psi_{jm}. \end{aligned} \tag{3.50}$$

The quantum number j can have any of the values $j = 0, \frac{1}{2}, 1, \frac{3}{2}, 2, \ldots$, that is any integer or half integer, and for a fixed j the permitted values of m are

$$m = j, j-1, j-2, \ldots., -j+1, -j. \tag{3.51}$$

In this purely formal theory the operatory $\hbar \underline{j}$ can represent either orbital or spin angular momentum, or as we shall see below, can even represent the sum of two angular momenta (Problem 3.11).

3.7. *Coupling of two angular momenta*

If $\underline{j}_1$ and $\underline{j}_2$ are operators which separately obey the commutation relations, equation (3.49), then it can be shown that the operator defined by

$$\underline{j} = \underline{j}_1 + \underline{j}_2 \tag{3.52}$$

also obeys the same commutation relations and is generally called the total angular momentum of the system. $\underline{j}_1$ and $\underline{j}_2$ may both be orbital momentum operators, or may both be spin operators or may be a mixture of the two. Thus once we have found eigenfunctions for the individual angular momenta $\underline{j}_1$ and $\underline{j}_2$, we then have the problem of constructing wavefunctions for a compound system of two angular momenta. These wave-functions are essential for the complete description of a single particle whose total angular momentum is a sum of orbital and intrinsic spin, as we shall see in section 3.8 below. They are also necessary for the discussion of many-electron atoms which we take up in section 3.9. Finally the results obtained in this section will be useful in the discussion of the selection rules for the emission and absorption of radiation which we give in Chapter 5.

Let the eigenfunctions of the angular momentum operators $\underline{j}_1$, $\underline{j}_2$ be $\psi_{j_1m_1}$ and $\psi_{j_2m_2}$ respectively, such that

$$\begin{aligned}\underline{j}_1^2\psi_{j_1m_1} &= j_1(j_1+1)\,\psi_{j_1m_1}\;; \quad j_{1z}\psi_{j_1m_1} = m_1\psi_{j_1m_1}\\ \underline{j}_2^2\,\psi_{j_2m_2} &= j_2(j_2+1)\,\psi_{j_2m_2}\;; \quad j_{2z}\psi_{j_2m_2} = m_2\psi_{j_2m_2}.\end{aligned} \tag{3.53}$$

We have used the same symbol ψ for both eigenfunctions even though they refer to different spaces, and may in the case of orbital and intrinsic spin wavefunctions have completely different forms. The simple direct-product wavefunctions

$$\psi_{j_1m_1j_2m_2} = \psi_{j_1m_1}\,\psi_{j_2m_2}$$

are simultaneous eigenfunctions of the operators $\underline{j}_1^2$, j_{1z}; $\underline{j}_2^2$, j_{2z} and form one possible set of basis functions for describing the combined system. This basis is called the uncoupled representation. We seek a new representation in which $\underline{j}^2$ and j_z together with $\underline{j}_1^2$ and $\underline{j}_2^2$ are diagonal, where $\underline{j}$ is the total angular momentum operator $\underline{j} = \underline{j}_1 + \underline{j}_2$ and j_z is its z-component. This coupled representation, denoted by $\psi_{j_1j_2jm}$, is obtained by taking suitable linear combinations of the uncoupled wavefunctions:

$$\psi_{j_1j_2jm} = \sum_{m_1m_2} C(j_1j_2j;m_1m_2m)\,\psi_{j_1m_1}\,\psi_{j_2m_2}. \tag{3.54}$$

The quantities $C(j_1j_2j;m_1m_2m)$ are simply numerical coefficients called Clebsch-Gordan or vector-coupling coefficients.

The three magnetic quantum numbers m_1,m_2,m are not all independent, as we discover by operating on both sides of equation (3.54) with $j_z = j_{1z} + j_{2z}$. We obtain

$$m\psi_{j_1j_2jm} = \sum_{m_1m_2} (m_1+m_2)C(j_1j_2j;m_1m_2m)\,\psi_{j_1m_1}\psi_{j_2m_2}$$

which may be rewritten as

$$\sum_{m_1 m_2} (m-m_1-m_2)C(j_1 j_2 j;m_1 m_2 m)\ \psi_{j_1 m_1}\ \psi_{j_2 m_2} = 0.$$

Now since the uncoupled wavefunctions $\psi_{j_1 m_1}\ \psi_{j_2 m_2}$ are linearly independent, this summation will only vanish if each term is identically zero, i.e.

$$(m-m_1-m_2)C(j_1 j_2 j;m_1 m_2 m) = 0.$$

Thus the Clebsch-Gordan coefficients are zero unless

$$m = m_1+m_2 .$$

The double summation in equation (3.54) therefore reduces to a single summation since $m_2=m-m_1$ is fixed once m_1 and m are specified. This condition can be incorporated in the Clebsch-Gordan coefficient and the third projection quantum number can be suppressed, giving

$$\psi_{j_1 j_2 jm} = \sum_{m_1} C(j_1 j_2 j;m_1,m-m_1)\ \psi_{j_1 m_1}\ \psi_{j_2 m-m_1}. \qquad (3.55)$$

The Clebsch-Gordan coefficients have the form of square roots of simple fractions and explicit relations for $C(j_1\ \tfrac{1}{2}\ j;m_1,m-m_1)$ and $C(j_1\ 1\ j;m_1,m-m_1)$ are given in Tables 3.3 and 3.4.

The range of the quantum numbers j and m must now be specified. Since $\underline{j}$ is an angular momentum operator we know that

$$-j \le m \le j \quad \text{and} \quad m_{max} = j.$$

Since $m=m_1+m_2$, the maximum value of m is obtained when m_1 and m_2 both have their maximum values. Thus

$$m_{max} = j_1+j_2.$$

This must also be the largest value of the quantum number j, thus $j_{max} = j_1+j_2$. The state in which $j=m=j_1+j_2$ is called the stretched state and there is a particularly simple relation between the coupled and the uncoupled representation in this case, since only one combination of m_1 and m_2 is possible. We have

$$\psi_{jm} = \psi_{j_1+j_2,j_1+j_2} = \psi_{j_1j_1}\psi_{j_2j_2}, \tag{3.56}$$

where the first two subscripts have been dropped from the notation of the coupled representation for the sake of clarity.

The next smaller value of m is $m=j_1+j_2-1$ which corresponds to the two uncoupled states $\psi_{j_1j_1-1}\psi_{j_2j_2}$ and $\psi_{j_1j_1}\psi_{j_2j_2-1}$. There are an infinite number of pairs of independent linear combinations of these states, one of which must represent the coupled state $\psi_{j_1+j_2,j_1+j_2-1}$ and the other must represent the coupled state $\psi_{j_1+j_2-1,j_1+j_2-1}$. Thus the allowed values of j are obtained as the series

$$(j_1+j_2),\ (j_1+j_2-1),\ (j_1+j_2-2),\ \ldots\ldots,\ j_{min}.$$

The smallest allowed value of j is obtained when the total number of linearly-independent coupled states equals the total number of uncoupled states from which they were formed, thus

$$\sum_{j_{min}}^{j_{max}} (2j+1) = (2j_1+1)(2j_2+1).$$

From this equation we find that $j_{min} = |j_1-j_2|$ and the allowed values of j are given by

$$j = (j_1+j_2),\ (j_1+j_2-1), \ldots\ldots, |j_1-j_2|. \tag{3.57}$$

If the numbers j_1, j_2, and j are represented as lengths, then from equation (3.57) we see that these three lengths must be capable of forming the sides of a triangle whose perimeter is an integer. If this triangular condition, represented symbolically by $\Delta(j_1j_2j)$, is not satisfied, then the Clebsch-Gordan coefficients in equation (3.55) are identically zero (Problem 3.12).

TABLE 3.3.

The Clebsch-Gordan coefficients $C(j_1\ \tfrac{1}{2}\ j; m_1, m-m_1)$

	$m_1 = m - \frac{1}{2}$	$m_1 = m + \frac{1}{2}$
$j = j_1+\frac{1}{2}$	$\left(\frac{j_1+m+\frac{1}{2}}{2j_1+1}\right)^{1/2}$	$\left(\frac{j_1-m+\frac{1}{2}}{2j_1+1}\right)^{1/2}$
$j = j_1-\frac{1}{2}$	$-\left(\frac{j_1-m+\frac{1}{2}}{2j_1+1}\right)^{1/2}$	$\left(\frac{j_1+m+\frac{1}{2}}{2j_1+1}\right)^{1/2}$

3.8. *Spin-orbit interaction and the vector model*

We now illustrate the usefulness of the coupled representation by considering the problem of spin-orbit interaction in one-electron atoms. The anomalous Zeeman effect shows that each electron with intrinsic spin angular momentum $\hbar\underline{s}$ possesses an associated magnetic moment given by

$$\underline{\mu} = -g_s\mu_B\underline{s} \tag{3.58}$$

where the spin g-factor has the value $g_s = 2(1{\cdot}001160 \pm 0{\cdot}000002)$ and $\mu_B = (9{\cdot}27410 \pm 0{\cdot}00006) \times 10^{-24}$ J T^{-1} is the value of the Bohr magneton. In an atom this magnetic moment experiences the effect of the magnetic induction $\underline{B}_l$ created by the orbital motion of the electron, and consequently there is an additional contribution to the Hamiltonian of the system of the form

$$\mathcal{H}_{s.o.} = -\underline{\mu}\cdot\underline{B}_l = \xi(r)\ \underline{s}\cdot\underline{l} \tag{3.59}$$

where $\xi(r)$ is given by

$$\xi(r) = \frac{\hbar^2}{2m^2c^2}\frac{1}{r}\frac{dV}{dr}\ . \tag{3.60}$$

TABLE 3.4.

The Clebsch-Gordan coefficients $C(j_1\ 1\ j;\ m_1,\ m-m_1)$

	$m_1 = m - 1$	$m_1 = m$	$m_1 = m + 1$
$j=j_1+1$	$\left\{\frac{(j_1+m)(j_1+m+1)}{2(2j_1+1)(j_1+1)}\right\}^{1/2}$	$\left\{\frac{(j_1-m+1)(j_1+m+1)}{(2j_1+1)(j_1+1)}\right\}^{1/2}$	$\left\{\frac{(j_1-m)(j_1-m+1)}{2(2j_1+1)(j_1+1)}\right\}^{1/2}$
$j=j_1$	$-\left\{\frac{(j_1+m)(j_1-m+1)}{2j_1(j_1+1)}\right\}^{1/2}$	$\frac{m}{\{j_1(j_1+1)\}^{1/2}}$	$\left\{\frac{(j_1-m)(j_1+m+1)}{2j_1(j_1+1)}\right\}^{1/2}$
$j=j_1-1$	$\left\{\frac{(j_1-m)(j_1-m+1)}{2j_1(2j_1+1)}\right\}^{1/2}$	$-\left\{\frac{(j_1-m)(j_1+m)}{j_1(2j_1+1)}\right\}^{1/2}$	$\left\{\frac{(j_1+m)(j_1+m+1)}{2j_1(2j_1+1)}\right\}^{1/2}$

This expression is only really useful in hydrogenic systems where the electrostatic potential V(r) is accurately known.

To obtain a simple description of one-electron atoms we must find wavefunctions which are simultaneous eigenfunctions of the main Hamiltonian $\mathcal{H}_0$, equation (3.9), and of the much smaller spin-orbit interaction, $\mathcal{H}_{s.o.}$, equation (3.59). An alternative way of stating this problem is that we are seeking a representation (i.e. a set of wavefunctions) in which the total Hamiltonian $\mathcal{H} = \mathcal{H}_0 + \mathcal{H}_{s.o.}$ is diagonal.

The wavefunctions of the uncoupled representation,

$$\psi_{nlm_l sm_s} = \psi_{nlm_l}\,\chi(m_s), \qquad (3.61)$$

which are obtained by taking direct products of the spatial wavefunctions, equation (3.43), with the spin wavefunctions, do not satisfy this requirement. The uncoupled representation would be a good description of the atom only if the spin-orbit interaction was negligible and in this case states with different values of m_l, m_s would be degenerate.

However, wavefunctions which are simultaneous eigenfunctions of the operators $\underline{l}^2$, $\underline{s}^2$, $\underline{j}^2$, and j_z where $\underline{j}=\underline{l}+\underline{s}$, do satisfy this requirement since then

$$\underline{j}^2 = (\underline{l}+\underline{s})^2 = \underline{l}^2 + \underline{s}^2 + 2\underline{l}.\underline{s}. \qquad (3.62)$$

In this representation the expectation value of the spin-orbit interaction has the well-defined value

$$\begin{aligned}\langle \mathcal{H}_{s.o.} \rangle &= \frac{\xi}{2}\langle nljm|\underline{j}^2-\underline{l}^2-\underline{s}^2|nljm\rangle \\ &= \frac{\xi}{2}\{j(j+1)-l(l+1)-s(s+1)\} \qquad (3.63)\end{aligned}$$

where

$$\xi = \frac{\hbar^2}{2m^2c^2}\left\langle \frac{1}{r}\frac{dV}{dr}\right\rangle \text{ (Problem 3.13).}$$

The wavefunctions $\psi_{nljm} \equiv |nljm\rangle$ are just those of the coupled representation given by equation (3.55). For a single electron $s=\frac{1}{2}$, and there are, according to equation (3.57), only two possible values of j given by $j=l\pm\frac{1}{2}$. The effect of the spin-orbit interaction is to split the previously degenerate states into two discrete energy levels separated by an amount $\Delta E = \xi(l+\frac{1}{2})$ as shown in Fig.3.3.

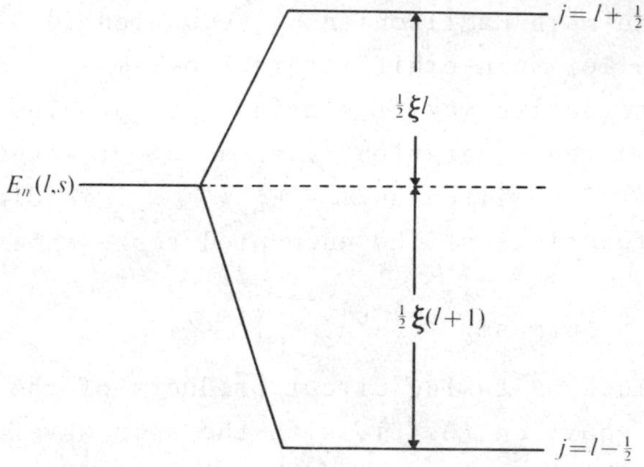

Fig.3.3. The effect of spin-orbit interaction in a one-electron atom.

The following explicit expressions for the wavefunctions representing a one-electron atom may be obtained by using the Clebsch-Gordan coefficients given in Table 3.3:

$$\begin{aligned}\psi_{nl\ j=l+\frac{1}{2},m} &= R_{nl}(r)\left\{\chi(+)\left(\frac{l+\frac{1}{2}+m}{2l+1}\right)^{\frac{1}{2}} Y_l^{m-\frac{1}{2}} + \right.\\ &\left. + \ \chi(-)\left(\frac{l+\frac{1}{2}-m}{2l+1}\right)^{\frac{1}{2}} Y_l^{m+\frac{1}{2}}\right\}\\ \psi_{nl\ j=l-\frac{1}{2},m} &= R_{nl}(r)\left\{\chi(+)\left(\frac{l+\frac{1}{2}-m}{2l+1}\right)^{\frac{1}{2}} Y_l^{m-\frac{1}{2}} - \right.\\ &\left. - \ \chi(-)\left(\frac{l+\frac{1}{2}+m}{2l+1}\right)^{\frac{1}{2}} Y_l^{m+\frac{1}{2}}\right\}.\end{aligned} \tag{3.64}$$

The angular momenta $\underline{l}$ and $\underline{s}$ are said to be coupled by the spin-orbit interaction to give a resultant total angular momentum $\underline{j}$. This coupling, which is represented analytically by equation (3.64), can be shown much more simply and vividly by means of the vector model. We represent the angular momentum operators $\underline{l}$, $\underline{s}$, and $\underline{j}$ as vectors which, because of the requirement $\Delta(lsj)$, must form the sides of a triangle as shown in Fig.3.4. The spin-orbit interaction produces a torque which causes the vectors $\underline{l}$ and $\underline{s}$ to precess about $\underline{j}$ with the angular velocity $\omega = \xi|\underline{j}|/\hbar$. This additional energy gives rise to the two fine-structure energy levels. Since expectation values in quantum mechanics are equivalent to time-averages in the vector model, the lengths $|\underline{l}|$, $|\underline{s}|$, and $|\underline{j}|$ are constant. However, the projections l_z and s_z on a z-axis fixed in space fluctuate due to the precessional motion. This corresponds to the fact that m_l, m_s are no longer good quantum numbers and the uncoupled representation becomes a poorer and poorer description of the atom as the spin-orbit precession gets faster and faster. By contrast, j and m are good quantum numbers since we assume that there are no external forces acting on the atom.

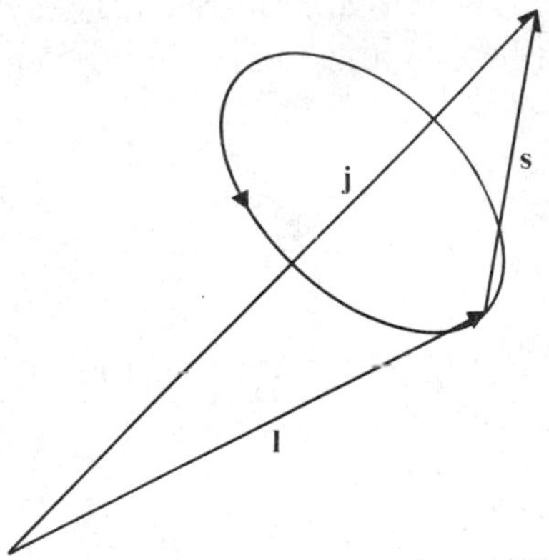

Fig.3.4. Vector model of a single-electron atom showing the precession of the angular momentum vectors $\underline{l}$ and $\underline{s}$ about their resultant $\underline{j}$.

3.9. Many-electron atoms

We wish now to introduce the more general case of an atom possessing N electrons and to derive the wavefunctions and quantum numbers which enable the energy levels of an atom to be labelled in a systematic way. The discussion will not be complete since the problem is considerably more difficult than the case of the one-electron atoms discussed previously.

3.9.1. The central-field approximation. The Schrödinger equation for the electrostatic interaction of N electrons in an atom with a nuclear charge Z is given by

$$\left\{ \sum_{i=1}^{N} \left(-\frac{\hbar^2}{2m}\nabla_i^2 - \frac{Ze^2}{4\pi\varepsilon_0 r_i}\right) + \sum_{i>j} \frac{e^2}{4\pi\varepsilon_0 r_{ij}} \right\} \Psi = E\Psi \qquad (3.65)$$

where the mutual repulsion term is summed over all pairs of electrons and Ψ now represents the time-independent N electron wavefunction. The mutual repulsion term prevents the direct solution of equation (3.65) using single-electron wavefunctions and it is necessary to find an approximate method for dealing with this equation. We therefore assume that each electron moves in a spherically-symmetric potential $U(r_i)$ and rewrite the Hamiltonian of equation (3.65) in the form

$$\mathcal{H} = \mathcal{H}_0 + \mathcal{H}_1 \qquad (3.66)$$

where

$$\mathcal{H}_0 = \sum_{i=1}^{N} \left\{-\frac{\hbar^2}{2m}\nabla_i^2 + U(r_i)\right\} \qquad (3.67)$$

and

$$\mathcal{H}_1 = \sum_{i>j} \frac{e^2}{4\pi\varepsilon_0 r_{ij}} - \sum_{i=1}^{N} \left\{\frac{Ze^2}{4\pi\varepsilon_0 r_i} + U(r_i)\right\}. \qquad (3.68)$$

We assume that $\mathcal{H}_1$, which contains the effect of the non-central forces, is small compared with $\mathcal{H}_0$. Then in the zeroth approximation in which $\mathcal{H}_1$ is neglected, we must solve the equation

$$\mathcal{H}_0\Psi = \sum_{i=1}^{N} \left\{-\frac{\hbar^2}{2m}\nabla_i^2 + U(r_i)\right\} \Psi = E\Psi . \qquad (3.69)$$

Substituting for Ψ a product of single-electron wavefunctions, we find that equation (3.69) is separable into a set of Schrödinger equations for each electron:

$$\{ -\frac{\hbar^2}{2m}\nabla_i^2 + U(r_i)\}\, \psi_{nlm_lm_s}(i) = E_{nl}(i)\, \psi_{nlm_lm_s}(i). \quad (3.70)$$

The total energy is given by the sum of the single-electron energy eigenvalues,

$$E = \sum_{i=1}^{N} E_{nl}(i),$$

and the individual electron wavefunctions or spin-orbitals have the form

$$\psi_\alpha(i) = \psi_{nlm_lm_s}(i) = R_{nl}(r_i)\, Y_l^{m_l}(\theta_i,\phi_i)\, \chi_{m_s}(i) \quad (3.71)$$

where α stands for the set of four quantum numbers n, l, m_l, m_s. Since $U(r_i)$ is no longer a simple Coulomb potential, states with the same n but different values of l may be widely separated in energy. When the values of n_i, l_i for all the N electrons have been given we have specified the electron *configuration*. In the central-field approximation the electrons of the configuration move completely independently of one another since the mutual interaction term, $\mathcal{H}_1$, has been neglected.

3.9.2. The Pauli exclusion principle. According to the Pauli exclusion principle, not more than one electron in a given atom can have a particular set of the four quantum numbers n, l, m_l, m_s. Electrons with the same n, l are called *equivalent* electrons since they have the same energy. Because of the degeneracy with respect to m_s and m_l, a maximum number of $2(2l+1)$ electrons can occupy a *shell* with a specified set of n and l. In the usual spectroscopic notation the value of n for a shell is given as a number, the value of l by a letter and the number of electrons in the shell as a numerical superscript. The letter code and the maximum number of electrons per shell are given by

$$\begin{aligned} l &= 0,\ 1,\ 2,\ 3,\ 4,\ 5 \\ &\ s,\ p,\ d,\ f,\ g,\ h \\ 2(2l+1) &= 2,\ 6, 10, 14, 18, 22 \end{aligned}$$

For example, the ground-state configurations of sodium (Z=11) and mercury (Z=80) are:

$$\text{Na} : 1s^2\ 2s^2\ 2p^6\ 3s$$

$$\text{Hg} : 1s^2\ 2s^2\ 2p^6\ 3s^2\ 3p^6(4s^2\ 3d^{10})4p^6(5s^2\ 4d^{10})5p^6$$
$$(6s^2\ 4f^{14}\ 5d^{10}).$$

The sequence in mercury gives the usual order in which the shells are filled as we go through the periodic system. The brackets indicate shells which have nearly the same energy and where the situation is often more complicated. The chemical properties of the elements are determined mainly by the least tightly-bound or valence electrons (3s in the case of sodium and $6s^2$ in the case of mercury). The spectra of atoms and ions in the visible and near-ultraviolet regions are due to the excitation of one of these valence electrons to states of higher energy.

The Pauli principle can be included in a more general symmetry principle, proposed by Heisenberg and Dirac, which states that electrons and other particles with half-integer spin must be represented by wavefunctions which are totally antisymmetric with respect to an interchange of the spin and coordinates of any two electrons. Slater showed that this requirement could be satisfied by a normalized determinantal product wavefunction of the form

$$\Psi(\alpha,\beta,\ldots,\nu) = (N)^{-1/2} \begin{vmatrix} \psi_\alpha(1) & \psi_\alpha(2) & \ldots\ldots\ldots & \psi_\alpha(N) \\ \psi_\beta(1) & \psi_\beta(2) & \ldots\ldots\ldots & \psi_\beta(N) \\ \cdot & \cdot & \cdot\ \cdot\ \cdot & \cdot \\ \cdot & \cdot & \cdot\ \cdot\ \cdot & \cdot \\ \psi_\nu(1) & \psi_\nu(2) & \ldots\ldots\ldots & \psi_\nu(N) \end{vmatrix} \tag{3.72}$$

since an interchange of two electrons is equivalent to an interchange of two columns of the determinant.

Pauli's principle is included since $\Psi(\alpha,\beta,\ldots,\nu)$ is zero if any pair of the sets of four quantum numbers are equal, e.g. $\alpha = \beta$. The Slater determinantal wavefunctions form an uncoupled representation where the good quantum numbers are the sets (n,l,m_l,m_s) for each electron. There are generally several states belonging to the same configuration corresponding to the different possible combinations of m_l,m_s quantum numbers. In the central-field approximation these states are all degenerate.

3.9.3. Corrections to the central-field approximation. Two important terms are omitted in the central-field approximation. The first is the residual electrostatic interactions represented by $\mathcal{H}_1$ (equation (3.68)). This gives rise to non-central forces which cause the motion of the electrons to be correlated with one another, rather than completely independent as in the central-field approximation. The second term is the magnetic spin-orbit interaction which was not included in equation (3.65):

$$\mathcal{H}_2 = \sum_{i=1}^{N} \xi(r_i)\, \underline{l}_i \cdot \underline{s}_i \,. \qquad (3.73)$$

We now assume that the energy intervals between one configuration and the next are very large. We need then only consider the results of $\mathcal{H}_1$ and $\mathcal{H}_2$ acting within a specified configuration. We shall find that the main effect of these interactions is to remove some of the degeneracy which exists in the central-field approximation.

It can be easily shown that the sum in $\mathcal{H}_2$ (equation 3.73)) is zero when taken over a closed or full shell. Thus we are concerned only with the effect of the spin-orbit interaction among the valence electrons. Similarly only the electrostatic repulsion term $\sum_{i>j} e^2/4\pi\varepsilon_0 r_{ij}$ in $\mathcal{H}_1$ (equation (3.68)) acting on the valence electrons leads to a splitting of the degenerate states belonging to the configuration. The effect of all the other terms in $\mathcal{H}_1$ taken over the valence electrons, or of $\mathcal{H}_1$ as a whole taken over closed shells, is simply to shift the total energy of the configuration without lifting the degeneracy.

We have therefore to consider the effect of the perturbation

$$\mathcal{H}_1' + \mathcal{H}_2' = \sum_{i>j}' \frac{e^2}{4\pi\varepsilon_0 r_{ij}} + \sum_i' \xi(r_i)\, \underline{l}_i \cdot \underline{s}_i \qquad (3.74)$$

where now the summations are taken only over the valence electrons of the configuration. The theory of complex spectra consists of finding the linear combinations of determinantal wavefunctions belonging to the configuration which diagonalize the perturbation given by equation (3.74).

3.9.4. L-S or Russell-Saunders coupling. In general the problem of diagonalizing $\mathcal{H}_1$ and $\mathcal{H}_2$ simultaneously is rather difficult, and only when either $\mathcal{H}_1$ or $\mathcal{H}_2$ is negligible is it possible to give a simple treatment. The most usual situation is that in which the electrostatic terms in equation (3.74) are much larger than the spin-orbit terms, i.e. $\mathcal{H}_1' \gg \mathcal{H}_2'$. This is called the Russell-Saunders case. The electrostatic interaction of one electron on another produces a torque which causes the individual orbital angular momenta $\underline{l}_i$ to precess as indicated in Fig.3.5. The orbital angular momenta $\underline{l}_i$ are coupled by the electrostatic interaction and a suitable representation is one in which the total orbital angular momentum $\underline{L} = \sum_i \underline{l}_i$ is a constant of the motion. Although $\mathcal{H}_1$ does not act in spin space, the exchange symmetry requirement which is built into the determinantal product wavefunctions means that spin and orbital angular momenta are closely linked. The result is that the individual spin angular momenta $\underline{s}_i$ are also coupled so that the total spin $\underline{S} = \sum_i \underline{s}_i$ is a constant of the motion, as shown in Fig.3.5.

The representation in which $\underline{L}$ and $\underline{S}$ are good quantum numbers is obtained by forming suitable linear combinations of the determinantal product wavefunctions, equation (3.72). The first-order shift in the energy produced by the electrostatic interaction is given by

$$\Delta E_1 = \langle LSM_LM_S \,|\, \sum_{i>j}' \frac{e^2}{4\pi\varepsilon_0 r_{ij}} \,|\, LSM_LM_S \rangle . \qquad (3.75)$$

It can be shown that ΔE_1 depends on the values of L and S,

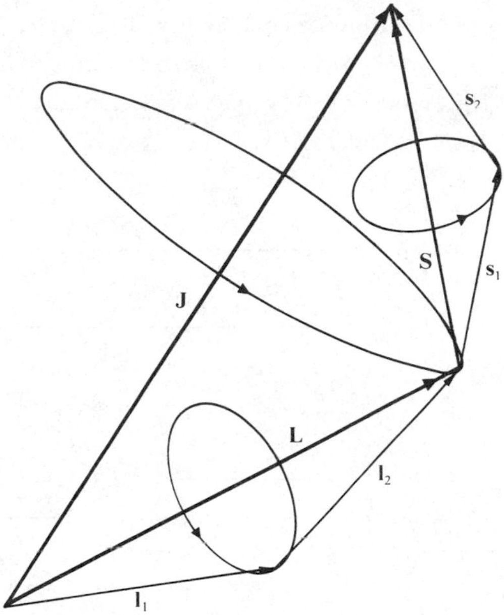

Fig.3.5. Vector model for the addition of the angular momenta of two electrons in L-S coupling.

a particular combination of L,S being called a *term*. Since ΔE_1 is independent of the values of the magnetic quantum numbers M_L and M_S, there is a (2L+1)(2S+1)-fold degeneracy with respect to M_L, M_S.

The spectroscopic notation for a term is obtained by representing the value of the quantum number L by a capital letter according to the code given in section 3.9.2. A numerical superscript is then added to indicate the value of the *multiplicity* given by 2S+1. Thus the term ^3D has L=2 and S=1. For the case of the configuration consisting of the valence electrons 4p3d the possible values of L and S are given by

$$L = 3,2,1$$
$$S = 1,0.$$

All combinations of these values of L and S are allowed and the levels corresponding to the different terms of this configuration are shown schematically in Fig.3.6. If the configuration consists of equivalent electrons, the Pauli exclusion principle considerably reduces the number of possible combinations of L,S (Problem 3.14).

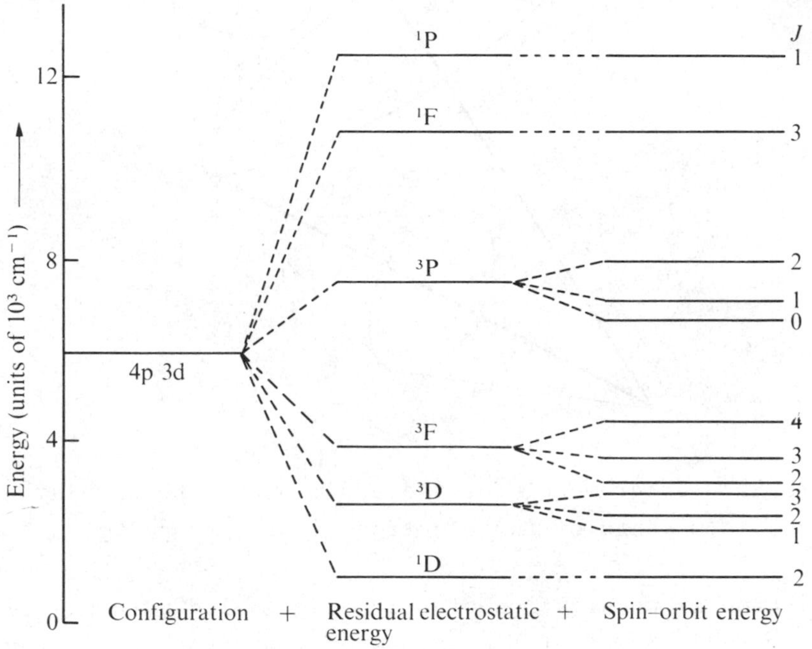

Fig.3.6. The terms and levels belonging to the configuration 4p3d in triply-ionized vanadium. (The spin-orbit splitting is not drawn to scale).

<u>3.9.5. Fine structure in L-S coupling.</u> We now consider the effect of the spin-orbit interaction

$$\mathcal{H}_2' = \sum_i{}' \xi(r_i)\, \underline{l}_i \cdot \underline{s}_i \tag{3.76}$$

If we assume that terms corresponding to different values of L,S are well separated in energy, then only the effect of $\mathcal{H}_2'$ within a single term need be considered. As in the one-electron case discussed in section 3.8, the perturbation is diagonal in a representation in which the total angular momentum $\underline{J}=\underline{L}+\underline{S}$ is a constant of the motion. This is another example of the coupled representation. In this case the wavefunctions are given by

$$\Psi(\gamma\ LSJM_J) = \sum_{M_L} C(LSJ;M_L,M_J-M_L)\ \Psi\ (\gamma\ LSM_LM_S) \qquad (3.77)$$

where γ represents all the other quantum numbers required to specify the configuration. The first-order shift in the energy produced by the spin-orbit interaction is given by

$$\Delta E_2 = \langle \gamma\ LSJM_J | \sum_i{}' \ \xi(r_i)\ \underline{l}_i \cdot \underline{s}_i | \gamma\ LSJM_J \rangle\ . \qquad (3.78)$$

By means of a rather lengthy calculation this expression may be transformed to

$$\begin{aligned} \Delta E_2 &= \langle \gamma\ LSJM_J | \zeta(L,S)\ \underline{L}.\underline{S} | \gamma\ LSJM_J \rangle \\ &= \tfrac{1}{2}\ \zeta(L,S)\ \{J(J+1) - L(L+1) - S(S+1)\} \qquad (3.79) \end{aligned}$$

where $\zeta(L,S)$ has the dimensions of energy and is constant within the levels of a given term.

Thus the spin-orbit interaction lifts the degeneracy of the term with respect to J and produces a fine-structure *multiplet* of closely spaced *levels*. Each energy level is specified by a set of quantum numbers (γ,L,S,J), the value of J being added as a subscript to the spectroscopic notation. Each level is still (2J+1)-fold degenerate with respect to the magnetic quantum number M_J and consequently a *state* specified by the set (γ,L,S,J,M_J) is sometimes referred to as a *sub-level*. Singlet terms, S=0, are not split by the spin-orbit interaction since in this case J=L and $\Delta E_2 = 0$. The effect of the spin-orbit interaction in the configuration 4p3d is shown at the extreme right of Fig.3.6.

The energy difference between two neighbouring levels in the same fine-structure multiplet is given by

$$\Delta E_2(J) - \Delta E_2(J-1) = \zeta(L,S)\ J \qquad (3.80)$$

and is proportional to the larger value of J. This is known as the Landé interval rule and is useful for determining the J-values of observed energy levels. Departures from the interval rule indicate a breakdown of one or more of the approximations we have made previously (Problem 3.15).

In the vector model the effect of the spin-orbit interaction is to cause the vectors $\underline{L}$ and $\underline{S}$ to precess about the direction of their resultant $\underline{J}$, as indicated in Fig.3.5. The L-S coupling approximation assumes that this precession due to the magnetic spin-orbit interaction is much slower than the precession of the vectors $\underline{l}_i, \underline{s}_i$ about $\underline{L}$ and $\underline{S}$ respectively, which is produced by the electrostatic interaction.

3.9.6. <u>The Zeeman effect in L-S coupling.</u> Finally in this review of quantum mechanics we recall that every circulating current I in classical electromagnetism possesses a magnetic moment, determined by the product of I and the area of the current loop. Similarly in the atom there exists a magnetic moment $\underline{\mu}_l$ associated with the orbital motion of the electron where $\underline{\mu}_l$ is given by the quantum mechanical operator $\underline{\mu}_l = -g_l \, \mu_B \underline{l}$. Although numerically $g_l \equiv 1$, this factor is introduced explicitly so that orbital and spin angular momenta are treated identically. It should be noted that the vectors $\underline{\mu}_l$ and $\underline{l}$ point in opposite directions due to the negative charge on the electron. As we mentioned in section 3.8, there also exists a magnetic moment associated with the electron spin angular momentum,

$$\underline{\mu}_s = -g_s \, \mu_B \, \underline{s}$$

where g_s is almost exactly twice the orbital g-factor. In a many-electron atom the individual orbital and spin magnetic moments combine to produce a total atomic magnetic moment given by

$$\begin{aligned} \underline{\mu} &= -\mu_B \sum_i (g_l \underline{l}_i + g_s \underline{s}_i) \\ &= -\mu_B (g_l \underline{L} + g_s \underline{S}). \qquad (3.81) \end{aligned}$$

When a magnetic field $\underline{B}$ is applied to the excited atoms in a gas discharge tube the spectral lines are observed to be split into a number of closely spaced components, a phenomenon known as the Zeeman effect. This splitting is caused by the additional energy acquired by the atomic magnetic moments in the applied magnetic field and is described by the perturbation operator

$$\mathcal{H}_1 = -\underline{\mu}\cdot\underline{B} = \mu_B(g_l\underline{L} + g_s\underline{S})\cdot\underline{B}\ . \tag{3.82}$$

In most experiments the Zeeman splitting is small compared with the fine-structure separations and the effect of the magnetic field can be obtained by calculating the diagonal matrix elements of the perturbation operator. It can be shown (Problem 3.16) that the energy of the sub-level $|\gamma\ LSJM_J\rangle$ is changed by an amount

$$\Delta E = \langle\gamma LSJM_J|\mathcal{H}_1|\gamma\ LSJM_J\rangle = g_J\mu_B BM_J \tag{3.83}$$

where the Landé g-factor is defined by

$$g_J = g_l\left\{\frac{J(J+1)+L(L+1)-S(S+1)}{2J(J+1)}\right\} + g_s\left\{\frac{J(J+1)+S(S+1)-L(L+1)}{2J(J+1)}\right\} \tag{3.84}$$

Thus only the component of the atomic magnetic moment parallel to $\underline{J}$ is observable and equation (3.81) may be replaced by the expression

$$\underline{\mu} = -g_J\ \mu_B\ \underline{J}. \tag{3.85}$$

By substituting the values $g_l=g_s/2=1$ into equation (3.84), an approximate expression for the Landé g-factor is obtained

$$g_J \cong \frac{3J(J+1)+S(S+1)-L(L+1)}{2J(J+1)} \tag{3.86}$$

which is sufficiently accurate for most purposes.

From equation (3.83) we see that the external magnetic field removes the degeneracy of the fine-structure states $|\gamma\ LSJM_J\rangle$. In weak magnetic fields a set of equally spaced magnetic sub-levels is produced separated by energies of $\hbar\omega_L$, where $\omega_L = g_J\ \mu_B B/\hbar$ is the Larmor angular frequency.

In the vector model of the atom the total angular momentum $\underline{J}$ is inclined at a constant angle to the direction of the field $\underline{B}$, determined by the magnetic quantum number M_J, and precesses about $\underline{B}$ at the Larmor frequency $\omega_L/2\pi$.

With the exception of the hyperfine structure and Zeeman effect of atoms with finite nuclear spin $\underline{I}$, which are considered separately in Chapter 18, this concludes our brief review of the quantum theory of atoms.

Problems

3.1. Construct the time-independent Schrödinger equation of a classical harmonic oscillator of mass m and restoring force constant k. Find the value of α for which

$$\psi(x) = A \exp(-\tfrac{1}{2}\alpha x^2)$$

is an eigenfunction of this equation. Calculate the corresponding eigenvalue in terms of the classical frequency of vibration $\omega_0/2\pi = (k/m)^{1/2}/2\pi$.

3.2. Calculate the expectation values of x, x^2, p_x and p_x^2 using the wavefunction given in Problem 3.1. Show that the product $(\Delta p)^2(\Delta x)^2 = \hbar^2$ where Δp and Δx are defined by $(\Delta p)^2 = \langle p_x^2\rangle - \langle p_x\rangle^2$ and $(\Delta x)^2 = \langle x^2\rangle - \langle x\rangle^2$. This is one example of the Heisenberg uncertainty principle. $\{ \int_{-\infty}^{\infty} \exp(-au^2)\, du = (\pi/a)^{1/2} \}$.

3.3. Construct the time-independent Schrödinger equation for the hydrogen atom in terms of the coordinates $\underline{r}_1$ and $\underline{r}_2$ of the electron and nucleus respectively. Introducing the coordinates of the centre of mass, $\underline{R}$, and of the electron relative to the centre of mass, $\underline{r}$, show that the Schrödinger equation may be separated into:

$$-\frac{\hbar^2}{2\mu}\nabla^2\psi + V(r)\psi = E\psi$$

and

$$-\frac{\hbar^2}{2(m+M)}\nabla^2\chi = E'\chi.$$

The first equation describes the motion of the electron relative to the centre of mass and the second describes the free motion of the centre of mass.

3.4. Verify that the spherical harmonics given in Table 3.1 satisfy the orthonormality relations of equation (3.25).

3.5. Plot the angular dependence of the functions $|Y_l^m(\theta,\phi)|^2$ on a polar diagram for $l=0,1,2$.

3.6. Verify that the spherical harmonics given in Table 3.1 satisfy the relation

$$Y_l^m(\pi-\theta,\phi+\pi) = (-)^l\, Y_l^m(\theta,\phi).$$

3.7. The orbital angular momentum operators are defined by $\underline{L} = \underline{r}\wedge\underline{p}$. Using $x = r\sin\theta\cos\phi$, $y = r\sin\theta\sin\phi$, $z = r\cos\theta$ express the cartesian components of $\underline{L}$ in terms of (r,θ,ϕ) and $(\frac{\partial}{\partial r}, \frac{\partial}{\partial\theta}, \frac{\partial}{\partial\phi})$. Hence verify that $\underline{L}^2$ is given by equation (3.33).

3.8. Using equations (3.31) and (3.33) verify that the spherical harmonics given in Table 3.1 are eigenfunctions of $\underline{l}^2$ and l_z. Show also that these functions are not eigenfunctions of the operators l_x and l_y.

3.9. The wavefunction of an electron in the ground state of atomic hydrogen is given by

$$\psi_{100}(r,\theta,\phi) = \left(\frac{1}{\pi a_0^3}\right)^{1/2} \exp(-r/a_0).$$

For this state calculate:

(a) the most probable value of r

(b) the expectation value of r

(c) the expectation value of the potential energy

(d) the expectation value of the kinetic energy.

[$\int_0^\infty x^n \exp(-ax)\, dx = n!/a^{n+1}$ where n is a positive integer and $a > 0$.]

3.10. By direct substitution in equation (3.13) calculate the energy associated with the hydrogenic wavefunction

$$\psi(r,\theta,\phi) = \frac{1}{8\sqrt{\pi}} \frac{1}{a_0^{3/2}} \left(\frac{r}{a_0}\right) \exp(-r/2a_0) \sin\theta \exp(i\phi).$$

Also calculate the expectation values of $\underline{l}^2$ and l_z for an atom in this state.

3.11. Show that the Pauli spin matrices

$$s_x = \frac{1}{2}\begin{pmatrix} 0 & 1 \\ 1 & 0 \end{pmatrix}; \quad s_y = \frac{1}{2}\begin{pmatrix} 0 & -i \\ i & 0 \end{pmatrix} \quad ; \quad s_z = \frac{1}{2}\begin{pmatrix} 1 & 0 \\ 0 & -1 \end{pmatrix}$$

satisfy the relations:

(a) $[s_x, s_y] = is_z$ and cyclic permutations

(b) $s_x^2 + s_y^2 + s_z^2 = \frac{3}{4}\underline{I}$ where $\underline{I}$ is the 2×2 unit matrix

(c) $[\underline{s}^2, s_z] = 0$.

3.12. If $\chi_+(i) = \begin{pmatrix} 1 \\ 0 \end{pmatrix}$ and $\chi_-(i) = \begin{pmatrix} 0 \\ 1 \end{pmatrix}$ show that $\chi_\pm(i)$ are eigenfunctions of $\underline{s}_i^2$ and s_{iz}. Using equation (3.55) and Table 3.3, construct the coupled spin wavefunctions for two spin 1/2 particles. Show that there are three possible states having a total spin angular momentum quantum number of unity which are eigenfunctions of $S_z = s_{1z} + s_{2z}$, having eigenvalues -1, 0, +1 respectively.

3.13. Show that the fine-structure splitting arising from spin-orbit interaction in the configuration nl $(l \neq 0)$ of a hydrogen-like ion of atomic number Z is

$$\Delta E \cong \frac{\alpha^2 Z^4}{n^3 l(l+1)} \text{ Rydbergs,}$$

where $\alpha = e^2/4\pi\varepsilon_0\hbar c$ is the fine-structure constant and for hydrogenic systems we have

$$\langle r^{-3} \rangle = \left\{ \left(\frac{a_0}{Z}\right)^3 n^3 l(l+\tfrac{1}{2})(l+1)\right\}^{-1} .$$

3.14. Find the allowed terms of the configuration np^2 in L-S coupling. Show that, for the configuration nl^x of x equivalent electrons each of orbital angular momentum l, the largest value of the total orbital angular momentum for terms of the highest multiplicity in L-S coupling is

$$L_{max} = x(2l+1-x)/2 \quad \text{for} \quad x \le 2l+1.$$

3.15. The spectrum of calcium, atomic number 20, contains a normal multiplet of six lines at 0, 14, 36, 106, 120, and 158 cm^{-1} above the lowest frequency line of the multiplet. From an analysis of these data in the L-S coupling approximation, what information can be obtained about the quantum numbers of the states involved in the transitions?

3.16. In the Zeeman effect the direction of the applied magnetic field is taken as the axis of quantization, 0z. Using the relations

$$L_z = \frac{\underline{L}\cdot\underline{J}}{J(J+1)} J_z \quad \text{and} \quad \underline{S}^2 = (\underline{J}-\underline{L})^2$$

which hold in L-S coupling, together with similar ones obtained by interchanging $\underline{L}$ and $\underline{S}$, expand the perturbation operator $\mathcal{H}_1$ given by equation (3.82) in terms of $\underline{J}^2$, $\underline{L}^2$, and $\underline{S}^2$. Hence show that the change in energy of the sub-level $|JM_J\rangle$ is given by $\Delta E = g_J \mu_B B M_J$ and obtain an expression for the Landé factor, g_J.

General references and further reading

A selection from the many texts available on atomic quantum mechanics and the quantum theory of angular momentum is given below. The condensed account presented in this chapter closely follows the treatment in Woodgate's excellent book.

Brink, D.M. and Satchler, G.R. (1962). *Angular momentum*, Oxford University Press, London.

Dicke, R.H. and Wittke, J.P. (1960). *Introduction to quantum mechanics*, Addison-Wesley, Reading, Mass.

Dirac, P.A.M. (1958). *The principles of quantum mechanics*, Clarendon Press, Oxford.

Messiah, A. (1962). *Quantum mechanics*, Vols. 1 and 2. North Holland Publishing Company, Amsterdam.

Rose, M.E. (1957). *Elementary theory of angular momentum*, Wiley, New York.

Woodgate, G.K. (1970). *Elementary atomic structure*, McGraw-Hill, London.

4

The spontaneous emission of radiation

Most of our knowledge about the energy levels of atoms and molecules, and the composition and physical processes occurring in the sun and stars comes from the study of radiation emitted in spontaneous transitions. In this chapter we shall be concerned only with spontaneous electric dipole transitions. We start by considering a classical model of the atom and then, by the application of the correspondence principle, show how the classical rate at which energy is radiated can be converted into a quantum-mechanical transition probability. This result is compared with that derived by the rigorous quantum theory of radiation and some difficulties which arise in the interpretation of quantum electrodynamics are discussed. We then show how the radiative lifetime of an excited atom can be calculated and introduce expressions for the oscillator strengths or f-values of spectral lines. These concepts are illustrated by a discussion of the f-values, transition probabilities, and radiative lifetimes of hydrogenic systems before the difficulties of calculating f-values for more complex atoms are briefly described.

4.1. The classical atomic model

We have seen in the previous chapter that the motion of electrons within an atom can only be correctly described by the methods of quantum mechanics. However it is often instructive to consider in detail a classical model of an atom since:

(a) the concepts and mathematical methods involved are simpler than those of quantum mechanics

(b) the model provides a physical picture which is a considerable aid in understanding many aspects of the interaction of light with atoms

(c) the results of calculations made using this model are often identical with those obtained by the full quantum mechanical treatment.

Indeed, during the early years of the twentieth century, the predictions of the classical Lorentz theory provided a guide to the development of a complete and rigorous quantum theory of the interaction of atoms and radiation.

The classical theory assumes that the 'atom' contains a massive nucleus, positive charge Ze, and Z negatively charged electrons. The electrons and nucleus each have dimensions of the order of 10^{-15}m and are assumed to occupy the same volume when the 'atom' is in its lowest energy level. If the 'atom' is excited, for example by a collision with an energetic particle, one or more electrons may be displaced from their equilibrium positions. They are then assumed to perform simple harmonic motion about the nucleus with characteristic frequencies given by $\omega_i/2\pi$. These frequencies are chosen to correspond to the wavelengths of light emitted by real atoms, $\lambda_i = 2\pi\, c/\omega_i$. The binding forces which are responsible for the harmonic oscillations are one of the postulates of the theory and cannot be derived from a consideration of the electrical forces within the 'atom'.

We now calculate the rate at which an excited 'atom' radiates energy. We assume, for the sake of simplicity, that it contains a single electron which is oscillating harmonically with amplitude z_0 and angular frequency ω along the direction of the z-axis:

$$z = z_0 \cos \omega t. \qquad (4.1)$$

The instantaneous electric dipole moment of the atom is given by

$$\underline{p}(t) = p_0 \cos \omega t \; \underline{\hat{k}} \qquad (4.2)$$

where $p_0 = -ez_0$. At a point $P(r,\theta,\phi)$ in the wavezone, the radiation fields obtained by substitution in equations (2.69) and (2.70) are given by

$$E_\theta = -\frac{p_0 k^2 \sin\theta}{4\pi\varepsilon_0 r}\cos(kr-\omega t) \tag{4.3}$$

$$H_\phi = -\frac{p_0 k^2 \sin\theta}{4\pi\varepsilon_0 r Z_0}\cos(kr-\omega t). \tag{4.4}$$

The emitted radiation is linearly polarized in the plane containing the z-axis and the point of observation, P. The time-averaged intensity at P is given by the Poynting vector, equation (2.46):

$$\underline{\overline{N}} = \frac{1}{2Z_0}\left(\frac{p_0 k^2}{4\pi\varepsilon_0}\right)^2 \frac{\sin^2\theta}{r^2}\,\hat{\underline{r}}. \tag{4.5}$$

The intensity decreases with the $1/r^2$ dependence that we expect for radiation from a localized source, and has a $\sin^2\theta$ angular dependence which is characteristic of radiation from linearly-polarized dipoles. The intensity is zero in the polar directions $\theta=0$ and $\theta=\pi$ and is independent of the angle ϕ. The oscillating electron is losing energy in the form of radiation at a rate given by

$$-\frac{dW}{dt} = \int_0^{2\pi} d\phi \int_0^{\pi} (\underline{\overline{N}}.\hat{\underline{r}})\, r^2 \sin\theta\, d\theta$$

$$= \frac{p_0^2 k^4 c}{12\pi\varepsilon_0}, \tag{4.6}$$

in agreement with equation (2.74).

4.2. *Radiative lifetime of a classical atom*

The total energy of the harmonically oscillating electron, W, is the sum of the kinetic and potential energies given by $W=m\omega^2 z_0^2/2$. The fraction of this energy lost by radiation in one period of the oscillation is

$$\frac{1}{W}\left(\frac{2\pi}{\omega}\frac{dW}{dt}\right) = \frac{e^2\omega}{3\varepsilon_0 mc^3}. \tag{4.7}$$

At optical frequencies, $\omega \approx 10^{15}\ s^{-1}$, this fraction is extremely small ($\approx 10^{-8}$), and we are therefore justified in writing the differential equation for the energy of the classical atom in the form

$$-\frac{dW}{dt} = \frac{e^2\omega^2}{6\pi\varepsilon_0 mc^3} W. \tag{4.8}$$

The solution of this equation has the decreasing exponential form which is characteristic of all types of spontaneous decay processes:

$$W(t) = W(0) \exp(-\gamma t) \tag{4.9}$$

where

$$\gamma = \frac{e^2\omega^2}{6\pi\varepsilon_0 mc^3} \tag{4.10}$$

is the classical spontaneous emission or radiative decay rate. The mean lifetime of the classical atom in the excited state is given by

$$\tau_{\mathrm{Cl.}} = \frac{1}{\gamma} = \frac{6\pi\varepsilon_0 mc^3}{e^2\omega^2} . \tag{4.11}$$

We now evaluate the classical lifetime, equation (4.11), for an oscillating electron representing: (a) an excited sodium atom in the 3 $^2P_{3/2}$ level, and (b) an excited mercury atom in the 6 $^1P_1$ level. The results are given in Table 4.1, together with the results of recent experimental determinations, $\tau_{\mathrm{expt.}}$ (Problem 4.1).

The classical lifetimes, 10^{-8}-10^{-9} s, are typical of the lifetimes of excited atoms which are able to decay by allowed electric dipole transitions. More surprisingly perhaps, the classical values are extremely close to the experimentally measured lifetimes, although this close agreement is only found in the case of the true resonance lines of atoms, i.e. allowed transitions connecting the first excited level to the ground state. Thus, although the motion of electrons in atoms can only be properly described by quantum mechanics, nevertheless a simple classical model is often capable of yielding accurate and useful results, es-

TABLE 4.1.

Comparison of classical lifetimes and experimental results in sodium and mercury

Level	Wavelength emitted (Å)	Classical lifetime (ns)	Experimental lifetime (ns)	Reference
$Na(3\ ^2P_{3/2})$	5890	15·6	16·6 ± 0·4	Kibble *et al.* (1967).
$Hg(6\ ^1P_1)$	1850	1·54	1·31 ± 0·08	Lurio (1965)

pecially when problems involving the interaction of atoms with resonance radiation are being considered. The classical radiative lifetime, equation (4.11), depends strongly on the angular frequency of the emitted radiation and we therefore expect that excited levels of nuclei which decay by emission of γ-rays might have lifetimes of the order 10^{-15}-10^{-17} s. Conversely, in the radio and microwave region, spontaneous emission by atoms and molecules is so slow that it is generally negligible under laboratory conditions and de-excitation in this case is due mainly to collisions. Only in the tenuous clouds of gas existing in interstellar space is the effect of collisions reduced to the level where spontaneous microwave emission becomes important.

4.3. *Spontaneous emission probability A_{ki}*

4.3.1. Inadequacy of simple quantum theory. The quantum theory of atoms outlined in Chapter 3 provides a time-dependent description of excited states of atoms. An excited hydrogen atom, for example, in the level n=2, l=1, m=0 is described by the time-dependent wavefunction:

$$\Psi_{210}(r,\theta,\phi,t) = \psi_{210}(r,\theta,\phi)\exp(-iE_2t/\hbar). \quad (4.12)$$

In the absence of external radiation, however, the theory predicts that if the atom is prepared in this state at time t=0 then it will remain in that state indefinitely. This is clearly incorrect, since hydrogen atoms in the 2p state are observed to decay rapidly to the ground state 1s with the spontaneous emission of Lyman α radiation, λ = 1216 Å.

Thus the simple quantum theory is inadequate and to account for the spontaneous emission of radiation we adopt alternative techniques: the first of these takes the classical result of section 4.2 and attempts to convert it into a quantum-mechanical form; the second develops a complete quantum theory in which not only the atoms but also the radiation fields are properly quantized. We examine each of these methods in the following sections.

4.3.2. Conversion of the classical result into quantum-mechanical terms. We start from the classical result for the time-averaged rate of radiation of energy by an oscillating electric dipole moment, equation (2.74):

$$-\frac{dW}{dt} = P_{E1} = \frac{\omega^4|\underline{p}_0|^2}{12\pi\varepsilon_0 c^3} \quad (4.13)$$

where

$$|\underline{p}_0|^2 = e^2\,(|x_0|^2 + |y_0|^2 + |z_0|^2).$$

We now identify the frequency of the classical oscillator with that of the spontaneous electric dipole transition from level k to level i:

$$\hbar\omega = \hbar\omega_{ki} = E_k - E_i.$$

(We use a convention in which the energy levels of an atom are labelled .....i,j,k,..... in ascending order from the ground state of the atom, but usually omit j to avoid confusion with the total angular momentum quantum number.) The mean rate of radiation can therefore be described in terms of the probability per unit time, A_{ki}, for the emission of a quantum of energy, $\hbar\omega_{ki}$, as follows

$$P_{E1} = \hbar\omega_{ki} A_{ki} = \frac{\omega_{ki}^4 |\underline{p}_0|^2}{12\pi\varepsilon_0 c^3} . \tag{4.14}$$

It is now necessary to replace $|\underline{p}_0|^2$ by a quantum-mechanical expression. A classical linearly oscillating electric dipole moment is given by

$$\underline{p}(t) = \underline{p}_0 \cos\omega t = \frac{1}{2}\underline{p}_0 (e^{i\omega t} + e^{-i\omega t}) \tag{4.15}$$

and a corresponding quantum-mechanical expression having the same frequency dependence as the second term in equation (4.15) would be

$$\int \Psi_i^* \underline{p} \Psi_k \, dv = \exp\{-i(E_k - E_i)t/\hbar\} \int \psi_i^* \underline{p} \psi_k \, dv. \tag{4.16}$$

We therefore postulate that the radiation field accompanying a spontaneous transition from a state k of higher energy to a state i of lower energy is the same, as far as the angular intensity distribution, polarization, and total power are concerned, as that given by the classical theory for a charge distribution whose electric dipole moment is given by

$$\left(\frac{\underline{p}_0 \exp(-i\omega t)}{2}\right)_{Cl.} \equiv \int \psi_i^* \underline{p} \psi_k \, dv \exp(-i\omega_{ki} t) \tag{4.17}$$

where $\underline{p} = -e\underline{r}$ is the electric dipole moment operator.

An explicit quantum-mechanical expression for the transition probability for spontaneous emission is obtained by substituting for $\underline{p}_0/2$ in equation (4.14), giving

$$A_{ki} = \frac{e^2\omega_{ki}^3}{3\pi\varepsilon_0\hbar c^3} \left| \int \psi_i^* \underline{r} \psi_k \, dv\right|^2. \tag{4.18}$$

Expressing $\underline{r}$ in terms of its cartesian components, $\underline{r} = x\hat{\underline{i}} + y\hat{\underline{j}} + z\hat{\underline{k}}$, the matrix element in equation (4.18) may be written in the form

$$|\langle i|\underline{r}|k\rangle|^2 = |\langle i|x|k\rangle|^2 + |\langle i|y|k\rangle|^2 + |\langle i|z|k\rangle|^2. \quad (4.19)$$

A rigorous derivation of A_{ki} from the classical theory is obviously impossible and the justification for the final result must rest on the agreement which exists between theoretical transition probabilities calculated from equation (4.18) and experimentally determined values. However, equation (4.18) is identical with the rigorous quantum electrodynamic result whose derivation is outlined in the next section.

4.4. *Spontaneous emission according to quantum electrodynamics*

The quantized field theory of radiation treats the radiation field as an ensemble of harmonic oscillators, each mode of the field being associated with an oscillator of angular frequency ω and having the allowed energies:

$$E_n = (n_{\underline{k},\underline{\varepsilon}} + \tfrac{1}{2})\,\hbar\omega \; ; \; n_{\underline{k},\underline{\varepsilon}} = 0,1,2,.... \quad (4.20)$$

where $n_{\underline{k},\underline{\varepsilon}}$ specifies the number of quanta in this mode of the field. The wave vector of the radiation is given by $\underline{k}$, and $\underline{\varepsilon}$ specifies the state of polarization. The number of quanta per mode is very small in the optical fields produced by conventional sources and only in laser beams does $n_{\underline{k},\underline{\varepsilon}}$ become appreciably greater than unity.

The theory now treats the combination of an atom plus the radiation field as a single system. If the system is in an initial state in which there are $n_{\underline{k},\underline{\varepsilon}}$ quanta in the radiation field and the atom is in an eigenstate labelled by k, then the probability per unit time that the atom will make a transition to a final state in which there are $(n_{\underline{k},\underline{\varepsilon}} + 1)$ quanta present and the atom is in a state of lower energy i is given by (Loudon(1973)):

$$P_{ki} = \frac{e^2\omega_{ki}^3}{3\pi\varepsilon_0\hbar c^3}\left|\int \psi_i^* \, \underline{r} \, \psi_k \, dv\right|^2 (n_{\underline{k},\underline{\varepsilon}} + 1). \quad (4.21)$$

This is independent of the vectors $\underline{k}$ and $\underline{\varepsilon}$ because an average has been performed over the directions of $\underline{k}$ and $\underline{\varepsilon}$ corresponding to the interaction of the atom with isotropic unpolarized radiation. The transition probability consists of two terms:

(a) the first term is proportional to the radiation density initially present in the field,

$$\rho(\omega) = \frac{\hbar\omega^3}{\pi^2 c^3} n_{\underline{k},\underline{\varepsilon}},$$

and corresponds to stimulated emission which we shall consider in detail in Chapter 9,

(b) the second term is finite even in the absence of external radiation ($n_{\underline{k},\underline{\varepsilon}} = 0$). This is the spontaneous emission process and the transition probability is identical with the result derived in the previous section, equation (4.18).

4.4.1. Comments on the quantum theory of spontaneous emission. Quantum electrodynamics is a sophisticated theory and as far as is presently known the results of calculations made with this theory are in agreement with experiment. However, the elegant mathematics of the theory does not answer the question: 'Is there a physical mechanism responsible for spontaneous emission?'. The usual answer to this is the following: even when there are no quanta present in the radiation field, there are still fluctuating electric and magnetic fields present which are associated with the zero-point energy of each mode, $\hbar\omega/2$. These zero-point fields are responsible for a small increase of the electron spin g-factor over the Dirac value of $g_s \equiv 2$, equation (3.58), and the Lamb shift in hydrogenic systems. It is therefore plausible to suppose that they are also responsible for spontaneous emission.

An alternative explanation of the mechanism of spontaneous emission has been proposed by Wheeler and Feynman (1945) in a theory which includes both the retarded and advanced electromagnetic potentials introduced in section 2.6. The difference between the two theories is most evident if we consider a hypothetical universe containing a single excited atom. Quantum electrodynamics predicts that the atom would immediately radiate spontaneously according to equation (4.21). Wheeler and Feynman claim that this result is incorrect and that, in order for spontaneous emission to occur, there must exist somewhere in the universe other atoms which are capable of absorbing the emitted radiation. The advanced electric fields of these absorbers act on the excited atom in such a way that emission occurs and the energy is conveyed by the retarded fields from the emitter to the absorber.

The radiative decay rate calculated by Wheeler and Feynman is identical to that given by equation (4.21), provided that there is sufficient matter in the universe to allow the emitted radiation to be completely absorbed. If this condition is not satisfied or if the distant matter is receding from the emitter with too great a velocity, then the rate of spontaneous emission will be less than that given by equation (4.21). Thus a connection can be made between spontaneous emission and cosmology which is discussed in detail by Hoyle and Narlikar (1968) and Pegg (1968).

This discussion has been included to show that our understanding of such an apparently simple phenomenon as spontaneous emission is not perhaps as complete as we would wish. Other difficulties which are connected with divergent expressions which occur in quantum electrodynamics are discussed by Leighton (1959, Ch. 20.12).

4.5. *Spontaneous transitions between degenerate levels*

The expression for the spontaneous transition probability, equation (4.18), refers to a transition between non-degenerate states k and i. Thus k and i are labels which should actually specify the magnetic quantum numbers, m_k and m_i, of the levels involved. Then the total decay rate of a non-degenerate sub-level $|km_k\rangle$ to a degenerate level $|i\rangle$ would be given by the sum of the transition probabilities taken over all the possible decay channels to the lower level:

$$A_{km_k,i} = \frac{e^2\omega_{ki}^3}{3\pi\varepsilon_0\hbar c^3} \sum_{m_i} |\langle im_i|\underline{r}|km_k\rangle|^2. \qquad (4.22)$$

However, if now the upper level is also degenerate we expect intuitively that the different degenerate states $|km_k\rangle$ will all decay at the same rate. If this were not so then there would be a preferred direction in space and the polarization of the emitted radiation would be observed to change with time. The following result must therefore hold (Problem 4.2):

$$\sum_{m_i} |\langle im_i|\underline{r}|km_k\rangle|^2 \quad \text{is independent of the value of } m_k.$$

Thus equation (4.22) also gives the radiative decay rate between degenerate levels k and i. The form of this result can be made more symmetrical by introducing an additional summation over m_k and dividing by the statistical weight $g_k = 2j_k+1$ of the upper level:

$$A_{ki} = \frac{e^2\omega_{ki}^3}{3\pi\varepsilon_0\hbar c^3} \frac{1}{g_k} \sum_{m_k m_i} |\langle im_i|\underline{r}|km_k\rangle|^2. \qquad (4.23)$$

In many-electron atoms, the operator in this equation should be replaced by $\underline{r} \to \sum_i \underline{r}_i$ where the summation extends only over the valence electrons provided that there is no configuration mixing.

The sum of the electric dipole matrix elements in equation (4.23) might, for a typical strong transition, have the value $\approx 20e^2a_0^2$. The resultant transition probability for a spectral line with $\lambda \approx 5000$ Å would be of the order of $10^8\ s^{-1}$ (Problem 4.3).

4.6. *Radiative lifetimes of excited atoms*

Let us consider an assembly of atoms in which $N_k(0)$ atoms per unit volume are excited into the level k at time t=0. In general these atoms will be able to decay by emission of radiation to a number of lower levels i. In the absence of effects such as collisional depopulation, repopulation by cascade from higher levels, etc., the population density in level k decreases at the rate given by

$$-\frac{dN_k}{dt} = N_k \sum_i{}' A_{ki}, \tag{4.24}$$

where the prime on the summation indicates that it extends only over those levels which have lower energy than that of level k. The population of the excited level decays exponentially:

$$N_k(t) = N_k(0)\exp(-t/\tau_k) \tag{4.25}$$

with radiative lifetime, τ_k, given by

$$\tau_k = 1/\sum_i{}' A_{ki}. \tag{4.26}$$

Since the transition probabilities for strong lines are of the order of $10^8\ s^{-1}$, the lifetime of a typical excited level would be $\approx 10^{-8}$s.

In the case of the lowest excited levels of many atoms, there is often only one term in the summation in equation (4.26). Thus a measurement of the lifetime of that level determines directly the transition probability of the resonance line. Techniques for the direct measurement of the radiative lifetimes of atoms and molecules are discussed in Chapter 6.

4.7. *Intensity of light emitted by optically thin sources*

In a gas discharge, atoms are usually excited isotropically by random collisions and the populations in different magnetic states of the same level can be assumed to be equal. This is known as natural excitation. The emitted radiation is unpolarized and of equal intensity in all directions. If the population density N_k in the excited level k is uniform throughout the volume V of the source, then $N_k V A_{ki}$ quanta per second will be emitted in the transition $k \to i$. The intensity of the source, which is the power radiated per unit solid angle, is given by

$$I_{ki} = \frac{N_k V \hbar \omega_{ki} A_{ki}}{4\pi} . \qquad (4.27)$$

This expression assumes that the source is optically thin, that is, every photon emitted is able to escape from the source without being re-absorbed by atoms in the lower level i. Self-absorption is usually only important if i is the ground state or is a metastable level.

The population density of excited atoms, N_k, is determined by the conditions existing in the source. The simplest situation is that in which the collisional excitation and de-excitation rates for some or all of the excited levels are much larger than the radiative decay rates. This is called local thermodynamic equilibrium (LTE) and for laboratory sources LTE generally exists only in high current arcs and shock tubes. In this case the density of excited atoms is determined by the Boltzmann distribution

$$\frac{N_k}{N} = \frac{g_k \exp(-E_k/kT)}{Z(T)}$$

where N is the total density of atoms and $Z(T) = \sum_i g_i \exp(-E_i/kT)$ is the partition function. The relevant temperature T is that which describes the distribution function of the exciting species. This is usually the electrons and T may be between 10^4 and 10^5 K. However, since electron-atom collisions are very inefficient in transferring momentum, the temperature describing the distribution of atomic kinetic energies will generally be much lower.

By contrast in low-current gas discharges, the equilibrium population is determined by a balance between excitation by electrons and decay by radiative and collisional processes. In this case detailed information is required about electron collision cross-sections and electric dipole transition probabilities before N_k can be calculated. However, the population in excited levels rarely exceeds 10^{-6} of the ground-state population. If we take $A_{ki} \approx 10^8\ s^{-1}$ and $\hbar\omega_{ki} \approx 4eV$ corresponding to emission in the blue-green region of the spectrum we find that the optical power radiated into unit solid angle at a typical source pressure of 1 Torr is of the order of 0.15 W cm^{-3}.

In some cases the excitation process is deliberately chosen to be anisotropic. For instance, the atoms may be excited by absorption using a unidirectional beam of light or by impact using a well-defined beam of electrons. The emitted radiation is then generally polarized and anisotropically distributed in space. This forms the basis of several extremely powerful techniques for the study of the Zeeman and hyperfine structure of atomic energy levels which are considered in detail in Chapters 15-17.

4.8. *Oscillator strengths*

We have seen in section 4.2 that the experimentally observed transition probabilities are often close to the classical radiative decay rate, γ, given by equation (4.10). It is therefore convenient to introduce a dimensionless quantity f, known as the oscillator strength or f-value, which describes what fraction of the energy of the classical oscillator should be ascribed to a given transition. For transitions from an upper level k to a lower level i we define the emission f-value, f_{ki}, by the relation

$$A_{ki} = -3\, f_{ki}\, \gamma, \tag{4.28}$$

and for an upward transition from level i to level k the absorption f-value, f_{ik}, is defined by

$$g_i\, f_{ik} = -\, g_k\, f_{ki} \equiv gf. \tag{4.29}$$

The f-values defined by equations (4.28) and (4.29) refer to transitions between degenerate levels and therefore include an average over all values of the magnetic quantum numbers as in equation (4.23).

Transitions which exhibit the classical Zeeman effect

i.e. $g_k=3$, $g_i=1$, come closest to behaving as classical oscillators. In such a transition if $A_{ki}=\gamma$, we would have an absorption f-value $f_{ik}=1$ and an emission f-value $f_{ki} = -1/3$. Transitions with oscillator strengths of this order of magnitude generally produce intense spectral lines and give rise to short lifetimes, in the range 10^{-8}-10^{-9} s. Transitions which are partially forbidden may have much smaller f-values, in the range 0·002-0·0004, and give rise to relatively long radiative lifetimes, 10^{-5}-10^{-6} s. The associated spectral lines may be so weak that they are difficult to detect.

Substituting for γ from equation (4.10) into equations (4.28) and (4.29), we obtain explicit expressions for the transition probability A_{ki} in terms of the emission and absorption oscillator strengths:

$$A_{ki} = \frac{e^2\omega_{ki}^2}{2\pi\varepsilon_0 mc^3}(-f_{ki}) = \frac{e^2\omega_{ki}^2}{2\pi\varepsilon_0 mc^3}\left(\frac{g_i}{g_k} f_{ik}\right). \qquad (4.30)$$

Finally the use of equation (4.23) enables the f-values to be related to the matrix elements of the electric dipole moment:

$$f_{ki} = -\frac{2m\omega_{ki}}{3\hbar g_k} \sum_{m_i m_k} |\langle km_k|\underline{r}|im_i\rangle|^2 . \qquad (4.31)$$

Again for many-electron atoms we must make the replacement $\underline{r} \rightarrow \sum_i \underline{r}_i$ in equation (4.31).

A knowledge of the f-values of spectral lines is important in many problems in the fields of spectroscopy, astrophysics, and plasma physics. Recently considerable effort has been made to compile tables of reliable f-values from the best available theoretical and experimental results. Two such compilations have been published by Wiese *et al.* (1966, 1969), the first containing data for the elements hydrogen to neon and the second for the elements sodium to calcium. These f-values are also useful in problems which are not directly connected with radiation processes. Thus expressions for the Van der Waals interaction between two neutral atoms involves a sum over f-values, and the refractive index of gases and the Stark shift of the energy levels of an atom subjected to an external electric field may be calculated in a similar way.

4.8.1. Oscillator strength sum rules. Sum rules for oscillator strengths are of considerable importance for checking the internal consistency of sets of calculations or measurements. It can be shown that if i is the lowest level in the energy-level system of a one-electron atom, then the absorption f-values obey the sum-rule (Problem 4.4)

$$\sum_k f_{ik} = 1. \tag{4.32}$$

This result may be derived intuitively since the sum of the different transitions should be equivalent to a single classical oscillator. A similar rule applies to the sum of the emission and absorption f-values for all the transitions from an excited level j:

$$\sum_{i<j} f_{ji} + \sum_{k>j} f_{jk} = 1. \tag{4.33}$$

The terms in this sum rule tend to cancel since the emission f-values f_{ji} are negative fractions. This is known as the Thomas-Kuhn-Reiche sum rule. In more complex atoms useful approximate sum rules can be obtained by replacing the value unity on the right side of equations (4.32) and (4.33) by z, where z is the number of equivalent electrons in the valence shell of the atom. This approximation breaks down if there is any appreciable interaction between different electron configurations.

In the central-field approximation more detailed sum rules can be derived for the jumping electron (Bethe and Salpeter 1957):

$$\sum_{n'} f_{nn'} = \left\{ \begin{array}{ll} -\frac{1}{3}\frac{l(2l-1)}{2l+1} & n,l\rightarrow n',l-1 \\ \frac{1}{3}\frac{(l+1)(2l+3)}{2l+1} & n,l\rightarrow n',l+1 \end{array} \right. \tag{4.34}$$

where n and l denote the principal and orbital quantum numbers respectively. Again the summation includes both emission ($n'<n$) and absorption ($n'>n$) processes. The sum rules show that for transitions in which $l \to l-1$, the downward transitions give the dominant contribution to the summation, whereas for $l \to l+1$ the upward transitions are more important.

4.9. The line strength, S_{ki}

The starting point of theoretical calculations of the oscillator strengths and transition probabilities of spectral lines are the wavefunctions of the energy levels involved. In these calculations it is often found to be convenient to compute first a quantity called the line strength:

$$S_{ki} = S_{ik} = \sum_{m_i m_k} |\langle km_k|e\underline{r}|im_i\rangle|^2. \qquad (4.35)$$

This is the quantum-mechanical analogue of the square of the classical electric dipole moment, $|\underline{p}_0|^2$ (equation (4.13)). The line strength is symmetrical with respect to the initial and final states and is often quoted in atomic units, $a_0^2 e^2 = 6{\cdot}459 \times 10^{-36}$ cm^2 esu^2 = $7{\cdot}188 \times 10^{-59}$ m^2 C^2. Once the line strength is known it is a simple matter to combine equations (4.29), (4.31), and (4.35) to obtain the oscillator strengths for absorption or emission:

$$g_i f_{ik} = -g_k f_{ki} = \frac{2m\omega_{ki}}{3\hbar} \frac{S_{ki}}{e^2}. \qquad (4.36)$$

4.10. Oscillator strengths in hydrogenic systems

Since explicit wavefunctions are available only for hydrogenic systems, it is in these cases alone that exact values for the f-values and transition probabilities can be calculated. To illustrate the use of the expressions derived in preceding sections, we consider the first line in the Lyman series, $2\,^2P \to 1\,^2S$, which in atomic hydrogen has a wavelength of 1216 Å (Problem 4.5). We ignore the fine structure since in hydrogenic systems it is so small that it can be observed only if high-resolution techniques are used. The statistical weights of the different terms are therefore given by $(2s+1)(2l+1) = 2(2l+1)$.

We start by evaluating the angular integrals appearing in the expression for the line strength, equation (4.35), using the detailed expressions for the spherical harmonics given in Table 3.1. The result is

$$S_{10}^{21} = 2e^2 \left| \int_0^\infty R_{21}(r)\, R_{10}(r)\, r^3\, dr \right|^2. \qquad (4.37)$$

The factor of 2 arises from the two possible values of $m_s = \pm \frac{1}{2}$, and it is also found that each m_l state of the ^2P term contributes equally to the sum in equation (4.35), as we expect on the grounds of symmetry. The radial integral in equation (4.37) may be evaluated by using the hydrogenic wavefunctions given in Table 3.2 with the result:

$$S_{10}^{21} = \frac{2^{16}}{3^9} \left(\frac{ea_0}{Z}\right)^2 = 3{\cdot}330 \left(\frac{ea_0}{Z}\right)^2.$$

This shows that the line strengths for hydrogenic systems with nuclear charge Z are related to those of hydrogen by

$$S_Z = S_H/Z^2.$$

From equations (3.42) and (3.46) the angular frequency of the Lyman transition is given by $\hbar\omega_{21} = 3Z^2e^2/8a_0$. Using this result and equation (4.37) in equation (4.36), we obtain the emission oscillator strength

$$f_{21} = 2^{13}/3^{10} = 0{\cdot}1387.$$

The oscillator strengths of hydrogenic systems are therefore independent of the nuclear charge,

$$f_Z = f_H,$$

as one might expect from the oscillator strength sum rules, equations (4.32) and (4.33).

The f-values of other transitions in hydrogenic systems can similarly be expressed in terms of the radial integrals (Problem 4.6)

$$\int R_{nl}(r)\, R_{n',l-1}(r)\, r^3\, dr. \qquad (4.38)$$

However, the evaluation of these integrals in the general case is rather difficult and we simply quote the final results of these calculations in Table 4.2. The table gives either the absorption or the emission oscillator strength, depending on the energies of the initial and final states of the transition. We see that the absorption f-value of the Lyman α resonance line accounts for over 40 per cent of the total oscillator strength in the Lyman series and that in all cases the oscillator strengths decrease rapidly with increasing value of the upper term principal quantum number, n. The second column of Table 4.2 illustrates the f-sum rule, equation (4.32), and shows that states in the continuum above the $n\ ^2P$ series limit make a significant contribution to this sum. In the columns where either $2\ ^2P$ or $3\ ^2P$ is the initial term, we see that the emission processes dominate in the f-value sums for the transitions $2\ ^2P \rightarrow n'\ ^2S$ and $3\ ^2P \rightarrow n'\ ^2S$, while absorption processes dominate for the transitions $2\ ^2P \rightarrow n'\ ^2D$ and $3\ ^2P \rightarrow n'\ ^2D$, as predicted by the detailed sum rules of equation (4.34). In both cases the total sum of the positive and negative contributions is zero.

From the results contained in Table 4.2 the transition probabilities and radiative lifetimes of hydrogenic energy levels can be calculated using equations (4.30) and (4.26) respectively. For the Lyman α transition in atomic hydrogen we obtain $A_{21} = 6{\cdot}265 \times 10^8\ s^{-1}$ and consequently $\tau(2\ ^2P) = 1{\cdot}60 \times 10^{-9}$ s. The results for other terms in hydrogen are given in Table 4.3. We see that all ^2P terms are short lived because there are always strong radiative transitions down to the ground state. By contrast the ^2S terms have remarkably long lifetimes due to the fact that all spontaneous transitions from these terms involve an increase in the l-quantum number and are therefore relatively weak, as predicted by the detailed sum rule, equation (4.34). Since the f-values of hydrogenic systems are independent of Z while the angular frequencies of the transitions are pro-

TABLE 4.2.

Absorption (or emission) oscillator strengths for hydrogen and the hydrogenic ions

Initial state	1s	2s	2p		3s	3p		3d	
Final state	n'p	n'p	n's	n'd	n'p	n's	n'd	n'p	n'f
n' = 1	—	—	-0·1387	—	—	-0.026	—	—	—
2	0.4162	—	—	—	-0.041	-0.145	—	-0.417	—
3	0.0791	0.435	0.0136	0.696	—	—	—	—	—
4	0.0290	0.103	0.0030	0.122	0.485	0.032	0.618	0.011	1.018
5	0.0139	0.042	0.0012	0.044	0.121	0.007	0.139	0.002	0.157
6	0.0078	0.022	0.0006	0.022	0.051	0.003	0.056	0.001	0.054
$\sum_{n'=1}^{\infty} f_{nn'}$	0.5650	0.649	-0.119	0.928	0.707	-0.121	0.904	-0.402	1.302
Continuous spectrum	0.4350	0.351	0.008	0.183	0.293	0.010	0.207	0.002	0.098
Total sum	1.0000	1.000	-0.111	1.111	1.000	-0.111	1.111	-0.400	1.400

portional to Z^2, the transition probabilities, equation (4.30), will increase rapidly with increasing nuclear charge:

$$A_Z = Z^4 A_H \quad \text{(Problem 4.7).}$$

TABLE 4.3.

Transition probabilities A_{ki} *and radiative lifetimes* τ_k *for atomic hydrogen*

Transition	Wavelength (Å)	g_i	g_k	A_{ki} (s^{-1})	τ_k (ns)
1s - 2p	1216	2	6	$6{\cdot}265 \times 10^8$	1.596
2p - 3s	6563	6	2	$6{\cdot}313 \times 10^6$	158·4
1s - 3p	1026	2	6	$1{\cdot}672 \times 10^8$	
2s - 3p	6563	2	6	$2{\cdot}245 \times 10^7$	5·273
2p - 3d	6563	6	10	$6{\cdot}465 \times 10^7$	15·47
2p - 4s	4861	6	2	$2{\cdot}578 \times 10^6$	
3p - 4s	18 751	6	2	$1{\cdot}835 \times 10^6$	226·6
1s - 4p	973	2	6	$6{\cdot}818 \times 10^7$	
2s - 4p	4861	2	6	$9{\cdot}668 \times 10^6$	
3s - 4p	18 751	2	6	$3{\cdot}065 \times 10^6$	
3d - 4p	18 751	10	6	$3{\cdot}475 \times 10^5$	12·31
2p - 4d	4861	6	10	$2{\cdot}062 \times 10^7$	
3p - 4d	18 751	6	10	$7{\cdot}037 \times 10^6$	36.16
3d - 4f	18 751	10	14	$1{\cdot}379 \times 10^7$	72.52

A selection of recent results of lifetime measurements in hydrogenic systems is given in Table 4.4. The agreement between the theoretical values and the experimental results is generally good. Measurements have also been made on other lines in the Balmer series n' → n=2. However, since the fine structure was not resolved in these experiments the measured lifetime is generally an average over the values for the different *l* states, weighted according to the initial population distribution. Since this is rarely a statistical equilibrium distribution, the interpretation of the results is difficult and for this reason we have not included them in Table 4.4.

TABLE 4.4.

Comparison of theoretical and experimental lifetimes in hydrogenic systems.

Atom or Ion	State	τ_{theory} (ns)	τ_{expt} (ns)	Reference
H	2p	1·596	1·600 ± 0·004	Bickel and
	3p	5·273	5·58 ± 0·13	Goodman (1966)
He^+	2p	0·100	0·097 ± 0·003	Heroux (1967)
	3p	0·330	0·31 ± 0·03	
Li^{++}	3s	1·96	2·1 ± 0·1	Bukow *et al*
	4s	2·80	2·9 ± 0·1	(1973)

4.11. *Theoretical oscillator strengths in complex atoms*

The calculation of line strengths and f-values for other atomic systems again involves the calculation of the angular and radial integrals appearing in equation (4.35). The angular integrals can generally be calculated exactly using the vector coupling techniques discussed in sections 3.7 and 3.9 , provided that L-S coupling is a good approximation.

However, the radial integrals can only be calculated using approximate radial wavefunctions. These are obtained by a variety of techniques such as the Hartree-Fock self-consistent field method and variational calculations. The accuracy of these wavefunctions is generally tested by calculating the expectation value of the Hamiltonian operator, $\mathcal{H}_0$, and comparing the result with the spectroscopically-determined energy levels. Agreement to within 1 per cent is often considered to be satisfactory. However, this procedure tests the wavefunction only over a limited range of the parameter r, since $\langle \psi|\mathcal{H}_0|\psi \rangle$ involves essentially an overlap integral of the wavefunction with itself. The expression for the oscillator strengths, equation (4.31), requires an integral which is a weighted overlap between different wavefunctions and the result is generally considerably less accurate than the energy eigenvalues.

It is therefore difficult to achieve an accuracy of better than 20 per cent in theoretical calculations of oscillator strengths unless a great deal of care and effort is taken. More importantly, the calculations are often only carried out for a few transitions in a given atom although for many applications in astrophysics and plasma physics results are required for many transitions in several different atoms or ions. In an attempt to remove some of these difficulties, Bates and Damgaard devised a technique in which the optically active electron was described by a hydrogenic wavefunction. This technique is relatively simple, it is easily applicable to a large number of different atoms, and often produces results which are surprisingly accurate. Unfortunately space does not permit us to discuss this method and the reader is referred to Griem (1964) for a detailed description.

Problems

4.1. Show that equation (4.11) for the mean lifetime of a classical oscillator may be rewritten in the form

$$\tau_{Cl.} = \left(\frac{3}{8\pi^2 r_0 c}\right) \lambda^2$$

where λ is the wavelength of the emitted radiation and $r_0 = e^2/4\pi\varepsilon_0 mc^2 = 2{\cdot}82 \times 10^{-15}$ m is known as the classical radius of the electron. Hence verify that the classical lifetimes for sodium and mercury given in Table 4.1 are correct.

4.2. Using the identities

$$j_- j_+ = j^2 - j_z(j_z+1)$$

and

$$j_+|jm\rangle = \{j(j+1) - m(m+1)\}^{1/2}|j\ m+1\rangle$$

where $j_{\pm} = j_x \pm ij_y$, prove that the summation

$$\sum_{m_i} |\langle km_k|\underline{r}|im_i\rangle|^2$$

is independent of the value of the magnetic quantum number, m_k.

4.3. An electron is confined to the region $0 \le x \le a$ by an infinitely deep one-dimensional square-well potential. The normalized wavefunctions for the electron are given by

$$\psi_m(x) = \left(\frac{2}{a}\right)^{1/2} \sin\left(\frac{m\pi x}{a}\right) \text{ where } m=1,\ 2,\ 3,\ldots,\infty.$$

Show that the matrix elements of the electric dipole moment operator, $p=e(x-a/2)$, taken between states with quantum numbers m and l are given by

$$p_{ml} = \frac{ea}{\pi^2}\left[\frac{\cos\{(m-l)\pi\}-1}{(m-l)^2} - \frac{\cos\{(m+l)\pi\}-1}{(m+l)^2}\right].$$

When $a=10a_0$, show that the first excited state decays with the emission of light of wavelength 3079 Å and that the transition probability for decay by the spontaneous emission of electric dipole radiation is $2{\cdot}25 \times 10^8\ s^{-1}$.

4.4. It can be shown that the matrix elements of the components of the electron position and momentum operators, $\underline{r}$ and $\underline{p}$ respectively, are connected by the equation

$$\frac{i}{\hbar}(E_k - E_i)\langle km_k|x|im_i\rangle = \langle km_k|\frac{p_x}{m}|im_i\rangle .$$

Use this relation to evaluate by matrix multiplication the diagonal matrix elements, $\langle im_i|C|im_i\rangle$, of the commutator $C = [p_x, x]$. Finally, by introducing the commutation relation, $[p_x, x] = -i\hbar$, show that

$$\sum_{k,m_k} \frac{2m}{3\hbar^2}(E_k - E_i)\,|\langle km_k|\underline{r}|im_i\rangle|^2 = 1$$

which is the sum rule for the absorption f-values, equation (4.32).

4.5. Using the hydrogenic wavefunctions for the states $n=1$, $l=0$, $m=0$, and $n'=2$, $l'=1$, $m'=0, \pm 1$ given in Tables 3.1 and 3.2 show that

$$|\langle 210|ez|100\rangle|^2 = \frac{1}{2}|\langle 2l\pm 1|e(x\pm iy)|100\rangle|^2 = \frac{1}{6}S_{10}^{21}$$

where the line strength S_{10}^{21} defined by equation (4.37) is equal to $2^{16}(ea_0)^2/3^9$. Hence evaluate the transition probability, A_{ki}, of the 1s - 2p Lyman α transition and show that the lifetime of the 2p level against spontaneous decay by electric dipole radiation is $\tau_k = 1{\cdot}596 \times 10^{-9}$ s.

4.6. Show that the radial integrals

$$|R_{nl}^{n',l-1}|^2 = |\int_0^\infty R_{n',l-1}\, R_{nl}\, r^3\, dr|^2$$

and line strengths, $S_{nl}^{n',l-1}$, for the fine-structure components of the Balmer α line of atomic hydrogen have the values given by:

2p-3s $\quad |R_{21}^{30}|^2 = \dfrac{2^{15}.3^8}{5^{12}}a_0^2 = \dfrac{1}{2}S_{21}^{30}$

2s-3p $\quad |R_{20}^{31}|^2 = \dfrac{2^{20}.3^7}{5^{12}}a_0^2 = \dfrac{1}{2}S_{20}^{31}$

2p-3d $\quad |R_{20}^{32}|^2 = \frac{2^{22}.3^8}{5^{12}} a_0^2 = \frac{1}{4} S_{20}^{32}$

Hence verify that the transition probabilities and lifetimes given in Table 4.3 are correct.

4.7. The lifetime of the 2p level of hydrogen against spontaneous decay by electric dipole radiation is $1{\cdot}596 \times 10^{-9}$ s. Estimate the radiative lifetimes of the 2p levels of both singly-ionized helium and doubly-ionized lithium.

References

Bickel, W.S. and Goodman, A.S. (1966). *Phys. Rev.* 148, 1.

Bukow, H.H., Buttlar, H.v., Haas, D., Heckmann, P.H., Holl, M., Schlagheck, W., Schürmann, D., Tielert, R., and Woodruff, R. (1973). *Nucl. Instrum. Meth.* 110, 89.

Heroux, L. (1967). *Phys. Rev.* 161, 47.

Hoyle, F. and Narlikar, J.V. (1968). *Nature, Lond.* 219, 340.

Kibble, B.P., Copley, G., and Krause, L. (1967). *Phys. Rev.* 153, 9.

Lurio, A. (1965). *Phys. Rev.* 140, A1505.

Pegg, D.T. (1968). *Nature, Lond.* 220, 154.

Wheeler, J.A. and Feynman, R.P. (1945). *Rev.Mod.Phys.* 17, 157.

General references and further reading

Bethe, H.A. and Salpeter, E.E. (1957). *Handbuch der Physik*, 35, p.88. Springer-Verlag, Berlin.

Condon, E.U. and Shortley, G.H. (1967). *The theory of atomic spectra*, Cambridge University Press, London.

Griem, H.R. (1964). *Plasma spectroscopy*, McGraw-Hill, New York.

Leighton, R.B. (1959). *Principles of modern physics*, McGraw-Hill, New York.

Loudon, R. (1973). *The quantum theory of light*, Clarendon Press, Oxford.

Wiese, W.L., Smith, M.W., and Glennon, B.M. (1966). *Atomic transition probabilities*, Vol.1. NSRDS - NBS4. U.S. Government Printing Office, Washington D.C.

__________, __________, and Miles, B.M. (1969). *Atomic transition probabilities*, Vol.2. NSRDS - NBS21, U.S. Government Printing Office, Washington D.C.

5

Selection rules for electric dipole transitions

5.1. Introduction

When the emission spectrum of an element is examined in detail it is soon realized that the number of observed lines is far smaller than the number one might expect if transitions between all pairs of levels were allowed. For example, in the spectrum of neutral helium, lines corresponding to the transitons $1\ ^1S_0 - n\ ^3P_{2,1,0}$ and $1\ ^1S_0 - n\ ^1D_2$ are not observed. These transitions are said to be forbidden by the electric dipole selection rules. These selection rules allow one to calculate quickly which transitions will occur once the angular momentum quantum numbers of the energy levels are specified. This is very helpful in the analysis of spectra, for instance in the transitions between the levels $^2D_{5/2,3/2}$ and $^2P_{3/2,1/2}$, the line corresponding to $J=5/2 \leftrightarrow J=1/2$ is forbidden and the $^2D-^2P$ multiplets consist of only three components. Similarly if one wishes to calculate the radiative lifetime of a given level the application of the selection rules usually reduces the summation in equation (4.26) to just a few non-zero terms.

5.2. One-electron atoms without spin

We consider first the case of atoms whose emission spectrum can be ascribed to a single electron. This includes all the hydrogenic systems, the alkalis and the singly-ionized alkaline earth elements, etc. For the sake of simplicity at this stage we use the uncoupled representation of equation (3.61):

$$\psi_{nlm_lm_s} = \psi_{nlm_l}\chi(m_s).$$

This would correspond to a situation in which $\underline{l}$ and $\underline{s}$ had been completely uncoupled by the application of a large external magnetic field (Paschen-Back effect). Since the electric dipole moment operator, equation (4.18), does not act directly on the spin wavefunctions we have in this case the selection rule

$$\Delta m_s = 0.$$

We can therefore neglect the effect of the spin and describe the electron by the spatial wavefunction alone:

$$\psi_{nlm_l}(r,\theta,\phi) = R_{nl}(r)\, Y_l^{m_l}(\theta,\phi). \qquad (5.1)$$

5.2.1. Selection rules for magnetic quantum number m_l. We are interested in the selection rules for transitions between an upper level labelled by the quantum numbers (n,l,m_l) and a lower level labelled by (n',l',m_l'). Since the degeneracy of the states with respect to the m_l quantum number has been lifted by the application of the external magnetic field, we can use equation (4.18) for the transition probability between non-degenerate states:

$$A_{E1}(nlm_l;n'l'm_l') = \frac{e^2\omega^3_{nl,n'l'}}{3\pi\varepsilon_0\hbar c^3}\left|\int \psi^*_{n'l'm_l'}\, \underline{r}\, \psi_{nlm_l}\, dv\right|^2 . \qquad (5.2)$$

In Cartesian components the vector $\underline{r}$ is given by

$$\underline{r} = x\hat{\underline{i}} \quad y\hat{\underline{j}} + z\hat{\underline{k}}.$$

However, since the wavefunctions are expressed in terms of spherical polar coordinates it is more convenient to expand $\underline{r}$ in terms of its components in the spherical basis

$$\hat{\underline{\varepsilon}}_{\pm 1} = \mp\frac{1}{\sqrt{2}}(\hat{\underline{i}} \pm i\hat{\underline{j}}) \;;\; \hat{\underline{\varepsilon}}_0 = \hat{\underline{k}}. \qquad (5.3)$$

These components are defined by the equations

$$\underline{r} = \sum_{q=-1}^{+1} (-)^q r_q\, \hat{\underline{\varepsilon}}_{-q} \qquad (5.4)$$

giving $\underline{r} = -\frac{1}{\sqrt{2}}\, r\sin\theta\, e^{-i\phi}\, \hat{\underline{\varepsilon}}_{+1} + r\cos\theta\, \hat{\underline{\varepsilon}}_0 + \frac{1}{\sqrt{2}}\, r\sin\theta\, e^{i\phi}\, \hat{\underline{\varepsilon}}_{-1}$

(5.5)

To derive the selection rules for m_l we need only consider the ϕ-dependence of the matrix elements given by

$$\int \psi^*_{n'l'm'_l} \, \underline{r} \, \psi_{nlm_l} \, dv = a \, \hat{\underline{\varepsilon}}_{+1} \int_0^{2\pi} \exp\{i(m_l - m'_l - 1)\phi\} \, d\phi + $$

$$+ \, b\hat{\underline{\varepsilon}}_0 \int_0^{2\pi} \exp\{i(m_l - m'_l)\phi\} d\phi + c\hat{\underline{\varepsilon}}_{-1} \int_0^{2\pi} \exp\{i(m_l - m'_l + 1)\phi\} d\phi \quad (5.6)$$

where a, b, and c involve integrals over r and θ. The integrals over the angle ϕ are identically zero unless the exponents of the exponential factors vanish. Thus the electric dipole transition probability is zero unless one of the selection rules

$$\Delta m_l = m_l - m'_l = +1, 0, -1$$

is satisfied.

<u>5.2.2. Polarization of the radiation.</u> The polarization of the radiation emitted when one of these selection rules is satisfied can be obtained by making use of the correspondence between the quantum-mechanical and classical dipole moments, equation (4.17), and the radiation field of a classical dipole oscillator. From equation (2.70) we see that the polarization of the electric field is determined by the component of $\underline{p}$ perpendicular to the direction of observation $\hat{\underline{r}}$. Thus the ellipse traced out by the electric vector $\underline{E}(r,\theta,\phi)$ has the same shape and orientation as the projection on to the plane perpendicular to $\hat{\underline{r}}$ of the ellipse described by the electron. In addition the motion of the $\underline{E}$ vector round its ellipse is in the same sense as the projected motion of the electron.

Introducing the time dependence explicitly we find that $\hat{\underline{\varepsilon}}_{+1}\exp(-i\omega t)$ represents anti-clockwise circular motion when viewed from the positive z-axis. Thus from equation (5.6) the wave emitted in transitions for which $\Delta m_l = +1$ is left circularly polarized when observed from the positive z-axis and is labelled σ^+, as shown in Fig.5.1. Similarly

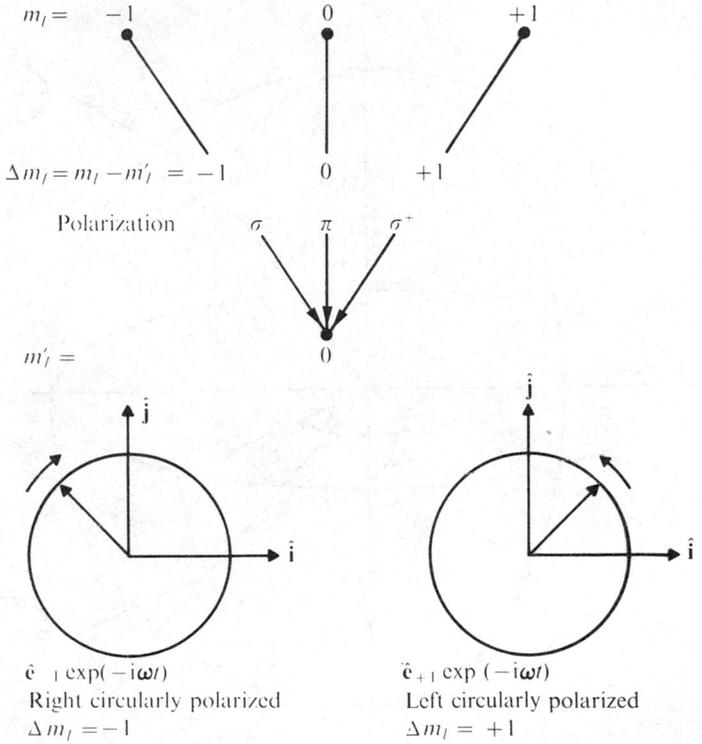

Fig.5.1. Selection rules for magnetic quantum number m_l and polarization of electric dipole radiation observed in the direction $\theta=0$.

transitions for which $\Delta m_l = -1$ in emission correspond to right circularly polarized radiation when observed from the direction $\theta=0$ and are labelled σ^-. The sense of rotation of the electric vector for σ^+ radiation is the same as that of the electric current in a coil generating the magnetic field.

In general the intensity and polarization of the detected radiation depend on the position of the observer (θ,ϕ) with respect to the atom. We denote by $(\hat{\underline{r}},\hat{\underline{\theta}},\hat{\underline{\phi}})$ a right-handed set of unit vectors in the directions of increasing (r,θ,ϕ) respectively, as shown in Fig.5.2. Then since an arbitrary vector $\underline{p}$ can be expressed in terms of its components parallel and perpendicular to the unit vector $\hat{\underline{r}}$ by

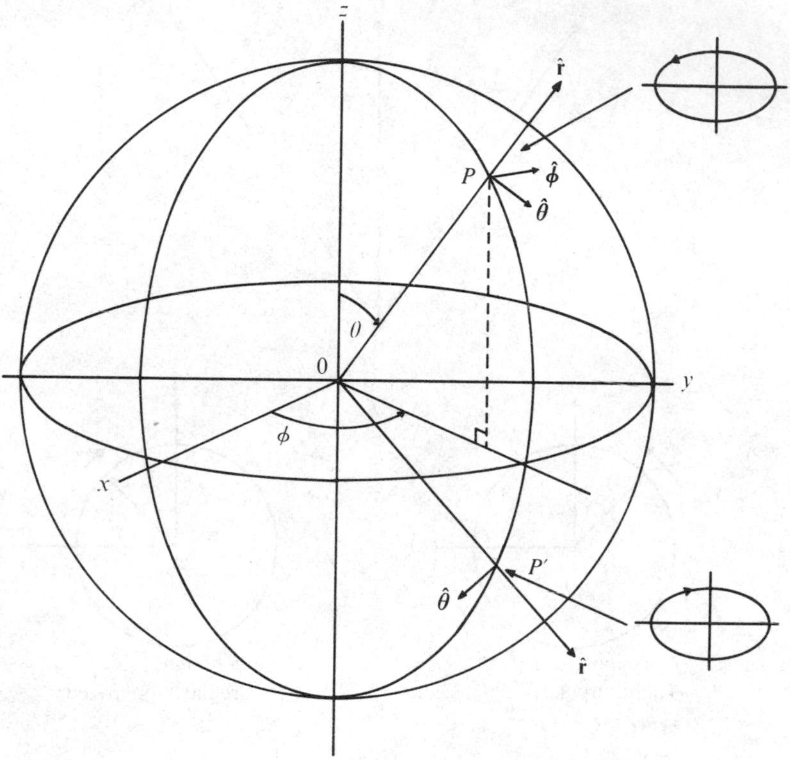

Fig. 5.2. Polarization of electric dipole radiation as a function of angle of observation for electric dipole transitions satisfying the selection rule $\Delta m_l = +1$.

$$\underline{p} = (\underline{p}.\hat{\underline{r}})\hat{\underline{r}} - \hat{\underline{r}} \wedge (\hat{\underline{r}} \wedge \underline{p}) \qquad (5.7)$$

we see, from equation (2.70), that the electric field is given by

$$\underline{E}(r,\theta,\phi) \propto \underline{p} - (\underline{p}.\hat{\underline{r}})\hat{\underline{r}} = p_\theta \hat{\underline{\theta}} + p_\phi \hat{\underline{\phi}}.$$

It is then a straightforward exercise (Problem 5.1) to express the field in terms of its linear and elliptically-polarized components:

$$\underline{E}(r,\theta,\phi) \propto \left\{ p_{-1} \frac{(\hat{\underline{\theta}}\cos\theta + i\hat{\underline{\phi}})}{\sqrt{2}} \exp(i\phi) - \right. \qquad (5.8)$$

$$\left. - p_0 \sin\theta \; \hat{\underline{\theta}} - p_{+1} \frac{(\hat{\underline{\theta}}\cos\theta - i\hat{\underline{\phi}})}{\sqrt{2}} \exp(-i\phi) \right\}$$

where

$$p_{\pm 1} = (-e)r_{\pm 1} = (-e)\{\mp \frac{1}{\sqrt{2}}(x \pm iy)\} \; ; \; p_0 = (-e)z.$$

The first term in equation (5.8) gives a contribution to the field which is left circularly polarized in the direction $\theta=0$ and becomes left elliptically polarized in directions $0<\theta<\pi/2$. In the plane $\theta=\pi/2$ this light is linearly polarized with the electric vector perpendicular to the direction of the applied magnetic field $\underline{B}$. Finally in the 'southern hemisphere' this radiation is observed to be right elliptically polarized as shown in Fig.5.2. The second term in equation (5.8) is responsible for radiation which is linearly polarized in a plane through the z-axis and the point of observation while the third term corresponds to right elliptically polarized light in the 'northern hemisphere'.

Combining equation (5.8) with the selection rules derived from equation (5.6), we see that the relative intensities of the different Zeeman components are given by

$$\left.\begin{aligned} I(\Delta m = +1) &= \frac{I_0}{2}(1+\cos^2\theta)|\langle\gamma'j'm-1|p_{-1}|\gamma jm\rangle|^2 \\ I(\Delta m = 0) &= I_0 \sin^2\theta|\langle\gamma'j'm|p_0|\gamma jm\rangle|^2 \\ I(\Delta m = -1) &= \frac{I_0}{2}(1+\cos^2\theta)\;|\langle\gamma'j'm+1|p_{+1}|\gamma jm\rangle|^2. \end{aligned}\right\} \quad (5.9)$$

In these expressions eigenfunctions of the total angular momentum $|jm\rangle$ have replaced the orbital angular momentum wavefunctions $|lm_l\rangle$ since, as we shall show below, the results of this paragraph have a general validity. The matrix elements of the components of $\underline{p}$ are given in Table 5.1 and enable the relative intensities of the Zeeman components to be calculated for any given transition (Problems 5.2-5.4). As expected, the angular distributions given by equation (5.9) are identical to those associated with the classical dipole oscillator which we discussed in section 2.8. In particular we notice that the intensity of π-polarized radiation, emitted in transitions for which $\Delta m=0$, is predicted

TABLE 5.1.

Matrix elements squared of the components of the electric dipole operator, $\underline{p}$.

For $j'=j$	$\lvert\langle\gamma'j\ m+1\lvert p_{+1}\rvert\gamma jm\rangle\rvert^2$	$= \frac{1}{2}(j-m)(j+m+1)\ A$
	$\lvert\langle\gamma'j\ m\lvert p_0\rvert\gamma jm\rangle\rvert^2$	$= m^2\ A$
	$\lvert\langle\gamma'j\ m-1\lvert p_{-1}\rvert\gamma jm\rangle\rvert^2$	$= \frac{1}{2}(j+m)(j-m+1)\ A$
For $j'=j+1$	$\lvert\langle\gamma'j+1\ m+1\lvert p_{+1}\rvert\gamma jm\rangle\rvert^2$	$= \frac{1}{2}(j+m+1)(j+m+2)\ B$
	$\lvert\langle\gamma'j+1\ m\lvert p_0\rvert\gamma jm\rangle\rvert^2$	$= \{(j+1)^2 - m^2\}\ B$
	$\lvert\langle\gamma'j+1\ m-1\lvert p_{-1}\rvert\gamma jm\rangle\rvert^2$	$= \frac{1}{2}(j-m+1)(j-m+2)\ B$
For $j'=j-1$	$\lvert\langle\gamma'j-1\ m+1\lvert p_{+1}\rvert\gamma jm\rangle\rvert^2$	$= \frac{1}{2}(j-m)(j-m-1)\ C$
	$\lvert\langle\gamma'j-1\ m\lvert p_0\rvert\gamma jm\rangle\rvert^2$	$= (j^2-m^2)\ C$
	$\lvert\langle\gamma'j-1\ m-1\lvert p_{-1}\rvert\gamma jm\rangle\rvert^2$	$= \frac{1}{2}(j+m)(j+m-1)\ C$

The constant factors

$A = \lvert\langle\gamma'j\Vert\underline{p}\Vert\gamma j\rangle\rvert^2,$

$B = \lvert\langle\gamma'j+1\Vert\underline{p}\Vert\gamma j\rangle\rvert^2,$

$C = \lvert\langle\gamma'j-1\Vert\underline{p}\Vert\gamma j\rangle\rvert^2$ are defined by equation (5.18)

to be zero in the directions $\theta=0$ and $\theta=\pi$, which agrees with one of the characteristic features of the observed Zeeman splittings of spectral lines.

5.2.3. Selection rules for orbital quantum number l. For an electric dipole transition to take place, we require that the matrix element of the electric dipole moment operator

$$\langle n'l'm_l'|-e\underline{r}|nlm_l\rangle = -e\int\psi^*_{n'l'm_l'}\,\underline{r}\,\psi_{nlm_l}\,dv \tag{5.10}$$

shall be non-vanishing. By inverting the coordinate system and calculating the new value of this matrix element using the parity properties of the spherical harmonics, equation (3.27), we obtain a result which is $(-1)^{l'+l+1}$ times the previous value. Thus only transitions between states of opposite parity are allowed for electric dipole radiation and we have

$$\Delta l = 1,3,5,\ldots\ldots .$$

A more detailed selection rule may be obtained by considering the angular integrals in the matrix element of equation (5.10). These involve the products $\cos\theta\, Y_l^{m_l}(\theta,\phi)$ and $\sin\theta\, Y_l^{m_l}(\theta,\phi)$ which can be transformed by the use of the recursion relations

$$\cos\theta\, Y_l^m = Y_{l+1}^m\left\{\frac{(l+1-m)(l+1+m)}{(2l+1)(2l+3)}\right\}^{1/2} + Y_{l-1}^m\left\{\frac{(l-m)(l+m)}{(2l+1)(2l-1)}\right\}^{1/2} \tag{5.11}$$

$$\sin\theta\, Y_l^m = -Y_{l+1}^{m+1}\left\{\frac{(l+m+1)(l+m+2)}{(2l+1)(2l+3)}\right\}^{1/2} + Y_{l-1}^{m+1}\left\{\frac{(l-m)(l-m-1)}{(2l-1)(2l+1)}\right\}^{1/2} \tag{5.12}$$

Now using the orthogonality property of the spherical harmonics we see that the electric dipole matrix elements of equation (5.10) will vanish unless

$$\Delta l = \pm 1. \tag{5.13}$$

<u>5.2.4. Intrinsic angular momentum of photons.</u> In a transition which obeys the selection rule of equation (5.13), for example the emission of Lyman α radiation $2\,^2P \rightarrow 1\,^2S$ the angular momentum of the atom decreases by one unit. The principle of conservation of angular momentum therefore requires that the emitted photon shall have an intrinsic angular momentum of $\hbar$. Similarly in the decay of the $m_l = +1$ sub-level of $2\,^2P$ the z-component of angular momentum of the atom decreases by $\hbar$. To compensate for this the σ^+

(left-circularly polarized) photon which is emitted (Fig. 5.1) has an angular momentum whose projection on the z-axis has the value $+1\hbar$. Similarly σ^- polarized photons possess a component of angular momentum along the z-axis of $-1\hbar$. These properties of the photon are consistent with the classical result discussed in section 2.9.

5.3. *One-electron atoms with spin*

The non-relativistic wavefunctions describing a one-electron atom with spin are given by equation (3.64), the good quantum numbers being (l,s,j,m). In an electric dipole transition from an initial state with total angular momentum quantum number j to a final state with quantum number j', a photon of unit intrinsic angular momentum is emitted. The selection rule for the j quantum number can now be derived by applying the principle of conservation of angular momentum to a vector model of the atom plus the radiation fields. It must be possible to create the initial angular momentum vector $\underline{j}$ by vector addition of $\underline{j}'$, representing the atom in its final state, and $\underline{\hat{l}}$ where $\underline{\hat{l}}$ is a unit vector representing the emitted photon. Applying the rules of the vector model we see that either $j=j'\pm1$ or $j=j'$. Therefore the selection rule for the total angular momentum quantum number is

$$\Delta j = j' - j = 0, \pm 1. \tag{5.14}$$

Similar considerations show that the magnetic quantum number m must obey the rule

$$\Delta m = 0, \pm 1. \tag{5.15}$$

The polarization and angular distribution of radiation satisfying a given Δm selection rule are the same as those discussed in section 5.2.2.

In the special case where the total angular momentum of the initial and final states is zero, $j=j'=0$, it is obviously impossible to satisfy the principle of conservation of angular momentum. These transitions are therefore strictly forbidden, a fact which can be expressed symbolically by

$$j=0 \nleftrightarrow j'=0. \tag{5.16}$$

This rule is responsible for the metastability of the lowest $^3P_0$ levels in mercury, cadmium, zinc, neon, argon, etc.

This derivation of the selection rules, equations (5.14), (5.15), and (5.16), has necessarily been rather qualitative and a more rigorous treatment is given in section 5.4.1 below. Meanwhile by using the explicit wavefunctions of equation (3.64) and the recursion relations (5.11) and (5.12), it is possible to evaluate the electric dipole matrix elements directly and to verify that these selection rules are in fact correct (Problem 5.5).

5.4. *Tensor properties of the electric dipole operator*

Unfortunately in order to derive the selection rules for many-electron systems and to give a rigorous derivation of the Δj selection rule it is necessary to employ the powerful techniques of the quantum theory of angular momentum which were discussed briefly in section 3.7. We start by expressing the electric dipole operator for a single electron in terms of its components in the spherical basis. Using equation (5.5) and Table 3.1 we have

$$\underline{p} = -e\underline{r} = er\left(\frac{4\pi}{3}\right)^{\frac{1}{2}}(Y_1^{-1}\underline{\hat{\varepsilon}}_{+1} - Y_1^0\underline{\hat{\varepsilon}}_0 + Y_1^1\underline{\hat{\varepsilon}}_{-1}). \tag{5.17}$$

It follows that the spherical components of the electric dipole operator will transform like spherical harmonics of order unity under any arbitrary rotation of the coordinate system. A quantity which transforms under rotations like the spherical harmonic $Y_k^q(\theta,\phi)$ is said to be an irreducible tensor operator T_q^k of rank k and projection q where the projection quantum number can take any integer value from -k to +k. Since any arbitrary function of θ and ϕ can generally be expanded as a sum of spherical harmonics, it is usually possible to express any physical operator in terms of irreducible tensor operators. For instance, the electric quadrupole moment operator defined by equation (4.45) can be shown to be a tensor of rank 2 (Problem 5.6).

5.4.1. The Wigner-Eckart theorem. In the theory of the interaction of radiation with atoms we are interested in the matrix elements of tensor operators taken between states of different quantum numbers. The very powerful Wigner-Eckart theorem states that these matrix elements are given by:

$$\langle \alpha' J' M' | T^k_q | \alpha\ J\ M \rangle = C(J\ k\ J';M\ q\ M') \langle \alpha' J' \| T^k \| \alpha\ J \rangle \quad (5.18)$$

where J,M are the total and magnetic quantum numbers of the states respectively, and α', α represent any additional quantum numbers required to specify the state. The matrix element therefore factorizes into a Clebsch-Gordan coefficient which contains all the information about the angular dependence of the states and of the operator involved, and a quantity $\langle \alpha' J' \| T^k \| \alpha\ J \rangle$ called the reduced matrix element which is independent of q,M, and M'.

The numerical value of the Clebsch-Gordan coefficient is zero unless the triangular condition $\Delta(J'\ k\ J)$ is satisfied. Thus we have

$$|J' - J| \le k \le J' + J. \quad (5.19)$$

The Clebsch-Gordan coefficient is also zero unless the magnetic quantum numbers satisfy the condition

$$M' = q + M. \quad (5.20)$$

These two conditions ensure essentially the conservation of the total angular momentum of the system and of the component of this angular momentum on the axis of quantization.

Since, as equation (5.17) shows, the single-electron electric dipole operator is a tensor operator of order unity, k=1, the triangular condition of equation (5.19) enables the selection rule for the j-quantum number

$$\Delta j = 0, \pm 1 \quad (5.21)$$

to be derived in a completely rigorous way. Similarly, using equation (5.20), we have the selection rule for the magnetic quantum number:

$$\Delta m = 0, \pm 1. \tag{5.22}$$

The reduced matrix element for the electric dipole operator is effectively the radial integral of r taken between the states involved (Problem 5.7). A more detailed discussion of the Wigner-Eckart theorem may be found in Messiah (1962, Vol.II, p.573 and Appendix C).

5.5. *Many-electron atoms*

5.5.1. Wavefunctions in L-S coupling. We recall that in the central field approximation, discussed in section 3.9, each electron in a many-electron atom is considered to move in a central potential independently of the others, at least in the first approximation. The ith electron may then be described by the function $\psi_\alpha(i)$ (equation (3.71)) in terms of a product of spatial and spin wavefunctions. If there are N electrons the Pauli principle requires that each of the sets of quantum numbers $\alpha, \beta, \gamma, \ldots, \nu$ should be different and a simple N-electron wavefunction would consist of the direct product of single-electron wavefunctions:

$$\psi_\alpha(1)\psi_\beta(2)\psi_\gamma(3) \ldots\ldots\ldots\ldots \psi_\nu(N). \tag{5.23}$$

A total wavefunction which is antisymmetric with respect to the exchange of the space and spin coordinates of any two electrons is then obtained by combining several direct product wavefunctions in the normalized Slater determinantal wavefunctions given by equation (3.72). These form an uncoupled representation in which the good quantum numbers are $(nlm_l m_s)$ for each electron. The residual Coulomb interaction, is diagonal in a representation in which the good quantum numbers are (LSM_LM_S). This representation is obtained by taking suitable linear combinations of the determinantal product wavefunctions and is called the L-S or Russell-Saunders coupling scheme.

$$\Psi(\gamma\ LSM_LM_S) = \sum_i a_i \Psi(\alpha, \beta, \ldots, \nu) \tag{5.24}$$

where a_i are the expansion coefficients and γ now represents all the quantum numbers which are required to specify the electron configuration. Finally the magnetic spin-orbit interaction is treated as a small perturbation which can be diagonalized by a transformation from the (LSM_LM_S) representation to the representation in which $(LSJM_J)$ are good quantum numbers. This transformation involves the use the Clebsch-Gordan coefficients:

$$\Psi(\gamma\ LSJM_J) = \sum_{M_LM_S} C(LSJ;M_LM_SM_J)\ \Psi(\gamma\ LSM_LM_S). \qquad (5.25)$$

We now derive the selection rules for electric dipole radiation using atomic wavefunctions in the L-S coupling approximation (equations (5.24) and (5.25)). We shall find that certain of the selection rules are rigorous but that others are only applicable if L-S coupling is a good description of the atomic system.

5.5.2. Selection rule for l. The matrix elements of the electric dipole operator for an N-electron atom have the form

$$\langle \gamma'S'L'J'M_J'|\sum_{i=1}^{N} -e\underline{r}_i|\ \gamma\ SLJM_J\rangle. \qquad (5.26)$$

The electric dipole operator, $\sum_{i=1}^{N} -e\underline{r}_i$, is odd under the parity transformation and therefore the parity of the electron configuration, given by $(-1)^{\sum_{i=1}^{N} l_i}$, must change in an electric dipole transition. This is a generalization of equation (5.13) and is known as the Laporte rule. Since the parity property of a free atom remains well defined whatever the coupling scheme, the Laporte rule remains valid even in complicated cases where it is no longer possible to assign a unique configuration to a given energy level. In a complicated spectrum it is therefore useful to characterize the energy levels as having odd or even parity, a term of odd parity being indicated by a superscript ^{o}, for example, npnd $^3D^o$.

Substituting equations (5.24) and (5.25) into equation (5.26), we see that the matrix element of the electric dipole operator eventually reduces to a sum over products of single-particle wavefunctions. A typical term in this summation has the form

$$\langle \psi_{\alpha'}(1)\psi_{\beta'}(2)\ldots\ldots\psi_{\nu'}(N)|e\underline{r}_i|\psi_\alpha(1)\psi_\beta(2)\ldots\ldots\psi_\nu(N)\rangle. \quad (5.27)$$

From the discussion in section 5.2 it is clear that for one particular electron we must have $\Delta l_i = \pm 1$ and that for the remaining electrons the initial and final quantum numbers must be identical, otherwise the orthogonality of the single-electron wavefunctions will ensure that all contributions to the dipole matrix element are zero. The selection rules for one-electron jumps,

$$\Delta l_i = \pm 1 \ ; \ \Delta l_{j \neq i} = 0 \quad (5.28)$$

rest on the assumption that the wavefunction is described exactly by a single-electron configuration. If there is strong configuration mixing then transitions involving the simultaneous change in the (nl) values of two electrons may occur and the rules of equation (5.28) will not apply. However, even for these two-electron jumps the parity selection rule must be obeyed.

<u>5.5.3. Selection rule for J.</u> The electric dipole operator $\sum_{i=1}^{N} -e\underline{r}_i$, is the sum of tensor operators of rank k=1 which operate on the spatial part of the wavefunctions of individual electrons. In other words, it operates in the orbital space (l_i) of the individual electrons. However, it also operates in the coupled orbital space (L) and the total space (J) because the (LSM_LM_S) and $(LSJM_J)$ representations are constructed from single-electron wavefunctions, equations (5.24) and (5.25). Applying the Wigner-Eckart theorem in the $(LSJM_J)$ representation and equations (5.19) and (5.20), we obtain the selection rules

$$\left.\begin{aligned} \Delta J &= 0, \pm 1 \\ J &= 0 \nleftrightarrow J' = 0 \\ \Delta M_J &= 0, \pm 1. \end{aligned}\right\} \qquad (5.29)$$

These selection rules are rigorous ones which are independent of the exact coupling scheme describing the individual spins and orbital angular momenta.

5.5.4. Selection rule for L. If we use the expansion of coupled wavefunctions given by equation (5.25), the matrix element of the electric dipole operator, equation (5.26), can be expressed as a sum over quantities of the type

$$\langle \gamma' L' S' M_L' M_S' | \sum_{i=1}^{N} -e\underline{r}_i | \gamma\, L S M_L M_S \rangle . \qquad (5.30)$$

Applying the tensor properties of the electric dipole operator and the Wigner-Eckart theorem to the space of the total orbital angular momentum (L), we derive the selection rules

$$\left.\begin{aligned} \Delta L &= 0, \pm 1 \\ L &= 0 \nleftrightarrow L' = 0 \\ \Delta M_L &= 0, \pm 1. \end{aligned}\right\} \qquad (5.31)$$

These selection rules are based on the assumption that L is a good quantum number and that Russell-Saunders coupling is an exact description of the atomic system. This is a good approximation in light atoms where the magnetic interactions are generally very much smaller than the residual electrostatic interactions. However, we may observe transitions which violate equations (5.31) in the spectra of heavier elements.

The selection rule $\Delta L = 0$ needs a little further explanation since transitions of this type are not observed in the spectra of most one- and two-electron atoms. For one-electron systems such as hydrogen and the alkalis, we have $L \equiv l_i$ and the selection rule $\Delta l_i = \pm 1$ effectively prevents $\Delta L = 0$ transitions. For simple two-electron systems such as helium, the alkaline earth elements, and zinc, cadmium, and mercury, the lowest configuration consists of two equivalent s electrons. The normal excited levels are almost entirely due to the promotion of just one of these electrons into a higher shell. Thus again $L \equiv l_i$ and transitions corresponding to $\Delta L = 0$ are not observed.

As an example of a more complex atom we consider the case of carbon which has a ground state configuration of $2\,p^2$ and excited configurations of the type 2pnp, 2pnd with $n \geq 3$ amongst others. The terms and their parities arising from these three configurations are given in Table 5.2. Transitions which simultaneously satisfy $\Delta l_i = \pm 1$, $\Delta L = 0$ and the parity selection rule are now possible and the multiplets

$$2p3d\ ^1D^o \rightarrow 2p^2\ ^1D \qquad 1481{\cdot}8\ \text{Å}$$

$$2p4d\ ^1P^o \rightarrow 2p3p\ ^1P \qquad 6587{\cdot}7\ \text{Å}$$

are observed in the spectrum of atomic carbon, CI.

TABLE 5.2

Terms and parities of selected configurations of the neutral carbon atom

Configuration	Parity	Terms
$2\,p^2$	even	^1S , ^3P , ^1D
2pnp	even	^1S , ^1P , ^1D
		^3S , ^3P , ^3D
2pnd	odd	$^1P^o$, $^1D^o$, $^1F^o$
		$^3P^o$, $^3D^o$, $^3F^o$

5.5.5. Selection rule for S. When the transition probability for electric dipole radiation is expressed in terms of the single-particle matrix elements of equation (5.27) it is obvious that the electric dipole operator does not act on the spin wavefunction of any of the electrons. We must therefore have $m_s = m_s'$ for all N electrons. Similarly the electric dipole operator does not act in the total spin space of the electrons (S) and the spin quantum numbers of the initial and final states in equation (5.30) must be the same, i.e.

$$\Delta S = 0. \tag{5.32}$$

This selection rule is not rigorous since it is again based on the assumption that L-S coupling is an exact description of the excited states of a given atom and transitions which violate equation (5.32) are well known.

An interesting example of the progressive breakdown of L-S coupling is provided by the group IIB metals Zn, Cd, and Hg. The lowest electron configuration of these elements is ns^2 resulting in an n $^1S_0$ ground state. The first excited configuration is nsnp which gives rise to the levels n $^1P_1$ and n $^3P_{2,1,0}$. Electric dipole transitions from the levels n $^3P_{2,0}$ are forbidden by the rigorous ΔJ selection rules (equation (5.20)) and these levels are metastable. If L-S coupling was an exact description of these levels, the transition n $^1S_0$-n $^3P_1$ would also be forbidden. However, this intercombination line is observed in the spectra of all three elements and in mercury it is one of the most intense lines in the spectrum.

To explain this fact we must assume that the true wavefunction describing the n $^3P_1$ levels is a mixture of the pure L-S coupled wavefunctions $\psi_{LS}(^1P_1)$ and $\psi_{LS}(^3P_1)$:

$$\psi(^3P_1) = \alpha\, \psi_{LS}(^3P_1) + \beta\, \psi_{LS}(^1P_1) \tag{5.33}$$

where α and β are numerical factors. Similarly the true wavefunction for the $^1P_1$ level is also a mixture of these states and since it must be orthogonal to $\psi(^3P_1)$, we may write it as

$$\psi(^1P_1) = -\beta\, \psi_{LS}(^3P_1) + \alpha\, \psi_{LS}(^1P_1). \tag{5.34}$$

We obviously hope that $\alpha \approx 1$ and $\beta \approx 0$, otherwise the L-S representation will be a very poor description of these energy levels. Assuming that the L-S wavefunctions are normalized and orthogonal, the normalization of $\psi(^1P_1)$ and $\psi(^3P_1)$ is achieved when

$$\langle \psi(^3P_1)|\psi(^3P_1)\rangle = \alpha^2 + \beta^2 = 1. \tag{5.35}$$

Since the levels $n\ ^1P_1$ and $n\ ^3P_1$ can each decay by only a single transition to the ground state $n\ ^1S_0$, we find from equations (4.23) and (4.26) that the ratio of the lifetimes is given by

$$\frac{\tau_1}{\tau_3} = \frac{\omega_3^3\ |\langle \psi(^3P_1)|e\underline{r}|\psi(^1S_0)\rangle|^2}{\omega_1^3\ |\langle \psi(^1P_1)|e\underline{r}|\psi(^1S_0)\rangle|^2} \tag{5.36}$$

Assuming that the ground-state wavefunction is given by a pure L-S state we have, using equations (5.33) and (5.34),

TABLE 5.3.

L-S coupling mixing parameters for Group IIB metals

Element	n	τ_1 (10^{-9} s)	λ_1 (Å)	τ_3 (10^{-7} s)	λ_3 (Å)	$(1-\alpha)$ (x 10^{-4})	β (x 10^{-3})
Zn	4	1·41	2138	200	3076	0·118	4·86
Cd	5	1·66	2288	23·9	3561	0·921	13·6
Hg	6	1·31	1850	1·18	2537	148	-171

$$\frac{\beta^2}{\alpha^2} = \frac{\tau_1\lambda_3^3}{\tau_3\lambda_1^3} . \tag{5.37}$$

Thus once the radiative lifetimes τ_1 and τ_3 of the singlet and triplet levels have been measured, the mixing parameters α and β can be determined with the aid of equations (5.35) and (5.37). The experimentally measured lifetimes for these levels in Zn, Cd, and Hg are given in Table 5.3 together with the resultant values of α and β.

We see that the value of β increases rapidly as we go from zinc to mercury, showing that magnetic spin-orbit interactions are becoming progressively more important. However, even for mercury the value of α is close to unity, showing that L-S coupling is still a useful description of the atomic system.

5.6. *Relative intensities in L-S coupling and forbidden transitions*

The relative intensities of a group of spectral lines which arise in electric dipole transitions between one L-S fine structure multiplet and another may be calculated from equation (5.26). The angular factors in this expression depend on the quantum numbers (S,L,J) and (S',L',J') of the levels involved and may be calculated by vector coupling techniques. Unfortunately the formulae so derived are themselves rather complicated and the reader is referred to standard works on atomic spectroscopy, for instance Condon and Shortley (1967), where tabulations of the relative intensities may be found.

Magnetic dipole and electric quadrupole transitions, which we have ignored so far, will be discussed in detail in Chapter 7. However, it should be noted that the selection rules governing these 'forbidden' types of radiation can be derived by the application of the techniques already discussed here. The results are summarized in Table 7.1 together with the rules for electric dipole radiation previously derived in this chapter.

Problems

5.1. In spherical polar coordinates show that the unit vectors $(\underline{\hat{r}},\underline{\hat{\theta}},\underline{\hat{\phi}})$ are related to the cartesian unit vectors $(\underline{\hat{i}},\underline{\hat{j}},\underline{\hat{k}})$ by

$$\underline{\hat{r}} = \underline{\hat{i}} \sin\theta \cos\phi + \underline{\hat{j}} \sin\theta \cos\phi + \underline{\hat{k}} \cos\theta$$

$$\underline{\hat{\theta}} = \underline{\hat{i}} \cos\theta \cos\phi + \underline{\hat{j}} \cos\theta \sin\phi - \underline{\hat{k}} \sin\theta$$

$$\underline{\hat{\phi}} = -\underline{\hat{i}} \sin\phi + \underline{\hat{j}} \cos\phi .$$

Hence show that $(p_\theta \underline{\hat{\theta}} + p_\phi \underline{\hat{\phi}})$ may be expressed in the form given in equation (5.8), where p_θ and p_ϕ are the components of the electric dipole moment operator $\underline{p}$ in the directions $\underline{\theta}$ and $\underline{\phi}$ respectively.

5.2. Using equations (3.86) and (5.6) verify that the transition $5\ ^1S_0 - 5\ ^1P_1$ at 2288 Å in the spectrum of atomic cadmium displays the classical Zeeman effect. Show that the magnitude of the splitting between the π and σ polarized components in a field of 1 T is $4{\cdot}669\ \text{cm}^{-1}$. Calculate the relative intensities of the Zeeman components viewed in directions parallel and perpendicular to the applied magnetic field using a classical argument and compare with the results obtained from equation (5.9) and Table 5.1.

5.3. Show that the Landé g-factors of the $3\ ^2S_{1/2}$, $3\ ^2P_{1/2}$, and $3\ ^2P_{3/2}$ levels of atomic sodium are 2, 2/3, and 4/3 respectively. Hence show that in a magnetic field of 0·25 T the D_1 line, $^2S_{1/2} - ^2P_{1/2}$, will be split into components at

$$\tilde{\nu}_1 \pm 2\Delta/3\ ;\ \tilde{\nu}_1 \pm 4\Delta/3$$

and the D_2 line, $^2S_{1/2} - ^2P_{3/2}$, will be split into components at

$$\tilde{\nu}_2 \pm \Delta/3\ ;\ \tilde{\nu}_2 \pm \Delta\ ;\ \tilde{\nu}_2 \pm 5\Delta/3$$

where $\Delta = \mu_B B/hc = 1{\cdot}167\ \text{cm}^{-1}$. The Zeeman pattern is assumed to be observed in a direction perpendicular to the applied field.

Using the selection rules for the magnetic quantum number and equation (5.9), determine the polarizations and relative intensities of the Zeeman components of the D_1 and D_2 lines viewed in directions both parallel and perpendicular to the applied field. The populations of the excited Zeeman sub-levels and the frequencies of the Zeeman components of the line may be assumed to be equal in this part of the calculation.

5.4. Show that the radiation emitted in electric dipole transitions between degenerate levels, equation (4.23), is unpolarized and has the same intensity in all directions. (Hint: in the direction (θ,ϕ) the relative intensities and polarizations of the light emitted in transitions between specified Zeeman sub-levels are given by equation (5.9)).

5.5. Making use of the coupled one-electron wavefunctions given by equation (3.64) and the recursion relations for the spherical harmonics, equations (5.1) and (5.12), show, by detailed calculation, that the matrix element of the z-component of the electric dipole operator between the states $\psi_{n,l,j=l+1/2,m}$ and $\psi_{n',l-1,j=l-3/2,m}$ is identically zero as required by the selection rules for j.

5.6. Show that the cartesian components of the electric quadrupole operator defined by

$$Q_{\alpha\beta} = e(3x_\alpha x_\beta - r^2 \delta_{\alpha\beta})$$

where $x_\alpha = x$, $x_\beta = y$, $x_\gamma = z$ can be expressed in terms of the spherical harmonics of order 2 given in Table 3.1. This proves that the quadrupole operator is a second rank tensor.

5.7. By considering the matrix elements of the components

$p_q = (-e)r_q$ of the electric dipole operator (a first rank tensor), show that the reduced matrix element for the 2p→1s Lyman α transition of atomic hydrogen is given by

$$|\langle 21 \| \underline{p} \| 10 \rangle|^2 = e^2 \left| \int_0^\infty R_{21}(r)\, R_{10}(r)\, r^3\, dr \right|^2 .$$

General references and further reading

Condon, E.U. and Shortley, G.H. (1967). *The theory of atomic spectra*, Cambridge University Press, London.

Messiah, A. (1962). *Quantum mechanics*, Vols. 1 and 2. North-Holland Publishing Company, Amsterdam.

6

Measurement of radiative lifetimes of atoms and molecules

We have shown in previous chapters that the f-values of spectral lines are important fundamental data which must be known before detailed calculations of the behaviour of gas discharges, plasmas, or stellar atmospheres can be undertaken. Since it is difficult, in many cases, to make theoretical calculations of f-values to an accuracy of better than 20 per cent, experimental measurements of these quantities are essential. A considerable number of different techniques have been developed for this purpose, many of them involving the determination of radiative lifetimes. In this chapter we discuss two such techniques, namely the beam-foil and the delayed-coincidence methods. In Chapter 8 we shall discuss the determination of the f-values of resonance lines by studies of the profiles of spectral lines and in Chapters 15 and 16 the use of the Hanle effect and optical double resonance methods.

From the relations given in Chapter 4 it can be seen that there are also a number of ways of measuring the f-values of electric dipole transitions which do not involve the measurement of lifetimes. For example, the intensity I_{ki} of the spontaneous emission in the spectral line corresponding to the transition from level k to level i is given by equation (4.27):

$$I_{ki} = \frac{A_{ki}N_k V\hbar\omega_{ki}}{4\pi} . \tag{6.1}$$

Thus an absolute measurement of I_{ki} will allow the transition probability and hence the f-value of this line to be determined, provided that the density of excited atoms, N_k, is known. It is possible to calculate N_k from thermodynamic arguments as explained in section 4.7 if arcs or plasmas are used in which LTE holds. Often, however, only relative f-values are obtained by this technique and they must be normalized by reference to lifetime measurements. Details of this method are given by Foster (1964). Another technique which we shall not have space to discuss but which has been widely used, especially in the U.S.S.R., is the Rhozdestvenskii hook method. This technique makes use of the anomalous dispersion of an atomic vapour which occurs close to an absorption line. It is discussed in detail by Ditchburn (1963) and Marlow (1967).

These alternative techniques usually require a knowledge of the density of atoms in a given level. Since this is often difficult to determine accurately, the absolute f-value measurements by these methods are generally inaccurate. This problem is avoided in lifetime measurements provided the data are taken at pressures low enough for radiation trapping and collision effects to be negligible. However, the radiative lifetime of a given level, equation (4.26), is usually determined by the sum of the transition probabilities of several different lines and in these situations the alternative techniques have an advantage since they give a direct measurement of the f-value of a single transition. Thus the two approaches are in a sense complementary.

We discuss first the beam-foil method for the measurement of lifetimes since conceptually and experimentally it is very simple. The technique is usually applied to the measurement of the lifetimes of excited states of atomic and molecular ions. The delayed-coincidence method which we treat in the second half of this chapter is experimentally more complex but it is perhaps the most accurate and widely applicable of the modern techniques.

6.1. The beam-foil method.

6.1.1. Early experiments using canal rays and atomic beams

The time-dependent exponential decay of excited atoms, equation (4.25), may be converted into a spatial variation of intensity by exciting a beam of fast-moving atoms at a given position. This is the basis of the canal ray method used by Wien (1927). Positive ions which have been accelerated in the cathode fall region of a low pressure glow discharge are allowed to pass through a narrow channel in the cathode, as shown in Fig. 6.1. The canal rays emerge into the evacuated space behind the cathode as excited atoms, having captured electrons by charge exchange collisions with atoms in the channel. The radiative decay of the moving atoms produces a faint glow extending beyond the cathode. The length of this glow enables the radiative lifetime to be measured, provided that the velocity of the atoms can be estimated. Unfortunately the lifetimes obtained by this technique were generally much too long due to repopulation of the excited states by radiative decay from higher levels.

Several similar methods were developed by the early workers in this field and these are described in Mitchell and Zemansky (1961). One technique, developed by Koenig and Ellett (1932), overcame the problem of repopulation by using optical excitation of a thermal atomic beam. However, this method was restricted to lifetimes greater than 10^{-6}s owing to the low velocity, $v \approx 10^5$ cm s^{-1}, of the atomic beam. In all of these experiments the calibration of the beam velocity led to considerable uncertainties in the final results. For this reason these techniques were little used after 1932.

6.1.2. Principle of the modern technique. Recently the concept of converting the time-dependent decay into a spatial variation of intensity has again become the basis of an important technique for the measurement of radiative lifetimes. This followed the discovery, made independently by Kay (1963) and Bashkin (1964), that ions in the beam of a Van de Graaff accelerator can be strongly excited by passing the beam through a thin carbon foil. The high velocity of the beam, $v \approx 10^8$ cm s^{-1}, means that the radiative decay usually extends over several centimetres and is easily measured. The velocity of the excited ions is generally known to an accuracy of 2 per cent or better, and therefore accurate lifetime measurements can be made in many cases.

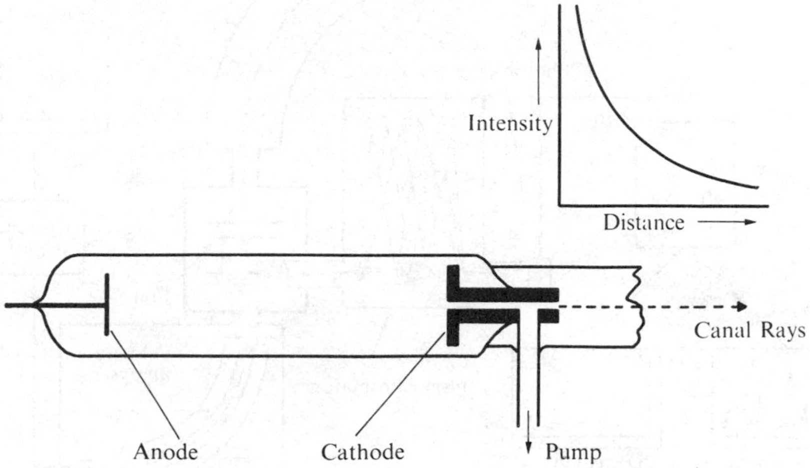

Fig.6.1. Decay of excited atoms and spatial variation of intensity in canal ray experiment.

Since the original discovery, this technique has been considerably developed both by Bashkin's group at the University of Arizona and by physicists in many other laboratories. The proceedings of the beam-foil spectroscopy conferences held in 1970 and 1972, (Martinson et al 1970, Bashkin 1973), provide useful collections of papers in this and related fields.

6.1.3. Apparatus for beam-foil experiments.

(a) General outline. The apparatus for a typical beam-foil experiment is shown in Fig.6.2. A beam of high-energy ions from the analysing magnet of a Van de Graaff accelerator is passed through a thin carbon foil. In the foil the ions interact strongly with the atoms of the foil and an appreciable number of the ions in the beam emerging from the downstream side of the foil are in excited levels. Usually there are several stages of ionization present in the excited beam, depending on the initial energy of the ions. It is

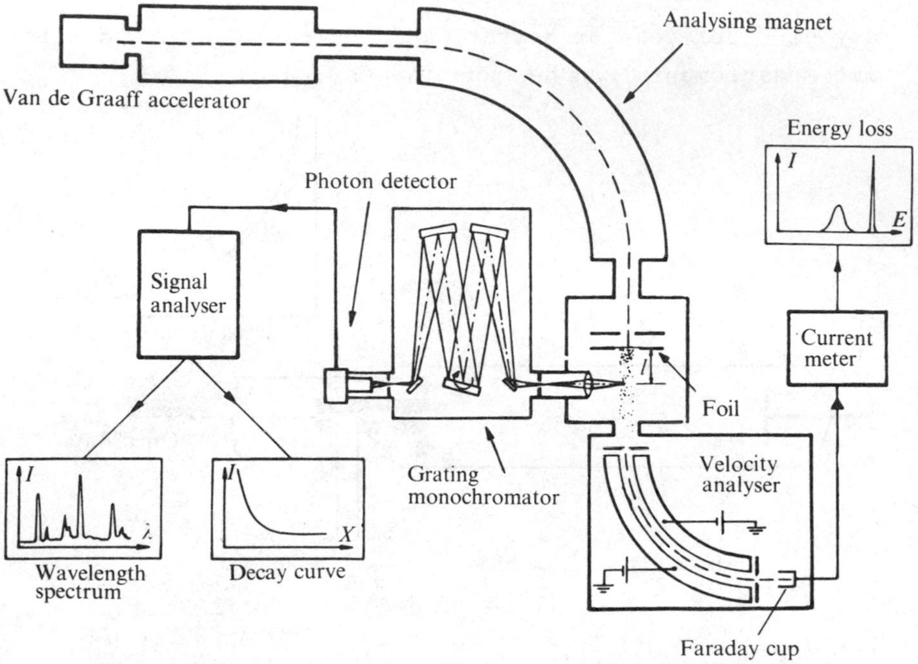

Fig.6.2. Experimental arrangement for studies of atomic spectra and the measurement of lifetimes of excited ions by the beam-foil technique. (After Martinson (1974)).

generally much more difficult to produce excited neutral atoms unless low beam energies are used. On the far side of the foil the beam is seen to glow, often with a faint blue light, as the excited atoms and ions decay by spontaneous emission. Owing to the very high velocity of the beam, the excited particles are able to move a considerable distance downstream before they decay. The intensity of the light emitted by the beam decreases downstream and a measurement of the intensity of the light from a given level of an ion as a function of distance from the position of the foil gives the lifetime of that level. In the apparatus shown in Fig. 6.2 the variation of intensity may be obtained as a plot on an X-Y recorder as the foil is translated away from the region of observation.

The excited beam has several important characteristics

as a light source:

(i) The de-excitation of the ions in the beam occurs in a drift tube under high vacuum, $\approx 10^{-6}$ Torr, and the effective density of the beam itself is very low. The ions therefore decay freely and are not influenced by the effects of collisions or radiation trapping.

(ii) Ions of most of the light and medium-heavy elements can be excited. Since, in many cases, very highly-ionized species are produced it is possible to study the lifetimes of levels which are not easily excited in other laboratory sources. The technique can also be applied to the study of the lifetimes of molecular ions.

(iii) The chemical purity of the excited beam is very high, which considerably simplifies the classification of spectral lines not previously observed in other sources.

(iv) Vacuum ultraviolet spectrographs may be easily coupled to the drift tube.

(b) Details of the Van de Graaff beam. In a typical beam-foil experiment energies of 2 MeV might be used for ions such as He^+, C^+, and N^+. The initial spread of beam energy is determined by the resolution of the analysing magnet and might be approximately $\pm 0{\cdot}02$ MeV. For N^+ this beam energy corresponds to a particle velocity of $5{\cdot}2 \times 10^8$ cm s^{-1}. The beams are excited by passing them through carbon, or occasionally aluminium, foils ranging in density from 5 - 40 μg cm^{-2}. These correspond to foil thicknesses of 250 Å and 2000 Å respectively. Thus the beam particles spend less than 10^{-13} s within the foil and the ions are effectively excited instantaneously. Gaseous targets have also been used to provide fast beams of excited atoms or molecules. In this method the Van de Graaff beam passes through a chamber filled with gas at 10^{-3} Torr before emerging into the observation chamber which is maintained at less than 10^{-5} Torr by differential pumping.

Beam currents are typically of the order of 1 μA, which gives 6×10^{12} ions s^{-1} passing through the foil for a singly-charged species. If we assume that the beam cross-section is 0·25 cm^2 we find that the particle density is only 5×10^4 ions cm^{-3}. If all the ions were in excited levels and if the mean radiative decay rate for the system were 10^8 s^{-1}, the total photon flux radiated into 4π solid angle would be of the order of 5×10^{12} cm^{-3} s^{-1}. Although this makes the beam easily visible in a darkened room, the number of photons on a single transition that can be collected within the solid angle of a given monochromator is often rather small.

The foils are prepared by vacuum evaporation from a carbon arc on to microscope slides which have first been coated with a thin film of soap or detergent. The carbon film can be floated off on water and pieces mounted on thin metal frames. The amount of light produced by the foils depends on their thickness but owing to surface contamination, foils of equal thickness do not necessarily produce equally intense excitation of the beam.

(c) Spectrographs and monochromators. For the preliminary identification of the lines produced by a given ion beam a fast spectrograph is useful, since the photographic plate provides a simple means of detection and long-term integration of the light emitted on all lines simultaneously. Alternatively a scanning photoelectric monochromator may be used, as shown in Fig.6.2. The lifetimes of excited levels can be obtained by photographic methods if the image of the ion beam is formed parallel to the slit of a stigmatic spectrograph. In these instruments there is a one-to-one correspondence between points in the image and points in the slit and therefore the plate records the spatial decay of intensity in each spectral line. However, this method gives very inaccurate results owing to the difficulty of calibrating the exposure of the photographic plate. Photoelectric detection is therefore essential for accurate lifetime measurements.

(d) Photon detection and signal measurement. Many of the spectral lines of interest in beam-foil experiments lie in the vacuum ultraviolet. In this region of the spectrum, the signal is often extremely small due to the restricted solid angle of the collection optics and the low efficiency of diffraction gratings. Consequently it is necessary to choose the most efficient detection method available and photon counting techniques are generally used. In this method the anode of the photomultiplier is usually operated at a high positive potential and each pulse of charge corresponding to a photoelectron emitted by the cathode is coupled capacitatively into a pulse amplifer. These pulses are counted by a scaler for a pre-selected time interval, of the order of 10-100 s. The number of counts recorded in this time might be typically 5×10^3 when the monochromator is set to receive light from the beam just as it leaves the foil.

In the visible region of the spectrum the signal level is often considerably higher than in the vacuum ultraviolet. Although the photon counting technique still has clear advantages, a large number of measurements have been made by recording the amplified photomultiplier current on a chart recorder as a function of the foil position.

As in all experiments, there is generally some unavoidable source of noise which contributes a background on which the signal is superimposed. In the vacuum ultraviolet this is due to energetic X-rays produced by the accelerator. In the visible region it is caused by thermal electrons randomly emitted by the cathode of the photomultiplier. In this case the so-called 'dark current' can be eliminated by cooling the photomultipler to $-40^{\circ}C$. An additional contribution to the background may be present if the pressure in the target chamber is not maintained below 10^{-6} Torr, for the ion beam may then excite the residual gas. In this case a further correction must be made to the observed signals.

(e) Calculation of lifetimes from the data. Once the spectral lines emitted by an ion beam have been identified, the decay of a given excited level, k, can be investigated by adjusting the monochromator to pass one of the radiative transitions from k to a lower level i. Let the number density of ions in the level k at the downstream face of the foil, x=0, be given by $N_k(0)$. Then if the beam remains of uniform cross-section S, and the ions have a uniform velocity v, the number density of excited ions at the position $x = l$ downstream is given by

$$N_k(l) = N_k(0)\exp(-t/\tau_k), \tag{6.2}$$

where $t = l/v$ is the time taken for an ion to travel from the foil to the observation region and τ_k is the mean radiative lifetime of the level k. The number of ions which decay radiatively in the beam between l and $l+\Delta l$ is given by

$$N_k(l,\Delta l) = N_k(0)\exp(-l/v\tau_k)\,[1-\exp(-\Delta l/v\tau_k)]\,S\,\Delta l. \tag{6.3}$$

The length of the observation region Δl is determined by the geometrical optics of the light-collecting system. The observed signal is proportional to $N_k(l,\Delta l)$, the constant of proportionality being determined by the solid angle of the collection optics, the combined efficiency of the monochromator and photomultiplier, and the branching ratio, $A_{ki}/\sum_i' A_{ki}$, for the observed spectral line. Typically the ratio of the number of photons detected to the number emitted in a given transition will range from 10^{-4} in the visible region to 10^{-6} in the vacuum ultraviolet.

The results of observations by Heroux (1967) on the $2s^2 2p^2\ ^3P - 2s2p^3\ ^3P^o$ transition of N^+ at 916 Å are shown in Fig.6.3. The counts observed in 100 s intervals are plotted semilogarithmically as a function of the distance from the foil. The data have been corrected for a constant background of 4 counts s^{-1}. The accuracy with which the data points lie on a straight line leads to the conclusion that the lifetime of the excited level may be derived directly. The result in this case is $\tau(^3P^o) = (0{\cdot}96 \pm 0{\cdot}03) \times 10^{-9}$ s.

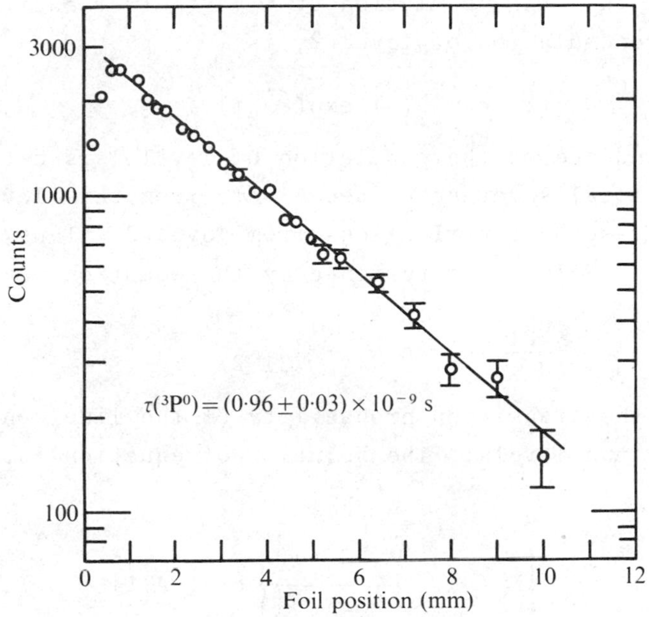

Fig.6.3. A semilogarithmic plot of the number of photons detected on the NII transition at 916 Å as a function of foil position. The data were obtained by Heroux (1967) using a N^+ beam energy of 1 MeV and a foil thickness of 500 Å. The transition is seen to be free of cascade.

6.1.4. Experimental difficulties.

(a) The effect of radiative cascade. Unfortunately the linear semilogarithmic plot of Fig.6.3 is not typical of most of the results obtained by the beam-foil technique. For many levels the plots show a pronounced curvature which is due to repopulation of the level being studied by radiative decay from higher-lying levels. This is only to be expected when one considers the unselective nature of the excitation process.

To study this problem in more detail we consider the

simplified energy-level diagram of a hypothetical atom shown in Fig.6.4. If the atoms in level 2 decay freely with a total rate A_2 then the population at time t after excitation, assuming no repopulation of level 2, is

$$N_2(t) = N_2(0) \exp(-A_2 t). \tag{6.4}$$

The time dependence of the population of level 1 is determined by the total spontaneous decay rate from this level, A_1, plus the cascade contributions from level 2. Thus the population of level 1 is determined by the equation

$$\frac{dN_1}{dt} = -A_1 N_1 + A_{21} N_2 \tag{6.5}$$

where A_{21} is the transition probability of the line connecting these two levels. The solution of equation (6.5) is given by

$$N_1(t) = \left\{ N_1(0) - \frac{A_{21} N_2(0)}{A_1 - A_2} \right\} \exp(-A_1 t) + \frac{A_{21} N_2(0)}{A_1 - A_2} \exp(-A_2 t) \tag{6.6}$$

where $N_1(0)$ is the population of level 1 at t=0. Thus in a beam-foil experiment the measurement of the emission from level 1 will yield a decay curve which consists of a sum or difference of two exponentials with lifetimes characteristic of levels 1 and 2.

In general, therefore, the data from beam-foil experiments must be analysed as a sum of exponential decay curves:

$$I_{ki}(l, \Delta l) = \sum_j C_j \exp(-l/v\tau_j). \tag{6.7}$$

A typical experimental curve showing the effect of cascade is given in Fig.6.5 for the resonance line 1^2S-2^2P of He^+ at 304 Å. The curve can be interpreted as the sum of a rapid decay due to the initial population created in the

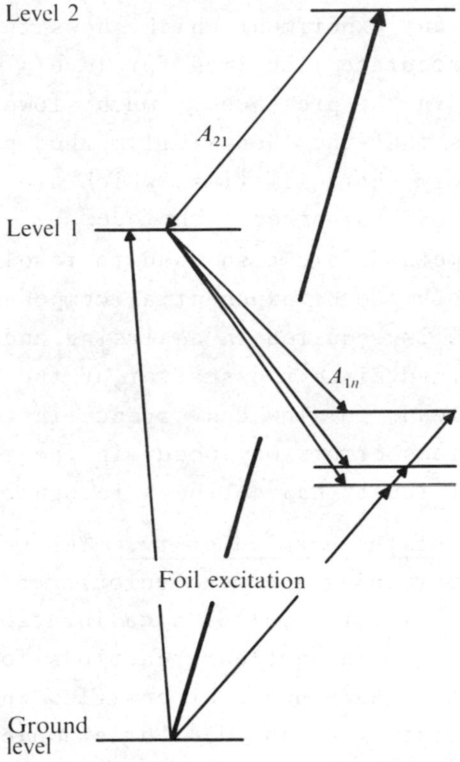

Fig.6.4. A simplified energy-level diagram showing the foil excitation and radiative cascade processes.

2^2P level and a slower decay produced by the repopulation of the 2^2P level by cascade. In this case the data were analysed by fitting a straight line to the tail of the curve, and by linear extrapolation the data at earlier times were corrected for the long-lived component. The result is that the fast decay exhibits a lifetime of

$$\tau(2^2P)_{\text{expt.}} = (0\cdot97 \pm 0\cdot03) \times 10^{-10} \text{ s.}$$

Since the helium ion is hydrogenic an accurate theoretical value for the lifetime is available:

$$\tau(2^2P)_{\text{theory}} = 0\cdot998 \times 10^{-10} \text{ s.}$$

This is an important experiment which shows that it is possible to obtain accurate lifetimes for levels exhibiting fast decays even in the presence of much slower components. It also indicates that the beam-foil method permits the measurement of very short lifetimes which are at present beyond the range of most other techniques

Numerical methods are also used to resolve the observed decay curves into a sum of exponential components. However, considerable care is required in analysing and interpreting the results obtained since the scatter of the data points at longer times leads to slow convergence in the fitting procedures. Serious errors may occur in the reported lifetimes if this difficulty has not been recognized.

(b) Calibration of the beam velocity. Two methods are available for determining the mean velocity of ions in the excited beam. The first involves a calibration of the accelerator voltage using nuclear reactions for which the cross-sections have sharp peaks at energies which have previously been accurately measured. For example, the reaction $^{7}Li(p,n)\,^{7}Be$ has a threshold at $1880{\cdot}60 \pm 0{\cdot}07$ keV. The heavy ions can either be accelerated at this calibration voltage or the bending magnet of the Van de Graaff may be calibrated over a small range. The energy loss of the ions passing through the foil must also be taken into account.

The second technique is to use a 90° electrostatic analyser in the target chamber. This enables the mean energy and also the energy distribution of the beam to be measured both before and after passing through the foil. This is the most satisfactory way of determining the energy loss of the ions passing through the foil, which may be quite large for ions of high atomic number moving at low velocities. The effect of the velocity distribution in the beam is usually negligible in comparison to other experimental errors.

(c) The effect of the Doppler shift. The excited ions are moving with extremely high velocities, $\approx 5 \times 10^{8}$ cm s^{-1}.

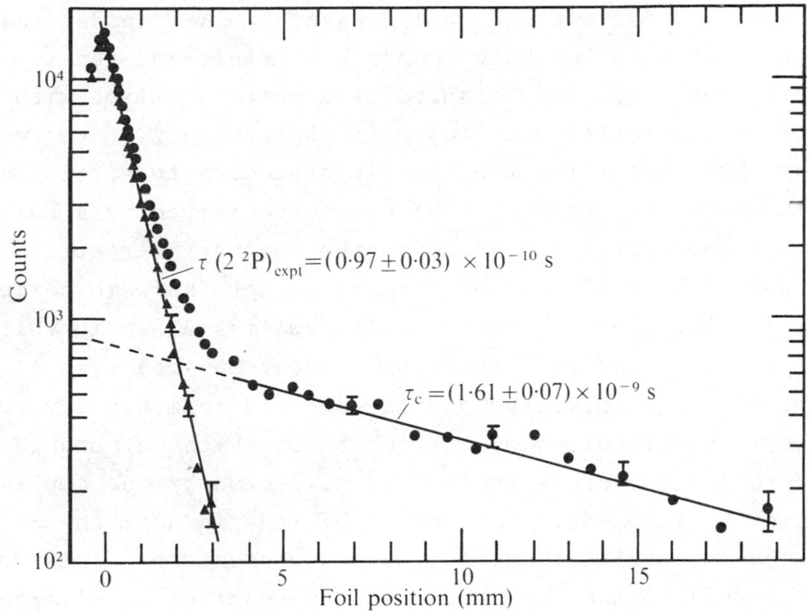

Fig.6.5. A semilogarithmic plot of the number of photons detected on the HeII transition at 304 Å as a function of foil position. The data were obtained using a He^+ beam energy of 1 MeV and a foil thickness of 500 Å. The transition is affected by cascade. •, data corrected for background; ▲, data corrected for background and cascade. (After Heroux (1967)).

Thus although the beam is generally viewed at right angles to the direction of motion, the finite range of the spectrometer acceptance angles gives rise to large Doppler shifts. When the spectrograph slit is perpendicular to the beam direction, this results in a Doppler broadening of the observed lines which may be of the order of 2 Å. This does not affect the lifetime measurements directly but it may well complicate the identification of the observed lines. When the spectrograph slit is parallel to the beam direction, Doppler shifts of ± 10 Å are possible.

(d) Beam fluctuations and foil breakage. The Van de Graaff beam current cannot be held completely stable and consequently the intensity measured at a position fixed with respect to the foil fluctuates. It is necessary to correct for this by normalizing the experimental data to a constant beam current. The beam current can be monitored by a Faraday cup. The charge collected by the cup may be recorded continuously as a current or integrated for the same period as the photomultiplier signal. Difficulties arise in this method owing to the fact that the Faraday cup measures the total charge collected and the excited beam usually consists of a distribution of charge states which may change as the foil ages. Thus the charge collected by the cup is not directly related to the number of ions in a given excited level. During the lifetime measurement the distance from the foil to the cup should remain constant to eliminate the effects of the divergence of the beam after passing through the foil.

A better method of obtaining the data necessary for normalization is to monitor the intensity of light in the transition of interest at some fixed distance from the foil. This overcomes most of the difficulties associated with changes in the populations of different charge states and different excited levels. This technique also enables the instant when a foil breaks to be detected. Foil breakage occurs frequently with 1 μA beam currents of heavy elements like Fe^+, and this information is necessary to avoid undue wastage of data.

6.1.5. Comparison of experimental and theoretical results. The usefulness of the results obtained by the beam-foil technique may be illustrated by considering the f-values of the resonance lines 2s ^2S-2p ^2P in the lithium isoelectronic sequence. In Fig.6.6 the sum of the absorption f-values of the transitions $^2S_{1/2}$- $^2P_{1/2}$ and $^2S_{1/2}$- $^2P_{3/2}$ calculated from the lifetime data using equations (4.26) and (4.30) are plotted together with the results of theoretical calculations using a variety of techniques. The data are dis-

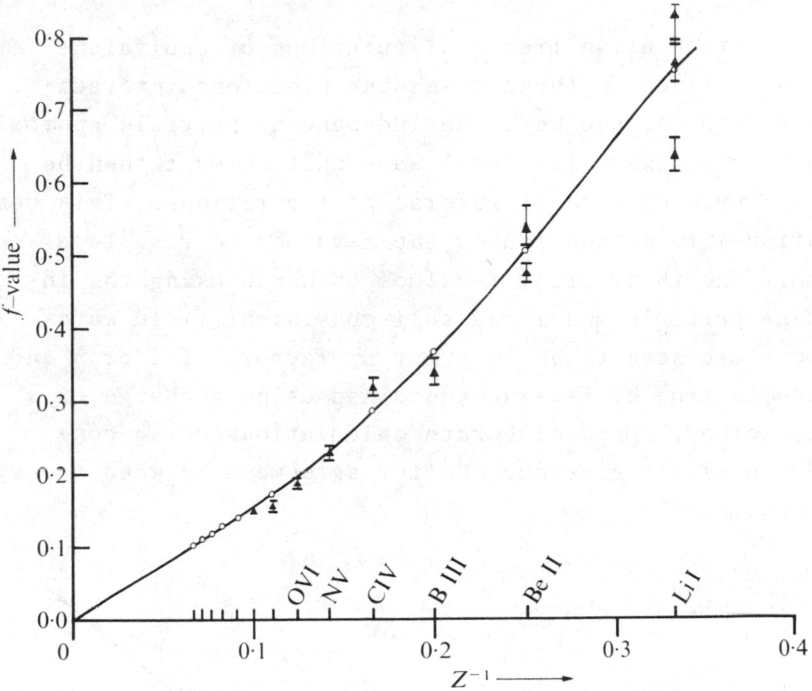

Fig.6.6. Variation of the sum of the absorption f-values for the 2s $^2S_{1/2}$- 2p $^2P^o_{1/2,3/2}$ transitions of the lithium isoelectronic sequence as a function of (nuclear charge, Z)$^{-1}$. ▲ , experimental values; o, theoretical values.

played as a function of Z^{-1}, where Z is the atomic number, since we expect the f-values for a given transition to vary smoothly as we move up an isoelectric sequence. There is very close agreement in this case between theory and experiment. In similar cases unknown f-values could be obtained by interpolation or extrapolation of other curves of this type.

Unfortunately this simple variation as a function of Z^{-1} does not always occur, as may be seen in Fig.6.7. This shows the f-values of the transition $2s^2 2p\ ^2P^o$ - $2s2p^2\ ^2D$ in the isoelectronic sequence of boron. The experimental data in this case include results from beam-foil and other lifetime measurements. Both the upper and lower levels of this transition arise from configurations of equivalent electrons. Since in these cases the electrons interact strongly with one another, the independent particle approximation breaks down. The level wavefunction must then be described by a mixture of several configurations. This configuration interaction causes the f-values to rise to a maximum. The theoretical f-values obtained using the independent particle model and self-consistent field wavefunctions are seen to be in error by factors of 2 or 3 and the same is true of f-values obtained using a charge expansion method. More elaborate calculations using configuration mixing give much better agreement between theory and experiment.

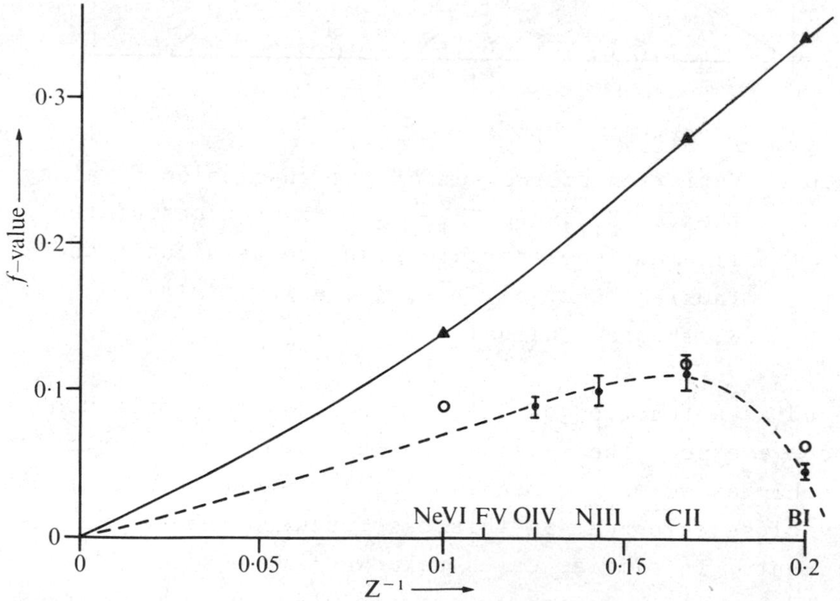

Fig.6.7. Variation of the sum of the absorption f-values for the $2s^2 2p\ ^2P^o_{1/2,3/2}$ - $2s2p^2\ ^2D_{3/2,5/2}$ transitions of the boron isoelectronic sequence as a function of (nuclear charge, Z)$^{-1}$. ▲, theoretical calculations using self-consistent field; o, theoretical calculations taking account of configuration interaction; •, experimental results obtained by a variety of techniques.

These two isoelectronic sequences clearly illustrate the usefulness of the lifetime measurements made by the beam-foil technique. Since the uncertainty in the beam velocity is often less than 2 per cent, the accuracy of the results for those levels which are free of cascade is usually limited by the photon counting statistics. However, for levels which do show cascade contributions uncertainties arise, especially if the decay rates of the different exponential components are of similar magnitude. It appears that these difficulties were not always realized in some of the early beam-foil experiments.

6.2. *Fast beam experiments using laser excitation*

The problems associated with radiative cascade in beam-foil experiments can be avoided by replacing the foil by selective optical excitation using one of the powerful lasers which are now available. In the first experiments of this kind reported by Andrä *et al.* (1973), the 4545 Å output of an argon ion laser was used to excite a 3 μA beam of Ba^+ accelerated to 300 keV. By adjusting the angle between the laser beam and the ion beam to approximately 23°, the laser was Doppler-tuned to the Ba^+ resonance line $6\ ^2S_{1/2}$-$6\ ^2P_{3/2}$ at 4554 Å. Good signals were obtained on the fluorescent light at 4554 Å with a maximum count rate of $10^4\ s^{-1}$. The decay curve was obtained as in a beam-foil experiment and was found to consist of a single exponential component. Analysis of the count rate over three decades gave a lifetime of $\tau(6\ ^2P_{3/2}) = (6{\cdot}20 \pm 0{\cdot}02) \times 10^{-9}$ s. Similar experiments have been performed on ion beams which have been excited by passing through gas target cells and which are then re-excited by a tunable dye laser. An accuracy of 1 part in 10^3 seems to be attainable with this new technique and many more results will appear in the near future.

6.3. The delayed-coincidence method using electron excitation

6.3.1. Principle of the technique. The delayed-coincidence technique, which is well known in nuclear physics, was first applied to the measurement of the lifetimes of excited atoms by Heron *et al.* (1954, 1956). However, it was not widely used until Bennett (1961) improved the method by incorporating a multichannel analyser into the experimental set up. Since then a large number of measurements on different elements have been made and the technique is now probably one of the most accurate and widely applicable of the methods currently available for the measurement of lifetimes.

In this technique a sample of atoms or molecules is excited to some level k, usually by a pulsed electron gun. The excitation is then cut off instantaneously at time t=0. The probability of detecting a photon on the decay transition from level k to level i in a time interval between t and $t+\Delta t$ after the removal of the excitation is given by

$$P(t, \Delta t) \propto A_{ki} N_k(0) \exp(-t/\tau_k) \Delta t. \qquad (6.8)$$

The emitted photons are detected by a photomultiplier. Each photo-electron produces a pulse of charge at the anode and the number of pulses received in a given time interval is counted by means of a fast scaler. A semilogarithmic plot of the number of counts accumulated as a function of the delay time t enables the radiative lifetime τ_k to be obtained.

6.3.2. The single channel method. The apparatus used by Heron *et al.* (1956) is shown in Fig.6.8. The excited atoms were produced by electron bombardment of helium at pressures in the range 5×10^{-3} to 5×10^{-2} Torr. The electron beam pulse was obtained by applying a positive voltage pulse of 2×10^{-8} s duration to the grid of the electron gun, which was normally held negative. The electron gun produced a current of 100 μA and was pulsed at a repetition rate of 10 kHz. The accelerating voltage was adjusted between 30 and 100 volts to give the maximum intensity for the line chosen for study.

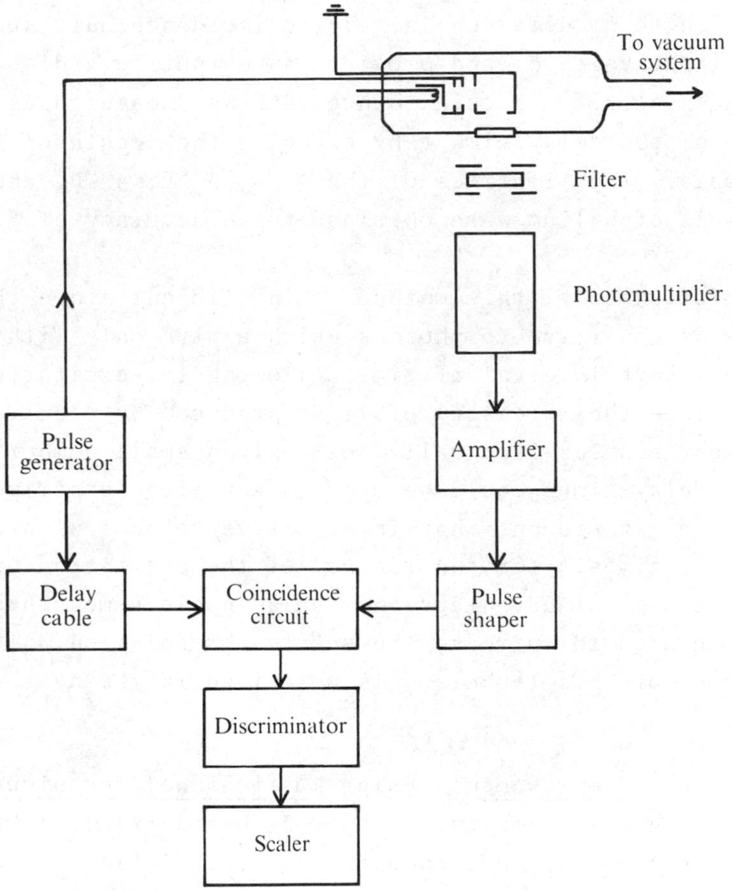

Fig.6.8. Schematic diagram of the single-channel delayed-coincidence apparatus used by Heron *et al.* (1956) for lifetime measurements in helium.

An interference filter placed between two collimating slits in front of the photomultiplier enabled a given spectral line from the excited level k to be selected. The arrival of a photon at the photomultiplier produced a sharp negative voltage pulse at the anode. This pulse was amplified and shaped before being applied to one side of a coincidence circuit. The pulses applied to the other input of the coincidence circuit were obtained from the electron gun pulse. These were delayed by a known amount by passing them down a fixed length of delay cable. The coincidence unit summed the two input voltages and provided an output to a discriminator and scaler. The coincidence rate was measured as a function of the delay time t by changing the length of the delay cable. The lifetimes of the 4 ^{3}S, 3 ^{3}P, 4 ^{3}P, and 3 ^{3}D levels of helium were obtained to an accuracy of 5 per cent.

Unfortunately, this method is inefficient since the detector is sensitive to photons which arrive only within a specified short interval of time following the excitation pulse. Since the intensity of light produced by electron bombardment sources is usually low, only a small number of discrete delay times could be used in any given experiment. Obviously an instrument that is sensitive to photons arriving at all times after the removal of the excitation pulse would have a great advantage and would enable a monochromator to be used to increase the spectral resolution. The development of this technique is described in the next section.

6.3.3. Recent improvements using multichannel techniques. The extension of the technique used by Heron *et al.* (1956) may be understood by reference to Fig.6.9. Again a sample of excited atoms is prepared by a pulsed electron gun. The gun voltage is applied simultaneously to the start input of a time-to-pulse-height converter. This initiates the charging of a capacitor by a constant current source. Some time later an atom in the sample may emit a photon on the transition of interest, which is isolated by the spectrometer and detected by the photomultipler. The resulting voltage pulse is amplified and applied to the stop input of the time-to-pulse-height converter. At this instant the capacitor stops charging and a voltage pulse, whose amplitude is proportional to the time interval which has elapsed between the excitation cut-off and the detection of the first photon, is sent to the multichannel pulse-height analyser. The nth address in the memory of the analyser corresponds to a voltage interval between V_n and $V_n + \Delta V$. If the incoming voltage pulse lies within this interval, one count will be added to the number stored in this particular address. Each address or channel of the analyser therefore corresponds to a definite time interval after the removal of the excitation pulse and the system has the advantage of being sensitive in all channels following this pulse. If no photon is detected before the capacitor reaches its peak charging voltage, corresponding to the last channel of the analyser, the time-to-pulse-height converter is reset for the start of a new excitation and decay cycle. When this sequence has been repeated many times the number of counts in a given channel is proportional to the number of excited atoms present at that time.

This multichannel delayed-coincidence technique was developed by Bennett (1961) and has since been applied to detailed studies of the lifetimes and collision-induced relaxation of excited levels of many different atoms, molecules, and ions. An excellent review of this method is given by Bennett *et al.* (1965). In the following sections we discuss various details of the experimental technique, methods used for the reduction of the data, and experimental difficulties.

(a) Electron gun design. Ideally the electron gun should produce a current pulse of several milliamperes since the cross-sections for electron impact excitation of atoms are very small, especially close to threshold. The width of the energy distribution of electrons in the beam should be less than 0·5 eV, otherwise many levels will be excited and radiative cascade will cause difficulties. In order to obtain good time resolution the electron beam must be cut off instantaneously and the transit time of electrons across the observation region should be as small as possible. Resolution times of 5×10^{-9}s are typical, although with more compact designs this can be reduced to approximately 1×10^{-9}s.

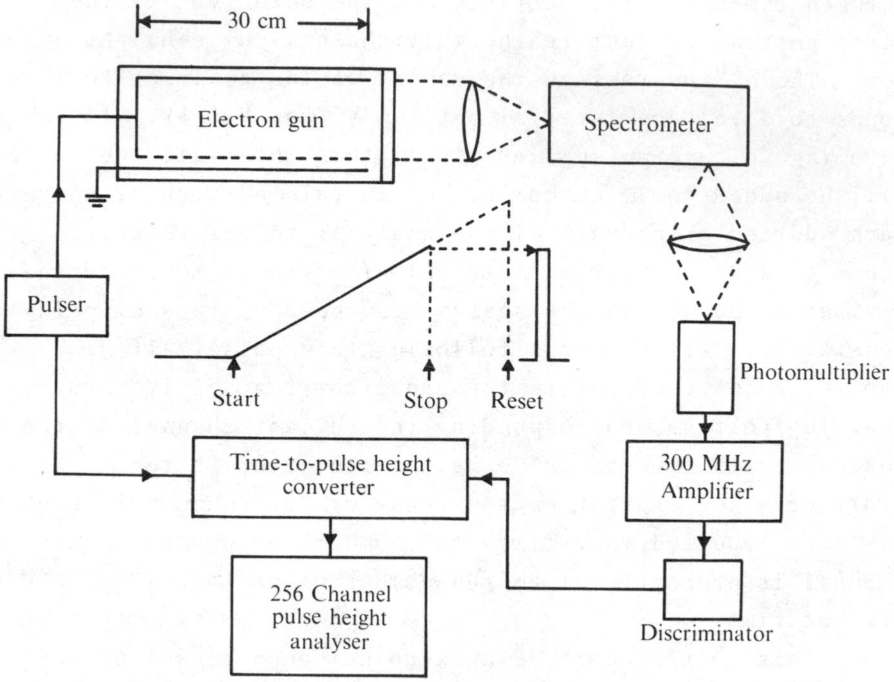

Fig.6.9. Schematic diagram of the multichannel delayed-coincidence apparatus developed for precision lifetime measurements by Bennett (1961).

(b) Photon detection. The use of a monochromator with a dispersion of the order of 16 Å/mm is essential for experiments on neon and other inert gases which have an abundance of closely spaced lines throughout the visible spectrum. It is desirable that the spectrometer should have a large numerical aperture so that the light-gathering power is high, and if the monochromator is to be used as efficiently as possible several gratings blazed for different spectral regions are required.

The photomultiplier should be chosen to have high sensitivity at the operating wavelength and low dark current. In the near infrared it is necessary to cool the photomultiplier to liquid nitrogen temperatures, 77 K, to reduce this dark current. In any photomultiplier there are fluctuations in the time taken for electrons to travel from the cathode to the anode which are caused by variations in the electron path between dynodes. This transit time spread may be of the order of 2×10^{-9} s and contributes to the finite resolving time of the system.

(c) Calibration. The multichannel analyser may be calibrated by introducing various lengths of delay cable before one input of the time-to-pulse-height converter. The time-delays so introduced are obtained from the resonant frequencies of the given cables, and the observed displacement of the peaks of the decay curves calibrates the time axis of the system.

(d) Differential non-linearity. In the ideal multichannel system the width of each channel is the same and remains constant over long periods of time. Any variation in channel width with channel number will lead to an excessively high or low number of counts being recorded in a given channel and to the appearance of spurious peaks and valleys in the experimental decay curves. This effect is known as differential non-linearity and is caused by departures from linearity in the saw-toothed waveform generated in the time-to-pulse-height converter. When this effect occurs the experimental data must be corrected before the final stage in the analysis is reached.

6.3.4. Experimental difficulties.

(a) The effect of radiative cascade. Close to threshold the electron excitation cross-sections are usually very small, $\approx 10^{-19}$ cm^2. Consequently the photon counting rates are often very low and to increase the signal it is frequently necessary to increase the mean electron energy above the threshold value. This usually results in the excitation of levels above the level being studied and radiative cascade occurs. Data typical of this situation are shown in Fig.6.10. In this plot the rapidly decaying component is due to the population directly excited into the $2p_6$ level of neon while the more slowly decaying tail is probably caused by radiative cascade from the 2s and 3s group of excited levels which lie between 1 and 2 eV above the $2p_6$ level, as shown in Fig.8.12.

The appearance of multiple components in the observed decay curves is an undesirable complication which must be avoided if at all possible. The only certain way of eliminating this problem is to use electron beams with very narrow energy distributions and to operate close to the excitation threshold of the level being studied. If this is not feasible the data must be analysed in terms of a multicomponent exponential as discussed in section 6.1.4.

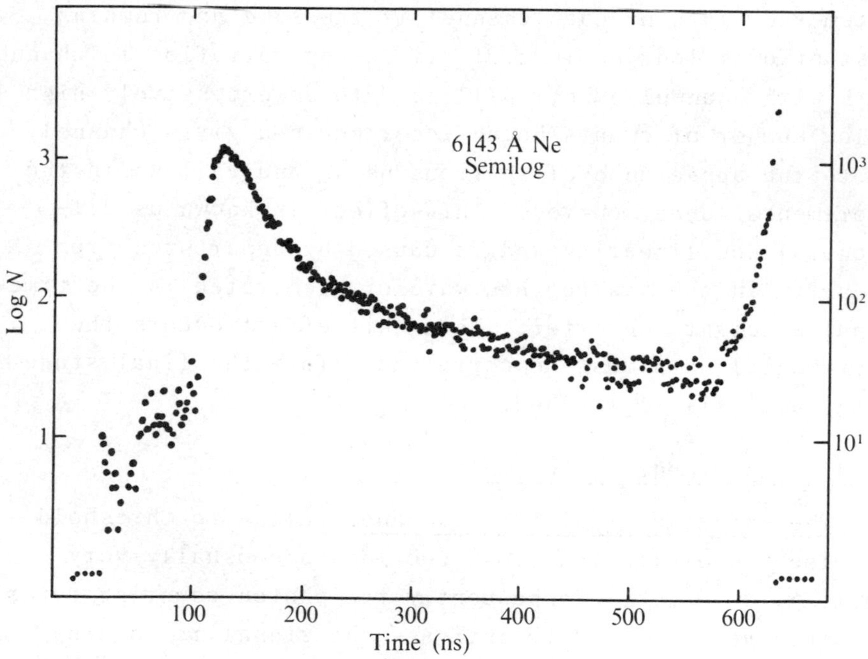

Fig.6.10. A plot of the log of the number of counts in a given channel of the multichannel analyser as a function of the delay time represented by the channel address. The data were obtained on the $1s_5$ - $2p_6$ transition of NeI at 6143 Å by Klose (1966) and illustrate the effect of cascade on the apparent lifetime of the $2p_6$ level.

(b) The effects of collisions. In lifetime measurements in helium it has been found that at pressures above 10^{-2} Torr the decay of excited singlet levels is increased by inelastic collisions with helium atoms in the ground level:

$$\begin{aligned} He(n\ ^1P_1) + He(1\ ^1S_0) &\rightarrow He(1\ ^1S_0) + He(n\ ^1F_3) + K.E. \\ &\rightarrow He(1\ ^1S_0) + He(n\ ^3F_{4,3,2}) + K.E. \end{aligned} \tag{6.9}$$

The initial effect of these collisions is to increase the decay rate from the spontaneous emission rate $\sum_i A_{ki}$ to

$$A'_k = \sum_i A_{ki} + N\sigma\overline{v}. \tag{6.10}$$

Here N is the density of atoms in the ground level, $\overline{v}$ is the mean relative velocity of the two colliding atoms, and σ is the mean collision cross-section which has been obtained by averaging over the velocity distribution of the two particles:

$$\sigma\overline{v} = \langle \sigma(v)\ v \rangle_{Av}. \tag{6.11}$$

This increase in the apparent decay rate due to atomic collisions is a very common phenomenon although the importance of the effect varies from level to level owing to the wide variation of the inelastic collision cross-sections. For near-resonant collisions, where very little energy needs to be transferred to or from the thermal kinetic energy of the system, the cross-sections tend to have very large values, of the order of 10^{-13} cm^2. If energy differences of the order of kT exist between the two sides of a reaction such as that given in equation (6.9), the observed cross-sections are of the same order as the gas kinetic cross-sections, $\approx 10^{-15}$ cm^2, while for energy differences greater than a few tenths of an eV the cross-sections tend to be very small, $\approx 10^{-20}$ cm^2, and are negligible for most purposes.

The numerical value of the collision-induced relaxation rate at the absolute temperature T is given by

$$N\sigma\bar{v} = 1{\cdot}404\ P\sigma\{(M_1+M_2)/M_1M_2T\}^{1/2} \times 10^{23}\ s^{-1}, \qquad (6.12)$$

where P is the pressure in Torr and the cross-section σ is measured in units of cm^2. M_1, M_2 are the atomic mass numbers of the two atoms involved in the collision. Thus at 300 K the collision-induced rate will be equal to a radiative rate of $10^8\ s^{-1}$ at a pressure of 39·0 Torr, assuming $M_1=M_2=20$ and $\sigma = 10^{-15}\ cm^2$.

When measurements are made of the apparent relaxation rate, equation (6.10) predicts a linear variation as a function of pressure. Results obtained by Bennett *et al.* (1965) for the decay of the $4\ ^1S_0$ level of helium are shown in Fig.6.11. The intercept of this graph yields a radiative lifetime of $\tau = 8{\cdot}40 \times 10^{-8}$s and the total inelastic cross-section, obtained from the slope of the linear portion of the curve, is $\sigma = 16 \times 10^{-16}\ cm^2$. It is obvious that at the highest pressures used the decay rate is not a linear function of pressure. The reason for this is that at these pressures the reverse reaction of equation (6.9) becomes significant, a process which is known as reciprocal transfer. It should be noted that the delayed-coincidence technique is probably the most direct and accurate method available for the measurement of inelastic collision cross-sections.

(c) Photon trapping. When the lifetime of a level which has a strong allowed transition to the ground level is measured, a marked increase of the lifetime with density is observed even at pressures which are so low that collisional effects are negligible. This is illustrated by measurements on the $3\ ^2P$ levels of sodium (see Fig.6.13) which are discussed in more detail in section 6.4 below. The effect is well understood and in this case it is caused by the reabsorption within the sample of photons in the $3\ ^2S_{1/2}$- $3\ ^2P_{3/2,1/2}$ transitions at 5890 Å and 5896 Å. This is called trapping or imprisonment of the resonance radiation. On average each reabsorption process increases the effective lifetime of the sample by approximately the natural atomic lifetime.

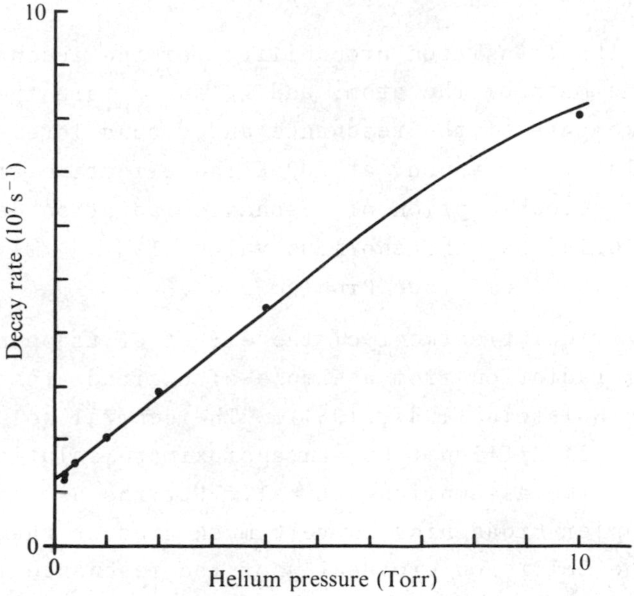

Fig.6.11. The decay rate of the 4 $^1S_0$ level of helium measured as a function of pressure by Bennett *et al.* (1965). The solid curve shows the result of a least-squares fit of the data to an equation of the form

$$A'_k = a_0 + a_1P + a_2P^2.$$

The departure of the curve from a straight line at the highest pressures is due to the effect of reciprocal transfer.

The size of the effect is due to the very large absorption coefficient, κ_0, for radiation at the centre of the Doppler-broadened resonance line, λ_0. The probability that a photon will be reabsorbed while travelling a distance L through a gas of density N_0 is given by x where

$$x = 1 - \frac{1}{\sqrt{\pi}}\int_{-\infty}^{\infty} \exp\{-\kappa_0 L \exp(-y^2)\} \exp(-y^2)\, dy. \quad (6.13)$$

The maximum value of the absorption coefficient κ_0 is given by

$$\kappa_0 = \frac{\lambda_0^3 N_0 g_k}{8\pi g_0} A_{k0} \left(\frac{M}{2\pi kT}\right)^{1/2} = N_0 \sigma_{Abs.} \qquad (6.14)$$

where A_{k0} is the transition probability for the resonance line, M is the mass of the atom, and g_k and g_0 are the statistical weights of the resonance and ground levels respectively. In sodium vapour at 300 K the effective cross-section for the reabsorption of resonance radiation defined by equation (6.14) has the enormous value of $\sigma_{Abs.} = 1{\cdot}2 \times 10^{-11}$ cm^2 (see Problem 9.6).

A theoretical treatment of the effect of trapping on the escape of radiation from a sample of excited atoms has been given by Holstein (1947, 1951). The general solution of the problem is difficult but an approximate solution was obtained using the assumption $\kappa_0 L \gg 1$. For the usual case where the Doppler broadening is very much greater than the natural or the collision broadening of the resonance line, the decay of a population of excited atoms which are initially distributed according to the lowest-order diffusion mode of an infinite cylinder of radius R (see Problem 7.10) is given by

$$N_k(t) = N_k(0) \exp\{-(gA_{k0} + \sum_{i \neq 0} A_{ki})t\} \qquad (6.15)$$

where g is an imprisonment factor given by

$$g = 1{\cdot}60[\kappa_0 R\{\pi \ln(\kappa_0 R)\}^{1/2}]^{-1}. \qquad (6.16)$$

Thus for levels with transitions which terminate on the ground level of the atom, accurate measurements over a wide pressure range are required in order to establish the true radiative lifetime of the excited level. For example, in helium measurements of the decay rates for the n $^1P_1$ levels at pressures in the range 10^{-3} to 10^{-2} Torr still give results which correspond to the complete trapping of the resonance lines, $g \approx 0$. In discharges photon trapping also occurs on transitions which terminate on metastable or resonance levels; however, this problem does not arise in delayed-coincidence lifetime measurements owing to the very low population density created in these levels.

6.3.5. Comparison of experimental results in neon. Table 6.1 gives the results of several recent determinations of the lifetimes of the levels of the $2p^5 3p$ configuration in neon obtained by the delayed-coincidence technique. The levels are labelled in the Paschen notation. This has the advantage of conciseness but gives no indication of the spectroscopic character of the levels which in this configuration can only be correctly represented in intermediate coupling. We see that the results for most of the levels are in good agreement. However, those of Bennett and Kindlmann (1966) are systematically shorter than those of the other authors and are generally considered to be the most accurate. A small uncorrected cascade contribution to the decay curves in the other experiments might account for this discrepancy. There is also close agreement between the experimental results and the sophisticated theoretical calculations of Feneuille *et al.* (1970).

A knowledge of the lifetimes and collision cross-sections for these levels is important since they are the lower levels of several important He-Ne laser transitions, for instance the well-known red transition $2p_4 - 3s_2$ at 6328 Å.

6.4. *Delayed-coincidence experiments using optical excitation*

It is also possible to apply the method of delayed coincidences to the measurement of the lifetimes of atoms which have been excited optically. This technique has the advantage that the very selective nature of the optical excitation process eliminates the effects of cascading discussed in section 6.3.4. In the past this method has been restricted to those levels which can be excited from the ground level by the absorption of intense resonance radiation. Thus only one or two levels could be studied

TABLE 6.1

Experimental and theoretical lifetimes (in ns) of the $2p^5 3p$ *configuration of neon*

	Delayed-coincidence measurements				Theory
Level	Klose (1966)	Bennett and Kindlmann (1966)	Bakos and Szigeti (1967)	Oshirovich and Verolainen (1967)	Feneuille *et al* (1970)
$2p_1$	14·7 ± 0·5	14·4 ± 0·3	14·8 ± 0·15	14 ± 1	13·65
$2p_2$	16·8 ± 1·4	18·8 ± 0·3	—	20 ± 1·6	17·82
$2p_3$	23·3 ± 4·8	17·6 ± 0·2	28·2 ± 0·85	18 ± 1·3	16·66
$2p_4$	22·4 ± 4·4	19·1 ± 0·3	—	24 ± 1·5	18·19
$2p_5$	18·9 ± 1·7	19·9 ± 0·4	—	23 ± 1·5	18·78
$2p_6$	22·0 ± 1·9	19·7 ± 0·2	22·0 ± 0·6	21 ± 1·4	18·77
$2p_7$	20·3 ± 1·6	19·9 ± 0·4	—	22 ± 1·3	18·62
$2p_8$	24·3 ± 2·0	19·8 ± 0·2	28·4 ± 1·2	25 ± 1·4	19·05
$2p_9$	22·5 ± 1·9	19·4 ± 0·6	19·5 ± 1·6	24 ± 1·5	18·52
$2p_{10}$	27·4 ± 2·9	24·8 ± 0·4	—	26 ± 1·5	23·65

in any one atom and the technique was further limited to those atoms whose resonance lines lie in or near the visible part of the spectrum. However, the recent development of pulsed dye lasers has removed some of these restrictions. These lasers are tunable in the range 3400 Å-12 000 Å and with frequency doubling wavelengths down to 2400 Å can be achieved. Their intense peak powers enable almost all levels to be reached either by direct or stepwise excitation.

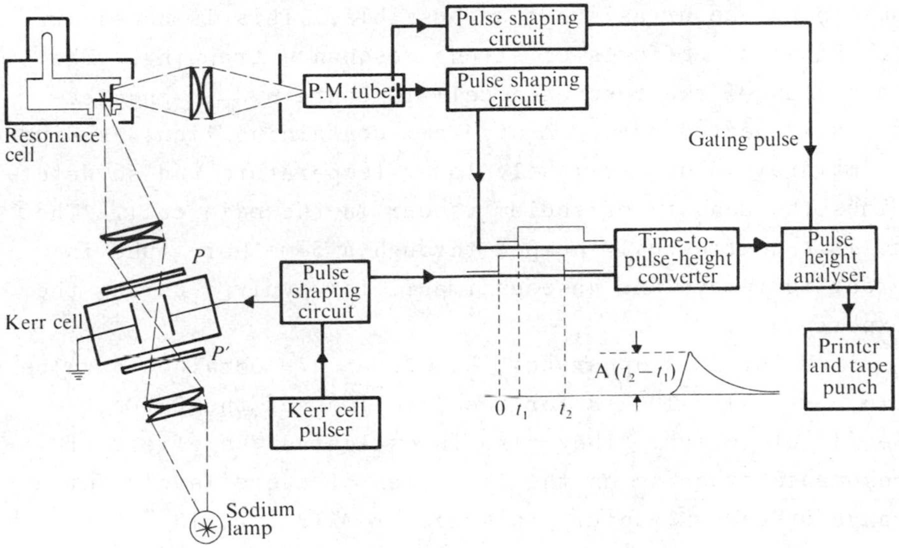

Fig.6.12. The multichannel delayed-coincidence apparatus used by Kibble *et al.* (1967) in the measurement of the lifetime of the 3 ^{2}P levels of sodium using optical excitation.

As an illustration of the optical excitation technique, we consider a detailed investigation made by Kibble *et al.* (1967) of the effect of the imprisonment of sodium resonance radiation on the measured lifetime of the 3 ^{2}P levels of sodium as well as the effect of molecular and atomic collisions on these levels. The apparatus used by Kibble *et al.* (1967) is indicated schematically in Fig.6.12. Atoms of the sodium vapour contained in the resonance cell are excited to the 3 $^2P_{1/2,3/2}$ levels by the absorption of light in the sodium D lines 5890 Å and 5896 Å emitted by an Osram sodium lamp. The excitation is pulsed by focusing the exciting radiation through a Kerr cell shutter placed between crossed polarizers. The Kerr cell is filled with nitrobenzene and requires a voltage of 15 kV to open it fully.

The resonance cell is specially designed to make the paths of the incident and fluorescent light within the absorbing vapour as short as possible. This is necessary to reduce the effects of strong resonance trapping. The main body of the resonance cell is contained in an electrically heated oven. A side arm containing liquid sodium is maintained at a slightly lower temperature and so determines the density of sodium vapour in the main cell. The cell is continuously pumped through a 3 mm bore tube in order to remove any gaseous impurities evolved during the experiments.

Using this apparatus, Kibble *et al.* obtained a value of $(16\cdot6 \pm 0\cdot4) \times 10^{-9}$ s for the lifetimes of the $3\,^2P_{1/2,3/2}$ levels of sodium. They also investigated the effect of resonance trapping on the lifetimes of these levels for a range of sodium vapour pressures from 10^{-7} to 10^{-4} Torr. Their results are given in Fig.6.13 and have been compared with the theories of Holstein (1947, 1951) and Milne (1926) by fitting the theoretical expressions to the data at the encircled point and at the low pressure limit of the observed lifetimes. Some uncertainty in this fitting is due to the fact that the contributions of the two separate levels $3\,^2P_{1/2}$ and $3\,^2P_{3/2}$ were not resolved in this experiment. This does not affect the value of the lifetime measured at the lowest pressures, but at higher densities the fact that the absorption oscillator strength for the $3\,^2S_{1/2}$-$3\,^2P_{3/2}$ transition is twice as large as that of the $3\,^2S_{1/2}$-$3\,^2P_{1/2}$ transition will cause stronger resonance trapping of the 5890 Å line. Thus in general, the observed decay curves should consist of two components with slightly different decay rates. It was not possible to resolve these in the analysis of the original data and it was therefore assumed that the observed lifetime was the mean of the effective lifetimes of the J=1/2 and J=3/2 levels. Milne's theory is seen to be in good agreement with the experimental values over the whole pressure range, while Holstein's expressions are in agreement only at pressures above

5×10^{-5} Torr where the condition $\kappa_0 L \gg 1$ assumed by Holstein is satisfied.

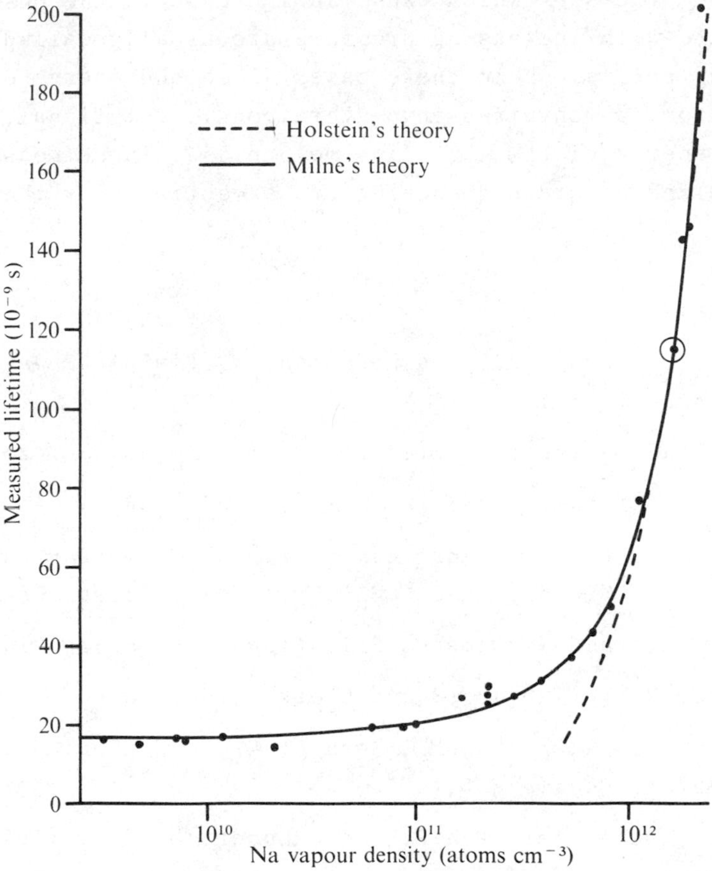

Fig.6.13. Variation of the measured lifetime of the 3 ^2P levels of sodium obtained by Kibble *et al.* (1967) as a function of sodium vapour density. The points are experimental, the solid curve is obtained from the theoretical predictions of Milne (1926) and the broken curve from those of Holstein (1947, 1951). The encircled point and the zero density lifetime of $\tau = 1{\cdot}66 \times 10^{-8}$s were used in the fit of the theoretical curves to the experimental data.

The same apparatus has been used to measure the decay

rate of sodium atoms in the 3 ^2P levels as a function of the pressure of the molecular gases N_2, H_2, HD, and D_2. The collision processes which cause an increase in the observed decay rate with increasing pressure are usually called quenching collisions in these cases since the energy of the sodium atom is converted into vibrational, rotational, and kinetic energy of the colliding molecules. These measurements have yielded accurate quenching cross-sections for these gases.

References

Andrä, H.J., Gaupp, A., and Wittmann, W. (1973). *Phys.Rev. Lett.* 31, 501.

Bakos, J. and Szigeti, J. (1967). *Optics Spectrosc.* 23, 255.

Bashkin, S. (1964). *Nucl.Instrum.Meth.* 28, 88.

Bennett, W.R. (1961). *Advances in Quantum Electronics*, (ed. J. Singer), p.28. Columbia Univ. Press, New York.

_____ and Kindlmann, P.J. (1966). *Phys.Rev.* 149, 38.

Ditchburn, R.W. (1963). *Light*. Blackie, London.

Feneuille, S., Klapisch, M., Koenig, E., and Liberman, S. (1970). *Physica* 48, 571.

Heron, S., McWhirter, R.W.P., and Rhoderick, E.H. (1954). *Nature*, Lond. 174, 564.

_____, _____ and _____ (1956). *Proc.R.Soc.* A234, 565.

Heroux, L. (1967). *Phys.Rev.* 153, 156.

Holstein, T. (1947). *Phys.Rev.* 72, 1212.

_____ (1951). *Phys.Rev.* 83, 1159.

Kay, L. (1963). *Phys.Lett.* 5, 36.

Kibble, B.P., Copley, G., and Krause, L. (1967). *Phys.Rev.* 153, 9.

Klose, J.Z. (1966). *Phys.Rev.* 141, 181.

Koenig, H.O. and Ellett, A. (1932). *Phys.Rev.* 39, 576.

Marlow, W.C. (1967). *Appl.Opt.* 6, 1715.

Milne, E.A. (1926). *J. Lond. Math.Soc.* 1, 40.

Oshirovich, A.L. and Verolainen, Ya.F. (1967). *Optics. Spectrosc.* 22, 181.

Wien, W. (1927). *Annln.Phys.* 83, 1.

General references and further reading

Bashkin, S. (ed.) (1973). *Beam-foil spectroscopy*, North-Holland, Amsterdam. (Nucl.Instrum.Meth. 110).

Bennett, W.R., Kindlmann, P.J., and Mercer, G.N. (1965). *Appl.Opt.Supplement on Chemical Lasers*, 2, 34.

Corney, A. (1970). *Adv.Electronics Electron Phys.* 29, 115. (A general review of lifetime measurements).

Foster, E.W. (1964). *Rep.Prog.Phys.* 27, 469.

Martinson, I., Bromander, J., and Berry, H.G. (eds.) (1970). *Beam-foil spectroscopy*, North-Holland, Amsterdam. (Nucl.Instrum.Meth. 90).

——— (1974). *Physica Scripta* 9, 281.

Mitchell, A.C.G. and Zemansky, M.W. (1961). *Resonance radiation and excited atoms*. Cambridge University Press, London.

7

Forbidden transitions and metastable atoms

We have seen in Chapter 5 that the transition probability for electric dipole radiation, equation (4.23), is only non-zero if certain selection rules are satisfied. In particular the initial and final states of the system must have opposite parity, where the parity of an electron configuration is given by $(-1)^{\sum_i l_i}$. In many atoms the lowest configuration consists of several equivalent electrons and so gives rise to a number of different terms; for instance, in atomic oxygen the lowest configuration is $2p^4$ and gives rise to the levels $^3P_{0,1,2}$, $^1D_2$, and $^1S_0$ as shown in Fig.7.1(a). Each of these levels has even parity, and thus, although the levels $^1D_2$ and $^1S_0$ are 2 eV and 4 eV respectively above the ground state $^3P_2$, decay by electric dipole radiation is strictly forbidden. In the absence of collisions these excited levels can decay only by the magnetic dipole and electric quadrupole transitions shown in Fig.7.1(a). As we shall show below, these processes are much slower than decay by electric dipole radiation, thus the $^1D_2$ and $^1S_0$ levels are metastable with radiative lifetimes of the order of 100 s and 1 s respectively.

Many other atoms possess metastable levels, e.g. mercury and the inert gases. In gas discharges these play an important role since the population in the metastable levels can be very high and these atoms are more easily ionized than atoms in the ground state. However, forbidden magnetic dipole and electric quadrupole lines are difficult to observe in the spectra of laboratory sources since they are extremely weak in comparison with the allowed electric dipole transitions. Also, the metastable atoms are often quenched by collisions with other atoms or with the walls of the discharge tube before radiation can occur. By contrast, in the upper atmosphere, the solar corona, and in gaseous nebulae the densities of atoms are so low that collision pro-

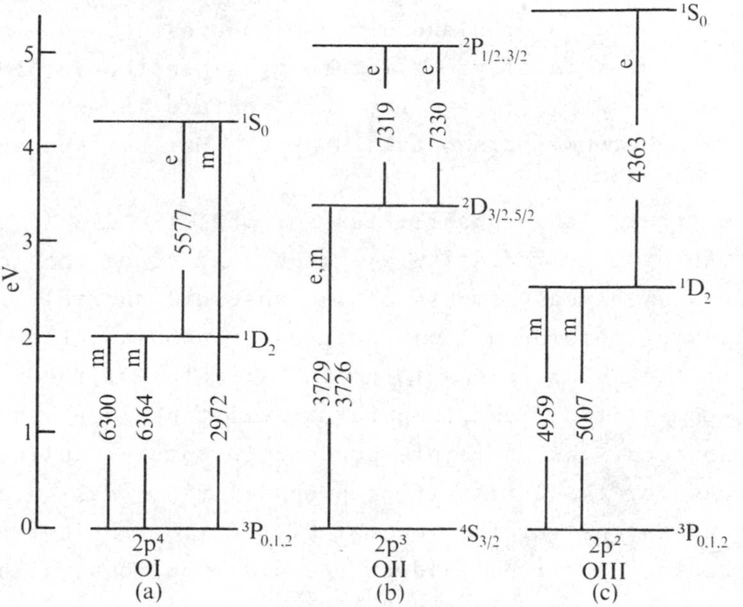

Fig.7.1. Energy levels of the ground configurations of OI, OII, and OIII showing the observed forbidden transitions. m indicates a magnetic dipole transition; e indicates an electric quadrupole transition; wavelengths are in Å.

cesses are often negligible (Problem 7.1). Consequently in the spectra of the aurora and the airglow the forbidden lines 5577 Å and 6300/6364 Å of atomic oxygen are prominent features. Similar transitions in the spectra of O^+ and O^{++}, shown in Fig.7.1(b) and (c), together with corresponding transitions in N^+ and N^{++}, are among the strongest lines in the spectra of gaseous nebulae. In the solar corona while the mean density is very low, the effective temperature is over 10^6 K and the visible spectrum is characterized by forbidden transitions of highly-stripped ions such as CaXII, FeXIII, NiXVI, etc.

Metastable levels also occur in the energy-level schemes of simple molecules, e.g. N_2, O_2, NO, and CO. These metastable molecules are responsible for prominent emission bands in the spectra of planetary atmospheres. Carbon monoxide in particular is of great interest, since the forbidden Cameron bands $a\,^3\Pi \rightarrow X\,^1\Sigma$ are the main feature in the spectra of the atmosphere of Mars obtained by the Mariner space probes.

The transition probabilities for magnetic dipole and electric quadrupole radiation are important since they can be combined with measurements of the absolute and relative intensities of forbidden lines emitted by nebulae, the aurora, or the solar corona to yield estimates of the number density, composition, and temperature existing in these various sources. We therefore proceed to obtain explicit expressions for these transition probabilities, making use of the expressions for the power radiated from the corresponding classical current and charge distributions which we obtained in sections 2.10 and 2.11.

7.1. Magnetic dipole transitions.

7.1.1. Breakdown of the electric dipole approximation. The expression for the transition probability for spontaneous emission of electric dipole radiation in a one-electron atom,

$$A_{E1} = \frac{\omega_{ki}^3}{3\pi\varepsilon_0\hbar c^3}\frac{1}{g_k}\sum_{m_k m_i}|\langle km_k|e\underline{r}|im_i\rangle|^2 \qquad (7.1)$$

was derived on the assumption that the spatial extent of the electron charge distribution in the atom was small compared with the wavelength of the light emitted. If this assumption is valid then phase differences in the contributions to the emitted electromagnetic wave from different parts of the atom can be neglected. This is known as the *electric dipole approximation* and when the matrix element in equation (7.1) is finite, the emitted radiation is called electric dipole (E1) radiation.

However, if the matrix element in equation (7.1) is identically zero for some reason, then it is necessary to introduce the factor

$$\exp(-i\underline{k}.\underline{r}') = 1 - i\underline{k}.\underline{r}' + \frac{1}{2}(i\underline{k}.\underline{r}')^2 - \ldots.. \tag{7.2}$$

from equation (2.61) which takes account of the relative phases of the contributions from different parts of the electron distribution. As shown in sections 2.10 and 2.11, the second term in the expansion of equation (7.2) can be separated into two distinct components. The first gives rise to magnetic dipole radiation and from equation (2.88) we find that the power radiated by a classical oscillating magnetic dipole is given by

$$P_{M1} = \frac{\mu_0 k^4 c}{12\pi} |\underline{\mu}|^2. \tag{7.3}$$

The quantum-mechanical expression for the magnetic dipole transition probability can be obtained from equation (7.3) in exactly the same way that the electric dipole transition probability was derived in section 4.3. We make the substitution

$$(\underline{\mu}/2)_{Cl.} = \int \psi_k^* \, \underline{\mu} \, \psi_i \, dv. \tag{7.4}$$

Then introducing the statistical weight g_k to take account of the degeneracy of the upper level as in section 4.5, we have

$$A_{M1} = \frac{\mu_0 \omega_{ki}^3}{3\pi\hbar c^3} \frac{1}{g_k} \sum_{m_k m_i} |\langle km_k|\underline{\mu}|im_i\rangle|^2 \tag{7.5}$$

where the quantum-mechanical magnetic moment operator is given by

$$\underline{\mu} = -\frac{e\hbar}{2m}(\underline{l} + 2\underline{s}) = -\mu_B(\underline{l} + 2\underline{s}). \tag{7.6}$$

For a many-electron atom, the single-electron operator in equation (7.5) must be replaced by a summation over the valence electrons and we have:

$$\underline{\mu} = -\mu_B(\underline{L} + 2\underline{S})$$

From equations (7.1) and (7.5) we find that the ratio

of the magnetic dipole to the electric dipole transition probability for an ion of atomic number Z is given approximately by

$$A_{M1}/A_{E1} \approx \frac{1}{c^2}\left|\frac{\mu}{e\underline{r}}\right|^2$$

$$= \left(\frac{Z\ \mu_B}{ea_0 c}\right)^2 = \left(\frac{Z\alpha}{2}\right)^2 \approx 10^{-5} \qquad (7.7)$$

where α is the fine structure constant $e^2/(4\pi\varepsilon_0\hbar c) \approx 1/137$. We can therefore usually neglect the effect of forbidden transitions on the lifetime of a given level if any of the electric dipole transition probabilities from that level are finite. Since a typical value for A_{E1} is $10^8\ s^{-1}$, equation (7.7) shows that levels which can decay only by the emission of magnetic dipole radiation in the visible spectral region are expected to have lifetimes of the order of 10^{-3} s or longer. These levels are therefore termed metastable.

7.1.2. Magnetic dipole radiation from atomic hydrogen. An interesting application of equation (7.5) arises in the case of the radio emission from atomic hydrogen at the frequency of 1420 MHz. This is the famous 21 cm wavelength line which has been used by radio astronomers to map the distribution of atomic hydrogen throughout the galaxy. It will be recalled that the ground state of atomic hydrogen has the electronic configuration 1s $^2S_{1/2}$ and that the proton possesses a nuclear spin of $I = \frac{1}{2}$. The ground-state hyperfine structure, shown in Fig.7.2, therefore consists of two levels F=1 and F=0 separated by an energy corresponding to the microwave frequency ν_{hfs} = 1420·40575 MHz. An atom in the upper hyperfine level can decay by spontaneous magnetic dipole radiation with a transition probability (Problem 7.2) given by

$$A_{M1} = \frac{\mu_0\omega_{hfs}^3}{3\pi\hbar c^3}\ \mu_B^2$$

$$= 2{\cdot}85 \times 10^{-15}\ s^{-1}. \qquad (7.8)$$

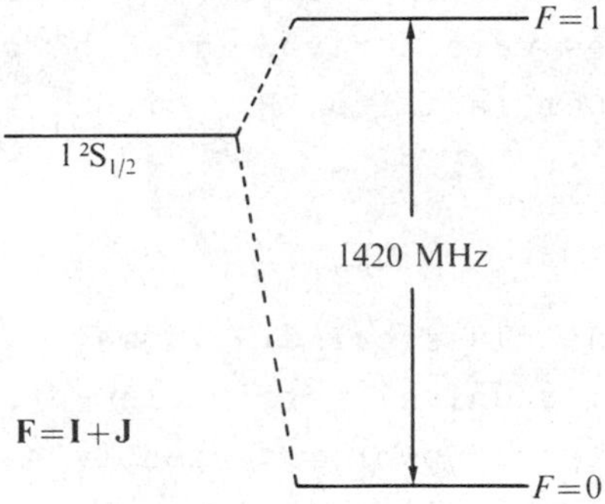

Fig.7.2. The hyperfine structure of the ground state of atomic hydrogen, $I = \frac{1}{2}$.

The spontaneous radiative lifetime for this electron spin re-orientation process is therefore $1{\cdot}14 \times 10^7$ years and only the great extent of the clouds of hydrogen gas makes the radio emission sufficiently intense to permit detection.

This calculation also shows that spontaneous magnetic dipole emission in the radio or microwave region will be quite unobservable in laboratory experiments. The techniques of radio frequency spectroscopy of atoms and nuclei in solids, liquids, and gases therefore depend on the stimulated emission and absorption processes which are discussed in detail in Chapter 9.

7.2. *Electric quadrupole radiation*

The second term in the expansion of the classical vector potential for an oscillating distribution of current and charge, equation (2.81), contains contributions from both magnetic dipole and electric quadrupole distributions, as shown in sections 2.10 and 2.11. We therefore expect these different distributions to radiate at similar rates. Thus, whenever the electric dipole transition probabilities from a given level are identically zero we must consider the possibility of decay by electric quadrupole radiation in addition to the magnetic dipole radiation discussed in the previous section.

The quantum-mechanical expressions for dipole radiation transition probabilities were derived from the classical expressions for the power radiated, P_{E1} or P_{M1}, by the relation

$$A_{ki} = \frac{4P}{\hbar\omega_{ki}} \tag{7.9}$$

while simultaneously the classical dipole magnitude was transformed into a quantum-mechanical matrix element. Applying equation (7.9) to the total power radiated by a classical electric quadrupole source, equation (2.100), we obtain the electric quadrupole transition probability in the form

$$A_{E2} = \frac{\omega_{ki}^5}{360\pi\varepsilon_0\hbar c^5}\,\frac{1}{g_k}\sum_{m_k m_i}\left|\langle km_k|-e\sum_{\alpha,\beta}(3r_\alpha r_\beta - r^2\delta_{\alpha\beta})|im_i\rangle\right|^2 \tag{7.10}$$

where r_α and r_β are the cartesian components of the electron position vector $\underline{r}$. For a many-electron system the electric quadrupole operator may be written in terms of the individual electron position vectors, $\underline{r}_i$, as

$$\underline{Q} = -e\sum_i(3\underline{r}_i\underline{r}_i - r_i^2\mathscr{I})$$

where $\mathscr{I}$ is the unit tensor.

The ratio of the electric quadrupole to the electric dipole transition probability is given by equations (7.1) and (7.10) as

$$A_{E2}/A_{E1} \approx \frac{3}{40}\left(\frac{\omega r}{c}\right)^2$$

$$= \frac{3}{40}\left(\frac{Z\alpha}{2}\right)^2 \approx 10^{-6} \tag{7.11}$$

Again the effect of forbidden electric quadrupole transitions on the lifetime of a given level will be negligible if any electric dipole transitions from that level are allowed. The lifetimes of levels which can decay only by the emission of electric quadrupole radiation are expected to be very long, of the order of 10^{-2} s or more (Problem 7.3).

For the visible region of the spectrum it is a good

approximation to neglect all emission of higher orders such as magnetic quadrupole and electric octupole radiation. However, in the X-ray and γ-ray region, this approximation is no longer valid since the wavelength of the radiation emitted is of the same order as the extent of the charge distribution involved in the emission process. Thus in nuclear spectroscopy, dipole and quadrupole radiation processes occur at similar rates and a study of the selection rules is necessary before the observed lines can be correctly assigned to a given decay mode. We shall also find such a study useful in the visible and ultraviolet regions of the spectrum.

7.3. *Selection rules for magnetic dipole and electric quadrupole transitions*

The selection rules for electric dipole radiation have already been considered in detail in Chapter 5 and are summarized in Table 7.1. Those for magnetic dipole and electric quadrupole transitions may be derived from equations (7.5) and (7.10) by the application of similar techniques. The task is therefore left as an exercise for the student (Problem 7.4); however, the comments which follow indicate the line of reasoning used and may prove helpful.

7.3.1. Rigorous selection rules. The rules given in Table 7.1 can be divided conveniently into two distinct sets: the rigorous selection rules and those which hold only when certain approximations are satisfied. Rules 1-3 are rigorous in the absence of nuclear spin and are derived from general symmetry arguments. Thus rule 1 comes from the fact that the photon possesses an intrinsic spin angular momentum, $\hbar$, and the application of the principle of conservation of angular momentum in the atom-photon system. The angular symmetry properties of atomic wavefunctions about the axis of quantization are used in the derivation of rule 2, while a reflection through the origin and the symmetry argument applied to states of well-defined parity leads to rule 3.

Of these rigorous selection rules perhaps the most im-

TABLE 7.1.

Selection rules for single photon transitions in atomic spectra

Rule	Electric dipole	Magnetic dipole	Electric quadrupole
1.	$\Delta J = 0, \pm 1$ $(0 \nleftrightarrow 0)$	$\Delta J = 0, \pm 1$ $(0 \nleftrightarrow 0)$	$\Delta J = 0, \pm 1, \pm 2$ $(0 \nleftrightarrow 0;\ \frac{1}{2} \nleftrightarrow \frac{1}{2};\ 0 \nleftrightarrow 1)$
2.	$\Delta M = 0, \pm 1$	$\Delta M = 0, \pm 1$	$\Delta M = 0, \pm 1, \pm 2$
3.	Parity change	No parity change	No parity change
4.	One electron jump $\Delta l = \pm 1$	No electron jump $\Delta l = 0,\ \Delta n = 0$	One or no electron jump $\Delta l = 0, \pm 2$
5.	$\Delta S = 0$	$\Delta S = 0$	$\Delta S = 0$
6.	$\Delta L = 0, \pm 1$ $(0 \nleftrightarrow 0)$	$\Delta L = 0$	$\Delta L = 0, \pm 1, \pm 2$ $(0 \nleftrightarrow 0;\ 0 \nleftrightarrow 1)$

portant is the parity rule. For most excited atomic levels there exist lower energy states of opposite parity, and the atom therefore decays by the emission of electric dipole radiation. Forbidden transitions are generally only observed in atoms or ions in which the lowest electronic configuration gives rise to several different terms, as in the case of the neutral and ionized oxygen level schemes shown in Fig.7.1.

7.3.2. Approximate selection rules. The last three selection rules in Table 7.1 are derived using explicit forms for the atomic wavefunctions. Since these can only be approximate solutions of Schrödinger's equation in the case of a many-electron atom, rules 4-6 in turn apply only approximately in many cases. Thus rule 4 is obeyed only if each of the states involved can be correctly described by means of a single configuration of electrons. Similarly rules 5 and 6 may be expected to apply only in Russell-Saunders coupling where the spin-orbit interaction is negligible compared with the electrostatic forces between the electrons.

In the Russell-Saunders approximation we see that magnetic dipole transitions are only possible between two states belonging to the same configuration and having the same values of L and S. Two such states must belong to the same fine structure multiplet and consequently their energy separation will be extremely small. The emitted radiation is then in the microwave or radio-frequency region and spontaneous transitions of this type have a negligibly small probability as we noted in section 7.1.2. We shall consider the selection rules for induced magnetic dipole transitions in more detail in Chapter 18 when we discuss hyperfine structure measurements by radio-frequency magnetic resonance.

As we discussed in section 5.5.5, many electric dipole transitions are known which violate the selection rule $\Delta S=0$, especially in the spectra of the heavier elements. A similar breakdown of this rule occurs also in magnetic dipole and electric quadrupole transitions, as Fig.7.1 shows.

7.3.3. Transition probabilities of forbidden lines. Accurate theoretical calculations of the transition probabilities of forbidden lines can be made without great difficulty for those lines which satisfy all the selection rules in Table 7.1. However, there are many cases of magnetic dipole or electric quadrupole lines where one or more of the approximate rules is broken. In these cases the atomic wavefunctions must be calculated in intermediate coupling and the resulting transition probabilities are consequently less reliable.

Unfortunately no generally applicable method of measuring the transition probabilities of forbidden lines has been developed. This is mainly because the lifetimes of the metastable levels which can decay only by magnetic dipole or electric quadrupole radiation in the visible region of the spectrum are of the order of 10^{-3}s or longer. In beam-foil experiments the velocities would be so great that no decay would be observed in any apparatus of convenient laboratory size. Similarly in the single-photon delayed-coincidence technique, the time required to obtain sufficient data would become quite prohibitive. The few reliable lifetime measurements that do exist have been made by the static afterglow technique. This was originally developed for experiments on the collisional destruction and diffusion of metastable atoms, which are discussed in detail in section 7.6. The difficulties encountered in the application of the afterglow and other methods to the experimental determination of the transition probabilities of forbidden lines have been reviewed by Corney (1973) and Corney and Williams (1972).

In this chapter so far we have been concerned with forbidden transitions between low-lying metastable levels belonging to a single electronic configuration. Recently interest has focussed on the energetic metastable levels of hydrogen-like and helium-like ions which emit radiation in the X-ray and vacuum ultraviolet region. Since these atoms are important in many atomic physics experiments, we discuss their properties in detail in the following sections.

7.4. Two-photon decay of hydrogenic systems

<u>7.4.1. Introduction.</u> The metastable $2\ ^2S_{1/2}$ level of hydrogen and its isoelectronic ions He^+, Li^{++}, etc.) forms an exceptionally interesting case. We show the relevant energy levels for singly-ionized helium in Fig. 7.3. since

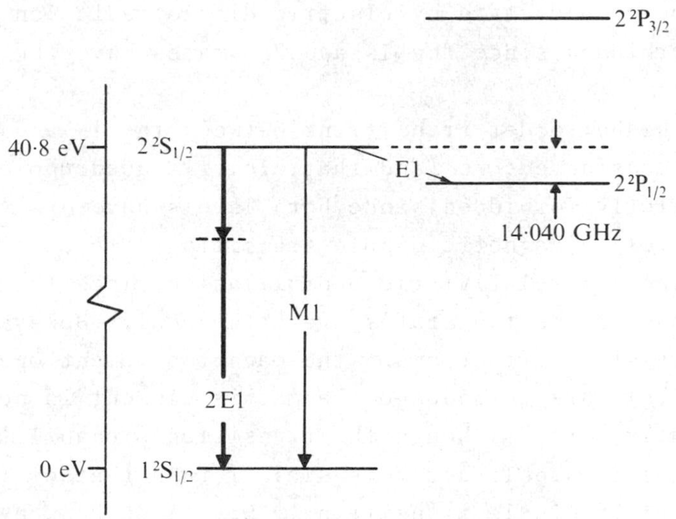

Fig. 7.3. Simplified energy level diagram for He^+ and possible decay modes of the $2\ ^2S_{1/2}$ state. (Not to scale).

many of the experiments that we shall discuss have been made on this particular system. In this ion, the metastable $2\ ^2S_{1/2}$ level lies 40·8 eV above the He^+ ground state, which in turn is 24·58 eV above the ground state of the neutral helium atom. On the Dirac theory the $2\ ^2S_{1/2}$ and $2\ ^2P_{1/2}$ levels are predicted to be degenerate. However, owing to the interaction of the electron with the quantized vacuum radiation fields, the $2\ ^2S_{1/2}$ level is displaced above the $2\ ^2P_{1/2}$ level by an amount known as the Lamb shift. Theory shows that the Lamb shift scales approximately as $(\alpha Z)^4$ and has the values 1·058 GHz and 14·040 GHz in hydrogen and ionized helium respectively.

Let us now consider in detail the metastability of the $2\ ^2S_{1/2}$ level. For an isolated atom or ion the only states of lower energy are the $1\ ^2S_{1/2}$ ground state and the Lamb-shifted $2\ ^2P_{1/2}$ level. Spontaneous electric dipole transitions to the $2\ ^2P_{1/2}$ level are allowed, but owing to the small value of the Lamb shift the transition probability is negligibly small, being about $2 \times 10^{-10}\ s^{-1}$ in hydrogen, which is equivalent to a lifetime of 163 years (Problem 7.5). Decay to the ground state by electric dipole radiation is strictly forbidden since the 1s and 2s states have the same parity.

When higher-order transitions between the 1s and 2s states are considered, we find that electric quadrupole radiation is strictly forbidden since both levels have $J = \frac{1}{2}$, (see Table 7.1). Magnetic dipole transitions are also forbidden in the non-relativistic approximation since the radial wavefunctions of the two states are orthogonal. However, when relativistic corrections to the magnetic moment operator, equation (7.6), are introduced the matrix element is no longer exactly zero, although the transition probability for the magnetic dipole decay is smaller than that of the allowed electric dipole transition $1\ ^2S_{1/2} - 2\ ^2P_{1/2}$ by a factor of $(Z\alpha)^6$. The explicit calculation by Drake (1971) gives

$$A_{M1}(2\ ^2S_{1/2}) = 2{\cdot}50 \times 10^{-6}\ Z^{10}\ s^{-1} \qquad (7.12)$$

From this we predict a lifetime for the $2\ ^2S_{1/2}$ level of hydrogen of about five days and this decay mode is therefore quite negligible for hydrogenic ions of low atomic number.

In fact the largest contribution to the radiative decay of the $2\ ^2S_{1/2}$ level comes from a process which cannot be predicted classically and which we have so far neglected. This is the simultaneous emission of two photons whose combined energies equal the energy difference between the $2\ ^2S_{1/2}$ and $1\ ^2S_{1/2}$ levels. In such a two-quantum process the angular momenta of the emitted photons can be coupled in such a way that the principle of conservation of angular momentum is also satisfied. The probability of this two-photon transition is smaller than that of an allowed transition by a quantity of order $\alpha(ka_0)^2 \approx \alpha(Z\alpha)^2$, and in hydrogen the decay rate turns out to be $8{\cdot}23\ s^{-1}$. The mean lifetime of the $2\ ^2S_{1/2}$ level is thus still extremely long compared with that of the $2\ ^2P_{1/2}$ level, a fact which was used to advantage in the first measurements of the Lamb shift. The study of two-photon decay is important for our understanding of the hydrogen atom and of multiple quantum processes. We therefore discuss the theoretical and experimental investigation of the phenomenon in some detail.

7.4.2. Theory of two-photon processes. The general theory of two-photon processes was first discussed by Göppert-Mayer (1931) and later applied by Breit and Teller (1940) to the $1\ ^2S_{1/2}-2\ ^2S_{1/2}$ transition in hydrogen. The probability per unit time for the spontaneous emission of two photons, one of which has an angular frequency between ω_1 and $\omega_1+d\omega_1$, is given by

$$A_{2E1}(\omega_1)d\omega_1 = \frac{8e^4}{(4\pi\varepsilon_0)^2\pi\hbar^2c^6}\,\omega_1^3\omega_2^3|\underline{M}_{ik}|^2_{Av}\,d\omega_1. \tag{7.13}$$

By conservation of energy, the sum of the energies of the two photons emitted equals the energy difference between the initial state k and the final state i:

$$\hbar\omega_1 + \hbar\omega_2 = \hbar\omega_{ki} = E_k - E_i$$

and the matrix element for the simultaneous emission of two electric dipole photons is (Breit and Teller 1940)

$$\underline{M}_{ik} = \sum_j \left\{ \frac{\langle i|\hat{\underline{\varepsilon}}_1\cdot\underline{r}|j\rangle\langle j|\hat{\underline{\varepsilon}}_2\cdot\underline{r}|k\rangle}{\omega_2+\omega_{jk}} + \frac{\langle i|\hat{\underline{\varepsilon}}_2\cdot\underline{r}|j\rangle\langle j|\hat{\underline{\varepsilon}}_1\cdot\underline{r}|k\rangle}{\omega_1+\omega_{jk}} \right\} \tag{7.14}$$

In equation (7.13) this matrix element must be averaged over the directions of propagation and the polarization vectors

$\hat{\underline{\varepsilon}}_1$ and $\hat{\underline{\varepsilon}}_2$ of the emitted photons independently. Finally, the summation in equation (7.14) must be carried out over all possible intermediate states, here denoted by $|j\rangle$, including the continuum as well as the discrete $n\ ^2P$ states.

<u>7.4.3. Angular correlation of the emitted photons.</u> It can readily be shown that the transition probability given by equation (7.13) is proportional to the square of the scalar product of the polarization vectors of the emitted photons, $(\hat{\underline{\varepsilon}}_1 \cdot \hat{\underline{\varepsilon}}_2)^2$. Averaging over the polarizations of both photons yields an angular distribution varying as $(1+\cos^2\theta)$ where θ is the angle between the propagation vectors of the two photons. Thus if the metastable atoms are monitored by two detectors which are insensitive to polarization, the photon-coincidence counting rate should vary as $(1+\cos^2\theta)$. The directions of emission of the photons are correlated as a consequence of the conservation of angular momentum in the total atom-photon system.

This angular correlation was first investigated experimentally by Lipeles *et al.* (1965) in the case of the $2\ ^2S_{1/2}$ level of He^+. A schematic diagram of their apparatus, further details of which are included in a review of two-photon decay processes presented by Novick (1969), is given in Fig.7.4. The apparatus consisted of an electron bombardment ion source, differential pumping chambers for pressure reduction, and a detection chamber. A slow beam of metastable and ground-state helium ions was focussed into the detection chamber by means of electrostatic lenses. Photons emitted were detected by two photomultipliers, one of which could be rotated about the beam axis. Although the pressure in the detection chamber was less than 1×10^{-8} Torr, a large photon flux was produced by collisions of the ground-state helium ions with the background gas. The two-photon signal was separated from this background radiation by photon-coincidence counting and also by selectively quenching the metastable ions. This was achieved by exciting the ions to the radiating $2\ ^2P_{1/2}$ level using

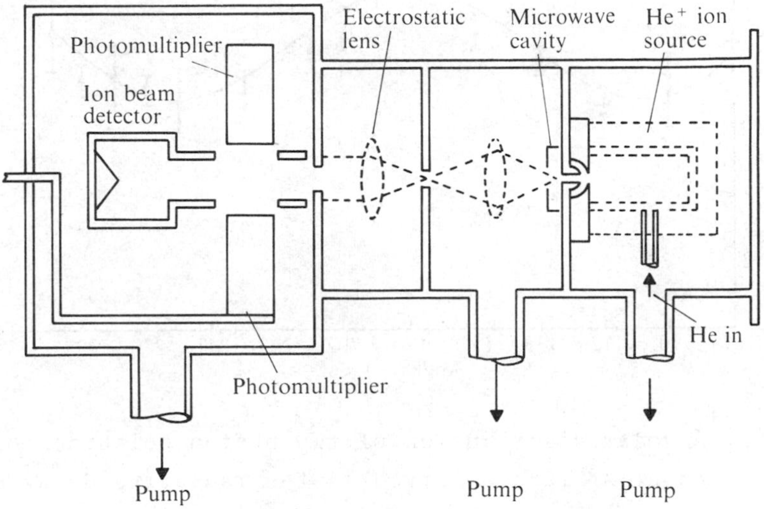

Fig.7.4. Experimental apparatus for the investigation of the two-photon decay of the metastable $2\ ^2S_{1/2}$ level of He^+. (After Novick (1969).)

stimulated absorption of microwave power at the 14 GHz Lamb shift frequency.

The two-photon coincidence counting rate arising from the metastable helium ions is shown in Fig.7.5 for various angular positions of the moveable phototube. The solid curve in Fig.7.5 was obtained by integrating the theoretical angular factor of $(1+\cos^2\theta)$ over the area of the photomultiplier cathode and over the length of the ion beam exposed to the detectors. The agreement of the experimental results with the predicted angular distribution gave the first unambiguous evidence that the $2\ ^2S_{1/2}$ level of He^+ decays by the simultaneous emission of two photons.

7.4.4. Spectral distribution of two-photon emission. From equation (7.14) we see that the spectral distribution is continuous and is also symmetric in ω_1 and ω_2, which is to be expected since both photons are emitted simultaneously.

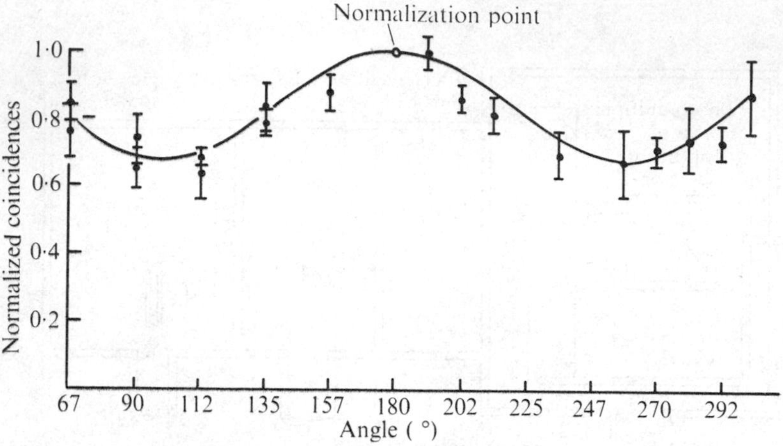

Fig.7.5. Angular distribution of two-photon coincidence counting rate observed in the radiative decay of the $2\ ^2S_{1/2}$ metastable level of He^+. (After Novick (1969).)

The spectral distribution $A_{2E1}(\omega_1)$ must therefore be symmetric about the angular frequency $\omega_{ki}/2=(E_k-E_i)/2\hbar$. In ionized helium this means that one photon of the two-photon pair will lie in the region between 304 Å and 608 Å while the other has a wavelength greater than 608 Å. The theoretical spectral distribution calculated by Spitzer and and Greenstein (1951) is shown in Fig.7.6 and displays a broad symmetric peak centred on the angular frequency $\omega_{ki}/2$, corresponding to a wavelength of 608 Å. Detailed experimental verification of this distribution has not been possible so far, although the results of Novick (1969) and his students show that there is no major discrepancy between theory and experiment.

In hydrogen as in ionized helium, the most intense part of the two-photon emission continuum lies in the vacuum ultraviolet. However, this two-photon decay of the $2\ ^2S_{1/2}$ state of hydrogen is thought to make a significant contribution to the continuous emission spectra of planetary nebulae in the visible region, although radiative recombina-

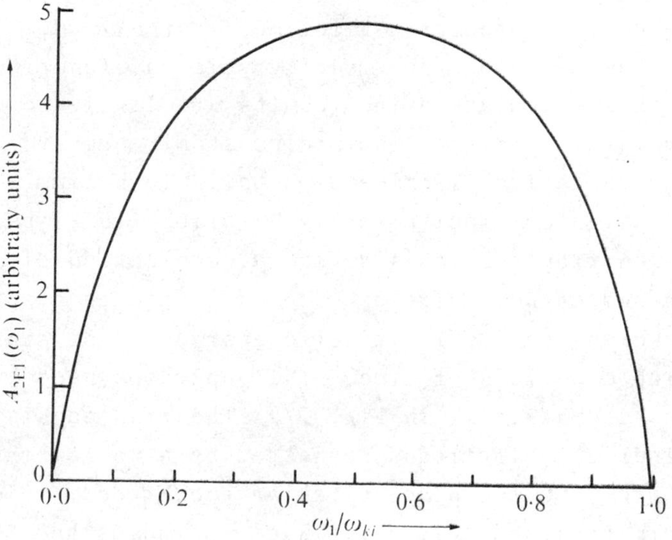

Fig.7.6. Theoretically predicted two-photon spectral distribution on an angular frequency scale. (After Novick (1969).)

tion of hydrogen and helium ions and free-free transitions are also important.

7.4.5. Radiative lifetimes of metastable helium ions. The total probability of two-photon emission, $A_{2E1}(2\ ^2S_{1/2})$, is obtained by integrating the spectral distribution of the two photons, $A_{2E1}(\omega_1)$, over all possible emission frequencies:

$$A_{2E1}(2\ ^2S_{1/2}) = \frac{1}{2}\int_0^{\omega_{ki}} A_{2E1}(\omega_1)\ d\omega_1. \qquad (7.15)$$

The factor of one-half in equation (7.15) arises because in the integration over the range from 0 to ω_{ki} each photon pair (ω_1,ω_2) is counted twice. The total theoretical transition probability was obtained by Schapiro and Breit (1959) in the form

$$A(1S,2S) = (8\cdot226 \pm 0\cdot001)\ Z^6\ s^{-1} \qquad (7.16)$$

which leads to a predicted lifetime for the 2 $^2S_{1/2}$ state of He^+ of 1·899 ms. Since non-relativistic wavefunctions were used in this calculation the results are likely to be accurate only for ions of low and medium atomic number, Z.

The radiative lifetime of metastable helium ions has been investigated experimentally by Prior (1972) using a Penning ion trap. In this device a combination of static electric and magnetic fields is used to create a potential well in which ions of low kinetic energy can be stored for considerable periods of time. The apparatus used by Prior is shown schematically in Fig.7.7. The trap consists of two cylindrical electrodes connected by a web of thin copper wires and two plane circular electrodes. Electrically these form a closed cylinder, 15 cm in radius and 30 cm long, whose ends are maintained at a positive potential of between 0·5-3·0 V with respect to the body. The electrostatic field limits the motion of the ions in the axial direction while a magnetic field of between 40-65 G applied coaxially with the cylinder limits the ion motion in the transverse direction.

The trap electrodes are enclosed in a high vacuum system where the background pressure is maintained below 5×10^{-8} Torr. During an experimental run, helium gas is admitted to the chamber through a needle valve and a small number of metastable helium ions are created at time t=0 by a pulsed electron beam. The ions are stored in the trap for a period of several lifetimes and during this time the exponential decay of the metastable ion population is monitored by photon counting. The signal is accumulated in a multichannel memory unit used in a multiscaling mode. In this mode of operation the memory address being accessed is advanced sequentially during the decay cycle by an internal or external pulse generator, and the photons detected are recorded as the counts stored in the different channels of the memory. By repeating the excitation and decay cycle many times a decay curve with a very low noise level can be built up in the memory. Fig.7.8 shows the

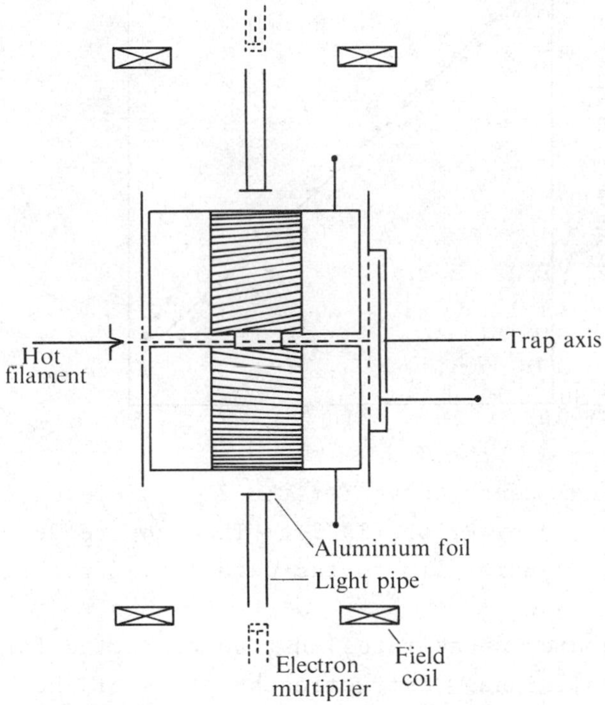

Fig.7.7. Schematic diagram of the Penning ion trap used by Prior (1972) in measurements of the radiative lifetime of He^+ ($2\ ^2S_{1/2}$). The rectangular shape in the centre of the figure represents a microwave horn located outside the trap. This enables ions in the $2\ ^2S_{1/2}$ state to be selectively quenched by inducing the $2\ ^2P_{1/2}$-$2\ ^2S_{1/2}$ electric dipole transition.

data accumulated after a run lasting 90 minutes. In this experiment it was again necessary to separate the two-photon decay from the radiation emitted by the background gas by selectively quenching the $2\ ^2S_{1/2}$ metastable ions.

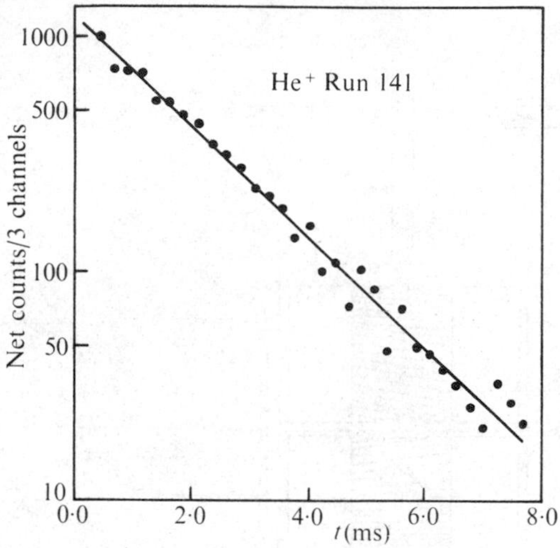

Fig.7.8. Typical decay curve for the 2 $^2S_{1/2}$ level of He^+ obtained by Prior (1972). The line represents the least-squares fit to the data giving τ = 1·866 ms.

The measured decay rates must be corrected for the effects of collisional destruction by atoms of the background gas and for the quenching caused by the electric and magnetic fields applied to the trap. These fields have the effect of mixing the metastable 2 $^2S_{1/2}$ state with the 2 $^2P_{1/2}$ state which then decays rapidly to the ground level. In these experiments the corrections amount to only 3 per cent and the final result is given in Table 7.2. The result is in agreement with a less accurate value obtained in experiments made by Kocher *et al.* (1972) in which the metastable beam intensity was measured as a function of distance from the source using the apparatus shown in Fig.7.4. There is also excellent agreement between the theoretical and experimental lifetimes given in Table 7.2 for the 2 $^2S_{1/2}$ level of He^+.

This close agreement has further interesting consequences. If a parity non-conserving interaction exists in hydrogenic systems, the wavefunction of the 2S level

TABLE 7.2.

Theoretical and experimental lifetimes of the $2\,^2S_{1/2}$ *level of hydrogenic systems*

Ion	τ_{expt}	τ_{theory}
H	—	0·1216 s
He^+	1·922 ± 0·082 ms	1·899 ms
S^{+15}	7·3 ± 0·7 ns	7·25 ns
Ar^{+17}	3·54 ± 0·25 ns	3·57 ns

would include a small mixture of the wavefunction of the 2P state, thus

$$\Psi(2S) = \alpha\psi(2S) + \beta\psi(2P) \qquad (7.17)$$

where the wavefunctions of well-defined parity are denoted by ψ. The mixed 2S state would then be able to decay to the ground state via the allowed Lyman α electric dipole transition, $1\,^2S_{1/2}$-$2\,^2P_{1/2}$. The uncertainty of the experimental lifetime given in Table 7.2 implies that in He^+ $|\beta| < 4\cdot 7 \times 10^{-5}$ (Problem 7.6), and places an upper limit on the strength of any parity non-conserving interaction.

7.4.6. Radiative decay rates of other hydrogenic ions. Although the lifetime of the $2\,^2S_{1/2}$ state in hydrogen and ionized helium is very long, equation (7.16) shows that in hydrogenic ions of atomic number $Z \approx 20$ the lifetime is reduced to a few nanoseconds and can therefore be measured by the beam-foil technique. In the apparatus used by Marrus and Schmieder (1972), shown in Fig.7.9, argon or sulphur

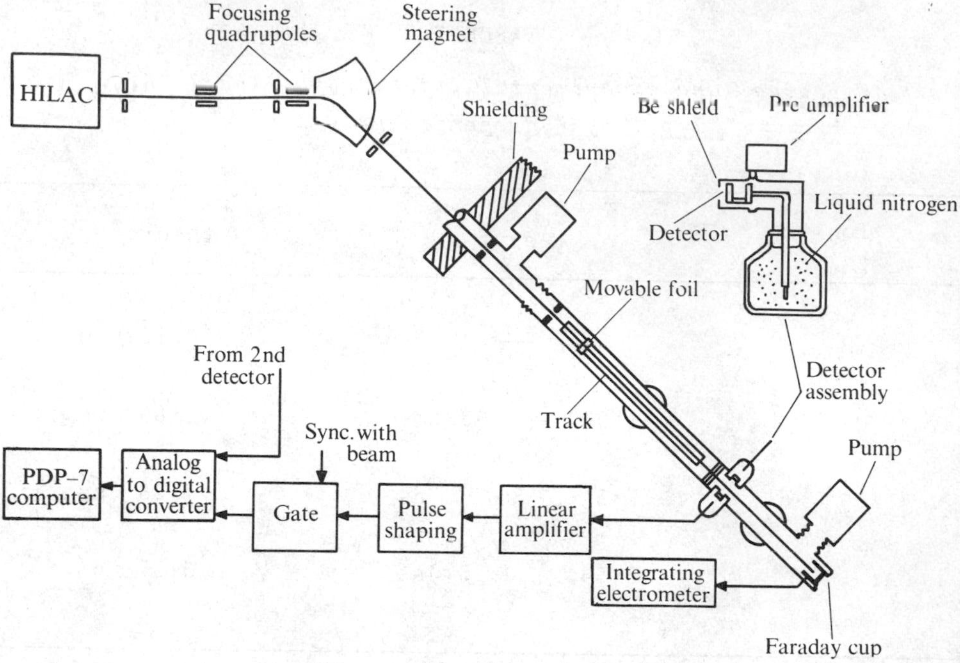

Fig.7.9. Experimental arrangement of Marrus and Schmieder (1972) for the investigation of the forbidden decay modes of one- and two-electron ions of high Z.

ions are accelerated to an energy of 10·3 MeV/nucleon by the Berkeley heavy ion accelerator, HILAC. For ^{40}Ar in the +14 charge state, this results in a final energy of 412 MeV and a velocity of $4{\cdot}4 \times 10^9$ cm s^{-1}. The beam is extracted from the HILAC by a steering magnet and passes into a long pipe containing a series of collimators and a 100μg/cm^2 beryllium foil mounted on a moveable track. After passing through the foil the argon beam consists approximately of 25 per cent Ar^{+16} (helium-like), 50 per cent Ar^{+17} (hydrogen-like), and 25 per cent Ar^{+18} (fully stripped). The ions in excited levels possessing allowed electric dipole transitions decay very close to the foil and the beam at the detectors consists only of ground state and 'metastable' ions. Ions in the metastable $2\ ^2S_{1/2}$ level decay by the emission of X-ray photons (Problem 7.8) which are detected by a pair

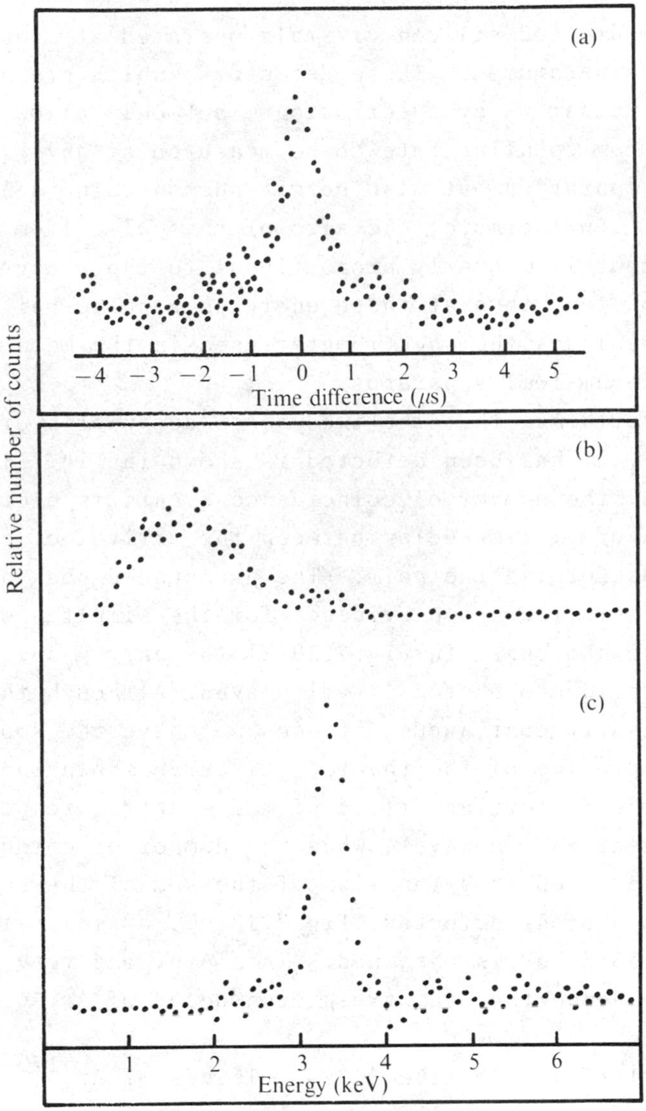

Fig.7.10. Delayed coincidence spectra from hydrogenic $Ar^{+17}(2\ ^2S_{1/2})$. (a) the number of counts versus time-delay between photons; (b) number of photons contributing to coincidence events as a function of photon energy; (c) number of events as a function of the sum of the energies of the two photons.

of lithium-drifted silicon crystals operated at liquid nitrogen temperatures. These detectors, which are normally used for nuclear γ-ray spectroscopy, not only allow the single-photon counting rate to be measured at any given foil-detector separation but also permit photon-coincidence experiments. Furthermore, the size of the pulse from the detector output is linearly proportional to the energy of the incident photon, thus at these energies a pulse-height analyser replaces the spectrometer used in the more conventional beam-foil apparatus.

The evidence that the two-photon decay of hydrogenic $Ar^{+17}(2\ ^2S_{1/2})$ has been detected is shown in Fig.7.10. In Fig.7.10(a) the number of coincidence events is plotted as a function of the time-delay between the arrival of each of the photons forming the pair. The pronounced peak at zero-time-delay provides good evidence for the simultaneous emission of two photons. In Fig.7.10(b) the energy spectrum of the coincidence photons is displayed. Although the spectrum is clearly continuous, it does not have the smooth symmetric profile of the theoretical curve shown in Fig.7.6 owing to the distorting effect of the spectral response of the detectors. However, when the number of coincidence events is plotted as a function of the sum of the energies of the two photons detected, Fig.7.10(c), a single line peaking at 3·3 keV is obtained. This provides very convincing evidence that the two-photon decay of Ar^{+17} is responsible.

The lifetimes of the $2\ ^2S_{1/2}$ levels of Ar^{+17} and S^{+15} were obtained by measuring the emitted intensity as a function of the separation between the foil and the detectors. The intensity was obtained by summing the counts recorded in an appropriate spectral range and then normalizing to the total integrated beam current as measured by the Faraday cup. After correction for a small background signal, the semi-logarithmic plots of the total number of counts as a function of foil-detector separation were linear and the radiative lifetimes could be determined accurately.

The results obtained are given in Table 7.2 together with the theoretical lifetimes calculated from equation (7.16). There is good agreement between theory and experiment for all the ions listed in Table 7.2., although when the relativistic magnetic dipole decay, equation (7.12), is taken into account the theoretical lifetimes of S^{+15} and Ar^{+17} should be reduced by 2 and 3 per cent respectively. It is clearly desirable that the decay rates of these highly-ionized atoms should be measured to an accuracy of one per cent or better. This would determine whether or not the discrepancy noted above was significant. It would also provide a more accurate check of the two-photon emission rates which so far have only been calculated using non-relativistic wavefunctions.

The successful detection of two-photon emission in hydrogenic systems prompted a more careful examination of the decay modes possible for metastable levels in two-electron atoms and ions. This has revealed further interesting features and we discuss the progress which has been made recently in the following section.

7.5. Forbidden transitions in helium-like systems

7.5.1. Introduction. The first few energy levels of the neutral helium atom are shown in Fig.7.11. We consider first the $2\ ^1S_0$ metastable level, the decay of the lower lying $2\ ^3S_1$ level being discussed briefly in section 7.5.4. Since the $2\ ^1S_0$ level lies above the $2\ ^3S_1$ level, decay by relativistically assisted magnetic dipole radiation is possible. However, the low frequency of this line combined with the $\Delta S=0$ selection rule makes decay by this route very improbable. Further consideration of the selection rules shows that the decay of the $2\ ^1S_0$ level by single-photon emission to the $1\ ^1S_0$ ground state is rigorously forbidden in all orders. Only by the simultaneous emission of two photons can the total angular momentum of the system be conserved and the $J=0 \rightarrow J=0$ transition accomplished. Theoretical calculations of the radiative lifetime of the

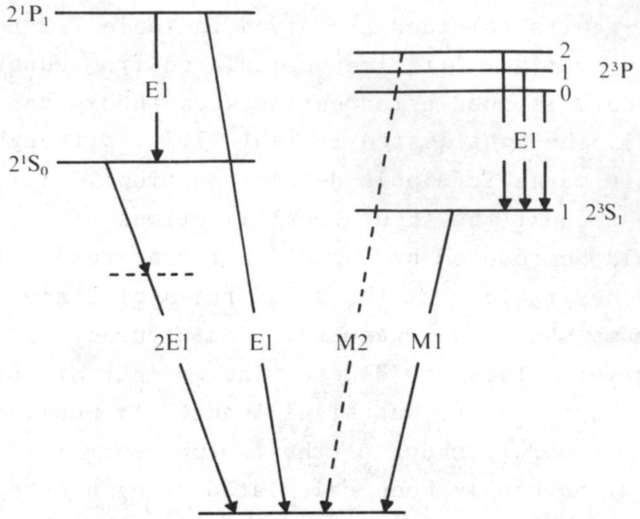

Fig.7.11. The principal decay modes of the low-lying energy levels of the neutral helium atom (not to scale).

$2\,^1S_0$ level by Drake *et al.* (1969) and by Jacobs (1971) gave a value of 19·5 ms. Since the calculations depend in a sensitive way on the wavefunctions of the states involved it is important to have confirmation of this result by an accurate measurement of the experimental lifetime. We now give details of an apparatus that was built for this purpose.

7.5.2. Thermal beam time-of-flight experiments.

(a) Details of the apparatus. The thermal beam time-of-flight technique was developed by Van Dyck, Johnson, and Shugart (1970, 1971) at the University of California in Berkeley especially for the measurement of the radiative lifetimes of metastable atoms and molecules. The principle of this technique may be easily understood from the schematic diagram of their vacuum apparatus shown in Fig.7.12. Atoms or molecules of the gas which it is desired to investigate pass through a cooled source chamber and effuse through a slot into a differential pumping region. They are next excited to the required metastable levels by a pulsed electron

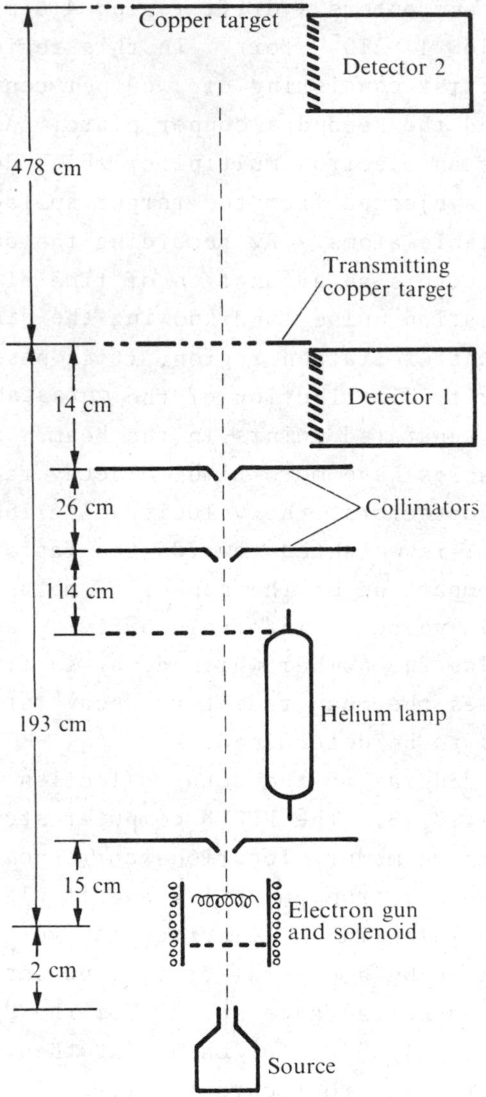

Fig.7.12. Schematic diagram of the thermal beam time-of-flight apparatus developed by Van Dyck, Johnson, and Shugart (1970,1971) for measurements of the radiative lifetimes of metastable atoms.

beam. Finally the beam of metastable atoms or molecules is collimated and enters a drift region 4·8 m long which is maintained below 1×10^{-8} Torr. In this region are two targets, the first consisting of a 60 per cent transparent copper mesh and the second a copper plate. Associated with each target is an electron multiplier which detects the Auger electrons ejected from the target surface by the impinging metastable atoms. By recording the output pulses from these detectors as a function of time after the electron gun excitation pulse, and knowing the distance of the targets from the excitation region, it is possible to obtain the velocity distribution of the metastable species at two spatially separated points in the beam. Since the slower metastables have more time to decay radiatively between the two detectors, the velocity distribution at the second detector is weighted towards the faster velocity classes. A comparison of the number of metastable atoms within specific velocity intervals arriving at the two detectors yields the number which decay in flight between them and enables the mean radiative decay rate of the metastable species to be determined.

A block diagram of the data collection system used is shown in Fig.7.13. The PDP-8 computer stores the Auger electron counts in memory locations corresponding to arrival times after the electron gun pulse and it also controls the magnetic field solenoid and electron gun voltages. The time base is generated by a crystal-controlled oscillator which supplies the channel advance pulses for the PDP-8. The electron gun is pulsed on for channel 0 then the following 199 channels are used to record the time of flight distributions. The dwell time per channel can be varied from 12·75 μs to 1 ms, depending on the radiative lifetime being measured. Data counts are collected at both detectors simultaneously. For good signals 10^7 metastable atoms would be counted during a 2 hour run; for weaker signals 4×10^5 counts in 12 hours would be typical. The data is punched on to cards for the final analysis.

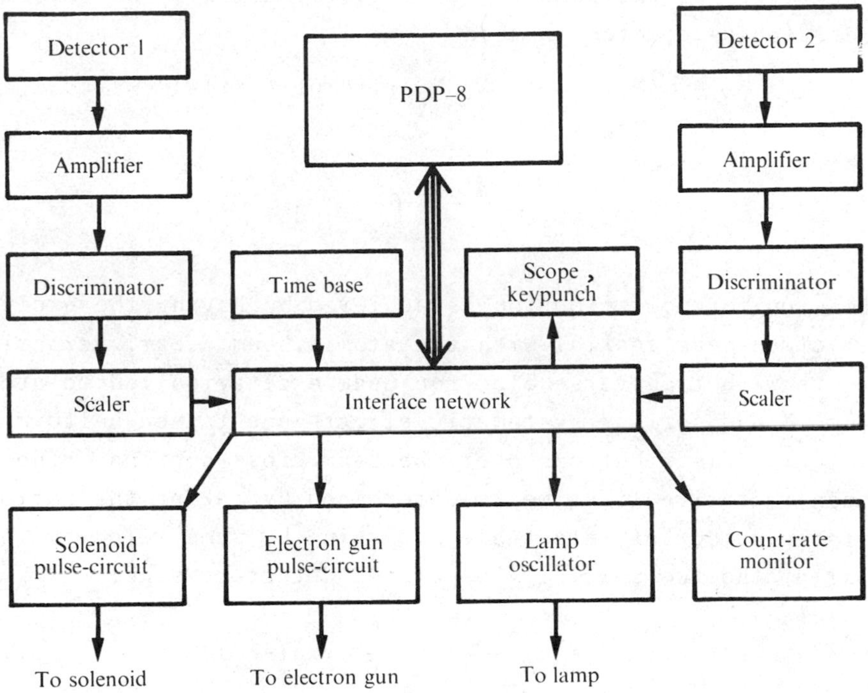

Fig.7.13. Block diagram of the data collection and storage system used in the time-of-flight experiments by Van Dyck *et al.* (1971).

(b) Time-of-flight data analysis. If we assume that the beam contains only one metastable state, the number of metastable atoms, $N_i(v)$, with velocity v that are counted at detector i is

$$N_i(v) = \int_{\text{surface}} \varepsilon_i \, N_0(v) \exp(-t_i/\tau) \, dS \qquad (7.18)$$

where t_i is the time-of-flight from the source to detector i,

τ is the radiative lifetime, and ε_i is the surface efficiency of detector i. If the initial velocity distribution of metastables is uniform across the beam, each point on the detector sees the same velocity distribution. The number of metastables counted is then

$$N_i(v) = C_i N_0(v) \exp(-t_i/\tau) \tag{7.19}$$

where

$$C_i = \int_{\text{surface}} \varepsilon_i \, dS.$$

This uniform distribution is achieved by having the exciting electron beam coaxial with the atomic beam. Early experiments with transverse electron beam excitation led to atom recoil and serious systematic effects due to non-uniform velocity distributions over the beam cross-section. The mean radiative lifetime is determined by taking the ratio of the number of metastable atoms in the same velocity interval at the two spatially separated detectors since

$$R = \frac{N_2(v)}{N_1(v)} = \frac{C_2}{C_1} \exp(-t/\tau) \tag{7.20}$$

where $t=t_2-t_1$ is the time-of-flight between the two detectors.

The time-of-flight distributions for the helium metastable level $2\,^1S_0$ are shown in Fig.7.14. The velocity intervals are determined by the channel width at detector 1, and the data for detector 2 have been averaged in such a way that the data plotted in Fig.7.14 give the number of metastable atoms in the same velocity intervals. Unfortunately, helium also has another long-lived metastable level, the $2\,^3S_1$, and it is necessary to find some way of distinguishing between this state and the $2\,^1S_0$ level. This is the purpose of the lamp shown in Fig.7.12. Metastable atoms in the $2\,^1S_0$ level can be excited by the absorption of radiation at 20 581 Å to the $2\,^1P_1$ level from which there is a fast allowed decay to the ground state. The metastables in the $2\,^1S_0$ can therefore be preferentially quenched while the

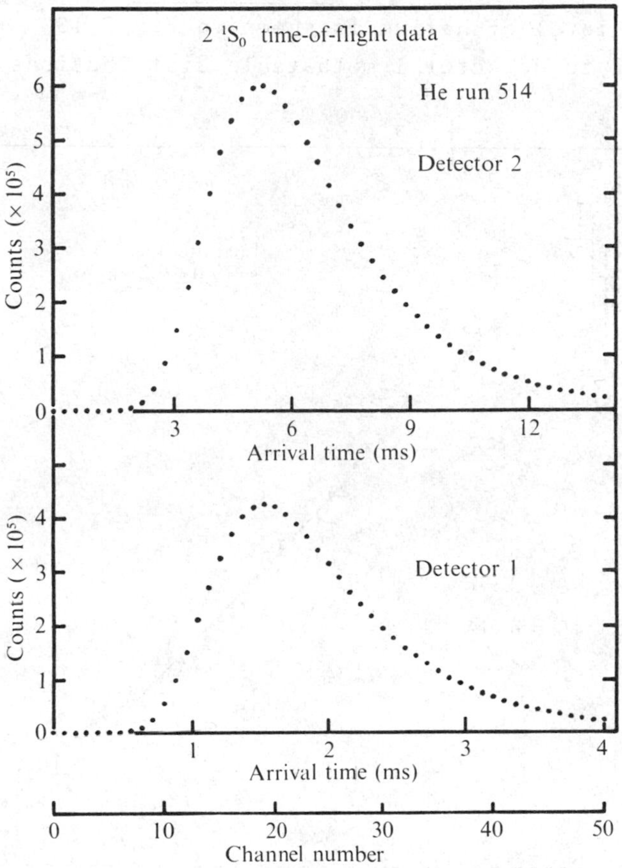

Fig.7.14. Time-of-flight distributions for He(2 $^1S_0$) obtained by Van Dyck *et al.* (1971). The channel numbers refer to detector 1. The data for detector 2 have been averaged over velocity intervals whose widths are determined by the channel width at detector 2.

2 $^3S_1$ metastables are unaffected. Data collection sweeps are performed with the quench lamp alternatively on and off and subtraction of the two sets of data allows the time-of-flight distributions for the 2 $^1S_0$ metastable alone to be determined.

(c) Results of the time-of-flight technique. The decay plot for the 2^1S_0 level of helium is shown in Fig.7.15. The ratio of detector 2 to detector 1 metastable distributions is

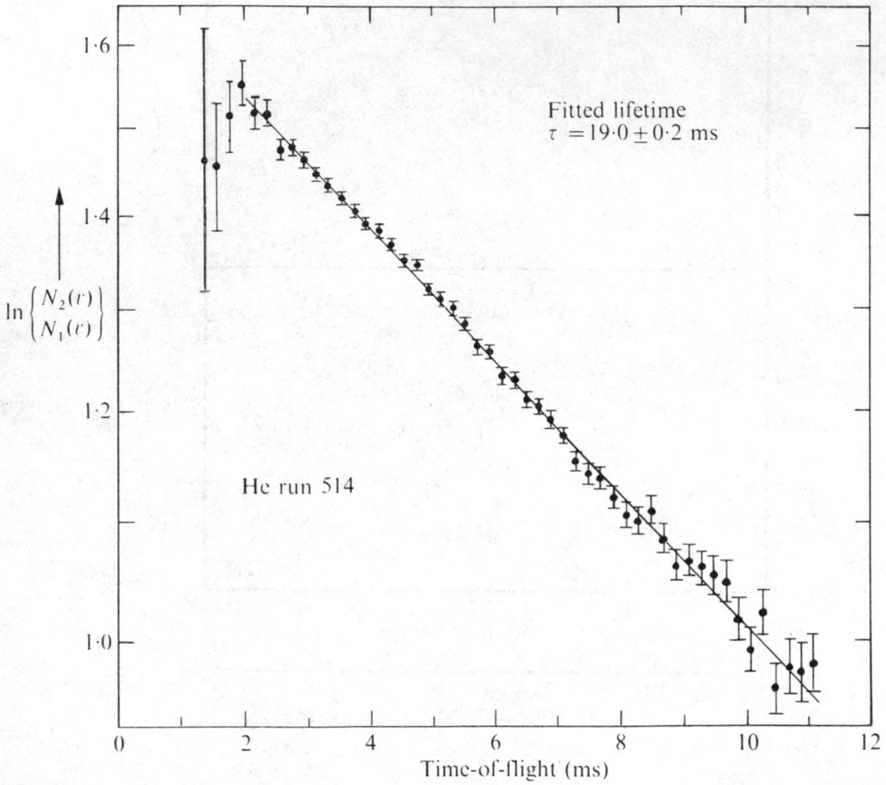

Fig.7.15. Decay plot for the He $2\ ^1S_0$ level obtained by Van Dyck *et al.* (1971). The mean lifetime is obtained from the slope of the least-squares fitted straight line. Data arising from points in the velocity distribution curves having less than 10 per cent of the peak values are not included in this fit.

plotted on a logarithmic scale as a function of the time-of-flight between the detectors, and the mean lifetime is obtained from the slope of the least-squares fitted straight line.

The final result for the experimentally determined lifetime of the $2\ {}^1S_0$ level of neutral helium is given in Table 7.3. We see that there is good agreement with the results of recent theoretical calculations in this particular case.

TABLE 7.3.

Experimental and theoretical lifetimes of the $2\ {}^1S_0$ level of helium-like systems

Ion	τ_{expt}	τ_{theory}
He	$19{\cdot}7 \pm 1{\cdot}0$ ms	$19{\cdot}5$ ms
Li^+	$0{\cdot}503 \pm 0{\cdot}026$ ms	$0{\cdot}513$ ms
Ar^{+16}	$2{\cdot}3 \pm 0{\cdot}3$ ns	$2{\cdot}35$ ns

<u>7.5.3. Two-photon decay of helium-like ions.</u> The two-photon decay of helium-like Li^+ in the $2\ {}^1S_0$ level has been measured by Prior and Shugart (1971) using the Penning ion-trap technique described in section 7.4.5. Their result, listed in Table 7.3, is also in good agreement with the theoretical calculations of Drake *et al.* (1969). The two-photon decay of helium-like Ar^{+16} has been confirmed by Marrus and Schmieder (1972) using the high energy beam-foil technique discussed in section 7.4.6. Although the accuracy of their lifetime result is not particularly high, the data collected in Table 7.3 provide a very convincing check of the theoretically computed decay rates over seven orders of magnitude.

We may therefore conclude that spontaneous decay by the simultaneous emission of two electric dipole photons is now reasonably well understood in both one- and two-electron systems. This knowledge could be important in the future since a closely related process involving the simultaneous absorption of two photons is rapidly becoming a favourite method of tunable dye laser spectroscopy.

7.5.4. Decay of the 2 $^3S_1$ level of helium-like systems. The lowest excited energy level of helium and its isoelectronic ions Li^+, Be^{++}, B^{+++} .. .. is the 2 $^3S_1$ level shown in Fig.7.11. Atoms in this state can decay radiatively only to the ground level and an examination of the selection rules given in Table 7.1 shows that decay by electric dipole radiation is strongly forbidden since the states have the same parity. The 2 $^3S_1$ level is therefore metastable. Furthermore decay by electric quadrupole radiation is also forbidden since this would violate the $J=1 \nleftrightarrow J=0$ rule. Although the transition to the ground state does seem to be possible by magnetic dipole radiation, this would violate the $\Delta S=0$ rule and since L-S coupling is an excellent approximation in helium, it was thought for a long time that the triplet metastable level would also decay preferentially by two-photon emission with a transition probability between 10^{-8}-10^{-9} s^{-1}.

Recently, however, Gabriel and Jordan (1969) suggested that certain solar X-ray lines should be identified as the 1 $^1S_0$-2 $^3S_1$ transition in the spectra of helium-like ions ranging from CV to MgXI. This single-photon emission was ascribed to spin-dependent relativistic corrections to the magnetic dipole operator, equation (7.5), of the type already discussed in connection with the decay of the 2 $^2S_{1/2}$ level of hydrogenic systems. Calculations then showed that in helium the 2 $^3S_1$ level decayed by this mode at a rate of $1{\cdot}27 \times 10^{-4}$ s^{-1} and that the transition probability scaled as $1{\cdot}66 \times 10^{-6}$ Z^{10} s^{-1} for ions of high atomic number.

In helium this single-photon decay mode has so far proved to be too slow to permit accurate experimental measurements although some evidence for the existence of a line at 625·56 Å in the spectrum of a helium afterglow has been obtained by Moos and Woodworth (1973). In helium-like Ar^{+16}, however, the decay rate is considerably faster and Marrus and Schmieder (1972) have succeeded in making measurements by the beam-foil technique. They obtained the value 172 ± 30 ns for the radiative lifetime, which is in approximate agreement with the theoretical value of 213 ns obtained by Drake (1971). Further experiments on other helium-like systems of medium and high atomic number are clearly required.

7.5.5. Radiative decay of other metastable species. In the light of the previous section it seems unlikely that we have, at present, anything like a complete understanding of the radiative decay of highly stripped ions. In these systems relativistic effects are becoming important and in addition the frequency dependence of the transition probabilities of higher orders of multipole radiation may cause these decay modes to compete effectively with decay by electric dipole radiation. This point has been emphasized by Marrus and Schmieder (1970) who observed a magnetic quadrupole line, $1\ ^1S_0 - 2\ ^3P_2$, in helium-like Ar^{+16} although in neutral helium the $2\ ^3P_2$ level decays by an allowed electric dipole transition to the $2\ ^3S_1$ metastable level. We are therefore likely to see continued developments in this field during the next few years.

The other noble gases Ne, Ar, Kr, and Xe all possess energetic metastable levels which in L-S coupling would be labelled as $n\ ^3P_0$ and $n\ ^3P_2$. Metastable levels with the same angular momentum quantum numbers also occur in the term schemes of the group IIB metals Zn, Cd, and Hg. Since these metastable states have opposite parity to the ground state, decay by the simultaneous emission of two photons is forbidden. Thus in all cases decay by very weak magnetic dipole and magnetic and electric quadrupole transitions is involved (Problem 7.9). Although Van Dyck, Johnson, and Shugart (1972) have been able to place a lower limit on the lifetimes in the case of the noble gases, no experiments are known which give an accurate measurement of the true radiative decay rate of these levels. We turn therefore to a discussion of the behaviour of these metastable atoms in gases at finite pressures.

7.6. Collision processes involving metastable atoms

7.6.1. Introduction. We have now shown in considerable detail that under high vacuum conditions atoms and ions in metastable levels decay by radiative processes which are generally very slow. Under more usual conditions such as exist in gas discharges at pressures in the range 10^{-2}-10^{1} Torr, these metastable species are rapidly destroyed by collisions and by diffusion to the walls of the container. The radiative transitions make an insignificant contribution to the effective decay rate and consequently we now examine some of the more important collision processes. Although in the active discharge electron-atom interactions are significant we do not have space to consider the many different aspects of such collisions. Instead we shall restrict ourselves to a consideration of the following atom-atom collision processes:

(a) Two-body quenching collisions in which the excitation energy is transferred into relative translational energy.

(b) Three-body collisions leading to the formation of excited molecules

$$A^* + 2A \rightarrow A_2^* + A. \tag{7.21}$$

(c) Diffusion to the walls followed by complete de-excitation.

(d) Collisional transfer to states from which optical transitions to the ground state can occur.

(e) Collisions involving a transfer of excitation to atoms of a different species

$$A^* + B \rightarrow A + B^*. \tag{7.22}$$

(f) Ionizing collisions (Penning ionization) with atoms of a different species

$$A^* + B \rightarrow A + B^+ + e. \tag{7.23}$$

One of the simplest and most accurate techniques for investigating the collision processes involving metastable atoms is the pulsed afterglow method. In this technique a sample of metastable atoms is created in a cylindrical or spherical cell by means of a pulsed discharge and the decay of the metastable density is observed following the termination of the excitation pulse. By measuring the decay rate as a function of gas pressure it is often possible to obtain enough information to determine the individual contributions to the observed decay rate made by each of the processes listed above.

7.6.2. The pulsed afterglow technique

(a) Typical experimental apparatus. The apparatus required

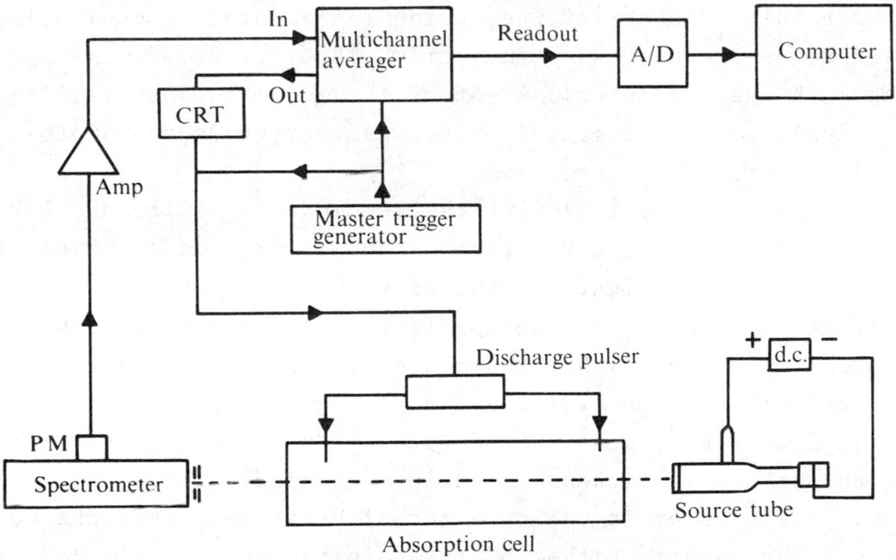

Fig.7.16. Schematic diagram of pulsed afterglow apparatus for the investigation of the collisional destruction of metastable atoms. (After Collins (1973).)

for a modern pulsed afterglow experiment is shown schematically in Fig.7.16. The heart of the system is the sample cell, usually cylindrical in shape, which is attached to a vacuum and gas handling system. In many experiments reproducible results can only be obtained if the cell is freed of adsorbed gases by careful bake-out and evacuation to 10^{-9} Torr before being refilled with specially purified gases. The discharge in the cell is created by a d.c. or r.f. power supply which is controlled by a master pulse generator. The metastable atoms created by the discharge are then detected by measuring the absorption of light in a transition connecting the metastable level with a higher radiating state. For instance, in helium afterglows the $2\ ^1S_0$-$3\ ^1P_1$ transition at 5016 Å and the $2\ ^3S_1$-$3\ ^3P_{0,1,2}$ multiplet at 3888 Å may be used to monitor the densities of the singlet and triplet metastables respectively. Light of the correct wavelength is obtained from a separate source tube containing the gas being investigated. Some of the precautions which must be observed when using the absorption monitoring technique are discussed in section 10.6.3. After passing through the sample cell light of the required spectral line is isolated with a monochromator and detected by a photomultiplier.

Immediately following the termination of the discharge pulse the metastable density and the absorption coefficient of the cell are high. As the density decays away in the afterglow, the absorption coefficient is reduced and the intensity of the light transmitted through the cell increases towards its maximum value. The afterglow signal may be recorded in real time either digitally by employing the photon counting and multichannel scaling technique described in section 7.4.5 or in analogue format by using a multichannel signal averager. Either of these instruments is simpler and makes more efficient use of the available signal than does the pulse-sampling method originally developed by Phelps and Pack (1955). By repetitively pulsing the discharge and triggering the channel address cycle of the memory device, a signal is accumulated with a relatively low noise level. Subsequent read-out and analysis of the data from the memory allows the effective decay rate of the metastable atoms to be determined. These data can then be interpreted in terms of various collision processes using the discussion which follows.

(b) Simplified model of the afterglow in pure gases. We consider first afterglows in gases consisting of a single atomic species. We assume that in the late afterglow the electron density is very small and we can therefore neglect the effects of collisions between metastable atoms and thermal electrons. We also assume that the density of metastable atoms created by the discharge is sufficiently low that we can neglect collisions involving a pair of metastable atoms. With these approximations the rate of change of the metastable atom density, $N_1(t)$, is given by

$$\frac{\partial N_1}{\partial t} = D_1 \nabla^2 N_1 - \alpha_0 N_0 N_1 - \beta_0 N_0^2 N_1. \qquad (7.24)$$

On the right-hand side of equation (7.24) the first term represents the loss by diffusion. D_1 is the diffusion coefficient for the metastable atoms and is inversely proportional to the pressure in the cell. The second term represents the destruction of the metastables by two-body collisions with ground state atoms of density N_0, which occurs at the rate $\alpha_0 N_0$. The third term gives the rate at which three-body collisions of the type given by equation (7.21) occur. Late in the afterglow it can be shown (Problem 7.10) that the metastable density decays exponentially with time according to

$$N_1(t) = N_1(0) \exp(-\Gamma_1 t). \qquad (7.25)$$

The effective decay rate, Γ_1, is given by

$$\Gamma_1 = D_1 \Lambda^{-2} + \alpha_0 N_0 + \beta_0 N_0^2 \qquad (7.26)$$

where Λ is the characteristic length of the lowest-order diffusion mode of the cell. For a cylindrical discharge cell of radius r and length l we have (McDaniel 1964, p.503)

$$\frac{1}{\Lambda^2} = \left(\frac{2\cdot 405}{r}\right)^2 + \left(\frac{\pi}{l}\right)^2 . \qquad (7.27)$$

It is now clear that the measurement of the effective decay rate Γ_1 over a suitable range of gas densities will allow the diffusion coefficient and the rate coefficients, α_0 and β_0, for two- and three-body collisions to be determined provided the dimensions of the discharge vessel are accurately known.

7.6.3. Afterglow experiments in pure helium. Some of the first systematic studies of discharge afterglows were made by Phelps and Molnar (1953) in argon, neon, and helium. Soon afterwards improvements in the techniques for measuring small, time-dependent absorption coefficients allowed Phelps (1955) to make a more detailed study of the collision pro-

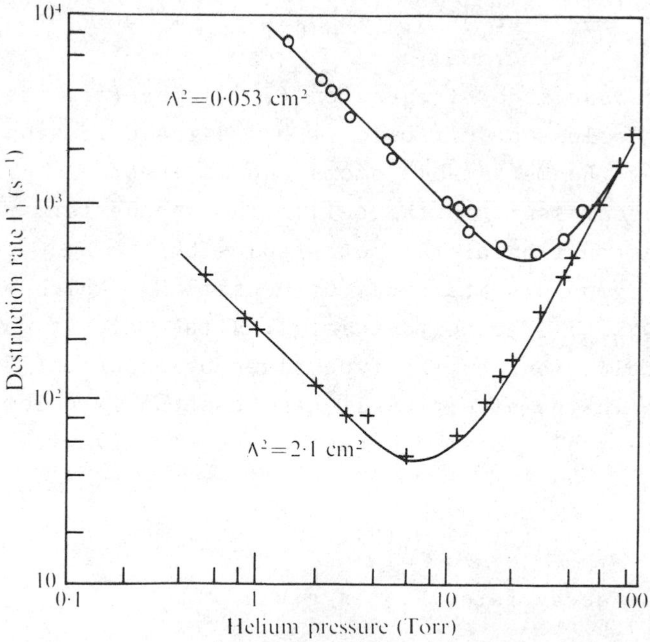

Fig. 7.17. Measured decay rate of the population of He(2 $^3S_1$) metastable atoms as a function of helium pressure at 300 K in two different absorption cells. (After Phelps (1955).)

cesses occurring in the helium afterglow. He found that late in the afterglow the decay of the $2\ ^1S_0$ metastable density was due to diffusion and two-body collisions while the collisional destruction of the $2\ ^3S_1$ metastable atoms occurred mainly in three-body processes. The effective decay rate of the triplet metastable level measured as a function of helium pressure is shown in Fig.7.17 for two different absorption cells. At pressures below 3 Torr the diffusion term in equation (7.26) dominates and the decay rate varies inversely proportionally with pressure, while above 30 Torr the collisional destruction term $\beta_0 N_0^2$ is the most important. A detailed analysis of Fig.7.17 and similar data for the $2\ ^1S_0$ level resulted in the values of the diffusion and collisional rate coefficients listed in Table 7.4. For the singlet metastables, the two-body rate coefficient can be expressed in terms of a velocity-averaged collision cross-section, $\alpha_0=\sigma(\bar{v})\bar{v}$, having the value $\sigma(\bar{v})=3\times10^{-20}$ cm^2 at 300 K. Measurements of the diffusion coefficients and collision cross-sections over a wide range of temperatures can give valuable information about the interatomic forces between metastable and ground-state

TABLE 7.4.

Metastable atom diffusion and collisional destruction coefficients in helium at 300 K

Helium metastable level	Diffusion coefficient D_1 (cm^{2}s^{-1} at 1 Torr)	Rate coefficient for two-body collisions, α_0 (cm^{3}s^{-1})	Rate coefficient for three-body collisions, β_0 (cm^{6}s^{-1})
$2\ ^1S_0$	440 ± 50	5×10^{-15}	—
$2\ ^3S_1$	470 ± 25	—	$2{\cdot}5\times10^{-34}$

atoms as Fitzsimmons *et al.* (1968) have shown.

It turns out that helium is a particularly simple system to understand since the nearest short-lived excited states, $2\ {}^3P_{0,1,2}$ and $2\ {}^1P_1$, lie 1·1 and 0·6 eV respectively above the $2\ {}^3S_1$ and $2\ {}^1S_0$ metastable levels. The transfer of metastable atoms into these radiating levels by thermal collisions is therefore extremely unlikely and can safely be ignored in helium. Unfortunately the effects of these collisions can no longer be neglected in systems like mercury and the other noble gases. There the radiating and metastable states are strongly coupled by collisions and in addition radiation trapping occurs in both the strong and weak emission lines. Although the afterglow technique is still applicable, the experimental results are considerably more difficult to interpret. We therefore move on to a discussion of collisions of the type given by equation (7.22), which are important in the helium-neon laser.

7.6.4. Afterglow experiments in helium-neon mixtures. When the afterglow of a helium-neon gas mixture is studied, it is found that the helium metastable population decays very much more rapidly than in pure helium at the same pressure. An examination of the helium and neon term diagram shown in Fig.11.3 quickly reveals that the $He(2\ {}^3S_1)$ metastable level lies very close to the 2s levels of neon. The destruction of the helium metastables is therefore accomplished mainly by excitation transfer reactions of the type

$$He(2\ {}^3S_1) + Ne \rightarrow He(1\ {}^1S_0) + Ne(2s). \qquad (7.28)$$

Since the rate of metastable destruction by neon atoms usually greatly exceeds the effect of helium-helium collisions, we neglect the last two terms in equation (7.24) and rewrite that equation as

$$\frac{\partial N_1}{\partial t} = D_1\ \nabla^2 N_1 - \sigma_T(\overline{v})\overline{v}\ n_0 N_1 \qquad (7.29)$$

where $\sigma_T(\overline{v})$ is the total destruction cross-section for collisions between helium metastables and ground state neon atoms of density n_0. It is again a straightforward exercise to show that the helium metastable density in the afterglow decays exponentially with an effective decay rate given by

$$\Gamma_1' = D_1 \Lambda^{-2} + \sigma_T(\bar{v})\bar{v}\, n_0. \qquad (7.30)$$

The density of excited neon atoms in one of the 2s levels, n_k, is determined by the competition between excitation transfer from the helium metastables and radiative decay to the lower 2p levels:

$$\frac{\partial n_k}{\partial t} = \sigma_k(\bar{v})\bar{v}\, n_0 N_1 - A_k n_k \qquad (7.31)$$

where $\sigma_k(\bar{v})$ is the velocity-averaged cross-section for the excitation transfer to the particular level k. In most cases the rate of decay by spontaneous emission, $A_k = \sum_i{}' A_{ki}$, is very much faster than the collisional excitation rate and the decay of excited neon atoms in the afterglow mirrors that of the helium metastable population:

$$n_k(t) \approx \frac{\sigma_k(\bar{v})\bar{v}\, n_0 N_1(0)}{A_k - \Gamma_1'} \exp(-\Gamma_1' t). \qquad (7.32)$$

The cross-section for the helium-neon excitation transfer process was determined by Javan *et al.* (1961) by observing the rate of decay of the densities of excited $He(2\,^3S_1)$ and Ne(2s) atoms in the afterglow of a pulsed r.f. discharge. The helium atoms were monitored by absorption on the $2\,^3S_1 - 2\,^3P_{0,1,2}$ infrared transition, while the light emitted in certain of the 2p-2s neon lines enabled the decay of the excited neon atoms to be recorded. It was found that the decay of the Ne(2s) population was identical to that of the helium triplet metastable population, as equation (7.32) predicts. Then by studying the decay rate of the Ne(2s) levels as a function of neon partial pressure, the total velocity-averaged cross-section for inelastic collisions was determined to be

$$\sigma_T(\bar{v}) = (3\cdot7 \pm 0\cdot5) \times 10^{-17}\ \text{cm}^2$$

at 300 K. Most of these collisions result in excitation transfer to the Ne(2s) levels and are responsible for the population inversion in the helium-neon laser systems discussed in detail in section 11.4.2.

7.6.5. Afterglow studies of Penning ionization collisions. Helium atoms in the 2 $^3S_1$ metastable level possess almost 20 eV of internal energy, an amount greater than the ionization potential of any other neutral atom. Consequently in collisions with atoms of another species there is always a certain probability that the energy will be used in an ionization process of the type described by equation (7.23). This is an example of a Penning ionization collision. If the collision involves atoms of metallic elements having low ionization potentials, e.g. zinc and cadmium, enough energy is available to raise the metallic ion simultaneously to an excited state:

$$He(2\ ^3S_1) + Zn(3d^{10}4s^2) \rightarrow He(1\ ^1S_0) + Zn^+(3d^9 4s^2) + e \tag{7.33}$$

Pulsed afterglow studies of Penning ionization excitation have been made by Collins *et al.* (1971 *a*,*b*) and are described in more detail in a review article by Collins (1973). The apparatus employed was similar to that shown in Fig.7.16, with the exception that the absorption tube contained zinc or cadmium pellets and was surrounded by an oven to maintain a suitable metal vapour density. The decay rate of the spontaneous emission from excited levels of the zinc ions in a helium-zinc afterglow was studied as a function of the zinc atom density. Typical experimental results obtained on the ZnII transitions at 5894 Å and 7478 Å are shown in Fig.7.18. The helium triplet metastable decay rate was measured in the same tube by absorption monitoring. Fig. 7.18 shows that the decay rates of the He(2 $^3S_1$) density and spontaneous ZnII emission at 5894 Å and 7478 Å are identical over a wide range of zinc atom density. This confirms that Penning ionization collisions

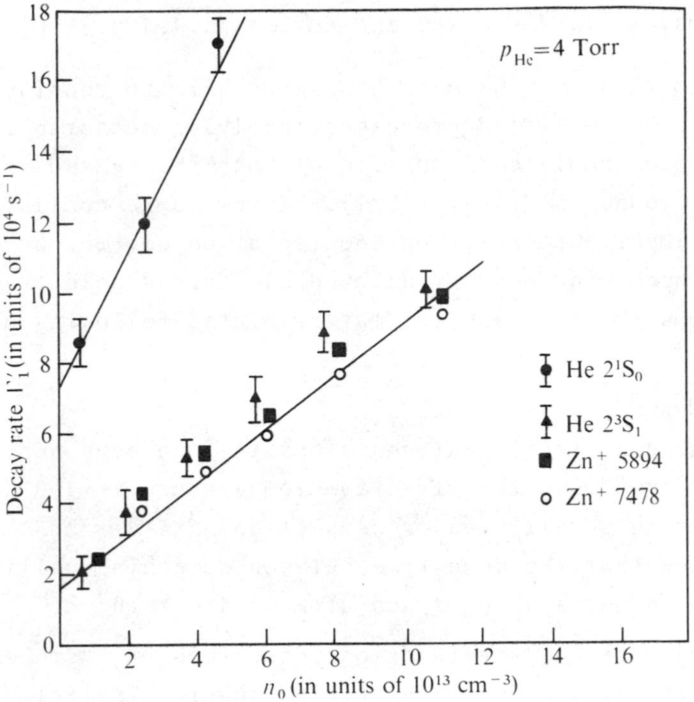

Fig.7.18. Decay rate of spontaneous emission from levels of the $3d^9 4s^2$ configuration of Zn^+ measured in a He-Zn afterglow as a function of zinc vapour density. The decay of the helium metastable atoms in the $2\ ^1S_0$ and $2\ ^3S_1$ levels were measured by absorption monitoring in the same afterglow. (After Collins (1973).)

with helium triplet metastables are providing the main source of excitation to the $3d^9 4s^2$ levels of the zinc ions in the afterglow. The experiments can therefore be analysed in terms of the modified afterglow equations discussed in section 7.6.4. From the slope of the curves in Fig.7.18 the total velocity-averaged cross-section for the destruction of He($2\ ^3S_1$) metastables by zinc atoms is determined to be

$$\sigma_{\text{Penning}}(\bar{v}) = 4 \times 10^{-15}\ \text{cm}^2.$$

Penning ionization collisions provide an efficient means of selective excitation and their importance in metal vapour laser systems is discussed in section 11.4.4.

7.6.6. Conclusion. We have had space here to consider only a few of the collision processes involving metastable atoms. A much more complete discussion of both theory and experiment is to be found in Massey (1971). Interatomic collisions also have an important effect on the radiation emitted by excited levels which can decay by allowed electric dipole transitions. This forms the main subject matter of the following chapter.

Problems

7.1. The density of hydrogen atoms in a gaseous nebula is 10^3 cm^{-3} and the effective temperature is 100 K. Assuming a collision cross-section of $1{\cdot}4 \times 10^{-18}$ cm^2, show that the mean time between quenching collisions for a metastable oxygen atom is $4{\cdot}8 \times 10^9$ s.

7.2. Generate the wavefunctions describing a hydrogen atom in the ground-state hyperfine sub-levels $|F=1,M_F\rangle$ and $|F=0,M_F=0\rangle$ using equation (3.55) and Table 3.3. Hence show that each of the matrix elements of the magnetic dipole moment operator between these sub-states has the value μ_B. Evaluate equation (7.8) and calculate the radiative lifetime, in years, of the upper hyperfine level.

7.3. Using the hydrogenic wavefunctions listed in Tables 3.1 and 3.2, show that the transition probability of the 1s - 3d electric quadrupole transition in atomic hydrogen has the value $2{\cdot}27 \times 10^3$ s^{-1}.

In practice the 3d level decays preferentially by allowed electric dipole radiation to the 2p level.

7.4. Using the fact that the electric quadrupole moment operator is a symmetric second rank tensor and the magnetic dipole moment operator transforms as an axial vector, derive the selection rules for magnetic dipole and electric quadrupole radiation given in Table 7.1.

7.5. Using Tables 3.1, 3.2, and equation (7.1) show that the transition probability for spontaneous electric dipole radiation on the $2\,^2S_{1/2}$-$2\,^2P_{1/2}$ transition of atomic hydrogen is $2{\cdot}67 \times 10^{-10}$ s^{-1}. The Lamb shift frequency in hydrogen is 1058 MHz.

7.6. The transition probability for the $1\,^2S_{1/2}$-$2\,^2P_{1/2}$ Lyman α line in He^+ is 1×10^{10} s^{-1}. Using the lower limit for the measured lifetime of metastable $He^+(2\,^2S_{1/2})$ given in Table 7.2., show that the magnitude of the parity-mixing coefficient β introduced in equation (7.17) is less than $4{\cdot}7 \times 10^{-5}$.

7.7. The $2\,^2S_{1/2}$ and $2\,^2P_{1/2}$ states of hydrogenic systems will also be mixed, as in equation (7.17), if the atoms are subjected to a finite static electric field $\underline{E}$. In the case of ionized helium show that the effect of this Stark mixing of the $2\,^2S_{1/2}$ state is to induce decay by single-photon electric dipole transitions at a rate given by

$$\Gamma_{\text{Stark}} = \frac{\Gamma_{2P}|V|^2}{\hbar^2\omega^2} = 6{\cdot}21 \times 10^{-3}\,E^2\,\text{s}^{-1}$$

where $\hbar\omega$ is the energy corresponding to the Lamb-shift frequency of 14·04 GHz and the perturbation matrix element is given by

$$\frac{V}{\hbar} = \frac{\langle 2S|e\underline{E}.\underline{r}|2P\rangle}{\hbar} = \frac{\sqrt{3}a_0 eE}{Z\hbar} = 6{\cdot}96 \times 10^4\,E.$$

(In these expressions E is expressed in Vm^{-1}.) Hence show that if Γ_{Stark} is to be less than 1 per cent of the theoretical two-photon decay rate of 527 s^{-1}, then $|\underline{E}|$ must be less than 29 V m^{-1}.

In neutral hydrogen the restrictions on $|\underline{E}|$ are very much more severe owing to the smaller values of both the Lamb shift and the two-photon decay rate.

7.8. Show that a photon emitted on the Lyman α line of hydrogenic Ar^{+17} has an energy of approximately 3·3 keV.

(Calculations including relativistic corrections show that the high-energy limit of the continuous two-photon emission spectrum of $Ar^{+17}(2\ ^2S_{1/2})$ lies at 3·34 keV).

7.9. The resonance and metastable levels of mercury, 6s6p $^1P_1$ and $^3P_{2,1,0}$ lie respectively 54 050, 41 179, 39 412, and 34 781 cm^{-1} above the ground state $6s^2\ ^1S_0$. The 6 $^3P_2$ level can decay by either magnetic dipole or electric quadrupole radiation to the 6 $^3P_1$ and 6 $^3P_0$ levels respectively. Show that the transition probabilities of these forbidden lines are of the order of 1·33 s^{-1} and $2 \times 10^{-3}\ s^{-1}$ respectively.

7.10. By writing the metastable density in equation (7.24) in the form

$$N_1(x,y,z,t) = \Psi(x,y,z)T(t)$$

show that the equation can be separated into:

$$\dot{T}(t) + \Gamma T(t) = 0$$

and

$$\nabla^2\Psi(x,y,z) + k^2\Psi(x,y,z) = 0$$

where $\Gamma = k^2D+\alpha_0 N_0+\beta_0 N_0^2$ and k is a constant. By further separation of the variables in cylindrical coordinates, show that the general solution for the spatial distribution of metastable atoms is given by

$$\Psi(r,\theta,z) = \sum_{m,n,q} C_{mnq}\ J_m(nr) \begin{matrix}\sin\\ \cos\end{matrix}\ m\theta \begin{matrix}\sin\\ \cos\end{matrix}\ qz$$

where $n^2=k^2-q^2$, $J_m(\xi)$ is the Bessel function of order m and C_{mnq} is a constant determined by the distribution at t=0. If $\Psi(r,\theta,z)$ is zero at the boundaries of a cylinder of radius r and length l, show that the characteristic length of the lowest-order diffusion mode is given by equation (7.27). ($J_0(\xi)=0$ at $\xi=2{\cdot}405$).

References

Breit, G. and Teller, E. (1940). *Astrophys.J.* 91, 215.

Collins, G.J. (1973). *J. Appl.Phys.* 44, 4633.

_____ , Jensen, R.C., and Bennett, W.R. Jr. (1971a). *Appl.Phys.Lett.* 18, 282.

_____ , _____ , _____ . (1971b). *Appl.Phys.Lett.* 19, 125.

Corney, A. (1973). *Nucl.Instrum.Meth.* 110, 151.

_____ and Williams, O.M. (1972). *J.Phys.B.* 5, 686.

Drake, G.W.F. (1971). *Phys.Rev.* A3, 908.

_____ , Victor, G.A., and Dalgarno, A. (1969). *Phys. Rev.* 180, 25.

Fitzsimmons, W.A., Lane, N.F., and Walters, G.K. (1968). *Phys.Rev.* 174, 193.

Gabriel, A.H. and Jordan, C. (1969). *Nature, Lond.* 221, 947.

Göppert-Mayer, M. (1931). *Annln.Phys.* 9, 273.

Jacobs, V. (1971). *Phys.Rev.* A4, 939.

Javan, A., Bennett, W.R. Jr., and Herriott, D.R. (1961). *Phys.Rev.Lett.* 6, 106.

Kocher, C.A., Clendenin, J.E., and Novick, R. (1972). *Phys. Rev.Lett.* 29, 615.

Lipeles, M., Novick, R., and Tolk, N. (1965). *Phys.Rev.Lett.* 15, 690.

Marrus, R. and Schmieder, R.W. (1970). *Phys.Rev.Lett.* 25, 1689.

_____ and _____ (1972). *Phys.Rev.* A5, 1160.

Moos, H.W. and Woodworth, J.R. (1973). *Phys.Rev.Lett.* 30, 775.

Novick, R. (1969). *Physics of the one- and two- electron atoms* (Eds.F.Bopp and H. Kleinpoppen) p.296. North-Holland, Amsterdam.

Phelps, A.V. (1955). *Phys.Rev.* 99, 1307.

_____ and Molnar, J.P. (1953). *Phys.Rev.* 89, 1202.

Phelps, A.V. and Pack, J.L. (1955). *Rev.Scient.Instrum.* 26, 45.

Prior, M.H. (1972). *Phys.Rev.Letts.* 29, 611.

_____ and Shugart, H.A. (1971). *Phys.Rev.Letts.* 27, 902.

Schapiro, J. and Breit, G. (1959). *Phys.Rev.* 113, 179.

Spitzer, L. and Greenstein, J.L. (1951). *Astrophys.J.* 114, 407.

Van Dyck, R.S., Johnson, C.E., and Shugart, H.A. (1970). *Phys.Rev.Letts.* 25, 1403.

_____, _____, and _____ (1971). *Phys.Rev.* A4, 1327.

_____, _____, and _____ (1972). *Phys.Rev.* A5, 991.

General references and further reading

A good review of forbidden transitions and their importance in astrophysics is that by

Garstang, R.H. (1962). *Atomic and molecular processes* (Ed. D.R. Bates) p.1. Academic Press, New York.

The properties of hydrogenic and helium-like systems are discussed in detail by

Bethe, H.A. and Salpeter, E.E. (1957). *Quantum mechanics of one- and two-electron atoms.* Springer-Verlag, Berlin.

The collision processes involving metastable atoms and molecules are considered in great detail by

Massey, H.S.W. (1971). *Electronic and ionic impact phenomena*, Vol.3. Oxford University Press, London.

and

McDaniel, E.W. (1964). *Collision phenomena in ionized gases.* Wiley & Sons Inc., New York.

8

The width and shape of spectral lines

So far we have assumed that the radiation emitted in a given transition is strictly monochromatic and that the spectral line recorded by an ideal spectrograph would be infinitely sharp. Normally, however, the lines observed with a simple spectrograph have an easily observable width which is directly related to the width of the entrance slit. If the width of this entrance slit is progressively reduced, the width of the spectral line eventually reaches a limiting value which is determined by diffraction or aberration effects in the instrument. The linewidth can be reduced still further by increasing the size of the grating, so increasing the resolving power, until, when instruments of the highest resolution are used, it is found that each spectral line consists of an intrinsic distribution of frequencies about the line centre, ω_{ki}. The profile of every spectral line has a finite width and characteristic shape which are determined by the conditions existing in the source, and it is this profile which we now wish to consider. In the sections which follow we shall assume that an instrument of infinite resolving power is being used unless otherwise stated.

We start this chapter by considering the width of spectral lines emitted by isolated stationary atoms. We again make use of the classical model of the atom and show how the finite lifetime of excited atoms leads to the so-called natural width. Although in the visible region of the spectrum the natural width of spectral lines is significant only in a few rather special cases, it does set a theoretical limit on the attainable accuracy of spectroscopic measurements. We next consider how interatomic forces affect the shape of the emitted line. These are responsible for the effect of pressure broadening which becomes significant at pressures above 10-100 Torr, and which is the dominant broadening mechanism in high temperature arcs and plasmas.

In the ultraviolet and visible region of the spectrum, however, most laboratory sources emit lines whose shape and width are determined mainly by the Doppler effect due to the random thermal motion of the emitting atoms.

The width and shape of spectral lines are of interest since:

(i) they can provide information about temperature, density, and composition existing in the source;

(ii) detailed calculations of the interaction of radiation with atoms require an accurate knowledge of the line profile;

(iii) the line profile is important in determining many of the characteristics of gas lasers.

In this chapter we shall be concerned only with the shape of lines emitted by an optically thin source. The change in the observed intensity distribution which can occur as a result of self-absorption by ground-state or metastable atoms will be discussed in Chapter 10.

8.1. The natural or radiative lineshape

<u>8.1.1. The radiation reaction force.</u> We return to a consideration of the classical atomic model which was introduced in sections 4.1 and 4.2. We found that there was a loss of energy in the form of radiation which occurred slowly over many cycles of the electron's motion. However, this loss of energy was not taken into account in the mechanical equation of motion of the electron. This situation can be remedied by introducing a radiation reaction force, F_r, such that the work done by the reaction force in one cycle of the oscillation is equal to the energy emitted into the radiation field:

$$\int F_r \, dz = \int F_r \, \dot{z} \, dt = -\frac{e^2}{6\pi\varepsilon_0 c^3} \int (\ddot{z})^2 \, dt \qquad (8.1)$$

where we have introduced the instantaneous rate of radiation obtained by using equation (4.3) and integrating the Poynting vector over the surface of a sphere, radius r. Now setting $\dot{z}=u$ and integrating the right-hand side of equation (8.1) by parts, we have

$$\int_{t_1}^{t_2} F_r\, u\, dt = -\frac{e^2}{6\pi\varepsilon_0 c^3}\left\{[u\dot{u}]_{t_1}^{t_2} - \int_{t_1}^{t_2} u\,\ddot{u}\, dt\right\}. \quad (8.2)$$

If the interval t_2-t_1 is an integral multiple of the period of the cyclic motion then the first term on the right-hand side of equation (8.2) is zero. Thus energy will be conserved on average if

$$F_r = \frac{e^2\ddot{u}}{6\pi\varepsilon_0 c^3}. \quad (8.3)$$

8.1.2. The classical lineshape. The modified equation of motion for the oscillating electron may now be written in the form

$$\ddot{z} + \omega_0^2 z = \frac{e^2}{6\pi\varepsilon_0 c^3 m}\dddot{z} \quad (8.4)$$

It can be easily shown (Problems 8.1 and 8.2) that the radiation reaction force is small compared with the binding force at the angular frequencies in which we are interested. Thus the equation of motion of the electron oscillator may, to a good approximation, be written in the form

$$\ddot{z} + \gamma\dot{z} + \omega_0^2 z = 0 \quad (8.5)$$

where $\gamma = e^2\omega_0^2/6\pi\varepsilon_0 c^3 m$ is the classical decay rate derived in section 4.2. The solution for $\gamma \ll \omega_0$ has the form of a damped oscillation:

$$z = z_0 \exp(-\gamma t/2)\exp(-i\omega_0 t). \quad (8.6)$$

It follows from equation (2.70) that the electric field radiated by the classical atom has a similar time-dependence:

$$E(t) = \begin{cases} E(0)\exp\{-i(\omega_0 - i\gamma/2)t\} & t \geq 0 \\ 0 & t < 0 \end{cases} \quad (8.7)$$

Since the emitted wave is no longer of infinite extent, it

is no longer strictly monochromatic and may be resolved into its different component frequencies by taking the Fourier transform:

$$E(\omega) = \left(\frac{1}{2\pi}\right)^{\frac{1}{2}} \int_{-\infty}^{\infty} E(t) \exp(i\omega t)\, dt \tag{8.8}$$

Substituting from equation (8.7) we have

$$\begin{aligned} E(\omega) &= \frac{E(0)}{\sqrt{(2\pi)}} \int_{0}^{\infty} \exp[i\{(\omega-\omega_0) + i\gamma/2\}t]\, dt \\ &= \frac{-E(0)/\sqrt{(2\pi)}}{i\{(\omega-\omega_0)+i\gamma/2\}} . \end{aligned} \tag{8.9}$$

The distribution of intensity, $I(\omega)$, that would be obtained using a spectrograph of infinite resolving power is proportional to $|E(\omega)|^2$ and is given by

$$\begin{aligned} I(\omega) &= I_0 \frac{\gamma/2\pi}{(\omega-\omega_0)^2+\gamma^2/4} \\ &= I_0\, \mathcal{L}(\omega-\omega_0,\gamma). \end{aligned} \tag{8.10}$$

The intensity distribution has been normalised by introducing I_0, the total energy radiated in the line, where

$$I_0 = \int_0^{\infty} I(\omega)\, d\omega.$$

This distribution is called the natural or radiative line-shape and is illustrated in Fig.8.1. It is a bell-shaped curve known as the Lorentzian distribution, $\mathcal{L}(\omega-\omega_0,\gamma)$, whose full width at half maximum intensity, (FWHM), is given by

$$\Delta\omega_{1/2} = \gamma = 1/\tau_{Cl}. \tag{8.11}$$

This is exactly the width which is predicted by the application of the uncertainty principle:

$$\Delta E.\ \Delta t = \hbar\Delta\omega.\tau = \hbar. \tag{8.12}$$

When expressed in terms of wavelength, the classical line width is a universal constant

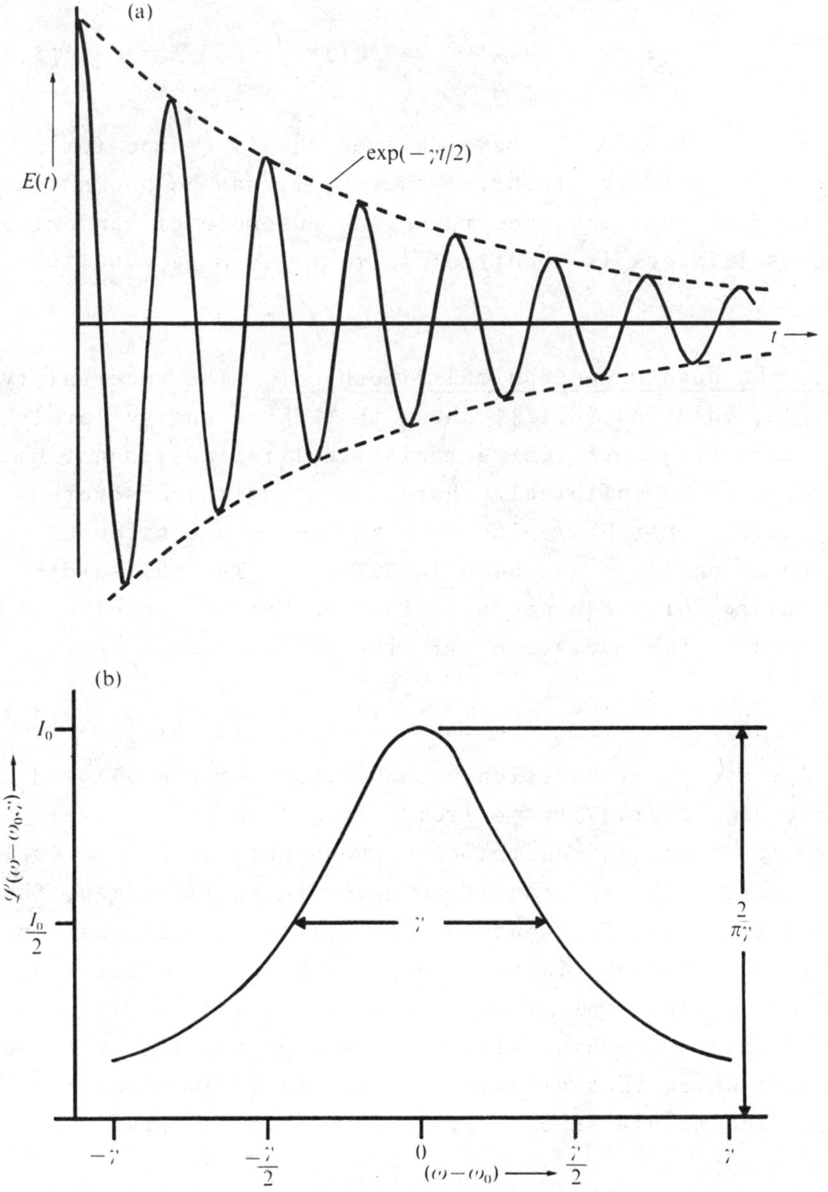

Fig.8.1. (a) Electric field radiated by classical dipole oscillator. (b) Spectral distribution of this radiation illustrating the normalized Lorentzian line profile defined by equation (8.10).

$$\Delta\lambda_{1/2} = \frac{2\pi c\Delta\omega_{1/2}}{\omega_0^2} = 1{\cdot}18 \times 10^{-4}\ \text{Å}. \qquad (8.13)$$

In this section we have derived the lineshape for a single classical oscillator. However, it can be shown (Problem 8.4) that the spectrum of an ensemble of randomly phased oscillators is identical to that given by equation (8.10).

8.1.3. The quantum-mechanical lineshape. The uncertainty principle, equation (8.12), shows that those energy levels of an atom which have finite radiative lifetimes cannot be considered to be infinitely sharp. Instead the discrete energy levels must be replaced by maxima in a continuous spectrum of energies as shown in Fig.8.2. The half-width of the energy or frequency distribution for the level k is determined by the radiative lifetime of that level

$$\Gamma_k = \Delta\omega_k = \Delta E_k/\hbar = 1/\tau_k = \sum_i{}' A_{ki} \qquad (8.14)$$

where A_{ki} are the transition probabilities of the allowed electric dipole transitions from k to all possible lower levels, i. Since in general both the upper and lower levels of an electric dipole transition have finite lifetimes, the uncertainty in the frequency of the emitted quanta must include the uncertainty in the energy of both the upper and the lower levels. This frequency distribution was first obtained quantum mechanically by Weisskopf and Wigner (1930) in a paper which is available in translation in Hindmarsh (1967). The result is once again a Lorentzian curve

$$I_{ki}(\omega) = I_0 \frac{\Gamma_{ki}/2\pi}{(\omega-\omega_{ki})^2+\Gamma_{ki}^2/4} \qquad (8.15)$$

with a half-width (FWHM) given by $\Delta\omega_{1/2}$ where

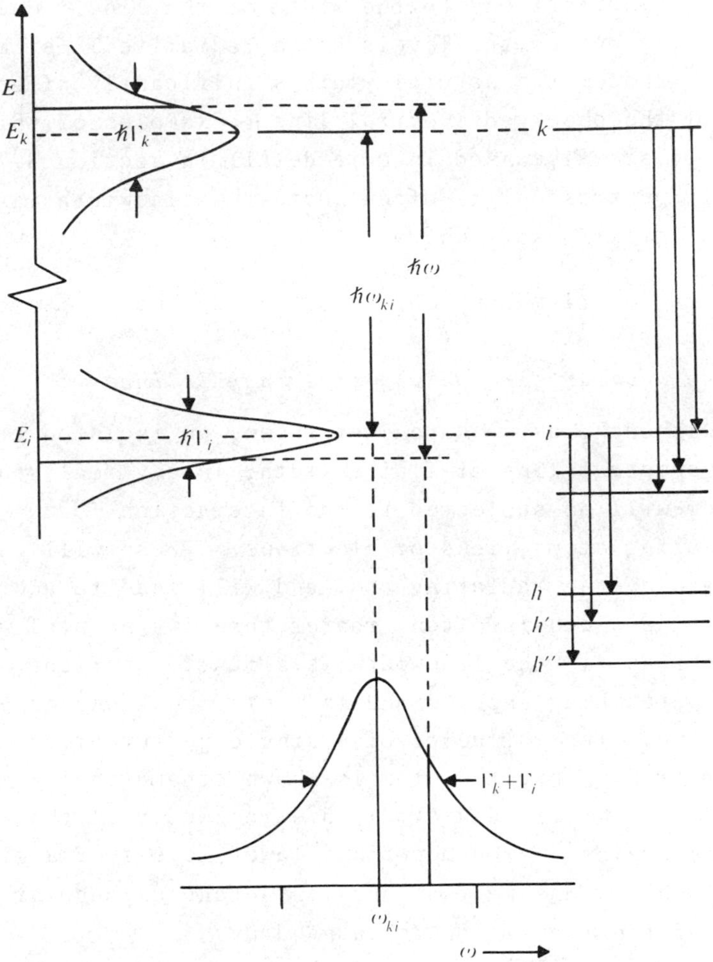

Fig.8.2. The finite width of excited atomic levels and the natural lineshape of radiation emitted spontaneously on the transition $k \to i$.

$$\Delta\omega_{1/2} = \Gamma_{ki} = \Gamma_k + \Gamma_i = \sum_j{}' A_{kj} + \sum_h{}' A_{ih}. \tag{8.16}$$

The natural linewidth of a typical transition, $\tau_k \approx 10^{-8}$ s, would be $\Gamma/2\pi \approx 16$ MHz which is generally small in comparison to other contributions to the width of the line. Only in cases where one of the levels has a radiative lifetime $\tau < 10^{-9}$ s does the natural width significantly affect the shape of the observed spectral line. Examples of this situation are discussed in more detail in section 8.7.3 below. Spectroscopists often quote the linewidth in units known as milliKaysers where

$$1 \text{ mK} = 10^{-3} \text{ cm}^{-1} \cong 30 \text{ MHz}$$

8.2. The pressure broadening of spectral lines

Although, as we have just shown, an isolated atom will emit a spectral line of finite width, in any real source the atom will be subjected to the interaction forces of neighbouring atoms, ions or electrons. These will perturb the state of the radiating atom and will lead to a broadening of the line which is often greater than its natural width. This increase in the linewidth is a function of the density of the perturbing species and is therefore known as pressure broadening. The influence of a single perturber at a distance r from an excited atom is shown schematically in Fig.8.3. If $\Delta V_k(r)$ and $\Delta V_i(r)$ are respectively the changes in the energies of the upper and lower levels of a given transition, we assume that the instantaneous angular frequency of the transition is changed by

$$\Delta\omega_{ki} = \{\Delta V_k(r) - \Delta V_i(r)\}/\hbar . \quad (8.17)$$

8.2.1. Interatomic forces. It is often possible to represent the long-range interaction between the excited atom and a perturber by a potential of the form

$$\Delta V_k(r) = C_n^k/r^n \quad (8.18)$$

where C_n^k is a constant which depends on the excited level involved and also on the perturbing species. The different values of the integer n which are of interest here are:

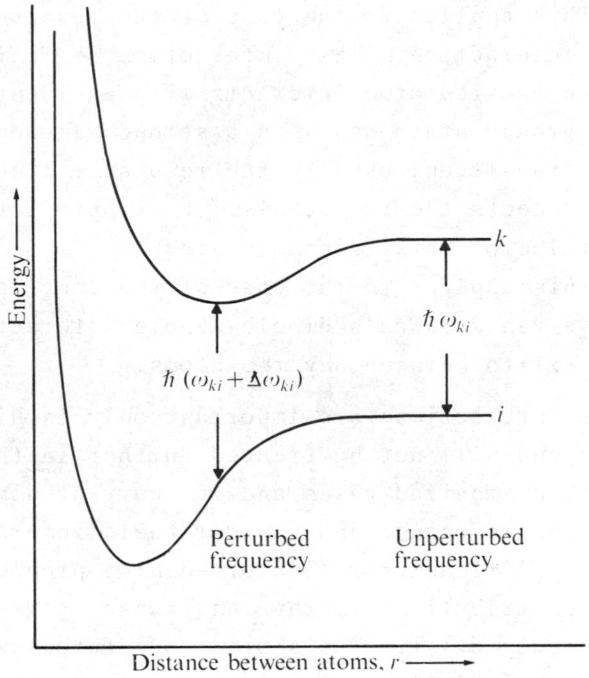

Fig.8.3. Energy levels of an atom as a function of the distance from a perturbing atom or molecule. The levels k and i do not remain parallel and the Franck-Condon principle predicts that the emitted radiation is shifted away from the unperturbed frequency ω_{ki}.

(i) n=2. This applies to the case of hydrogen and hydrogenic ions in the electric fields produced by other ions or electrons. These give rise to a Stark shift of the energy levels which depends linearly on the electric field strength.

(ii) n=4. This describes the Stark broadening in helium and other systems where the splitting is a quadratic function of the electric field of the perturbers.

(iii) n=3. This applies to the case of the resonance dipole-dipole interaction. This interaction is finite only when the excited atom interacts with an identical atom in the ground state and when a strong allowed electric dipole transition, usually the resonance line of the atom, connects the two levels. C_3^k is proportional to the f-value of this resonance line.

(iv) n=6. This applies to the case of the long-range attractive van der Waals dipole-dipole interaction which always exists between any two atoms.

The first two interactions are important only in highly-ionized gases and will not be treated further in this book. Turning now to un-ionized gases and low current discharges, we find that the resonance and van der Waals interactions are the most significant for line broadening problems, although other contributions to the long-range attractive forces undoubtedly exist. The short-range forces which are responsible for chemical bonding or interatomic repulsion have been omitted from the list because, unlike the long-range forces, they are difficult to calculate and cannot be expressed by any general formula. However, in the case of spectral lines broadened by collisions with hydrogen, helium, and neon, these repulsive forces play a very significant part.

8.2.2. The quasi-static and the impact approximations. The representation of the interatomic forces by equation (8.18) refers to the simple case of a binary system of an excited atom and a single perturber having a fixed separation and orientation. In general, however, the excited atom may interact with several perturbers and it is necessary to average over the different orientations and paths of these perturbers. This averaging can only be performed satisfactorily in two limiting cases known as the quasi-static and the impact or phase shift approximations respectively. We will now derive qualitative criteria which define the range of validity of these two approximations.

From the properties of the Fourier transform, equation (8.8), we see that the shape of the line profile at a frequency separation $\Delta\omega=\omega_0-\omega$ from line centre is determined by the wavetrain emitted during the time interval Δt, where $\Delta t \approx 1/\Delta\omega$. If we imagine the atom to emit a wavetrain until perturbed by a strong collision, the time of interest, Δt, will be the mean time between collisions, T_c. This time must be compared with the duration of one collision t_c. It is often a valid approximation to consider the perturber moving past the emitting atom on a classical trajectory. For neutral species these paths would be approximately straight lines, whereas for ions or electrons they would be hyperbolae. One can then consider the radiating atom to be perturbed only during the duration of the collision, defined by

$$t_c = \rho/\bar{v} \tag{8.19}$$

where ρ is the distance of closest approach and $\bar{v}$ is the mean velocity at this point.

When the duration of a collision is much greater than the time between collisions, the line broadening is produced while the perturber moves only a short distance. It is therefore possible to ignore the motion of the perturbers completely and we have the quasi-static approximation:

$$\text{Quasi-static approximation} \quad t_c \gg 1/\Delta\omega \approx T_c. \tag{8.20}$$

The excited atom is assumed to emit a wave shifted in frequency by equation (8.17) and the lineshape for the sample as a whole is obtained by averaging over the probability distribution of the stationary perturbers. This approximation is used to calculate the Stark broadening produced by ions in a plasma and for neutral atom broadening at pressures $>$ 100 Torr. From equations (8.19) and (8.20) we see that the quasi-static approximation is likely to be better at high densities, where T_c is short, and at low temperatures, where $\bar{v}$ is small.

At the other extreme, where the duration of the collision is short compared with the time between collisions, the impact approximation can be used:

$$\text{Impact approximation} \quad t_c \ll 1/\Delta\omega \approx T_c. \tag{8.21}$$

The phase shift produced by one collision is calculated and the result is averaged over all impact parameters. The impact approximation is used to describe the line broadening at pressures from 0-100 Torr. From equations (8.19) and (8.20) we see that the impact approximation should work best at low pressures, where T_c is long, and high temperatures, where $\bar{v}$ is large.

However, it should be remembered that the time of interest is determined by the frequency displacement from line centre, $\Delta\omega$. It is therefore possible that the wings of the line profile have to be calculated by the quasi-static approximation while the profile near the line centre is described by the impact approximation.

<u>8.2.3. Line broadening in the quasi-static approximation.</u> We consider first the interaction of an excited atom at r=0 with a single stationary perturber. The probability that the perturber lies in the spherical shell between r and r+dr is given by

$$P(r)\, dr = \frac{4\pi r^2 dr}{V} \tag{8.22}$$

where V is the volume of the enclosure. From equations (8.17) and (8.18) the excited atom will emit a wave whose frequency is shifted from the line centre by an amount

$$\Delta\omega = C'_n/r^n \tag{8.23}$$

where $C'_n = (C^k_n - C^i_n)/\hbar$. Substituting for r in equation (8.22), we find that wings of the line profile are given by the probability distribution

$$P(\Delta\omega)\, d(\Delta\omega) = \frac{4\pi C'^{3/n}}{nV(\Delta\omega)^{(n+3)/n}}\, d(\Delta\omega). \tag{8.24}$$

This result can be easily modified to describe an assembly of N perturbing particles provided that the actual frequency shift is attributed only to the effect of the nearest neighbour. However, this is only satisfactory for those interactions which have a relatively short range, i.e. $n \geq 6$. For Stark broadening, $n=2$ and 4, it is necessary to include the simultaneous effect of all the perturbers, which greatly complicates the calculation. Experimental results illustrating the application of the quasi-static theory will be discussed in section 8.7.5 below.

8.2.4. The development of impact approximation theories.

(a) The Lorentz theory. The impact approximation (Problem 8.5) was introduced in its simplest form by Lorentz (1906). He assumed that an atom excited at the instant $t=0$ would continue to radiate according to equation (8.7) until the wave-train was terminated at the instant of collision, $t=T_i$. The spectral distribution of this truncated damped harmonic oscillation may be obtained using the technique developed in section 8.1.2 with the result:

$$I(\omega)_i \propto \frac{1+\exp(-\gamma T_i)-2\exp(-\gamma T_i/2)\cos(\omega_0-\omega)T_i}{(\omega-\omega_0)^2+\gamma^2/4} \qquad (8.25)$$

For $\gamma \approx 0$ this distribution is identical to the Fraunhofer diffraction pattern of a single slit. This spectrum must now be averaged over the distribution of times between collisions, T_i. The probability that the atom will suffer a collision in a distance dl after having travelled a distance l without collision is

$$P(l)\,dl = \exp(-l/\lambda)\,dl/\lambda \qquad (8.26)$$

where λ is the mean free path for the excited atom. The distribution of times between collisions is therefore given by

$$P(T_i)\,dT_i = \exp(-T_i/T_c)\,dT_i/T_c \qquad (8.27)$$

where $1/T_c$ is the mean number of collisions per unit time,

$$1/T_c = \bar{v}/\lambda. \tag{8.28}$$

The spectral distribution of radiation, obtained by averaging equation (8.25) over the distribution of equation (8.27), is given by

$$I(\omega) = I_0 \frac{(\gamma/2+1/T_c)}{\pi} \frac{1}{(\omega-\omega_0)^2+(\gamma/2+1/T_c)^2}. \tag{8.29}$$

The line profile is again predicted to be a Lorentzian with a width at half intensity (FWHM) which is increased to

$$\Delta\omega_{1/2} = \gamma + \frac{2}{T_c}$$

$$= \gamma + 2N\sigma\bar{v} \tag{8.30}$$

where N is the density of perturbers, $\bar{v} = \{8kT(M_1+M_2)/\pi M_1M_2\}^{1/2}$ is the mean relative velocity and σ is the collision cross-section. In terms of the kinetic theory, the collision cross-section is given by $\sigma = \pi\rho^2$ where the collision radius ρ is taken as the distance between the centres of the atoms at the moment of collision, or alternatively as the sum of the radii of the colliding particles.

The Lorentz theory is in qualitative agreement with the experimental result that the increase in the linewidth at low pressures is proportional to the density of perturbers. However, the values of the collision cross-section derived from the observed pressure broadening by neutral atoms, $\sigma \approx 10^{-14}$-10^{-15} cm^2, are considerably larger than gas kinetic collision cross-sections, $\sigma \approx 10^{-16}$ cm^2. This discrepancy could be explained by the fact that the effective radii of excited atoms are always greater than those of ground-state atoms. However, the Lorentz theory also requires the radiation to be completely quenched by the collision. Although quenching collisions occur in certain situations, they generally have cross-sections which are much too small to explain the observed pressure broadening. Moreover many examples of interatomic collision processes are known, e.g. excited sodium atoms in the 3 $^2P_{1/2,3/2}$ levels perturbed by inert gases, where broadening of the lines is observed in the complete absence of quenching. We shall now consider some improvements on the simple Lorentz theory.

(b) The Weisskopf theory. Weisskopf (1933) showed that the assumption of a complete disruption of the radiation is not essential since the broadening may be explained by optical collisions in which the change of phase is sufficient to cause the light waves emitted before and after the collision to be completely incoherent. The change in the instantaneous angular frequency $\Delta\omega$ during a single collision in which the perturber follows the straight line path shown in Fig.8.4(a) is plotted in Fig.8.4(b). The phase shift for this collision is then given by the analytical expression

$$\eta(\rho) = \int_{-\infty}^{\infty} \Delta\omega \, dt = \int_{-\infty}^{\infty} \frac{C_n' \, dt}{(\rho^2+\bar{v}^2t^2)^{n/2}} \qquad (8.31)$$

where in this case ρ is the impact parameter. Using the substitution $\bar{v}t = \rho\tan\theta$, the integral may be expressed as

$$\eta(\rho) = \frac{C_n' a_n}{\bar{v}\rho^{n-1}} \qquad (8.32)$$

where a_n is a standard integral of the form

$$a_n = \int_{-\pi/2}^{\pi/2} \cos^{n-2}\theta \, d\theta. \qquad (8.33)$$

For the cases of interest here we have $a_2=\pi$, $a_3=2$, $a_4=\pi/2$, $a_6=3\pi/8$. Weisskopf assumed arbitrarily that a collision which produced a phase change equal to or greater than unity caused the wavetrain which followed to be completely incoherent with that which preceded the collision. This produces a collision in the Lorentz sense without, however, quenching the atom. From equation (8.32) the optical collision radius, or Weisskopf radius, is given by

$$\rho_W = \left(\frac{C'_n a_n}{\bar{v}}\right)^{1/(n-1)} \tag{8.34}$$

This theory again predicts a Lorentzian line profile in which the pressure broadening produced by the optical collision cross-section, $\sigma = \pi\rho_W^2$, is given

$$\Delta\omega_{1/2}(\text{collision}) = 2N\pi\left(\frac{C'_n a_n}{\bar{v}}\right)^{2/(n-1)} \bar{v}. \tag{8.35}$$

For van der Waals broadening, n=6, the values of collision radius calculated from equation (8.34) are of the order of 5-10 Å while for the strong resonance interaction, n=3, they are considerably larger, 35-50 Å (Problem 8.6). Thus the significant difference in the magnitude of van der Waals and resonance broadening can be explained. However, since the assumption that a collision has occurred when $\eta(\rho) \geq 1$ is quite arbitrary, the agreement between the theoretical linewidth calculated from equation (8.35) and the experimentally observed broadening is only approximate. In contrast, the dependence of the linewidth on the mean relative velocity $\bar{v}$ is expected to be correct. For resonance broadening, n=3, $\Delta\omega_{1/2}$ is predicted to be independent of temperature, while for van der Waals broadening, n=6, we find that there is a small temperature dependence:

$$\Delta\omega_{1/2}(\text{collision}) \propto \bar{v}^{3/5} \propto T^{3/10}$$

Thus by observing the dependence of the pressure broadening on the temperature of the source important information can be obtained about the interatomic forces which are responsible for the broadening of the line.

The Weisskopf theory, however, is unable to account for the shift in the frequency of the centre of the line which is observed in both van der Waals and Stark broadening. In order to improve the theory it is necessary to include the effect of weak, distant collisions for which $\eta(\rho) \leq 1$ and in addition to consider in detail the phase changes which occur in strong collisions.

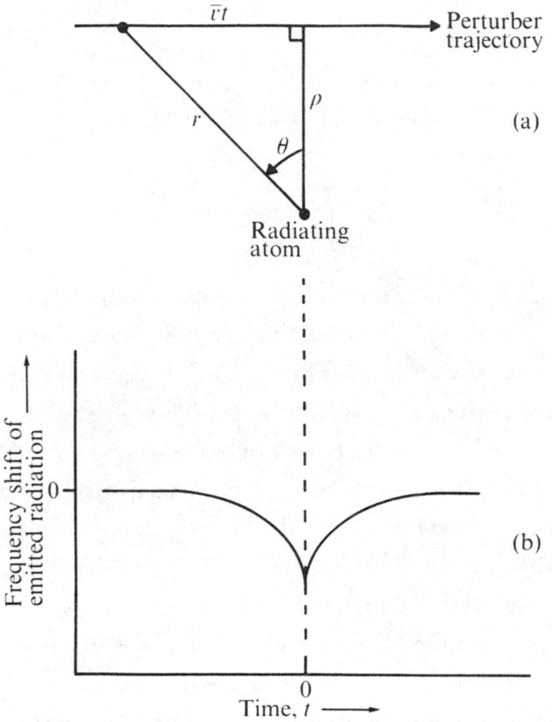

Fig.8.4. Interatomic collisions and the broadening of spectral lines. (a) Trajectory of perturber moving past a radiating atom. (b) Frequency shift of emitted radiation as a function of time.

(c) The adiabatic impact theory. A method of including phase changes smaller than unity by considering the Fourier transform of the instantaneous electric dipole moment of the ensemble was introduced by Lindholm (1941) and Foley (1946). The result of this calculation is again a Lorentz profile given by

$$I(\omega) = I_0 \frac{(\Gamma_{ki}/2 + N\sigma_r\bar{v})}{\pi} \frac{1}{(\omega - \omega_0 + N\bar{v}\sigma_i)^2 + (\Gamma_{ki}/2 + N\bar{v}\sigma_r)^2} \quad (8.36)$$

The width of the line is determined by the real part of the optical collision cross-section defined by (Problem 8.7)

$$\sigma_r = 2\pi \int_0^\infty \{1 - \cos\eta(\rho)\}\rho \, d\rho \qquad (8.37)$$

and the shift of the line centre is determined by the imaginary part

$$\sigma_i = 2\pi \int_0^\infty \sin\eta(\rho)\rho \, d\rho. \qquad (8.38)$$

The integrands of equations (8.37) and (8.38) are plotted in Figs.8.5 (a) and (b) respectively as functions of ρ^2. The real and imaginary parts of the optical collision cross section are therefore represented by the areas beneath the curves. For $\rho < \rho_W$, the integrand of equation (8.38) is a rapidly oscillating function which averages to zero. The main contribution to the line shift therefore comes from the distant collisions, $\rho > \rho_W$. On the other hand, the integrand of equation (8.37) shown in Fig.8.5(a) is always positive and the broadening of the line is determined almost entirely by collisions for which $\rho < \rho_W$.

The adiabatic theory enables quite accurate theoretical estimates of the broadening and shift of spectral lines to be made in simple cases. For interaction potentials of the form C_n/r^n it predicts a unique ratio of linewidth to line shift, $2\sigma_r/\sigma_i$. In the case of the van der Waals interaction this has the value of 2·76, while for resonance broadening the shift in the line centre should be identically zero. However, the adiabatic theory assumes that the perturbation causes no change in the internal state of the system. It is well known that inelastic collisions of atoms and molecules can cause transitions between degenerate or nearly degenerate levels. Modern developments of the theory have attempted to include these non-adiabatic effects and have led to very precise information about the shape and width of pressure broadened lines, as explained in section 8.7.

Finally it is useful to note that the radiative and collision broadening processes have the same effect on all atoms in the sample. They are consequently known as *homogeneous broadening* mechanisms. By contrast, the Doppler

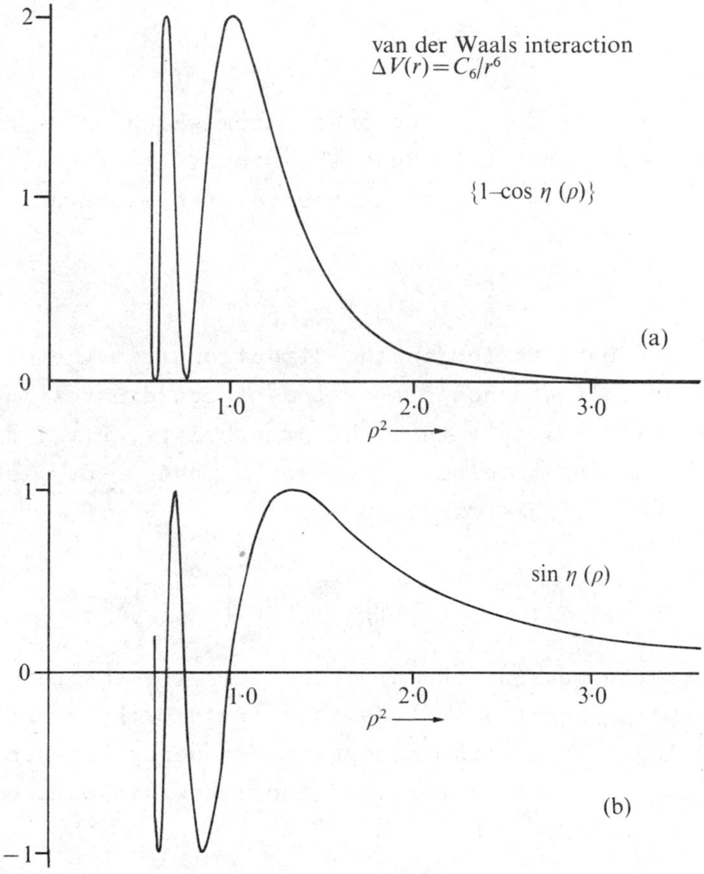

Fig.8.5. Impact theory of line broadening. Integrands of (a) the real and (b) the imaginary parts of the optical collision cross-section for a two-body collision as a function of (impact parameter, $\rho)^2$.

frequency shift associated with moving atoms is an example of a broadening mechanism which varies from one atom to another. It is therefore referred to as an *inhomogeneous broadening* mechanism. Although collision broadening is important in high pressure or highly-ionized gases the shape and width of the spectral lines emitted by most low pressure, low current discharge lamps are determined by the Doppler effect, as we now explain.

8.3. Doppler broadening

Let us consider a stationary atom which emits a wave of angular frequency ω_0. When the atom is moving with a velocity $\underline{v} \ll c$ relative to an observer the observed frequency is given by

$$\omega_0' = \omega_0 (1 - \frac{\underline{v}.\hat{\underline{r}}}{c}) \tag{8.39}$$

where $\hat{\underline{r}}$ is a unit vector in the direction from the observer to the atom. If we choose a system of coordinates in which the z-axis lies along $\hat{\underline{r}}$ then the probability, $P(v_z)\, dv_z$, of an atom having a velocity between v_z and v_z+dv_z is given by the Maxwellian distribution

$$P(v_z)\, dv_z = \left(\frac{M}{2\pi kT}\right)^{1/2} \exp\left(-\frac{Mv_z^2}{2kT}\right) dv_z \tag{8.40}$$

where M is the mass of the atom and T is the absolute temperature. Using equations (8.39) and (8.40), the probability of detecting a wave with an angular frequency between ω_0' and $\omega_0' + d\omega_0'$ is given in terms of the linewidth parameter $\Delta = 2(\omega_0/c)(2kT/M)^{1/2}$ by

$$\begin{aligned} P(\omega_0')\, d\omega_0' &= \frac{2}{\Delta\pi^{\frac{1}{2}}} \exp\{-4(\omega_0'-\omega_0)^2/\Delta^2\}\, d\omega_0' \\ &= \mathscr{G}(\omega_0'-\omega_0,\Delta)\, d\omega_0'. \end{aligned} \tag{8.41}$$

Thus the profile of a Dopper-broadened spectral line is given by the normalized Gaussian distribution $\mathscr{G}(\omega_0'-\omega_0,\Delta)$ shown in Fig.8.6. The half width (FWHM) is given by

$$\begin{aligned} \Delta\omega_{1/2}(\text{Doppler}) &= \Delta(\ln 2)^{1/2} = 2\,\frac{\omega_0}{c}\left[\frac{2kT}{M}\ln 2\right]^{1/2} \\ &= 7{\cdot}16 \times 10^{-7}\omega_0 \left(\frac{T}{A}\right)^{1/2} \end{aligned} \tag{8.42}$$

where A is the atomic or molecular weight of the radiating particle. As an example the Doppler width of the red lines of neon, $\lambda \approx 6000$ Å, A=20, is $16\cdot6 \times 10^{-3}$ Å $\equiv 46\cdot2$ mK at T=300 K. This is about two orders of magnitude greater than the natural width of spectral lines discussed in section 8.1. The large Doppler width of spectral lines, particularly in the case of light elements such as hydrogen and helium, has severely limited the resolution attainable in optical spectroscopy and has delayed fundamental discoveries such as the Lamb shift.

It is important to note that the Doppler width is proportional to the observed frequency, ω_0. Thus one powerful technique of improving the resolution in experimental studies of fine or hyperfine structure is to use microwave or radio frequencies. These methods are discussed in detail in Chapters 16-18. Similarly from equation (8.42) we see that $\Delta\omega_{1/2}$(Doppler) $\propto \sqrt{T}$. Thus the Doppler width at optical frequencies can be reduced by cooling the source, but it is often necessary to go to liquid nitrogen temperatures before any useful increase in resolution is obtained. Even in cooled discharges, however, the effective temperature of the radiating atoms is generally higher than the walls of the tube since they are excited in energetic collisions with ions and electrons. Consequently it is also necessary to limit the current through the cooled discharge to a low value ≈ 1 mA. Experiments involving the use of cooled discharge tubes are discussed in section 8.7 below.

The Doppler width at optical frequencies can be considerably reduced by using a well-collimated beam of atoms, as described by Odintsov (1961) and Stanley (1966). The radiation emitted or absorbed is observed in a direction at right angles to the beam and the effective velocity in the line of sight can be reduced indefinitely by increasing the collimation of the beam (Problem 8.8). These sources are, however, considerably more difficult to construct than conventional discharge tubes or absorption cells and have very low intensities and absorption coefficients. In addition the width observed in emission in the case of light atoms is always larger than that estimated from the beam collimation and oven temperature because of the momentum imparted by electron impact excitation. For these reasons the technique has not been widely used.

The recent invention of tunable dye lasers seems likely to overcome some of these problems. Moreover these devices have already led to the development of other powerful methods of eliminating the Doppler effect, some of which are discussed in Chapters 13 and 14.

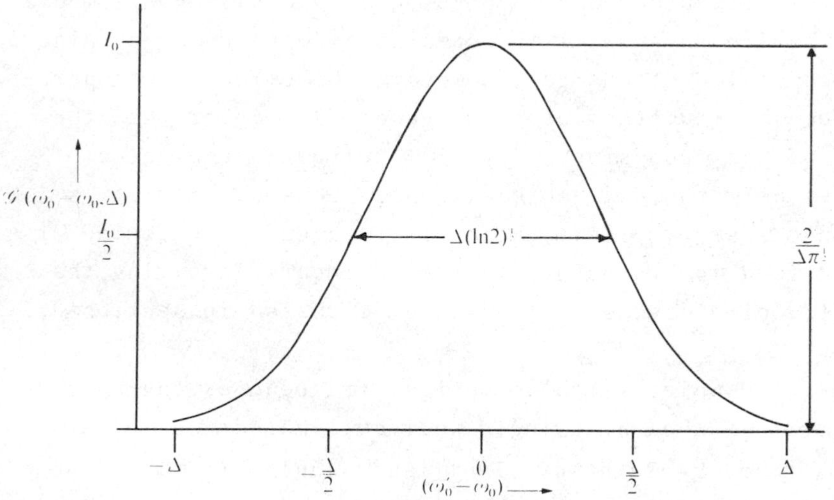

Fig.8.6. Normalized Gaussian distribution defined by equation (8.41) describing the Doppler broadening of spectral lines.

8.4. *Comparison of Doppler, collision, and natural widths*

In order to compare the different contributions to the width of a spectral line, let us take as an example an atom of atomic mass A=100 having an excited level with a life-time of 10^{-8}s, which radiates at 5000 Å. The gas temperature is assumed to be 300 K and the optical collision cross-section section is taken as $\sigma_r = 5 \times 10^{-15}\ cm^2$. The various contributions to the linewidth are evaluated in Table 8.1, for a gas pressure of 10 Torr. We see that under these conditions, which are typical of a low pressure glow discharge, the collisional width and the natural width are of the same order of magnitude but are smaller by a factor of nearly 10^2 than the Doppler width. The instrumental resolving power which would be required to investigate the Doppler profile in detail would be $R \geq (\lambda/\Delta\lambda_D) \approx 10^6$. To attain this kind of resolution a spectrograph would generally require a very large grating combined with a collimating system of several metres focal length. Instruments of this type are found in only a few research laboratories. The scanning Fabry-Perot étalon is much more compact and for a given resolution has a much larger light-gathering power. Consequently this is the instrument generally used for line profile studies.

TABLE 8.1.

A comparison of natural, collisional, and Doppler widths

Natural width	Collisional width	Doppler width	Units
1	0·573	118	$10^8\ s^{-1}$
15·9	9·12	744	MHz
0·53	0·304	24·8	$10^{-3}\ cm^{-1}$
0·133	0·076	6·2	mÅ

8.5. Voigt profiles

Although we have considered the various contributions to the width of spectral lines separately, it is obvious that in any experimental situation several effects will usually be acting simultaneously. Consequently the observed lineshape will not have either a simple Lorentzian or Gaussian profile. To investigate this we consider a moving atom whose resonance frequency is observed to be at ω_0'. Due to collisions or to the finite radiative lifetime the emitted radiation consists of a distribution of frequencies ω about the centre frequency ω_0' which is given by the Lorentzian function

$$\mathcal{L}(\omega-\omega_0',\Gamma) = \frac{\Gamma/2\pi}{(\omega-\omega_0')^2+\Gamma^2/4} \tag{8.43}$$

where Γ is the full width of the distribution at half intensity. The spectral profile emitted by an ensemble of moving atoms is obtained by averaging equation (8.43) over the thermal distribution of equation (8.41) giving

$$\begin{aligned} I(\omega_0-\omega,\Gamma/\Delta) &= \int_0^\infty \frac{\Gamma/2\pi}{(\omega-\omega_0')^2+\Gamma^2/4} \cdot P(\omega_0')\, d\omega_0' \\ &= \int_0^\infty \frac{\Gamma/2\pi}{(\omega-\omega_0')^2+\Gamma^2/4}\, \frac{2}{\Delta\sqrt{\pi}} \exp\left\{-\frac{4(\omega_0'-\omega_0)^2}{\Delta^2}\right\} d\omega_0' \end{aligned} \tag{8.44}$$

The resulting line profile has the form of a folding integral of the Lorentzian and Gaussian distributions:

$$I(\omega_0-\omega,\Gamma/\Delta) = \int_0^\infty \mathcal{G}(\omega_0'-\omega_0,\Delta)\, \mathcal{L}(\omega-\omega_0',\Gamma)\, d\omega_0' \tag{8.45}$$

which is known as the Voigt profile. The shape of this profile is determined uniquely by the ratio Γ/Δ, and in Fig.8.7 the Lorentzian, Gaussian, and Voigt profiles are compared for the case $\Gamma = \Delta(\ln 2)^{1/2}$. Unfortunately the Voigt profile cannot be expressed in an analytic form. However, it can be evaluated numerically and several tabulations of this and related functions have been published, including those by Davies and Vaughan (1963), Fried and Conte (1961), and Hummer (1965).

For the example considered in section 8.4 we have $\Gamma/\Delta \ll 1$ and the line profile would be Gaussian out to three Doppler widths from the line centre before changing over to the more slowly decreasing Lorentzian shape. For the opposite case, $\Gamma/\Delta \gg 1$, the profile would be approximately Lorentzian over the entire lineshape. When $\Gamma/\Delta \approx 1$ it is possible to determine both Γ and Δ from a detailed comparison of the experimental and theoretical line profiles as is explained in more detail in section 8.7 below.

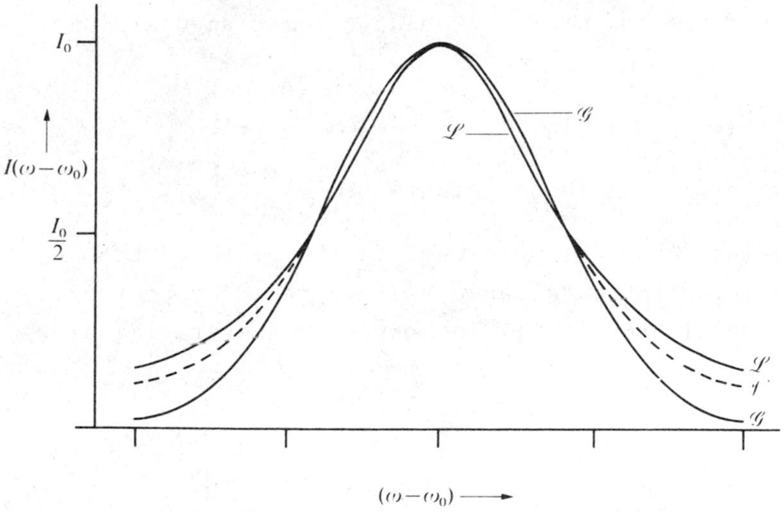

Fig.8.7. Comparison of the shapes of the Lorentzian distribution $\mathscr{L}$, the Gaussian distribution $\mathscr{G}$, and the Voigt line profile $\mathscr{V}$ for which $\Gamma = \Delta(\ln 2)^{1/2}$. Profiles are normalized to the same peak intensity and have the same half width.

8.6. *Effect of the instrumental profile*

In the analysis of the shape of spectral lines it is often necessary, as we have seen, to use spectroscopic instruments of the highest possible resolving power. Even then the width of the instrumental profile is frequently comparable with the width of the line which it is desired to examine. We must therefore consider how this affects the shape of the observed spectrum.

The instrumental function is most conveniently specified in terms of the angular frequency, ω_P, of the monochromatic wave which produces a maximum intensity on a detector located at the point P. In a grating spectrograph, for example, ω_P would vary as the detector P was moved along the focal plane. If the instrument is now illuminated by monochromatic light of angular frequency ω the instrumental profile is given as the transmission function, $T(\omega-\omega_P)$. For a grating spectrograph with a relatively wide entrance slit the instrumental profile would be a rectangular function

$$T(\omega-\omega_P) = T_0 \qquad \omega - \frac{\Delta\omega}{2} \le \omega_P \le \omega + \frac{\Delta\omega}{2}$$

$$= 0 \qquad \text{otherwise}$$

where the width $\Delta\omega$ is proportional to the width of the entrance slit. If the entrance slit is made very narrow, the instrumental profile will be given by the grating diffraction pattern

$$T(\omega-\omega_P) = T_0 \left[\frac{\sin\{(\omega-\omega_P)/\Delta\omega\}}{N \sin\{(\omega-\omega_P)/N\Delta\omega\}}\right]^2$$

where N is the total number of grooves on the grating and $\Delta\omega$ is now inversely proportional to the width of the grating.

When the instrument is illuminated with light whose intrinsic spectral distribution is represented by $I(\omega)$, the spectrogram obtained as the detector position is changed is given by the folding integral:

$$S(\omega_P) = \int_0^\infty T(\omega-\omega_P)\, I(\omega)\, d\omega \qquad (8.46)$$

The effect of this convolution is illustrated in Fig.8.8 for the case of a spectrograph with a rectangular transmission function illuminated by light having a broad symmetric spectral profile.

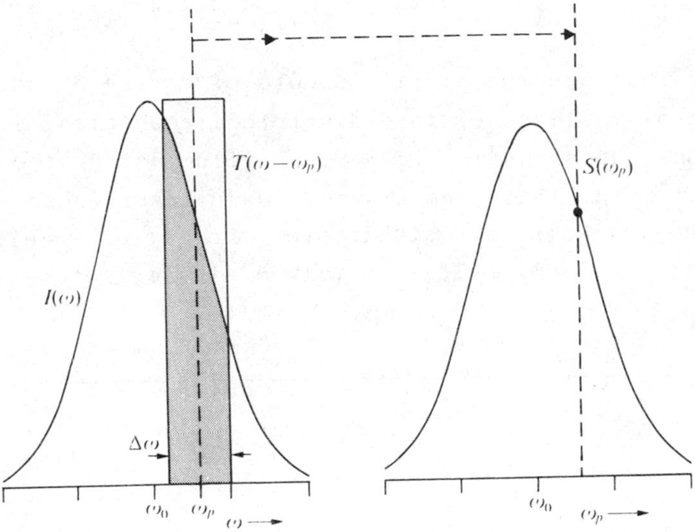

Fig.8.8. The relation between the intrinsic spectral distribution of the source I(ω) and the observed profile $S(\omega_p)$ for a spectrograph having a rectangular transmission function $T(\omega-\omega_p)$.

It is now necessary to find a method of determining the intrinsic line profile, I(ω), from the shape of the observed spectrum, $S(\omega_p)$. If the data are available in digital form it is possible to compute the theoretical folding integrals for a number of assumed profiles I(ω) until there is satisfactory agreement between the theoretical and experimental spectra. Alternatively, it may be possible to assume that the instrumental profile is approximately described by a Lorentzian or Gaussian distribution and to make use of the following theorems:

(i) The folding integral of two Lorentzian distributions whose widths are characterized by Γ_1 and Γ_2 respectively is a Lorentzian whose width is given by $\Gamma = \Gamma_1 + \Gamma_2$.

(ii) The folding integral of two Gaussian functions whose widths are characterized by Δ_1 and Δ_2 is a Gaussian whose total width is given by

$$\Delta\omega_{1/2} = \Delta(\ln 2)^{1/2} = (\Delta_1^2 + \Delta_2^2)^{1/2} (\ln 2)^{1/2}.$$

The instrumental profile should of course be checked experimentally at one or more discrete frequencies. This can be done, as shown in Fig.8.9, by illuminating the instrument with light from a single mode laser whose frequency and intensity are stable over the period required to obtain the spectrum. Alternatively a single isotope lamp

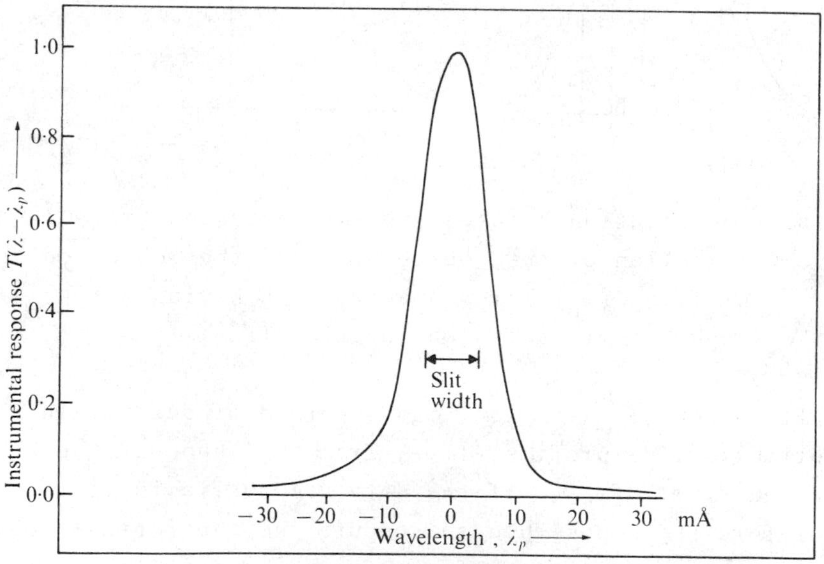

Fig.8.9. Central part of the instrumental profile of the echelle grating spectrometer of the Department of Astrophysics, University of Oxford. The grating has 300 grooves mm^{-1} and the ruled area is $256 \times 128\ mm^2$. (After Collins (1970).)

run at low current and cooled to an accurately measured temperature may be used. It is then assumed that the lamp emits a Doppler-broadened line corresponding to the bath temperature.

8.7. *Line profile measurements at low pressures and temperatures*

At low pressures the impact theory of line broadening should accurately predict the shape and width of the central part of the line profile, as discussed in section 8.2.2. To study this region of the spectral line it is necessary to employ the techniques of high-resolution spectroscopy described by Jacquinot (1960) and Steel (1967), and in addition to reduce the Doppler width as far as possible by cooling the light source. The inert gases are very suitable for this type of high-resolution spectroscopy since pure samples of the even isotopes can be readily obtained. This eliminates the difficulties caused by isotope shifts or the hyperfine structure associated with the odd isotopes. Discharges can be run in the inert gases over a wide range of temperature and the pressure of the gas may be determined accurately. These factors are important in the line broadening measurements, as will be seen in the following discussion.

8.7.1. Apparatus for high-resolution studies of line profiles

A schematic diagram of the apparatus used by Kuhn and Vaughan (1963), Vaughan (1966a), Kuhn and Lewis (1967), and Vaughan (1968) in a series of detailed studies of spectral line profiles in helium, neon, argon, and krypton is shown in Fig.8.10. The source was a low current d.c. discharge tube cooled in a bath of liquid nitrogen or ice. The light was taken from a constricted region of the discharge to reduce the effects of self absorption. Typical discharge currents were 0·5 mA at pressures in the range 0·1-10 Torr.

An f/16 Littrow spectrograph using Bausch and Lomb precision replica gratings isolated the spectral line which

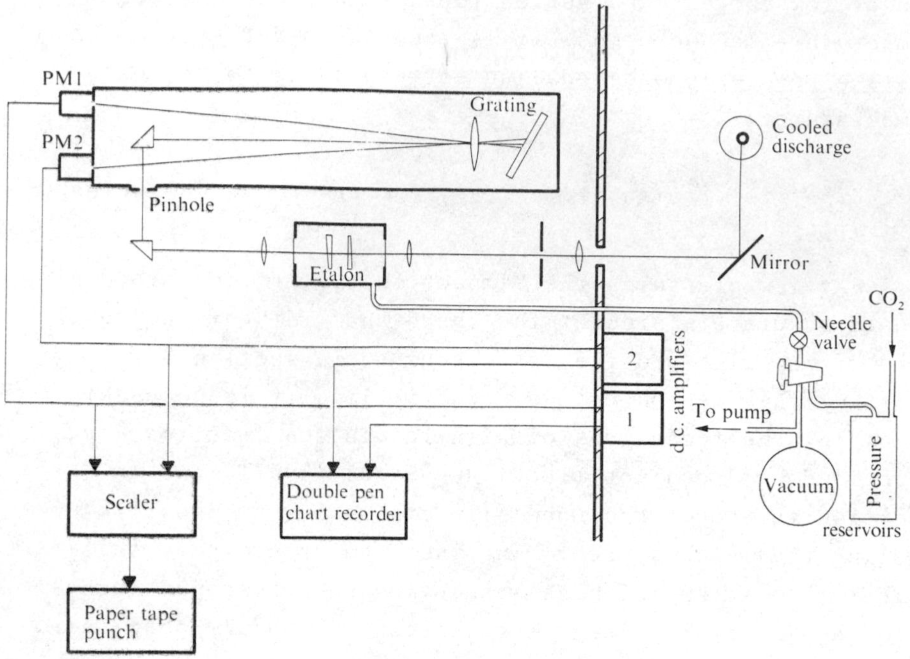

Fig.8.10. A schematic diagram of the apparatus used for high-resolution studies of spectral profiles of the inert gases. (After Rebbeck (1970).)

it was desired to study. This spectrograph was combined with a Fabry-Perot étalon to obtain the necessary high resolution. The étalon plates, which were flat to better than $\lambda/150$, were coated with high-reflectivity films of silver, aluminium or multilayer dielectrics depending on the wavelength of the line being studied. The size of the étalon spacer was chosen so that overlapping of orders due to the finite width of the spectral line was avoided. The central spot of the Fabry-Perot fringe pattern was imaged on to a pinhole which replaced the entrance slit of the spectrograph.

To record the profile of a spectral line, the Fabry-Perot fringe pattern was scanned across the pinhole by slowly changing the pressure of carbon dioxide in a chamber surrounding the étalon. In the early work the intensity of the light received by a photomultiplier placed behind a similar pinhole in the focal plane of the spectrograph was recorded as a function of pressure in analogue form on an X-Y recorder. The pressure scale was linear and could be converted directly into a wavenumber scale since the peaks of successive orders are separated by (1/2d) cm^{-1} where d is the étalon spacing. Recently the apparatus has been improved by using the photomultipliers in the photon counting mode and recording the data digitally on punched paper tape. A typical experimental trace of the helium line $2\ ^1P_1 - 4\ ^1S_0$ at 5047 Å obtained by this technique is shown in Fig.8.11.

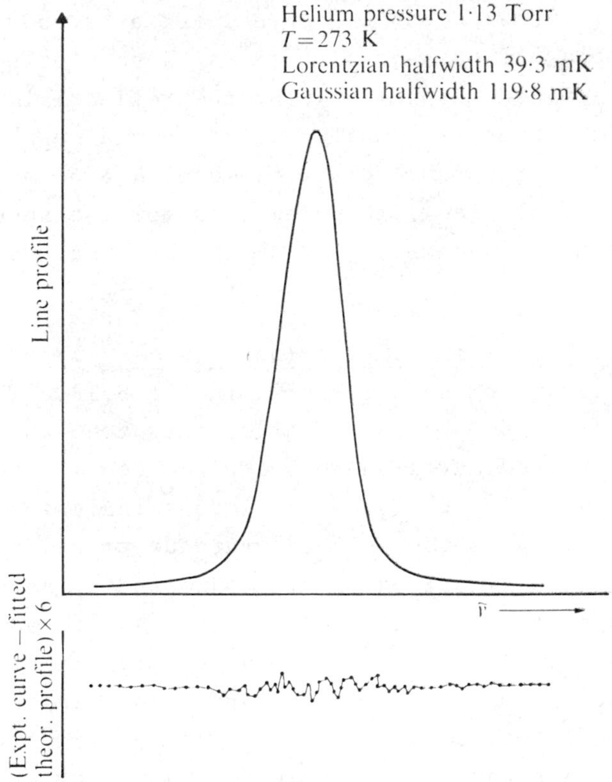

Fig.8.11. Profile of the $2\ ^1P_1 - 4\ ^1S_0$ line of helium at 5047 Å recorded using a scanning Fabry-Perot étalon and photon counting. (After Malvern (1975).)

8.7.2. Analysis of the experimental curves. The digital output enables the line profiles to be computer analysed, which is considerably more accurate than the graphical analysis used by Kuhn and Vaughan (1963) and others. The different orders of the experimental recording are superimposed to produce an averaged experimental trace which is treated as a Voigt profile. This is compared with computer-generated Voigt functions until there is agreement between the theoretical and experimental curves, thus determining the Lorentzian and Gaussian parameters (Γ,Δ) of the observed spectrogram. It can be shown that the instrumental profile of the Fabry-Perot étalon can be represented as a convolution of Lorentzian and Gaussian functions whose widths (Γ_2,Δ_2) are determined respectively by the reflectivity of the étalon and the imperfect parallelism and flatness of the étalon plates. The theorems given in section 8.6 may then be used to obtain the linewidth parameters for the spectral line (Γ_1,Δ_1) from the parameters (Γ,Δ) of the observed spectrogram.

8.7.3. Resonance broadening in the inert gases. A partial energy-level diagram for neon is shown in Fig.8.12. The level structure is also typical of argon, krypton, and xenon. The first four excited levels have been labelled in the L-S notation although the spacings of the levels indicate that intermediate coupling is the only accurate description. The $^1P_1$ levels of all the inert gases are connected by strong resonance lines to the ground level and have short radiative lifetimes in the range 5×10^{-10}-1×10^{-9}s. The intercombination lines from the $^3P_1$ levels are weaker in argon and neon and the lifetimes of these levels are slightly longer. All these transitions lie in the vacuum ultraviolet so that it is impossible to study them directly at high resolution. However, the widths of both the $^1P_1$ and the $^3P_1$ levels may be obtained by high-resolution studies of lines which terminate on these levels. Some of these transitions, which in neon are in the red region of the spectrum, are shown in Fig.8.12. They start on levels whose radiation

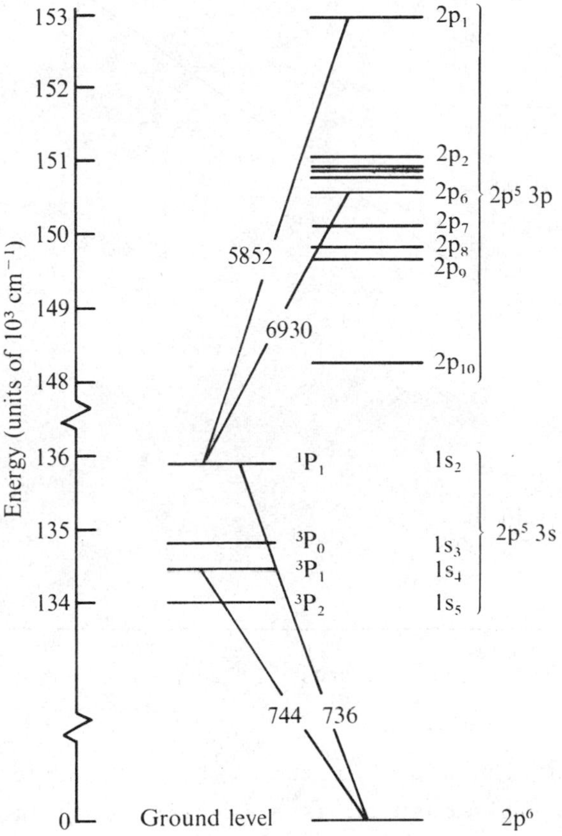

Fig.8.12. A partial energy-level diagram for neon.

widths are generally very small and which can be accurately calculated from the measured lifetimes listed in Table 6.1.

Detailed analysis has proved that at pressures between 0-20 Torr and temperatures between 85-300 K the profiles of these lines are accurately represented by Voigt functions. The Lorentzian widths, Γ_1, are found to be a linear function of the density of the inert gas in the discharge, as shown in Fig.8.13. The pressure broadening is due mainly to the broadening of the lower levels of these transitions through the resonance interaction discussed in section 8.2.1. Under these conditions the impact theory of pressure broadening is expected to be valid, and indeed measurements at different

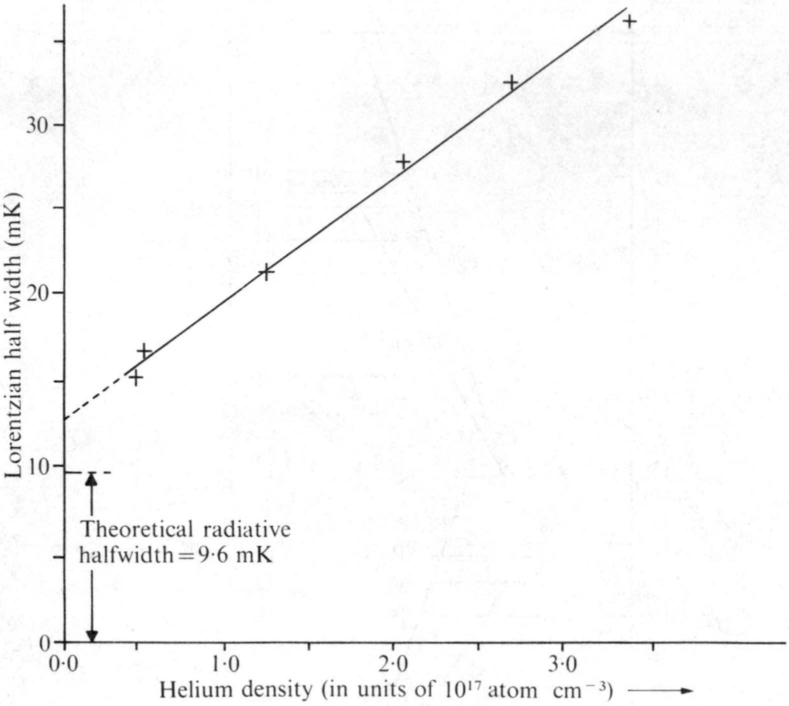

Fig.8.13. Lorentzian width of the 2 $^1P_1$-3 $^1S_0$ line of helium at 7281 Å measured as a function of helium density. The lower level of this transition is broadened by collisions through the resonance dipole-dipole interaction. (After Vaughan (1966a).)

gas temperatures have shown that the broadening is independent of temperature (as predicted by equation (8.35) for the resonance interaction, n=3) and that there is no shift of the line centre.

The impact theory of resonance broadening (Breene 1964, p.48) predicts that

$$\Delta\omega_{1/2}(\text{collision}) = \frac{k_{JJ'}\pi e^2 f_{J'J} N}{8\varepsilon_0 m\omega} \qquad (8.47)$$

where N is the density of perturbing atoms, $f_{J'J}$ and ω are the absorption oscillator strength and angular frequency of the resonance line, and $k_{JJ'}$ is a constant which depends on the total angular momentum quantum numbers J,J' of the upper and lower levels of the resonance transition. Values of $k_{JJ'}$ obtained by several recent calculations are given in Table 8.2 for the $^1S_0-^1P_1$ resonance lines. The last four values quoted in the table are expected to be accurate to better than 2 per cent.

TABLE 8.2

Theoretical values of the resonance broadening constant, $k_{JJ'}$

k_{10}	Reference
1·45	Omont (1965)
1·53	Omont and Meunier (1968)
1·532	Stacey and Cooper (1969)
1·54	Berman and Lamb (1969)
1·532	Carrington, Stacey, and Cooper (1973)

Unfortunately, accurate theoretical f-values for the resonance lines of the inert gases are difficult to calculate except in the case of helium where a precise value for the absorption f-value of the $1\ ^1S_0-2\ ^1P_1$ resonance line at 548 Å ($f_{01} = 0{\cdot}2761 \pm 0{\cdot}0001$) has been obtained by Schiff and Pekeris (1964). This has enabled Vaughan (1966b) to obtain the value $k_{10} = 1{\cdot}44 \pm 0{\cdot}09$ from his observations of the resonance broadening of the 7281 Å and 6678 Å lines in helium. The agreement between this result and the theoretical values listed in Table 8.2 is within the experimental precision and is an indication of the accuracy of the modern line broadening calculations.

Alternatively the theoretical value, $k_{10} = 1{\cdot}53$, may be accepted and the experimental measurements of resonance broadening may be combined with equation (8.47) to obtain the f-values of the resonance transitions in the inert gases. These values are given by Corney (1970) where they are compared with f-values obtained by other experimental and theoretical techniques.

8.7.4. Resonance broadening in the alkali metals. The technique described in section 8.7.2 has been applied to a study of resonance broadening in potassium by Lewis *et al.* (1971). In principle the shape of the resonance lines of the alkali metals can also be studied in absorption as described by Chen and Phelps (1968). However, most absorption experiments have been performed with only moderate spectroscopic resolution and the observations are confined to the far wings of the lines. The widths of the absorption lines are almost entirely determined by the effect of collision broadening, but unfortunately it is not possible to apply the results of the impact theories since these apply only to that part of the line profile satisfying $\Delta_\omega \ll t_c^{-1}$, as explained in section 8.2.2.

8.7.5. Foreign gas broadening. We shall now consider briefly the case of non-resonant interactions. This includes the broadening of spectral lines by neutral foreign gases and also the self-broadening of lines not connected to the resonance levels. It is observed that most of the perturbing gases produce a shift of the line centre towards the red end of the spectrum. Since the dominant long-range force is the attractive van der Waals interaction, this effect is adequately explained by equation (8.38). Of the noble gases only helium and occasionally neon produce shifts towards the violet end of the spectrum. These perturbers have such small polarizabilities that the short-range repulsive forces have a larger effect on the spectral line than the relatively weak van der Waals attraction.

Typical of experiments performed under conditions where the impact approximation is valid are those in krypton made by Vaughan and Smith (1968) using the techniques of high-resolution spectroscopy described previously. The most accurate measurement of the width-to-shift ratio for self-broadening gave the value $-2{\cdot}96 \pm 0{\cdot}2$ which is in good agreement with the theoretical value of $-2{\cdot}76$ expected for the van der Waals interaction. The ratio of the broadening at T=295 K and T=90 K was found experimentally to be $1{\cdot}48 \pm 0{\cdot}12$ whereas the ratio predicted by equation (8.35) for van der Waals broadening is $1{\cdot}43$. There is therefore excellent agreement between theory and experiment in this case.

However, when the broadening of krypton by other noble gases was examined, widely different temperature coefficients and width-to-shift ratios were obtained. This confirmed a conclusion reached earlier by Smith (1967): an adequate theory of foreign gas broadening in the impact approximation requires the use of a potential containing both attractive and repulsive components. Unfortunately such potentials are extremely difficult to calculate theoretically, nor is it possible to unambiguously derive an empirical potential, $V(r)$, from measurements made under conditions where the impact theory is valid.

On the other hand in the far wings of the line the observed spectral profile is non-Lorentzian and depends strongly on temperature. Gallagher and his co-workers (Hedges *et al.* 1972, Carrington and Gallagher 1974) have shown that in this region the quasi-static theory of line-broadening is applicable and that interatomic potentials can be determined without ambiguity. Experiments have been carried out in cesium and rubidium broadened by noble gases at densities of the order 10^{19} cm^{-3} and potentials for the $X^2\Sigma$, $A^2\Pi$ and $B^2\Sigma$ states of the alkali-inert gas molecules have been determined. Unfortunately the theoretical analysis is somewhat involved and the reader should refer to the original papers for further details.

Problems

8.1. A classical electron oscillator emits light of wavelength 5000 Å. Show that in this system the radiation reaction force is approximately 10^8 times smaller than the harmonic binding force, $\omega_0^2 z$.

8.2. A more accurate description of the radiation from a classical electron oscillator may be obtained by assuming a solution of the form $z=z_0 \exp(-\alpha t)$ in equation (8.4) Show that α is given by the roots of the equation

$$\alpha^3 \gamma/\omega_0^2 + \alpha^2 + \omega_0^2 = 0.$$

To terms of order $(\gamma/\omega_0)^2$, show that the complex roots of this cubic equation may be expressed in the form:

$$\alpha_{\pm} = \frac{\gamma}{2} \pm i(\omega_0 + \Delta\omega_0)$$

where $\Delta\omega_0$ is the radiative level shift given by $\Delta\omega_0 = -5\gamma^2/8\omega_0$. This is the classical counterpart of the well-known Lamb shift in hydrogenic systems.

8.3. A stationary atom of mass M emits a photon of energy $\hbar\omega$ and momentum $\hbar\underline{k}$ where $|\underline{k}| = \omega/c$. Applying the principles of conservation of energy and momentum, show that the relative shift in the angular frequency of the photon due to recoil of the atom is $(\omega-\omega_0)/\omega_0 \cong -\hbar\omega_0/2Mc^2$.

Estimate the ratio of the recoil angular frequency shift to the natural width of the transition for an atom of ^{20}Ne emitting:

(a) a photon of visible light, $\hbar\omega \approx 2$ eV, and

(b) a γ-ray photon, $\hbar\omega \approx 15$ keV.

Owing to these recoil shifts resonant absorption of γ-rays by nuclei cannot generally take place. Those nuclei which demonstrate the Mössbauer effect form an exception to this rule.

8.4. Consider the electromagnetic wave emitted by an assembly of a large number, N, of stationary classical electron oscillators. The individual oscillators, each having the resonance frequency $\omega_0/2\pi$, are assumed to be excited with random phases at arbitrary instants of time. Show that the frequency distribution of the radiation emitted by the ensemble is identical to the spectrum of a single oscillator, equation (8.10), but that the intensity of the radiation is greater by a factor of N.

8.5. By neglecting radiation damping and the Doppler effect show that the emission spectrum of an ensemble of colliding, classical electron oscillators is given by

$$I(\omega) = \frac{I_0}{\pi} \frac{1/T_c}{(\omega-\omega_0)^2+(1/T_c)^2}$$

where T_c is the mean time between collisions. It may be assumed that each collision completely quenches the radiating atom and that the times between collisions are governed by a Poisson distribution.

8.6. The interaction constant for resonance broadening is given approximately by $C_3' = e^2 f_{ik}/4\pi\varepsilon_0 m\omega_{ki}$ where ω_{ki} and f_{ik} are respectively the angular frequency and absorption f-value of the atomic resonance line. Hence calculate the Weisskopf optical collision radius, equation (8.34), for the case of resonance broadening of:

(a) the sodium D_2 line at 5890 Å, and

(b) the mercury intercombination line at 2537 Å,

given that the absorption f-values of these lines are 0·650 and $2{\cdot}45 \times 10^{-2}$ respectively.

Estimate the collision-broadened width of these spectral lines in their own atomic vapours at 0·1 Torr pressure.

8.7. Using equations (8.32) and (8.37) show that, according

to the impact theory of pressure broadening, the half widths of the emission lines in gases where the atoms interact with one another either through the resonance dipole-dipole interaction or by means of van der Waals forces are

$$\Delta\omega_{1/2}(\text{Resonance}) = 8\pi C_3' N I_1$$

and

$$\Delta\omega_{1/2}(\text{van der Waals}) = 8\pi\left[\frac{3\pi C_6'}{16\bar{v}}\right]^{2/5} \bar{v} N I_2$$

where I_1 and I_2 are given by

$$I_1 = \int_0^\infty x \sin^2\left(\frac{1}{x^2}\right) dx = 0{\cdot}392; \; I_2 = \int_0^\infty x \sin^2\left(\frac{1}{2x^5}\right) dx = 0{\cdot}301.$$

Hence calculate the collision-broadened width of the 7601 Å line of krypton at 300 K and a krypton pressure of 50 Torr given that $C_6' = 5{\cdot}10 \times 10^{-31}$ cm^6 s^{-1}.

8.8 An atomic beam light source is formed by placing a circular collimating aperture, diameter 2a, at a distance l from a fine hole in the source oven. After passing through the collimating aperture the atoms are excited by electron bombardment and collisional recoil effects are assumed to be negligible. Show that light emitted in a direction at right angles to the axis of the beam has a Gaussian spectral profile

$$I(\omega_0')\, d\omega_0' = I_0 \exp\left\{-\frac{4(\omega_0'-\omega_0)^2}{\Delta^2(1+l^2/a^2)}\right\} d\omega_0'$$

where $\Delta = (2\omega_0/c)(2kT/M)^{1/2}$. The Doppler width of light emitted by the beam is therefore smaller by a factor of $(1+l^2/a^2)^{1/2} \approx l/a$ than that emitted by atoms in random motion at the temperature of the source.

A beam of helium atoms is excited to the $2\,^1P_1$ level ($E_k = 21{\cdot}2$ eV, $\tau_k = 5{\cdot}55 \times 10^{-10}$ s) by electron bombardment. Estimate the effect of line broadening due to atomic recoil and compare this with the natural line width of the $2\,^1S_0 - 2\,^1P_1$ line at 2·058 μm.

References

Berman, P.R. and Lamb, W.E. (1969). *Phys.Rev.* 187, 221.

Carrington, C.G., Stacey, D.N., and Cooper, J. (1973). *J.Phys. B.* 6, 417.

_____ , and Gallagher, A. (1974). *Phys.Rev.A.* 10, 1464.

Chen, C.L. and Phelps, A.V. (1968). *Phys.Rev.* 173,62.

Corney, A. (1970). *Adv.Electronics Electron Phys.* 29, 115.

Collins, B.S. (1970). D.Phil.Thesis, University of Oxford.

Davies, J.T. and Vaughan, J.M. (1963). *Astrophysics.J.* 137, 1302.

Foley, H.M. (1946). *Phys.Rev.* 69, 616.

Fried, B.D. and Conte, S.D. (1961). *The plasma dispersion function.* Academic Press, New York.

Hedges, R.E.M., Drummond, D.L., and Gallagher, A. (1972). *Phys.Rev. A.* 6, 1519.

Hindmarsh, W.R. (1967). *Atomic spectra.* Pergamon Press, Oxford.

Hummer, D.G. (1965). *Mem.R.astr.Soc.* 70, 1.

Jaquinot, P. (1960). *Rep.Prog.Phys.* 23, 267.

Kuhn, H.G. and Lewis, E.L. (1967). *Proc.R.Soc.* A299, 423.

_____ and Vaughan, J.M. (1963). *Proc.R.Soc.* A277, 297.

Lewis, E.L., Rebbeck, M.M., and Vaughan, J.M. (1971). *J.Phys. B.* 4, 741.

Lindholm, E. (1941). *Ark.Mat.Astr.Fys.* 28B, No.3.

Lorentz, H.A. (1906). *Proc.Acad.Sci.Amst.* 8, 591.

Malvern, A.R. (1975). D.Phil.Thesis, University of Oxford.

Odintsov, V.I. (1961). *Optics.Spectrosc.* 10, 202.

Omont, A. (1965). *J.Phys.Radium, Paris,* 26, 26.

_____ and Meunier, J. (1968). *Phys.Rev.* 169, 92.

Rebbeck, M.M. (1970). D.Phil.Thesis, University of Oxford.

Schiff, B. and Pekeris, C.L. (1964). *Phys.Rev.* 134, A638.

Smith, G. (1967). *Proc.R.Soc.* A297, 288.

Stacey, D.N. and Cooper, J. (1969). *Phys.Letters* A30, 49.

Stanley, R.W. (1966). *J.opt.Soc.Am.* 56, 350.

Steel, W.H. (1967). *Interferometry.* Cambridge University Press, London and New York.

Vaughan, J.M. (1966a). *Proc.R.Soc.* A295, 164.

——— (1966b). *Phys.Letters* 21, 153.

——— (1968). *Phys.Rev.* 166, 13.

——— and Smith, G. (1968). *Phys.Rev.* 166, 17.

Weisskopf, V. (1933). *Phys.Z.* 34, 1.

——— and Wigner, E. (1930). *Z. Phys.* 63, 54.

General references and further reading

Breene, R.G. (1964). *Handbuch der Physik*, Vol.27, p.1. Springer-Verlag, Berlin.

This article reviews the basic theory of the pressure broadening of spectral lines.

Panofsky, W.K.H. and Phillips, M. (1962). *Classical electricity and magnetism.* Addison-Wesley, Reading, Mass.

Thorne, A. (1974). *Spectrophysics.* Chapman & Hall, London.

Stone, J.M. (1963). *Radiation and optics.* McGraw-Hill, New York.

9

The absorption and stimulated emission of radiation

So far in this book we have emphasized the spontaneous emission of radiation. This occurs at a constant rate, independent of any external influences. Now we wish to turn our attention to the processes of stimulated emission and absorption of radiation which are induced by the presence of an external electromagnetic wave. In this chapter we prepare the foundations for an understanding of both the formation of spectral lines, which is considered in detail in Chapter 10, and the physics of gas lasers which is discussed in Chapters 11-13.

We commence by deriving the absorption cross-section of a classical electric dipole oscillator. The result should be similar to that obtained on the basis of the quantum theory and is of further interest since the frequency dependence of the cross-section is predicted in a simple way. Next we obtain the relations between the spontaneous emission transition probability, A_{ki}, and the Einstein coefficients for absorption and stimulated emission, denoted by B_{ik} and B_{ki} respectively. The expressions for B_{ik} and B_{ki} are then confirmed by means of quantum mechanics using time-dependent perturbation theory. This enables the probability of stimulated emission and absorption of radiation to be given in terms of the oscillator strengths of spectral lines. Finally we show that there is close agreement between the classical and quantum-mechanical expressions for the total absorption cross-section and explain how the atomic frequency response may be introduced into the quantum-mechanical results.

9.1. *Classical description of absorption by electric dipole oscillator*

We consider first the absorption of radiation by a classical 'atom', represented by an harmonically oscillating electron, interacting with the electric field, $\underline{E}(t)$, of an incident electromagnetic wave. The equation of motion for the oscillator is obtained by writing equation (8.5) in vector form and including the additional force experienced by the electron:

$$\ddot{\underline{r}} + \gamma\dot{\underline{r}} + \omega_0^2\underline{r} = -\frac{e}{m}\underline{E}(t). \tag{9.1}$$

To solve this equation we make a Fourier expansion of both $\underline{r}(t)$ and $\underline{E}(t)$:

$$\underline{r}(t) = \left(\frac{1}{2\pi}\right)^{\frac{1}{2}} \int_{-\infty}^{\infty} \underline{r}(\omega) \exp(-i\omega t)\, d\omega \tag{9.2}$$

$$\underline{E}(t) = \left(\frac{1}{2\pi}\right)^{\frac{1}{2}} \int_{-\infty}^{\infty} \underline{E}(\omega) \exp(-i\omega t)\, d\omega. \tag{9.3}$$

Since both $\underline{r}(t)$ and $\underline{E}(t)$ are real, the negative frequency components of their transforms must equal the complex conjugates of the positive frequency components:

$$\begin{aligned} \underline{r}(-\omega) &= \underline{r}^*(\omega) \\ \underline{E}(-\omega) &= \underline{E}^*(\omega). \end{aligned} \tag{9.4}$$

Substituting equations (9.2) and (9.3) into the equation of motion we obtain

$$\underline{r}(\omega) = -\frac{e}{m}\frac{\underline{E}(\omega)}{\omega_0^2 - \omega^2 - i\omega\gamma} \tag{9.5}$$

just as for a monochromatic wave.

The energy absorbed by the oscillator from the radiation field, ΔU, can be obtained by calculating the time integral of the rate of work done by the field on the oscillator:

$$\Delta U = \int_{-\infty}^{\infty} -e\, \dot{\underline{r}}.\underline{E}(t)\, dt. \tag{9.6}$$

Using equations (9.2) and (9.3) together with the reality conditions, equation (9.4), and the Fourier representation of the Dirac delta function:

$$\int_{-\infty}^{\infty} \exp\{i(\omega-\omega')t\}dt = 2\pi\delta(\omega-\omega'),$$

we can transform equation (9.6) to (Problem 9.1)

$$\Delta U = 2e \, \mathrm{Re} \int_0^\infty i\omega \underline{r}(\omega).\underline{E}^*(\omega) \, d\omega. \tag{9.7}$$

Substituting from equation (9.5), the energy absorbed by the electron is given by

$$\Delta U = \frac{2e^2}{m} \int_0^\infty |\underline{E}(\omega)|^2 \frac{\omega^2\gamma}{(\omega_0^2-\omega^2)^2+\omega^2\gamma^2} \, d\omega. \tag{9.8}$$

Since $\gamma \ll \omega_0$, the integrand is sharply peaked around $\omega=\omega_0$ and the approximations $(\omega_0^2-\omega^2) \approx 2\omega_0(\omega_0-\omega)$ and $\omega^2 \approx \omega_0^2$ are valid. Thus we obtain

$$\Delta U = \frac{e^2}{m} \int_0^\infty |\underline{E}(\omega)|^2 \frac{\gamma/2}{(\omega_0-\omega)^2+\gamma^2/4} \, d\omega. \tag{9.9}$$

The energy absorbed can also be written in terms of the product of the energy flow per unit area and an absorption cross-section. From equations (2.44) and (9.3) the energy flux per unit angular frequency range in a medium where the refractive index may be assumed to be unity is given by

$$\frac{d|\underline{N}|}{d\omega} = 2\varepsilon_0 c \int_0^\infty |\underline{E}(\omega)|^2 \, d\omega. \tag{9.10}$$

Therefore, if the spectral distribution of intensity, $|\underline{E}(\omega)|^2$, changes slowly compared with the frequency response of the oscillator, we may interpret equation (9.9) in terms of a frequency-dependent absorption cross-section given by

$$\sigma(\omega) = \frac{\pi e^2}{2\varepsilon_0 mc} \cdot \frac{\gamma/2\pi}{(\omega_0-\omega)^2+\gamma^2/4} . \tag{9.11}$$

As might be expected, the absorption cross-section has the Lorentzian frequency dependence which is characteristic of the classical oscillator. The intensity of a collimated beam of radiation, $I_\omega(x)$, propagating through a gas of stationary classical 'atoms', density N per unit volume, would vary as

$$I_\omega(x) = I_\omega(0)\exp\{-N\sigma(\omega)x\}.$$

The total energy absorbed per unit length is proportional to the integral of the absorption cross-section over the atomic frequency response. This has a constant value given by

$$\int_0^\infty \sigma(\omega)\,d\omega = \frac{\pi e^2}{2\varepsilon_0 mc} = 2\pi^2 r_0 c, \tag{9.12}$$

where we have introduced the classical radius of the electron given by

$$r_0 = \frac{e^2}{4\pi\varepsilon_0 mc^2} = 2\cdot 82 \times 10^{-15}\ \text{m}, \tag{9.13}$$

and thus established a classical sum rule for absorption processes.

In this calculation the phases of the different frequency components of the electric field were assumed to be randomly distributed, corresponding to the radiation field of a conventional thermal source. However, when the incident radiation is monochromatic the average energy absorbed may be either positive or negative, depending on the phase of the atomic oscillations with respect to those of the perturbing electric field. Energy may now be transferred either to the atom from the radiation field, or from the atom to the field. The latter case provides a classical analogue of the stimulated emission process introduced by Einstein and discussed in detail in the following section.

9.2. *Einstein's treatment of stimulated emission and absorption*

We must now consider how the stimulated emission and absorption of radiation are treated in the case of a real atom. The relations between the probability of stimulated and spontaneous transitions were derived by Einstein (1917) using arguments based on statistical mechanics. This paper, which appeared well before the development of a complete quantum theory, is available in translation in ter Haar

(1967). Einstein considered a gas of identical stationary atoms interacting with radiation inside a cavity maintained at the temperature T. The atoms are assumed to have quantized energy levels $E_i, E_j, E_k, \ldots$ with statistical weights $g_i, g_j, g_k, \ldots$. In thermal equilibrium the total number of atoms, N, will be distributed over the available levels according to the Boltzmann distribution

$$N_i = \frac{N g_i \exp(-E_i/kT)}{\sum_i g_i \exp(-E_i/kT)} . \tag{9.14}$$

This distribution is maintained by a dynamic equilibrium in which atoms continually make upward and downward transitions between the levels. We consider in particular the transitions between the levels i and k, shown in Fig.9.1, where $E_k > E_i$. An atom can make an upward transition from the

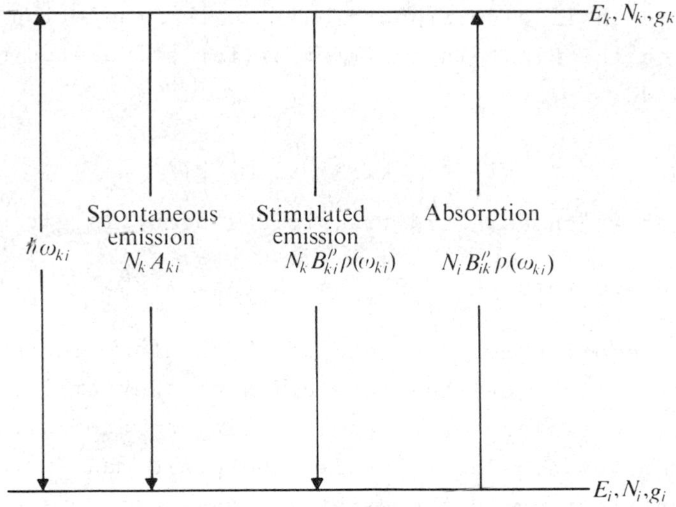

Fig.9.1. Spontaneous and stimulated emission and absorption of radiation. The rate at which transitions take place between the levels k and i is given for atoms interacting with radiation whose energy density per unit bandwidth is $\rho(\omega)$.

level i by the absorption of a photon of energy $\hbar\omega_{ki}$. The transition probability per unit time, P_{ik}, for absorption will be proportional to the energy density of radiation per unit bandwidth, $\rho(\omega_{ki})$, at the angular frequency ω_{ki}, in accord with the results of the classical calculation of section 9.1. Introducing the Einstein coefficient for absorption B^{ρ}_{ik}, we have

$$P_{ik} = B^{\rho}_{ik}\,\rho(\omega_{ki}).$$

The rate at which upward transitions are occurring is given by

$$\text{Upward rate} = N_i P_{ik} = N_i\, B^{\rho}_{ik}\,\rho(\omega_{ki}).$$

The transition probability for downward transitions from level k to level i, P_{ki}, is the sum of the spontaneous emission probability, A_{ki}, and a stimulated emission probability which is also proportional to the radiation density. Introducing the Einstein coefficient for stimulated emission, B^{ρ}_{ki}, we have

$$P_{ki} = A_{ki} + B^{\rho}_{ki}\,\rho(\omega_{ki}).$$

The rate at which downward transitions occur is given by

$$\text{Downward rate} = N_k P_{ki} = N_k\{A_{ki} + B^{\rho}_{ki}\,\rho(\omega_{ki})\}.$$

In thermal equilibrium the rate at which atoms leave level i must obviously equal the rate at which they arrive in that level, but since we are considering an atom with a large number of available levels, there are many more processes contributing to these rates than those considered here. However, in statistical mechanics the principle of detailed balancing states that in equilibrium the rate of transfer from i to k must equal the rate from k to i for every contributing transfer process. Thus we have

$$N_i B^{\rho}_{ik}\,\rho(\omega_{ki}) = N_k\{A_{ki} + B^{\rho}_{ki}\,\rho(\omega_{ki})\} \tag{9.15}$$

which leads to

$$\frac{N_k}{N_i} = \frac{B^{\rho}_{ik}\,\rho(\omega_{ki})}{A_{ki} + B^{\rho}_{ki}\,\rho(\omega_{ki})} \tag{9.16}$$

An alternative expression for the ratio of the populations in thermal equilibrium may be obtained from equation (9.14)

$$\frac{N_k}{N_i} = \frac{g_k}{g_i}\exp(-\hbar\omega_{ki}/kT). \tag{9.17}$$

From equations (9.16) and (9.17) we obtain an expression for the energy density of radiation in the cavity:

$$\rho(\omega_{ki}) = \frac{g_k A_{ki}\exp(-\hbar\omega_{ki}/kT)}{g_i B^{\rho}_{ik} - g_k B^{\rho}_{ki}\exp(-\hbar\omega_{ki}/kT)}. \tag{9.18}$$

This expression should be compared with the result obtained by Planck for black-body radiation:

$$\rho(\omega) = \frac{\hbar\omega^3}{\pi^2 c^3}\cdot\frac{1}{\exp(\hbar\omega/kT)-1}. \tag{9.19}$$

If equations (9.18) and (9.19) are to be compatible at all temperatures we require (Problems 9.2 and 9.3)

$$g_i B^{\rho}_{ik} = g_k B^{\rho}_{ki}$$

and (9.20)

$$B^{\rho}_{ik} = \frac{\pi^2 c^3}{\hbar\omega^3_{ki}}\cdot\frac{g_k}{g_i}A_{ki}.$$

Introducing the explicit expression for A_{ki} in terms of the atomic matrix elements, equation (4.23), we obtain the absorption B-coefficient for electric dipole radiation

$$B^{\rho}_{ik} = \frac{\pi e^2}{3\varepsilon_0\hbar^2}\frac{1}{g_i}\sum_{m_k,m_i}|\langle km_k|\underline{r}|im_i\rangle|^2. \tag{9.21}$$

The relations between the spontaneous emission coefficient and the coefficients for stimulated emission and absorption, given by equation (9.20), do not include the frequency response of the atom. They have been derived for an electromagnetic field which is assumed to be isotropic and unpolarized, and which is described in terms of a density of radiation, $\rho(\omega)\, d\omega$, which is assumed to be a slowly varying function of the angular frequency ω. Different expressions for the B-coefficients will be obtained if the radiation field is characterized by its intensity, as will be explained in section 9.4 below.

The Einstein derivation assumes an idealized situation in which the excitation and de-excitation processes are controlled by the radiation in the cavity. This situation is achieved in reality only inside stars or very high temperature plasmas. In a low pressure discharge there is usually no thermal equilibrium between the atoms and the radiation because the light readily escapes from the source. Consequently we observe a line spectrum rather than a continuum. Collisions play the dominant role in maintaining the equilibrium populations in a low pressure discharge. However, this does not invalidate the arguments used here since the A- and B-coefficients depend only on the internal structure of the atoms and not on their external environment.

An important feature of the stimulated emission process which is not obvious from the Einstein treatment is the fact that the radiation which is emitted in a stimulated transition $k \rightarrow i$ has the same frequency and direction of propagation as that which stimulated the transition. This aspect is most easily visualized on the basis of the classical oscillator model section 9.1 since the phase of the amplified wave must have a definite relationship with that of the incident wave. This coherent amplification process forms the basis of all laser systems.

9.3. The semi-classical treatment of absorption and induced emission

We shall now examine the interaction of atoms with

electro-magnetic radiation in more detail using the semi-classical theory developed in most standard texts on quantum mechanics, e.g. Schiff (1968). In this method the atoms are described quantum mechanically while the electromagnetic field is described classically. We consider a plane, linearly polarized wave propagating in the direction $\underline{k}$ and represented by

$$\underline{E}(t) = \underline{E}_\omega \cos(\underline{k}.\underline{r} - \omega t). \qquad (9.22)$$

This will later be taken as the Fourier component at the angular frequency ω of a more general time-dependent field. It is convenient to represent this electric field in the complex notation

$$\underline{E}(t) = \tfrac{1}{2} \underline{E}_\omega \left[\exp\{i(\underline{k}.\underline{r}-\omega t)\} + \exp\{-i(\underline{k}.\underline{r}-\omega t)\}\right]. \qquad (9.23)$$

The wave is assumed to interact with a stationary, one-electron atom whose instantaneous dipole moment is given by $\underline{p} = -e\underline{r}$. For transitions involving the absorption or emission of a single electric dipole photon it is sufficient to take the classical interaction energy $-\underline{p}.\underline{E}(t)$ as the perturbation operator. The small additional contribution to the Hamiltonian of the system is thus given by

$$\begin{aligned} \mathcal{H}_1 &= -\underline{p}.\underline{E}(t) \\ &= \frac{e\underline{r}.\underline{E}_\omega}{2} \left[\exp\{i(\underline{k}.\underline{r}-\omega t)\} + \exp\{-i(\underline{k}.\underline{r}-\omega t)\}\right]. \end{aligned} \qquad (9.24)$$

The wavefunction, $\Psi(\underline{r},t)$, describing the atom perturbed by the incident radiation, is now a solution of the time-dependent Schrödinger equation

$$i\hbar \frac{\partial \Psi}{\partial t} = \mathcal{H}\Psi \qquad (9.25)$$

where

$$\mathcal{H} = \mathcal{H}_0 + \mathcal{H}_1$$

and $\mathcal{H}_0$ is the Hamiltonian for the unperturbed atom, equation (3.9). We seek solutions of equation (9.25) of the form

$$\Psi(\underline{r},t) = \sum_n a_n(t)\, \psi_n(\underline{r}) \exp(-iE_n t/\hbar) \qquad (9.26)$$

where the spatial wavefunctions are eigenstates of the time-independent equation

$$\mathcal{H}_0 \psi_n(\underline{r}) = E_n \psi_n(\underline{r}). \qquad (9.27)$$

For the moment these wavefunctions are to be understood to represent nondegenerate eigenstates.

If the atom is in state i at t=0 then $a_i(0)=1$ and $a_{n\neq i}(0)=0$. We now wish to calculate the probability, $|a_k(t)|^2$, that after a time t the atom has made an upward transition and is in the state k. Substituting equation (9.26) into equation (9.25), we have

$$\sum_n (i\hbar \dot{a}_n + E_n a_n)\psi_n \exp(-iE_n t/\hbar) = (\mathcal{H}_0 + \mathcal{H}_1) \sum_n a_n \psi_n \exp(-iE_n t/\hbar) \qquad (9.28)$$

Since $\psi_n(\underline{r})$ is an eigenfunction of $\mathcal{H}_0$, equation (9.28) becomes

$$\sum_n i\hbar \dot{a}_n \psi_n \exp(-iE_n t/\hbar) = \sum_n a_n \mathcal{H}_1 \psi_n \exp(-iE_n t/\hbar). \qquad (9.29)$$

The radiation field is now assumed to be so weak that the probability amplitudes $a_n(t)$ do not change significantly with time. We may therefore, as an approximation, use the initial values of a_n on the right-hand side of equation (9.29) with the result:

$$\sum_n i\hbar \dot{a}_n \psi_n \exp(-iE_n t/\hbar) = \mathcal{H}_1 \psi_i \exp(-iE_i t/\hbar). \qquad (9.30)$$

To obtain $\dot{a}_k$ we multiply equation (9.30) by $\psi_k^* \exp(iE_k t/\hbar)$ and integrate over the spatial coordinates, giving

$$i\hbar \dot{a}_k = \int \psi_k^* \mathcal{H}_1 \psi_i \, dv \exp\{i(E_k - E_i)t/\hbar\} . \qquad (9.31)$$

Introducing the explicit time dependence of $\mathcal{H}_1$ from equation (9.24), we have

$$i\hbar\dot{a}_k = \tfrac{1}{2}\langle k|e\underline{r}.\underline{E}_\omega \exp(i\underline{k}.\underline{r})|i\rangle \exp\{i(\omega_{ki}-\omega)t\} +$$

$$+ \tfrac{1}{2}\langle k|e\underline{r}.\underline{E}_\omega \exp(-i\underline{k}.\underline{r})|i\rangle \exp\{i(\omega_{ki}+\omega)t\} \quad (9.32)$$

where $\omega_{ki} = (E_k - E_i)/\hbar$. The probability amplitude, $a_k(t)$, may now be obtained by integrating equation (9.32) using the initial condition $a_k(0)=0$:

$$a_k(t) = \tfrac{1}{2} e \langle k|\underline{r}.\underline{E}_\omega \exp(i\underline{k}.\underline{r})|i\rangle \left[\frac{1-\exp\{i(\omega_{ki}-\omega)t\}}{\hbar(\omega_{ki}-\omega)}\right]. \quad (9.34)$$

The contribution to the integral from the second term in equation (9.32) is negligible in the region $\omega \approx \omega_{ki}$ where absorption takes place due to the factor $(\omega_{ki}+\omega)$ which appears in the denominator. Thus the probability of finding the atom in the excited state k at time t is given by

$$|a_k(t)|^2 = \frac{e^2}{4\hbar^2} |\langle k|\underline{r}.\underline{E}_\omega \exp(i\underline{k}.\underline{r})|i\rangle|^2 . \frac{\sin^2\{(\omega_{ki}-\omega)t/2\}}{\{(\omega_{ki}-\omega)/2\}^2} . \quad (9.35)$$

For most problems involving the interaction of atoms with optical-frequency radiation we are concerned with radiation from broad-band rather than from perfectly monochromatic sources. We therefore integrate equation (9.35) with respect to ω and assume that the spectral distribution of the incident radiation varies slowly compared with that of the atomic frequency response. The result is

$$|a_k(t)|^2 = \frac{e^2}{4\hbar^2} |\langle k|\underline{r}.\underline{E}_\omega \exp(i\underline{k}.\underline{r})|i\rangle|^2 \int_0^\infty \frac{\sin^2(\omega_{ki}-\omega)t/2}{\{(\omega_{ki}-\omega)/2\}^2} d\omega$$

$$= \frac{\pi e^2}{2\hbar^2} |\langle k|\underline{r}.\underline{E}_\omega \exp(i\underline{k}.\underline{r})|i\rangle|^2 t \quad (9.36)$$

where we have used the standard integral

$$\int_{-\infty}^{\infty} \frac{\sin^2 x}{x^2} dx = \pi$$

and the magnitude of the Poynting vector of the incident radiation is now given by equation (9.10). The absorption transition probability per atom per unit time is therefore

$$P_{ik} = \frac{\pi e^2}{2\hbar^2} |\langle k|\underline{r}.\underline{E}_\omega \exp(i\underline{k}.\underline{r})|i\rangle|^2 . \qquad (9.37)$$

Since $|\underline{r}|$ is of the order of 10^{-8}cm and $|\underline{k}|$ is of the order of 10^5 cm^{-1}, $\underline{k}.\underline{r}$ is small compared to unity and we may expand the exponential factor in equation (9.37) as

$$\exp(i\underline{k}.\underline{r}) = 1 + i\underline{k}.\underline{r} + \ldots\ldots$$

and retain only the first term of this expansion. The transition probability then involves only the matrix elements of the electric dipole operator:

$$P_{ik} = \frac{\pi e^2}{2\hbar^2} |\langle k|\underline{r}.\underline{E}_\omega|i\rangle|^2 . \qquad (9.38)$$

Introducing the polarization vector $\hat{\underline{\varepsilon}}$ through the relation $\underline{E}_\omega = \hat{\underline{\varepsilon}}\, E_\omega$ and using equation (2.47) to express the electromagnetic wave in terms of its radiation density per unit angular frequency interval,

$$\rho(\omega) = \frac{1}{2}\varepsilon_0 E_\omega^2 , \qquad (9.39)$$

the transition probability becomes

$$P_{ik} = \frac{\pi e^2}{\varepsilon_0\hbar^2} |\langle k|\underline{r}.\hat{\underline{\varepsilon}}|i\rangle|^2 \rho(\omega). \qquad (9.40)$$

This gives the transition probability between nondegenerate states induced by the absorption of radiation from a linearly polarized beam. If the radiation is isotropic and unpolarized we must average equation (9.40) over all possible directions of the polarization vector giving

$$P_{ik} = \frac{\pi e^2}{3\varepsilon_0\hbar^2} |\langle k|\underline{r}|i\rangle|^2 \rho(\omega) \qquad (9.41)$$

where $|\langle k|\underline{r}|i\rangle|^2 = |\langle k|x|i\rangle|^2 + |\langle k|y|i\rangle|^2 + |\langle k|z|i\rangle|^2$.

Comparing equation (9.41) with the definition of the Einstein coefficient for absorption given in section 9.2 leads to the result

$$B^{\rho}_{ik} = \frac{\pi e^2}{3\varepsilon_0 \hbar^2} |\langle k|\underline{r}|i\rangle|^2. \qquad (9.42)$$

The transition probability for stimulated emission from k to i is obtained by using the second term in equation (9.32) (Problem 9.4). For nondegenerate levels we obtain

$$B^{\rho}_{ik} = B^{\rho}_{ki}.$$

This result, which also follows directly from the Hermitian property of the perturbation operator $\mathcal{H}_1$, is an example of microscopic reversibility which provides the quantum mechanical basis for the principle of detailed balancing used in section 9.2.

The results given by equations (9.41) and (9.42) are valid only for nondegenerate states. Thus the labels i and k should be replaced by m_i, m_k representing nondegenerate Zeeman states of the levels i and k. A result valid for transitions between degenerate levels is obtained by summing equation (9.42) over the final states and averaging over the initial states, giving

$$B^{\rho}_{ik} = \frac{\pi e^2}{3\varepsilon_0 \hbar^2} \frac{1}{g_i} \sum_{m_i, m_k} |\langle k m_k|\underline{r}|i m_i\rangle|^2 \qquad (9.43)$$

which is in agreement with the result obtained by the Einstein derivation, equation (9.21). For a many-electron atom the electric dipole moment operator in equation (9.43) $\underline{p} = -e\underline{r}$ must be replaced by $\underline{p} = -e \sum_i \underline{r}_i$.

9.4. Einstein B-coefficients defined in terms of intensity

The isotropic, unpolarized radiation field which induces stimulated transitions between the levels i and k may also be described in terms of its intensity $I(\omega)$. Since the probability of a transition can be expressed in terms of either $\rho(\omega)$ or $I(\omega)$, we introduce a new B-coefficient through the definition

$$P_{ik} = B^{\rho}_{ik}\,\rho(\omega) = B^{I}_{ik}\,I(\omega). \qquad (9.44)$$

For isotropic radiation, $I(\omega)$ and $\rho(\omega)$ are related by

$$I(\omega) = \frac{c}{4\pi}\rho(\omega) \tag{9.45}$$

for a medium whose refractive index is unity. Consequently the absorption B-coefficient when the radiation is expressed in terms of its intensity becomes

$$B^{I}_{ik} = \frac{4\pi^2 e^2}{3\varepsilon_0 \hbar^2 c} \frac{1}{g_i} \sum_{m_i, m_k} |\langle km_k|\underline{r}|im_i\rangle|^2 . \tag{9.46}$$

Additional expressions, differing by factors of 2π from those of equations (9.43) and (9.46), will be obtained if the radiation field is expressed in terms of normal frequency rather than angular frequency units, since we require that

$$\rho(\omega)\, d\omega = \rho(\nu)\, d\nu.$$

It is often convenient to have expressions for the probability of transitions stimulated by an approximately undirectional beam of unpolarized light which fills the solid angle $d\Omega$. The transition rate is then $d\Omega/4\pi$ of the transition rate for isotropic radiation as may be verified from equation (9.40) (Problem 9.5).

9.5. Relations between Einstein B-coefficients and f-values

It is also convenient to have expressions for the Einstein B-coefficients in terms of the oscillator strengths of spectral lines introduced in section 4.8. Substituting equations (4.30) and (4.31) into equations (9.43) and (9.46) for the absorption coefficients we obtain

$$B^{\rho}_{ik} = \frac{\pi e^2 f_{ik}}{2\varepsilon_0 \hbar m \omega_{ki}} \tag{9.47}$$

$$B^{I}_{ik} = \frac{2\pi^2 e^2 f_{ik}}{\varepsilon_0 \hbar m \omega_{ki} c} . \tag{9.48}$$

Expressions for the coefficients for stimulated emission follow directly by using the relation $g_k B_{ki} = g_i B_{ik}$ given in equation (9.20).

9.6. The integral of the total absorption cross-section

We now consider a beam of radiation, intensity $I(\omega)$, uniformly filling the solid angle $d\Omega$. The power absorbed from this beam by one atom is given by the energy of one photon multiplied by the transition probability:

$$\text{Power absorbed} = \hbar\omega_{ik}P_{ik} = \hbar\omega_{ik}B^{I}_{ik}\, I(\omega)\,\frac{d\Omega}{4\pi}$$

$$= I(\omega)\, d\Omega\, \frac{\pi e^2 f_{ik}}{2\varepsilon_0 mc} \qquad (9.49)$$

where use has been made of equation (9.48). The power absorbed may also be expressed in terms of the integral of the total absorption cross-section per atom introduced in section 9.1 as follows:

$$\text{Power absorbed} = I(\omega)\, d\Omega \int_0^\infty \sigma(\omega)\, d\omega. \qquad (9.50)$$

From equations (9.49) and (9.50) we obtain an expression for the integral of the total absorption cross-section associated with the transition i to k:

$$\int_0^\infty \sigma(\omega)\, d\omega = \frac{\pi e^2 f_{ik}}{2\varepsilon_0 mc} = 2\pi^2 r_0 c f_{ik} \qquad (9.51)$$

This result should be compared with the classical expression, equation (9.12). We see that the f-value of an absorption line can be interpreted as that fraction of the integral of the classical total absorption cross-section which is to be associated with the given transition.

9.7. Introduction of the atomic frequency response

The Einstein B-coefficients discussed in sections 9.2-9.6 apply to the case of atoms interacting with a radiation field whose energy density per unit angular frequency interval, $\rho(\omega)$, is a slowly varying function of ω. The transition probabilities have effectively been averaged over the frequency response of the atoms in the ensemble. However, in many applications and particularly in the case of lasers, we require the transition rates induced by a monochromatic wave of angular frequency ω. These may be obtained by multiplying the frequency-averaged B-coefficients by the appropriate normalized line shape function, $g(\omega)$, discussed in Chapter 8. Thus for the case of a line whose shape is determined by natural or collision broadening, $g(\omega)$ will be the normalized Lorentzian distribution

$$g(\omega) = \mathcal{L}(\omega-\omega_{ki};\Gamma) = \frac{\Gamma/2\pi}{(\omega-\omega_{ki})^2+\Gamma^2/4} \tag{9.52}$$

More frequently $g(\omega)$ will be given by the Doppler-broadened lineshape (Problem 9.6)

$$g(\omega) = \mathcal{G}(\omega-\omega_{ki};\Delta) = \frac{2}{\Delta\sqrt{\pi}} \exp\{-4(\omega-\omega_{ki})^2/\Delta^2\}. \tag{9.53}$$

or alternatively one of the Voigt profiles discussed in section 8.5.

This lineshape factor was not included in the quantum-mechanical calculation of section 9.3 since the wavefunction was expanded in terms of the stationary states of $\mathcal{H}_0, \psi_n(\underline{r})$, whose energy uncertainty, ΔE_n, is identically zero. The classical calculation discussed in section 9.1 has the advantage that $g(\omega)$ is included explicitly, as may be seen by referring to equation (9.11).

Problems

9.1. Using the Fourier expansions defined in equations (9.2)-(9.4), prove that the energy absorbed by an oscillating electric dipole $\underline{p} = -e\underline{r}(t)$ from an electromagnetic field $\underline{E}(t)$ having a wide spectral distribution is given by

$$\Delta U = 2e \operatorname{Re} \int_0^\infty i\omega\, \underline{r}(\omega).\underline{E}^*(\omega)\, d\omega.$$

9.2. Show that the relation between the Einstein coefficients for spontaneous and stimulated emission,

$A_{ki}/B^{\rho}_{ki} = \hbar\omega^3_{ki}/\pi^2c^3$, is consistent with the requirement, equation (4.21), that the ratio of the stimulated and spontaneous transition probabilities is equal to the number of quanta per mode of the radiation field.

9.3. An atom is placed in a cavity and allowed to interact with black-body radiation. For radiative transitions:

(a) in the visible region at a frequency of 10^{15} Hz, and

(b) in the microwave region at a frequency of 10^9 Hz,

show that the probability of stimulated emission will equal that for spontaneous emission at temperatures of $6{\cdot}93 \times 10^4$ K and $6{\cdot}93 \times 10^{-2}$ K respectively.

This illustrates why resonant cavities for amplification by stimulated emission in the visible region (lasers) need to have very much greater Q values than those for the amplification of microwaves (masers).

9.4. An atom interacting with a beam of radiation is initially in an excited state $|i\rangle$ such that $E_i > E_k$. Show that the term oscillating as $\exp\{i(\omega_{ki}+\omega)t\}$ in equation (9.32) is now responsible for stimulated emission and hence prove that the probabilities for absorption and stimulated emission involving two non-degenerate levels are equal.

Making use of the results contained in section 5.2.2 or otherwise, show that the atomic transition probability per unit time induced by the absorption of unpolarized, isotropic radiation of energy density $\rho(\omega)$ is given by equation (9.41).

9.5. Prove that the transition probability, P_{ik}, for absorption induced by an approximately undirectional beam of unpolarized light of intensity $I(\omega)$ which uniformly fills the solid angle $d\Omega$ is given by

$$B^I_{ik}\ I(\omega)\ \frac{d\Omega}{4\pi} = \frac{4\pi^2e^2}{3\varepsilon_0\hbar^2c}\ \frac{1}{g_i}\sum|\langle km_k|\underline{r}|im_i\rangle|^2\ \frac{I(\omega)\,d\Omega}{4\pi}$$

9.6. In two cells containing helium gas and sodium vapour respectively, the atomic densities at 300 K are sufficiently small that the absorption line profiles are dominated by Doppler broadening. Show that the maximum absorption cross-sections per atom for:

(a) the sodium D_2 line at 5890 Å, $f_{ik} = 0{\cdot}650$, and

(b) the helium resonance line, $1\,{}^1S_0 - 2\,{}^1P_1$ at 584 Å, $f_{ik} = 0{\cdot}276$

are $1{\cdot}232 \times 10^{-11}$ cm^2 and $2{\cdot}164 \times 10^{-13}$ cm^2 respectively.

References

Einstein, A. (1917). *Phys.Z.* 18, 121.

ter Haar, D. (1967). *The old quantum theory*. Pergamon Press, Oxford.

General references and further reading

Jackson, J.D. (1972). *Classical electrodynamics*, Wiley, New York.

Loudon, R. (1973). *The quantum theory of light*. Clarendon Press, Oxford.

Panofsky, W.K.H. and Phillips, M. (1962). *Classical electricity and magnetism*. Addison-Wesley, Reading, Mass.

Schiff, L.I. (1968). *Quantum mechanics*. McGraw-Hill, New York.

10
Radiative transfer and the formation of spectral lines

We shall now apply the concepts developed in Chapter 9 to a discussion of the emission and absorption of radiation by an excited gas. We start by deriving the equation of radiative transfer in terms of the volume emission and absorption coefficients for line radiation and consider simple solutions for the case of uniformly excited sources. The terms source function and optical thickness are defined and the effect of self absorption and self reversal in optically thick sources is outlined.

We then turn to a consideration of the absorption lines produced by the transmission of a continuous spectrum through an absorbing vapour. The concept of equivalent width is explained and we show that the equivalent width of an absorption line is determined by the product $N_i f_{ik} L$. Details are given of the measurement of relative oscillator strengths by the absorption technique using a King furnace.

Absorption spectrophotometry also provides a sensitive technique for the measurement of the density of atoms. We consider the applications of this technique to the determination of the abundance of the elements in the solar atmosphere, to spectrochemical analysis, and to the measurement of metastable and ground-state densities in gases or vapours.

10.1. Derivation of the equation of transfer

We shall now derive a simple equation which determines the change of intensity $I_\omega(x)$ of an approximately collimated beam of radiation confined to the solid angle $d\Omega$ as it propagates in the direction $0x$ through a gas of excited atoms. The radiation is assumed to be unpolarized and its angular frequency ω is chosen to lie close to that of an atomic absorption line, ω_{ki}. The absorption and emission coefficients of the gas will then be determined mainly by the

Einstein A- and B-coefficients associated with the transition $k \to i$. The transfer equation is obtained by considering the change in the radiant energy contained within the angular frequency interval $d\omega$ as the beam passes through a cylindrical volume-element of cross-section dS and length dx, as shown in Fig. 10.1.

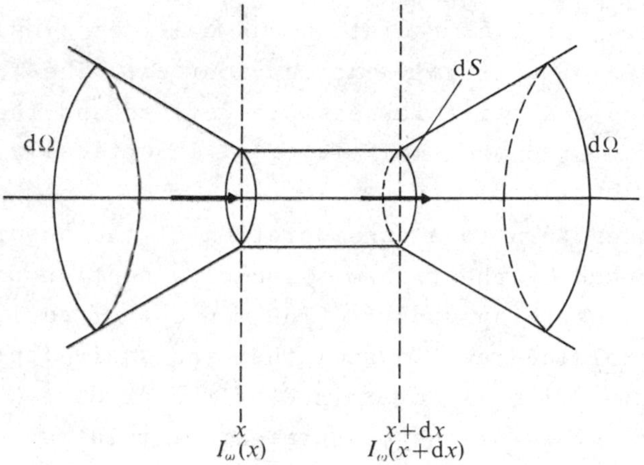

Fig.10.1. The geometric volume-element considered in the derivation of the equation of radiative transfer for a beam of light confined to the solid angle $d\Omega$.

As $dx \to 0$ in Fig.10.1 we have

$$\{I_\omega(x+dx) - I_\omega(x)\}\, dS\, d\omega\, d\Omega\, dt = \frac{dI_\omega}{dx}\, dx\, dS\, d\omega\, d\Omega\, dt. \quad (10.1)$$

The energy absorbed from the beam as it passes through the volume-element is given by the energy of one photon multiplied by the number of upward transitions occurring in the time interval dt. If there are N_i atoms per unit volume in the level i we have

$$\text{Energy absorbed} = \hbar\omega N_i\, dx\, dS\, B^I_{ik}\, g(\omega)\, d\omega\, \frac{d\Omega}{4\pi}\, I_\omega(x)\, dt. \quad (10.2)$$

The lineshape factor, $g(\omega)$, has been introduced to describe the atomic frequency response, as discussed in section 9.7, while the term $d\Omega/4\pi$ takes account of the fact that we are considering transitions induced by a collimated beam of radiation confined to the solid angle $d\Omega$. Atoms in the excited level k give energy to the beam by both spontaneous and stimulated emission of radiation into the solid angle $d\Omega$ and frequency interval $d\omega$. The energy emitted in the time interval dt is given by

$$\text{Energy emitted} = \hbar\omega N_k \, dx \, dS \, A_{ki} \, g(\omega) \, d\omega \, \frac{d\Omega}{4\pi} \, dt + $$
$$+ \hbar\omega N_k \, dx \, dS \, B^I_{ki} \, g(\omega) \, d\omega \, \frac{d\Omega}{4\pi} \, I_\omega(x) \, dt. \quad (10.3)$$

Using equations (10.2) and (10.3) in equation (10.1), the equation of radiative transfer becomes

$$\frac{dI_\omega}{dx} = \frac{\hbar\omega}{4\pi} \{A_{ki}N_k + (B^I_{ki}N_k - B^I_{ik}N_i)I_\omega\} \, g(\omega). \quad (10.4)$$

Introducing the relation $g_i B^I_{ik} = g_k B^I_{ki}$, we find that the equation of transfer may be written more concisely as

$$\frac{dI_\omega}{dx} = \varepsilon_\omega - \kappa_\omega I_\omega \quad (10.5)$$

where ε_ω and κ_ω are the volume emission and absorption coefficients respectively defined by

$$\varepsilon_\omega = \frac{\hbar\omega}{4\pi} A_{ki} N_k g(\omega) \quad (10.6)$$

and

$$\kappa_\omega = \frac{\hbar\omega}{4\pi} B^I_{ik} N_i \left(1 - \frac{g_i N_k}{g_k N_i}\right) g(\omega). \quad (10.7)$$

In the M.K.S. system the absorption coefficient, κ_ω, has units of m^{-1}. In most physical situations and particularly in the examples considered in this chapter, the upper and lower level population densities satisfy $g_i N_k / g_k N_i \ll 1$. Consequently the absorption coefficient, equation (10.7), is positive. However, there are a few systems where conditions can be chosen so that a population inversion exists for some of the possible transitions. In these cases the absorption coefficient is negative and an electromagnetic wave travelling through the medium will be amplified rather than attenuated. Some examples of these systems will be discussed in Chapters 11-13.

Once the emission and absorption coefficients of the excited gas are specified as functions of frequency and position, the equation of transfer, equation (10.5), may in principle be solved to obtain the intensity at any point. In practice this equation is difficult to solve in all but the simplest possible situations, some of which are considered in the next section. Nevertheless, solutions of the general equation of radiative transfer are essential to a complete understanding of the behaviour of stellar and planetary atmospheres.

10.2. *Solution of the transfer equation for uniformly excited sources*

We now assume that the volume emission and absorption coefficients are independent of x, i.e. the atomic densities N_k and N_i and the lineshape factor $g(\omega)$ are constant. The equation of transfer, equation (10.5), may then be solved by multiplication by the integrating factor $\exp(\kappa_\omega x)$, giving

$$\frac{d}{dx}\{I_\omega(x)\ \exp(\kappa_\omega x)\} = \varepsilon_\omega\ \exp(\kappa_\omega x). \qquad (10.8)$$

If the excited gas is taken to be in the form of a slab bounded by the planes x=0 and x=L, the intensity of radiation at the angular frequency ω that would be observed at the face x=L is given by

$$I_\omega(L) = I_\omega(0)\ \exp(-\kappa_\omega L) + \frac{\varepsilon_\omega}{\kappa_\omega}\{1 - \exp(-\kappa_\omega L)\} \qquad (10.9)$$

where $I_\omega(0)$ is the intensity of the beam of radiation incident on the sample at x=0.

The ratio $S_\omega=\varepsilon_\omega/\kappa_\omega$ appearing in equation (10.9) is usually referred to as the *source function* and the quantity $\tau_\omega(L) = \kappa_\omega L$ is known as the *optical thickness* of the source. It determines the amount by which a collimated beam of intensity $I_\omega(0)$ is attenuated as it passes through the medium. When the absorption coefficient is a function of position, a more general expression for the optical thickness is

$$\tau_\omega(z) = \int_0^z \kappa_\omega(z')\, dz' \tag{10.10}$$

where z=L-x is measured from the face x=L into the medium. In this case $\tau_\omega(z)$ would more correctly be termed the *optical depth*. The distance z at which $\tau_\omega(z) \approx 1$ represents the effective depth within the sample from which most of the light of frequency $\omega/2\pi$ is emitted. For a commercial sodium discharge lamp this depth will vary from a fraction of a mm at the centre of one of the resonance lines to more than a cm in the wings of the line, owing to the rapid variation of κ_ω with ω as determined by the line profile, $g(\omega)$.

10.2.1. Optically thin sources, $\tau_\omega(L) \ll 1$. To investigate the intensity of radiation emitted by a uniformly excited column of gas of length L, we set the incident intensity, $I_\omega(0)$, equal to zero and equation (10.9) reduces to

$$I_\omega(L) = \frac{\varepsilon_\omega}{\kappa_\omega}\{1 - \exp(-\kappa_\omega L)\}. \tag{10.11}$$

If the effective optical depth is very small,

$$\tau_\omega(L) = \kappa_\omega L \ll 1,$$

the source is said to be optically thin. Expanding the exponential factor in equation (10.11) we have

$$I_\omega(L) \approx \varepsilon_\omega L = \frac{\hbar\omega}{4\pi} A_{ki}\, N_k\, g(\omega)\, L. \tag{10.12}$$

In this case the intensity is proportional to the density of excited atoms, N_k, to the spontaneous emission transition probability, A_{ki}, and to the length of the column of excited gas, L. The spectral distribution of the emitted radiation is identical to the intrinsic line profile, $g(\omega)$. This particular relation was an important postulate underlying the discussion of the shape of spectral lines presented in Chapter 8. The measurement of the intensities of spectral lines emitted by optically thin sources is the basis of two widely-used techniques: the spectrochemical analysis of metals and other materials and the determination of oscillator strengths of atomic and molecular transitions. Details of the first are given in Sawyer (1963) and of the second by Foster (1964). A comprehensive table of f-values determined by the emission method has been published by Corliss and Bozmann (1962).

10.2.2. Optically thick sources, $\tau_{\omega}(L) \gg 1$. In the case of spectral lines which terminate on the ground, metastable, or resonance levels of an atom the assumption that $\tau_{\omega}(L) \ll 1$ is frequently invalid. Substantial populations accumulate in these levels in a gas discharge and consequently the emitted radiation has a large probability of being re-absorbed before it leaves the source. When the optical depth is of the order of or greater than unity, it is convenient to substitute for ε_{ω} and κ_{ω} in equation (10.11) the explicit expressions given by equations (10.6) and (10.7). The ratio A_{ki}/B^{I}_{ik} which occurs can be expressed in terms of fundamental constants by using equation (9.20) and recalling that $B^{I}_{ik} = 4\pi B^{\rho}_{ik}/c$. Finally the emitted intensity is given by

$$I_{\omega}(L) = \frac{\hbar\omega^3}{4\pi^3 c^2} \frac{\{1 - \exp(-\kappa_{\omega}L)\}}{\{(N_i g_k/N_k g_i) - 1\}} \tag{10.13}$$

As the number density of atoms or the length of the column of excited gas is increased, there is initially a proportional increase in the emitted intensity $I_{\omega}(L)$. However, when the optical depth, $\kappa_{\omega}L$, is of the order of unity self absorption becomes appreciable and the intensity increases more slowly. Simultaneously the profile of the emitted radiation becomes broader and then starts to flatten off as shown in Fig.10.2. Thus self absorption tends to level out intensity differences between spectral lines. When this is undetected it can cause serious errors to be made in the relative oscillator strengths of spectral lines determined from intensity measurements. Self absorption may also distort the intensity distributions predicted in fine or hyperfine structure multiplets. For instance, the intensity ratio of the sodium D lines emitted by a sodium discharge lamp is usually observed to be approximately 1:1 instead of the ratio 2:1 predicted theoretically.

Eventually, when $\kappa_\omega L \gg 1$, self absorption leads to an equilibrium between the radiation and the collision processes responsible for the excitation of atoms in the gas. If the populations of the levels i and k are maintained in thermal equilibrium at the electron temperature T_e then the limiting intensity, obtained by using $\kappa_\omega L \gg 1$ in equation (10.13), is given by the black-body distribution

$$I_\omega(L) = \frac{\hbar\omega^3}{4\pi^3 c^2}\{\exp(\frac{\hbar\omega}{kT_e}) - 1\}^{-1} . \tag{10.14}$$

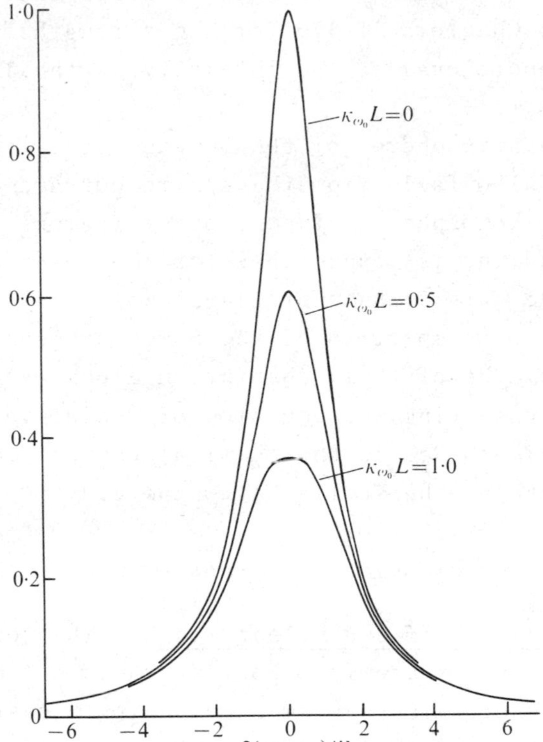

Fig.10.2. The effect of self-absorption on the profiles of spectral lines emitted by a uniformly excited column of gas.

Thus the measurement of the absolute intensity of an optically-thick line would enable the effective temperature of the system to be determined.

10.3. Non-uniform sources

In most laboratory sources the excitation temperature decreases towards the boundary of the discharge region. Consequently the absorption in this part of the source is increased and the profile of the emitted radiation is not only appreciably broadened but often shows a pronounced dip at the centre of the line. This effect is known as *self reversal* and it has been studied in detail by Cowan and Diecke (1948). It is particularly important that self reversal is avoided in the lamps used in the resonance fluorescence experiments described in Chapters 15-17, for the strengths of the signals are proportional to the intensity at the line centre frequency $\omega_{ki}/2\pi$.

In the atmospheres of the Sun and stars the excitation temperature also falls from the centre outwards to the boundary of the photosphere. In the outer regions this temperature gradient is responsible for the absorption line spectrum (the dark Fraunhofer lines) which is superimposed on the continuous spectrum of the Sun. By measuring the equivalent widths of these absorption lines (see section 10.4), and by solving the equation of radiative transfer, it is possible to deduce the chemical composition and physical state of the stellar atmosphere.

10.4. Equivalent widths of absorption lines

10.4.1. Definition of equivalent width. The source function, $\varepsilon_\omega/\kappa_\omega$, for a gas or atomic vapour which is not excited by an electric discharge is very small even for a system in thermodynamic equilibrium at 2000 K (Problems 10.1 and 10.2). Spontaneous and stimulated emission can therefore often be neglected. Consequently, the intensity of radiation of angular frequency ω transmitted by a column of absorbing gas whose temperature and density is uniform is given by equation (10.9) as

$$I_\omega(L) = I_\omega(0)\exp(-\kappa_\omega L). \tag{10.15}$$

If the incident radiation comes from a tungsten lamp or high-pressure xenon arc, $I_\omega(0)$ will be essentially constant in the region of an absorption line. The transmitted intensity will then have a dip centred on ω_{ki} and the spectral line is observed in absorption, as shown in Fig.10.3. The *equivalent width*, W_ω, of the absorption line is defined as the width of a rectangular strip of height $I_\omega(0)$ which has the same area as that of the absorption line. This equality is represented by the two shaded areas in Fig.10.3 and leads to the expression:

$$W_\omega I_\omega(0) = \int \{I_\omega(0) - I_\omega(L)\}\, d\omega \tag{10.16}$$

where the integral is taken over the profile of the absorption line. From equation (10.16), the equivalent width is given by

$$W_\omega = \int \left\{1 - \frac{I_\omega(L)}{I_\omega(0)}\right\} d\omega \tag{10.17}$$

where W_ω has units of angular frequency. In astrophysical applications the intensity is usually measured as a function of the wavelength λ, and the equivalent width W_λ, which has units of length, is defined by

$$W_\lambda = \int \left\{1 - \frac{I_\lambda(L)}{I_\lambda(0)}\right\} d\lambda = \frac{\lambda_{ki}^2}{2\pi c} W_\omega. \tag{10.18}$$

In older references (e.g. Mitchell and Zemansky 1966) the equivalent width is often referred to as the *total absorption* since the total power absorbed from the beam is given by $W_\omega I_\omega(0)$. Thus it is not essential to resolve the absorption line and the equivalent width may be measured using a spectrometer of only moderate resolving power. However, the results should be corrected for the effect of light scattered by optical components in the spectrometer

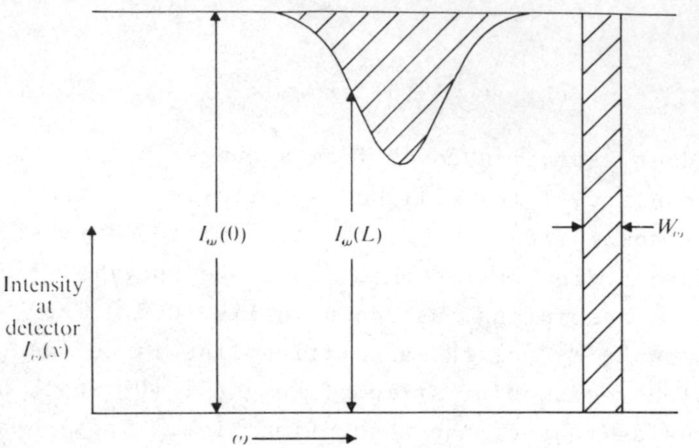

Fig.10.3. Definition of the equivalent width of an absorption line, W_ω, in angular frequency units. The areas of the shaded sections are equal.

and serious systematic errors may occur if the instrumental profile possesses extensive wings. It is therefore preferable to employ a high-resolution instrument, if this is available, and to obtain W_ω by making measurements of transmitted intensity at a number of discrete points on the profile of the absorption line.

The equivalent width is given in terms of the absorption coefficient by substituting equation (10.15) into equation (10.17), giving

$$W_\omega = \int \{1 - \exp(-\kappa_\omega L)\}\, d\omega. \tag{10.19}$$

Referring to equation (10.7) we see that the equivalent width of an absorption line is determined by the product $B^I_{ik}N_iL$, or alternatively $f_{ik}N_iL$ when the oscillator strength is introduced. Thus the measurement of equivalent widths is an important method for determining either oscillator strengths or atomic densities and further details of these techniques are given in the sections which follow.

10.4.2. Equivalent width for optically-thin samples, $\kappa_\omega L \ll 1$.

For a column of absorbing gas in which the optical depth at the centre of the line is small compared to unity, $\kappa_\omega L \ll 1$, we may expand the exponential factor in equation (10.19) and to a good approximation we have

$$W_\omega \approx \int \kappa_\omega L \, d\omega$$

$$= \frac{\hbar\omega_{ki}}{4\pi} B^I_{ik} N_i L\left(1 - \frac{g_i N_k}{g_k N_i}\right) \int g(\omega)\, d\omega. \qquad (10.20)$$

The integral in equation (10.20) is equal to unity since the lineshape function $g(\omega)$ is normalized. Also for most transitions in the visible region, the ratio of upper to lower state populations in thermal equilibrium satisfies

$$\frac{g_i N_k}{g_k N_i} = \exp(-\hbar\omega_{ki}/kT) \ll 1 \qquad (10.21)$$

at all temperatures which are attainable in the laboratory. Thus equation (10.20) reduces to

$$W_\omega = \frac{\hbar\omega_{ki}}{4\pi} B^I_{ik} N_i L. \qquad (10.22)$$

Introducing the absorption f-value for the transition from equation (9.48) we have finally (Problems 10.3-10.5)

$$W_\omega = \frac{\pi e^2}{2\varepsilon_0 mc} f_{ik} N_i L. \qquad (10.23)$$

In the optically-thin case the f-value of an absorption line can be obtained directly from a measurement of the equivalent width provided that the density of absorbing atoms, N_i, is known. Experiments in which this technique has been used to determine oscillator strengths of astrophysical interest are described in section 10.5. Alternatively, if the f-value of the line is known, a measurement of W_ω enables the density of absorbing atoms to be obtained. This forms the basis of the experimental techniques discussed in section 10.6.

10.4.3. Equivalent width for optically-thick samples, $\kappa_\omega L \gg 1$. When the approximation $\kappa_\omega L \ll 1$ is no longer valid, the complete expression for the equivalent width, equation (10.19), must be used. The frequency dependence of the absorption coefficient, equation (10.7), is given by the normalized line shape $g(\omega)$ which is usually taken to be a Voigt function. However, the shape of the Voigt profile depends on the ratio of the Lorentzian to the Gaussian line-width parameters, Γ/Δ, which is not usually accurately known. In this case it is therefore difficult to obtain the oscillator strength of an absorption line directly from a measurement of its equivalent width. Consequently values of W_ω are measured as a function of the density of atoms in the absorbing gas and a graph of log W_ω is plotted against log N_i. This is a *curve of growth*. Theoretical curves of growth are also plotted by evaluating equation (10.19) using a range of different parameters $\alpha = 2\Gamma/\Delta$ as shown in Fig.10.4(a). At low densities the equivalent width is proportional to N_i, in agreement with equation (10.23), but as N_i increases the curve of growth flattens out and eventually at high densities it can be shown (Problem 10.6) that

$$W_\omega = \left(\frac{\pi e^2 f_{ik} N_i L\Gamma}{\varepsilon_0 mc} \right)^{1/2} \qquad (10.24)$$

The comparison of the theoretical and experimental curves of growth enables the best values of Γ/Δ and f_{ik} to be determined.

Experimental results for the resonance line of copper, $4\ ^2S_{1/2}-4\ ^2P_{3/2}$, at 3247 Å obtained by Bell and Tubbs (1970) are shown in Fig.10.4(b). The full line is the theoretical curve of growth which gives the best fit to the experimental data points. The absorption f-value obtained by this investigation is $f_{ik} = 0{\cdot}43 \pm 0{\cdot}02$ and is in good agreement with the results obtained by other techniques.

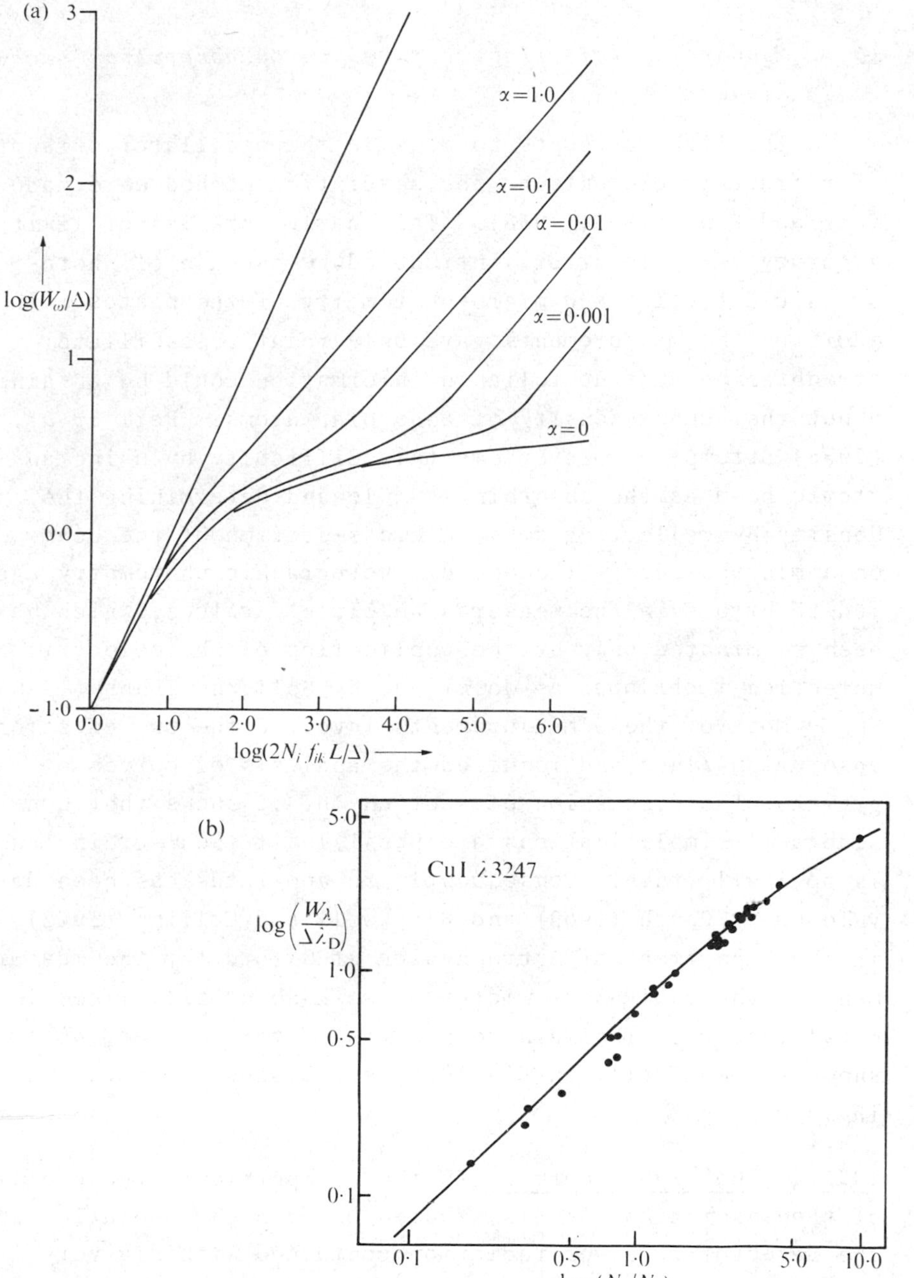

Fig.10.4. Equivalent width as a function of the product $N_i f_{ik} L$ in optically thick samples. (a) Theoretical curves of growth for different values of $\alpha=2\Gamma/\Delta$. (b) Experimental curve of growth for the resonance line of copper, $4\ ^2S_{1/2}-4\ ^2P_{3/2}$, at 3247 Å obtained by Bell and Tubbs (1970).

10.5. Measurement of relative f-values by absorption techniques

The first attempts to measure the oscillator strengths of refractory elements by the absorption method were made by King and King (1935, 1938). This early work was of limited accuracy owing to errors introduced by the use of photographic detection and micro-photometry of the plates. In addition the measurements gave only relative oscillator strengths because no reliable information could be obtained about the number density of absorbing atoms. Bell *et al.* (1958) attempted to overcome this difficulty by using an atomic beam as the absorbing sample and determining the density by collecting a known cross-section of the beam on a microbalance. The use of photographic photometry again led to errors in the measured absolute f-values, which have been eliminated only by the application of photoelectric detection techniques as described by Bell and Tubbs (1970).

Most of these measurements involved the use of strong absorption lines and required the analysis of curves of growth. The discussion of section 10.4.2 shows that considerable simplifications are possible if the absorbing sample is optically thin. Consequently an apparatus has been developed by Peach (1969) and Blackwell and Collins (1972) in the Department of Astrophysics at Oxford for the measurement of the equivalent widths of weak absorption lines in metal vapours. The main components of this system, which is shown schematically in Fig.10.5, are discussed in the following sections.

10.5.1. The light source. In these experiments the profile of the absorption line is studied under high resolution and the detector receives radiation contained within a very small wavelength or frequency interval. Moreover, the relatively small solid angle of the spectrometer allows only a small fraction of the light emitted by the source to be collected. Consequently a source of high intensity is required and in these experiments the continuum emission from

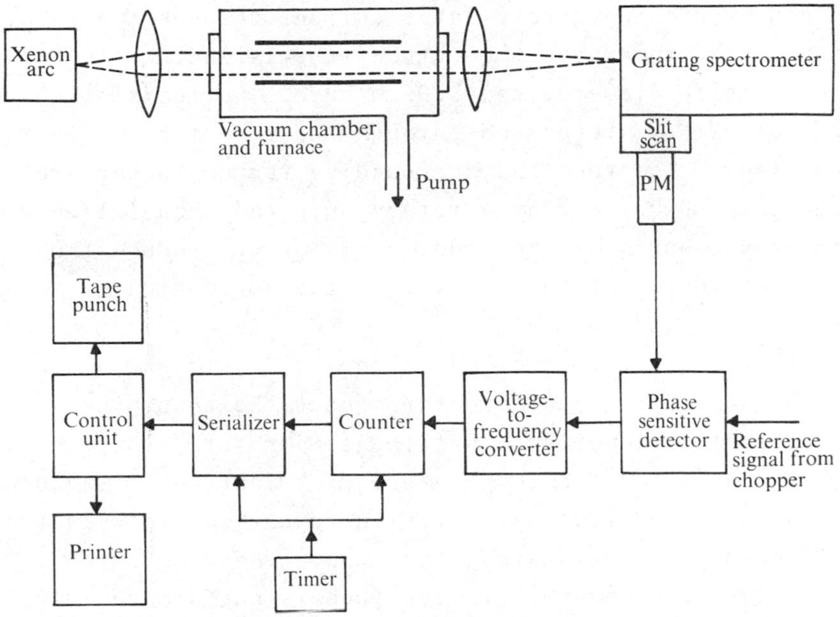

Fig.10.5. Schematic diagram of King furnace and spectrometer for measurements of relative f-values by the absorption technique. (After Collins (1970).)

a high pressure xenon arc is used.

10.5.2. The absorption cell. A few elements, such as the alkali metals, have appreciable vapour pressures at easily attainable temperatures and may be studied by using sealed absorption cells of glass or quartz. However, for astrophysical applications there is considerable interest in the elements of the first transition series, e.g. manganese, iron, cobalt, and nickel. For these elements suitable vapour pressures are only attained at temperatures above 2000 K and the standard technique involves the use of a carbon furnace introduced by King (1922) as the absorption cell. At high temperatures, carbon is one of the few suitable materials since it reacts only slowly with most metals and has a low vapour pressure.

A King furnace consists basically of a carbon tube

mounted in a vacuum enclosure. The high temperature is attained by passing a very large current through the tube. In the Oxford apparatus the carbon tube is 122 cm long, 7.6 cm outside diameter and 5.0 cm bore. Currents of up to 5000 A at 42·5 V are passed through the tube from a 200 kW transformer. The tube and surrounding vacuum jacket are evacuated to 0·02 Torr by a rotary pump and then filled with argon at pressures of the order of 5 Torr to reduce the rate of diffusion of the metal from the hot region of the tube.

In the derivation of equation (10.23) it was assumed that the density was uniform throughout the absorbing vapour. It is therefore important that the temperature along the tube is constant. The temperature distribution is measured by sighting an optical pyrometer on to carbon blocks which are fixed at intervals along the tube. These measurements show that the temperature in the tube is uniform to ± 10 K over a length of 80 cm.

10.5.3. The spectrometer and detection system. The intrinsic width of a weak absorption line is determined mainly by the Doppler width characteristic of the furnace temperature. Thus to obtain detailed measurements across the line profile it is necessary to employ a spectrometer of high resolving power. The Oxford instrument has an échelle grating with 316 grooves mm^{-1}, blazed at 63° and used in a Littrow mounting. The grating is 256 × 128 mm, giving a theoretical resolving power of $6{\cdot}93 \times 10^5$ in 9th order. A resolution of 95 per cent of the theoretical value has been achieved in practice. The collimator lens has a focal length of 9·60 m giving a dispersion of 6·4 mm/Å in the focal plane. The small free spectral range of the main instrument makes it necessary to use a small prism monochromator as a pre-disperser.

Light transmitted by the spectrometer is detected by an EMI 9558 photomultiplier. This is combined with a selective chopper and phase-sensitive detector to generate a signal proportional to the difference $\{I_\omega(0)-I_\omega(L)\}$. As the exit slit of the spectrometer is stepped across the absorption line profile, the output from the phase-sensitive detector is converted into digital form and punched on paper tape for detailed analysis, as indicated in Fig.10.5. The system enables long integration times to be used and has the advantage that the absorption line profile is measured from a zero base-line. In addition, source fluctuations are compensated by a feedback loop which controls the high voltage applied to the photomultiplier. This low noise detection system is capable of measuring the profiles of very weak absorption lines and so enables experiments to be performed on the linear portion of the curve of growth.

A typical scan of an absorption line in the spectrum of iron is shown in Fig.10.6. The detection system has been calibrated by placing in the beam coated glass plates whose attenuation has been measured in separate experiments.

<u>10.5.4. Discussion of some typical results.</u> The equivalent width of an absorption line, such as that shown in Fig.10.6, may be obtained by integrating the area under the absorption profile. To obtain the oscillator strength from W_ω (equation (10.23)) it is necessary to know the density of absorbing atoms. Unfortunately in the King furnace the vapour density cannot be deduced from the measured temperature since the system is not in thermal equilibrium. There is a constant diffusion metal vapour as atoms evaporate from the hot central section of the furnace and then condense on the cooler outer sections. Absolute f-values are therefore very difficult to obtain by this method. However, the ratios of f-values of several absorption lines which all start from the same lower level i can be obtained since in this case the unknown atomic density, N_i, is eliminated. Some of the results for the product $g_i f_{ik}$ obtained by Blackwell and Collins (1972) for lines in the spectrum of neutral manganese are given in Table 10.1. The results have been normalized to the gf value for the 4030·75 Å line which is arbitrarily taken as 100. There is good agreement

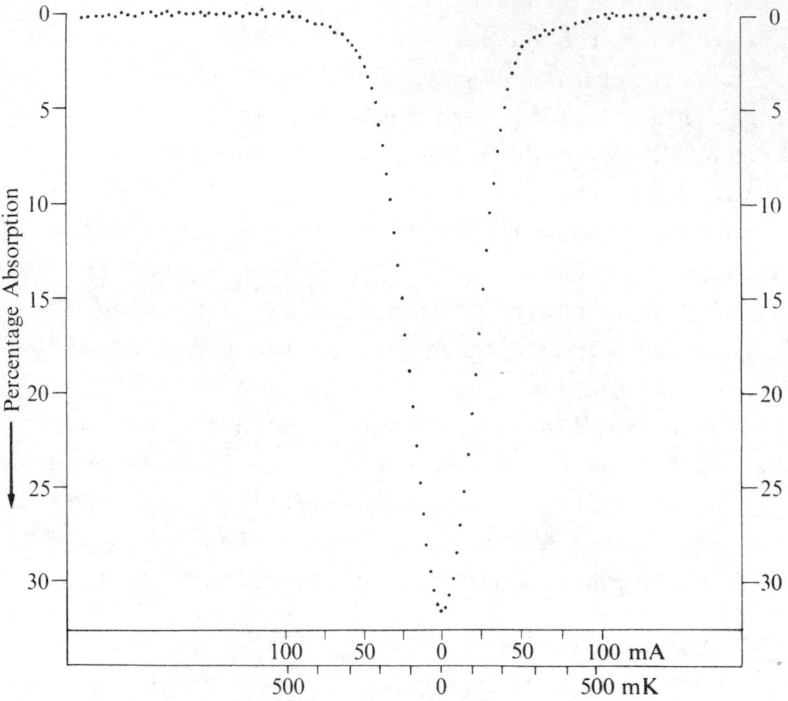

Fig.10.6. Absorption line of iron at 4427 Å obtained using the King furnace and scanning spectrometer shown in Fig.10.5. The integration time was 1 s per point and the spectrometer slit was adjusted to give a resolution $\lambda/\Delta\lambda = 1{\cdot}1 \times 10^5$. (After Collins (1970).)

with the results of Ostrovsky and Penkin (1957) obtained by the hook-method, but the result for the 4034·49 Å line obtained by Corliss and Bozmann (1962) by the emission method is in serious error. This is undoubtedly due to the effects of self absorption in the source which were emphasized in section 10.2.2. The relative f-values obtained by the absorption technique for lines which originate on the same lower level are probably accurate to better than 5 per cent.

The relative f-values of absorption lines which start on different lower levels i and i' can be obtained if the ratio of the populations $N_i/N_{i'}$ is assumed to be given by

TABLE 10.1.

Relative oscillator strengths for the transitions $a\ ^6S_{5/2} - z\ ^6P_{3/2,5/2,7/2}$ *in the spectrum of neutral manganese.*

Wavelength (Å)	Blackwell and Collins (1972) gf	Ostrovsky and Penkin (1957) gf	Corliss and Bozmann (1962) gf	L-S coupling gf
4030·755	100	100	100	100
4033·073	70·42 ± 1·4	71·43	69·6	75
4034·490	46·63 ± 1·3	48·26	25·3	50

the Boltzmann distribution. This assumption is probably valid, provided that i and i' are not separated by more than 1-2 eV. In this case we have from equation (10.23)

$$\frac{g_i f_{ik}}{g_{i'} f_{i'k'}} = \frac{W_\omega(ik) L_{i'k'}}{W_\omega(i'k') L_{ik}} \exp\{(E_i - E_{i'})/kT\} \qquad (10.25)$$

where L_{ik} and $L_{i'k'}$ are the effective absorbing path lengths for the two lines concerned. These are not equal due to the fall in temperature from the centre to the ends of the absorption tube. The ratio $L_{i'k'}/L_{ik}$ may be estimated by assuming that the vapour density at any point is proportional to the vapour pressure of the pure substance at the local temperature and that the same temperature determines the distribution among the excited levels. Results obtained by Peach (1969) for lines in the spectrum of neutral iron are given in Table 10.2. The large error estimate for the 5167·49 Å line is due to the uncertainties in the effective path correction and the measured temperatures.

TABLE 10.2.

Relative f-values for lines in the spectrum of Fe I *obtained by measurement of equivalent widths*

Wavelength (Å)	Transition	Energy of lower level (eV)	gf
5166·29	a $^5D_4$ - z $^7D^o_5$	0	1
5168·90	a $^5D_3$ - z $^7D^o_3$	0·05	1·61 ± 0·03
5167·49	a $^3F_4$ - z $^3D^o_3$	1·48	906 ± 136

10.6. *Determination of chemical composition and atomic densities by absorption techniques*

<u>10.6.1. Abundance of elements in the solar atmosphere.</u> Using the King furnace absorption technique, Blackwell and Collins (1972) measured the relative f-values of 20 lines in the spectrum of neutral manganese. These were then converted to an absolute scale using results previously obtained by Ostrovsky and Penkin (1957). Many of the same transitions appear as absorption lines in the solar spectrum, a section of which is given in Fig.10.7. This spectrum was taken with the low-noise spectrometer described in section 10.5 combined with a coelostat. Since the f-values of these lines are now known it should be possible to obtain the solar abundance of manganese from a measurement of the equivalent widths of the observed lines.

Unfortunately the inversion procedure is rather involved. In order to calculate the theoretical equivalent

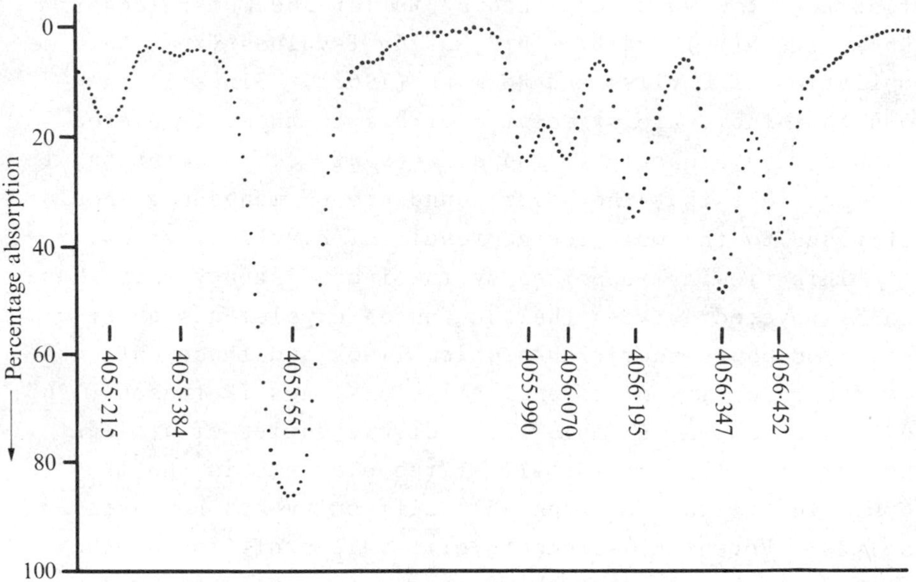

Fig.10.7. High-resolution scan of the solar spectrum showing absorption lines of neutral manganese. (After Collins (1970).)

width of a solar absorption line it is necessary to first construct a model of the Sun's atmosphere giving, amongst other things, the abundance of important elements and the variation of temperature and pressure with depth. Using this model atmosphere the line and continuum absorption coefficients must be calculated. Finally the equation of radiative transfer must be solved to give theoretical line profiles for assumed values of the product $g_i f_{ik} N_i$, where N_i is proportional to the solar abundance of the element under consideration.

Calculations along these lines have been performed by Blackwell *et al.* (1972) using experimentally determined gf-values. When the results were compared with the measured equivalent widths of the manganese solar absorption lines they obtained a solar abundance of $\log N(\mathrm{Mn}) = 5{\cdot}4 \pm 0{\cdot}2$ referred to an assumed hydrogen abundance of $\log N(\mathrm{H}) = 12{\cdot}0$. This result is significantly larger than the value obtained by Müller and Mutschlecner (1964), log N(Mn) = 4·81 ± 0·2, using f-values from the compilation of Corliss and Bozmann (1962). Since it is known that there are systematic errors in these tabulated f-values, due probably to the effects of self absorption, it seems possible that the solar abundance of manganese should be revised to the more recent result of Blackwell *et al.*

This view is supported by the inconsistency which has been discovered between the abundances of elements determined from photospheric absorption lines and those obtained from the intensity of coronal emission lines (Pottasch (1963, 1964)). As shown in Table 10.3, discrepancies of at least a factor of ten exist for all of the elements in the iron group. In the case of iron this difficulty has now been resolved. Recent measurements of f-values of iron by the beam-foil technique have shown conclusively that previous f-values for iron obtained by the emission method contained serious systematic errors and there has been a consequent revision of the abundance of iron in the Sun towards the coronal value.

10.6.2. Atomic absorption in spectrochemical analysis. Atomic absorption spectrophotometry is widely applied in the metallurgical, medical, agricultural, and industrial fields for the determination of trace concentrations of many elements. The technique is based on the measurement of the absorbance, $\ln\{I_0(0)/I_0(L)\}$, of an optically-thin atomic vapour at the peak of the absorption line. From equations (10.7), (10.15), and (9.48) we obtain

$$\text{Absorbance} = \ln\left\{\frac{I_0(0)}{I_0(L)}\right\} = \frac{\pi e^2}{2\varepsilon_0 mc} f_{ik} N_i L\ g(\omega_0). \quad (10.26)$$

Thus the absorbance is directly proportional to the atomic concentration N_i which it is desired to measure.

A typical atomic absorption spectrophotometer is shown schematically in Fig.10.8. It consists of a hollow-cathode lamp emitting an intense, narrow, and unreversed

TABLE 10.3

Comparison of solar abundances of iron group elements derived from photospheric absorption and coronal emission lines

Element	log N		
	Photosphere		Corona
	Goldberg *et al.* (1960)	Müller and Mutschlecner (1964)	Pottasch (1964)
Hydrogen	12·0	12·0	12·0
Chromium	5·36	5·07	6·0
Manganese	4·90	4·81	5·7
Iron	6·57	6·70	7·87
Cobalt	4·64	4·41	5·6
Nickel	5·91	—	6·72

line of the element which it is desired to measure. Usually the resonance line of the element is chosen. The absorbing vapour is formed by spraying a solution of the material for which the analysis is required into the flame of a carefully designed burner. A mixture of air and acetylene burning at 2325°C is most commonly used, but for elements which form refractory compounds it may be necessary to employ the nitrous-oxide/acetylene flame which burns at 2700°C. In the flame the solvent rapidly evaporates and chemical compounds undergo thermal dissociation to form an atomic vapour containing the element of interest.

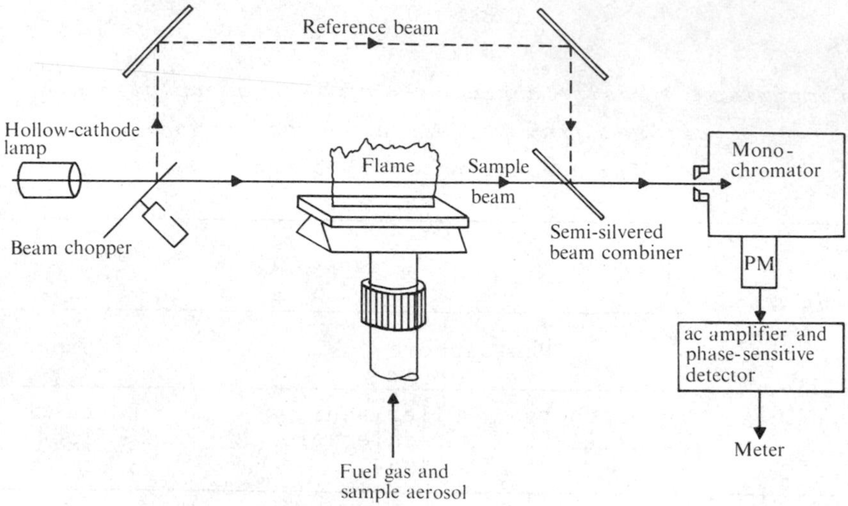

Fig.10.8. Schematic diagram of an atomic absorption spectrophotometer used for chemical analysis.

Light from the hollow-cathode lamp is focussed into the flame and then refocussed on to the slit of a monochromator. The monochromator serves simply to isolate the spectral line chosen for analysis and thus only a moderate resolving power is required. The light is detected by a photomultiplier and the absorption signal is obtained by phase-sensitive detection using the double beam and chopper arrangement shown in Fig.10.8. This eliminates background emission from the flame and enables a.c. amplification to be used. The absorbance may be obtained directly by incorporating a logarithmic amplifier and the signal is displayed either digitally or on a moving-coil meter.

The instrument is usually calibrated by measuring the absorbance produced by a range of standard solutions and then plotting an empirical working curve. In this way atomic absorption spectrophotometry may be used for the analysis of some 68 elements and in many cases the limit of sensitivity is as low as 0·1 parts per million. Unfortunately the assumption that the linewidth of the source is negligible compared with the linewidth of the absorption coefficient of the flame is not in fact correct and the linear relation predicted by equation (10.26) holds only at the lowest concentrations. For direct absolute measurements of atomic densities these line profiles must be considered in detail. This forms the subject of the following section.

10.6.3. Measurement of atomic densities by optical absorption

A knowledge of the densities of atoms in the ground and metastable levels is often essential for the interpretation of data in atomic collision experiments. This is particularly true for the alkali metals where optical pumping experiments are frequently used to determine collision cross-sections. Since data obtained from vapour pressure tables are unreliable, the required densities are often obtained by measuring the absorption of the gas or vapour when illuminated by light of the resonance or other suitable lines. The f-values of the resonance lines of the alkalis are known from direct lifetime and Hanle effect experiments and it should therefore be possible to obtain accurate vapour densities.

Absorption techniques which require no spectral resolution and in which the profile of the resonance lamp must be assumed rather than determined experimentally are described by Mitchell and Zemansky (1966). These methods were used in early experiments such as the work on the population of metastable levels in inert gas discharges reported by Ladenburg (1933). Unfortunately systematic errors are difficult to avoid in these experiments since the profile of the resonance lamps is usually considerably distorted by self absorption or even self reversal. Both of these effects vary noticably with lamp design and operating conditions.

Most of these uncertainties may be avoided by measuring the lamp and the absorption profiles using a pressure-scanned Fabry-Perot étalon and the techniques of high-resolution spectroscopy described in section 8.7. The gradual improvements that have been made in this method are described by Jarrett and Franken (1965), Gibbs and Hull (1967), and Gallagher and Lewis (1973). The apparent spectrum of the lamp is given by the convolution of the actual source spectrum $S(\omega')$ with the instrumental transmission function $T(\omega-\omega')$:

$$I_\omega(0) = \int_{-\infty}^{\infty} S(\omega')T(\omega-\omega')\, d\omega'. \qquad (10.27)$$

Similarly, after passing through the absorption cell of length L, the apparent absorption spectrum is

$$I_\omega(L) = \int_{-\infty}^{\infty} S(\omega')\, \exp(-\kappa_{\omega'} L)T(\omega-\omega')\, d\omega'. \qquad (10.28)$$

In the limit that the instrumental profile is very much narrower than the emission and absorption profiles, the measured absorbance, $\ln\{I_\omega(0)/I_\omega(L)\}$, would approach the value given by equation (10.26) with $g(\omega_0)$ replaced by $g(\omega)$. This situation is difficult to achieve in practice owing to the narrow widths of the Doppler-broadened lines. Thus equations (10.27) and (10.28) must be evaluated using the known instrumental function $T(\omega-\omega')$ and assumed values of the lamp profile and the absorption coefficient until there is agreement with the observed emission and absorption spectra. This procedure determines the absorption coefficient κ_ω and hence the atomic density.

Results obtained by Gallagher and Lewis (1973) for the vapour density of rubidium are shown in Fig.10.9. Also shown is a plot of the density N in atom cm^{-3} calculated from the expression

$$\log_{10}(N) = -\frac{A}{T} - (B+1)\log_{10}T + C + DT + 18{\cdot}985. \qquad (10.29)$$

which was obtained by Nesemeyanov (1963) from a weighting of various vapour pressure data. The excellent agreement is perhaps fortuitous since some of the individual data used by Nesemeyanov differ by up to 60 per cent from the results of Gallagher and Lewis (1973), thus indicating the

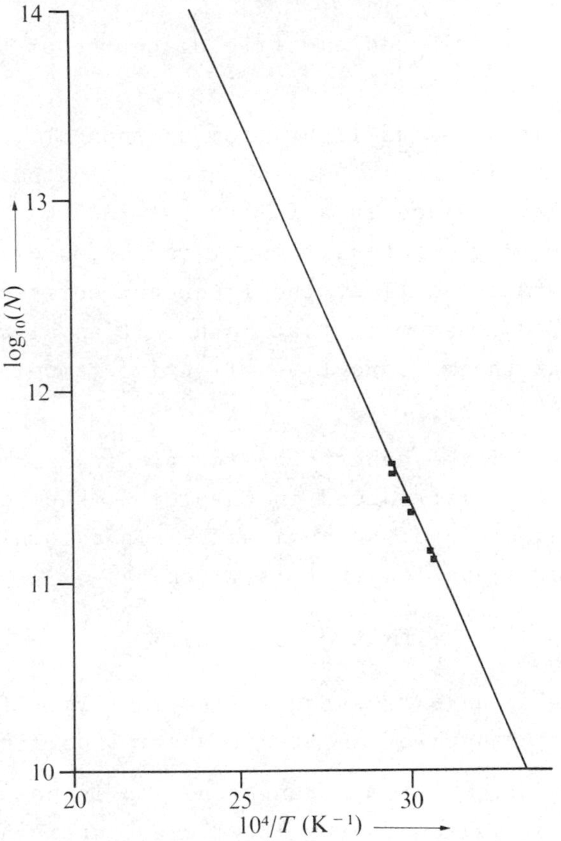

Fig.10.9. Vapour density of rubidium as a function of temperature. ■ , experimental measurements of Gallagher and Lewis (1973). The line plots equation (10.29) with A=4529·6, B=2·991, C=15·8825, and D=0·00059.

need for accurate measurements by the absorption technique.

Problems

10.1. Show that the ratio of the emission coefficient to the absorption coefficient for a gas in thermal equilibrium at 2000 K is of the order of 10^{-13} Jm^{-2} at a wavelength corresponding to $\omega = 4 \times 10^{15}$ s^{-1}. Hence in the visible region of the spectrum spontaneous emission from gases is quite negligible unless they are excited, either by an electric discharge or by some other means.

10.2. A collimated beam of light from a xenon arc, effective temperature 10^4 K, is passed through an atomic absorption cell contained in a furnace at 2500 K. Calculate the ratio of the intensities of radiation emitted by the arc and the cell at the frequency corresponding to $\hbar\omega = 3$ eV, assuming that both emit as black-body sources at their respective effective temperatures. (Ans: $2 \cdot 85 \times 10^{-5}$).

10.3. Making use of the concept of the atomic absorption cross-section introduced in Chapter 9, show that the equivalent width of an optically-thin column of gas of density N and length L is given by

$$W_\omega = N L \int_0^\infty \sigma(\omega)\, d\omega.$$

Hence derive an expression for W_ω in terms of the oscillator strength of an atomic absorption line.

10.4. Collimated light from a tungsten lamp is passed through a column of sodium vapour 1 cm long containing 10^9 sodium atoms cm^{-3} at a temperature of 500 K. The wavelength of the sodium D_1 line, $3\,^2S_{1/2}$-$3\,^2P_{1/2}$, is 5896 Å and its absorption oscillator strength is 0·325. Show that the column of vapour is optically thin at this wavelength and that the equivalent width of the absorption line formed by the vapour is $5 \cdot 42 \times 10^7\ s^{-1}$.

10.5. An absorption spectrum is formed by passing a collimated beam of light from a tungsten filament lamp through a column of iron vapour 10 cm long containing $3 \cdot 5 \times 10^{13}$ atoms cm^{-3} at a temperature of 1600 K. Calculate the oscillator strength of the absorption line at 4430 Å whose equivalent width is measured as $W_\lambda = 8 \cdot 13 \times 10^{-3}$ Å. (Ans: $f_{ik} = 1 \cdot 337 \times 10^{-4}$).

10.6. At high densities the atomic lineshape function, $g(\omega)$, may be represented by a Lorentzian distribution of half width, Γ. When, under the same conditions, the equivalent width of a spectral line is measured in an optically-thick sample, show that

$$W_\omega = \left(\frac{\pi e^2 f_{ik} N_i L \Gamma}{\varepsilon_0 mc} \right)^{1/2} .$$

(Hint: replace the continuous intensity distribution of the lamp by a Gaussian emission profile, $I(\omega) = I(\omega_{ki})\exp\{-a^2(\omega-\omega_{ki})^2\}$, and calculate the limit of $W_\omega(a)$ as $a \to 0$).

References

Bell, G.D., Davis, M.H., King, R.B., and Routly, P.M. (1958). *Astrophys.J.* 127, 775.

__________ and Tubbs, E.F. (1970). *Astrophys.J.* 159, 1093.

Blackwell, D.E. and Collins, B.S. (1972). *Mon.Not.R.ast.Soc.* 157, 255.

__________, Collins, B.S., and Petford, A.D. (1972). *Solar Phys.* 23, 292.

Collins, B.S. (1970). D.Phil. Thesis, University of Oxford.

Corliss, C.H. and Bozmann, W.R. (1962). *N.B.S. Monograph No. 53.* U.S. Govt. Printing Office, Washington D.C.

Gallagher, A. and Lewis, E.L. (1973). *J.Opt.Soc.Am.* 63, 864.

Gibbs, H.M. and Hull, R. (1967). *Phys.Rev.* 153, 132.

Goldberg, L., Müller, E.A., and Aller, L.H. (1960). *Astro-phys.J.Suppl.* 5, No.45, p.1.

Jarett, S.M. and Franken, P.A. (1965). *J.opt.Soc.Am.* 55, 1603.

King, A.S. (1922). *Astrophys.J.* 56, 318.

__________ and King, R.B. (1935). *Astrophys.J.* 82, 377.

__________ and __________ (1938). *Astrophys.J.* 87, 24.

Ladenburg, R. (1933). *Rev.Mod.Phys.* 5, 243.

Müller, E.A. and Mutschlecner, J.P. (1964). *Astrophys.J. Suppl.* 9, No.85, p.1.

Nesemeyanov, A.N. (1963). *Vapour pressures of the chemical elements*. Elsevier, New York.

Ostrovsky, V.I. and Penkin, N.P. (1957). *Optical transition probabilities* (ed. I. Meroz) Vol.I. p.332, Israel Program for Scientific Translations, S. Monson, Jerusalem.

Peach, J.V. (1969). *Mon.Not.R.ast.Soc.* 144, 171.

Pottasch, S.R. (1963). *Astrophys.J.* 137, 945.

____________ (1964). *Mon.Not.R.ast.Soc.* 128, 73.

General references and further reading

Cowan, R.D. and Diecke, G.H. (1948). *Rev.Mod.Phys.* 20, 418.

Foster, E.W. (1964). *Rep.Prog.Phys.* 27, 469.

Jefferies, J.T. (1968). *Spectral line formation*. Blaisdell, Waltham, Mass.

Mitchell, A.C. and Zemansky, M.W. (1966). *Resonance radiation and excited atoms*. Cambridge University Press, London.

Sawyer, R.A. (1963). *Experimental spectroscopy*. Dover, New York.

11
Population inversion mechanisms in gas lasers

Conventional light sources rely either on thermal emission from incandescent solids at high temperatures, e.g. lamps with tungsten filaments operating at 2000-3000 K, or on spontaneous emission from atoms and molecules excited by collisions in electric discharges, e.g. fluorescent discharge tubes, mercury and sodium arc lamps. It is found that the widths of the spectral lines emitted by such sources are quite large, ≈ 1000 MHz. Also, at two spatially separated points on the same wavefront, the phases of the electric fields are completely uncorrelated unless the points considered are very close to one another. These sources are therefore said to be spatially and temporally incoherent and consequently are of very limited utility in holography and long path-difference interferometry. Moreover, the intensity of radiation emitted by such sources cannot exceed that of a black body at the effective temperature of the source, as we have seen in section 10.2.2.

In contrast the light emitted by a laser is generated by stimulated emission in an optical resonator. As a result the radiation has high spatial and temporal coherence which is not found in light emitted by any other source. In addition, the energy densities of radiation in laser beams are equivalent to those emitted by black-body sources at temperatures in the range 10^{15}-10^{25} K. These unique properties of laser radiation have proved a great stimulus to fundamental and applied research over the last fifteen years. In this and the two subsequent chapters we shall discuss the basic physics of gas lasers and show how they have led us to a deeper understanding of the interaction of light and atoms (Problem 11.1).

11.1. Introduction

A laser amplifier consists of a gas, liquid or solid

sample in which the intensity of an electromagnetic wave is increased by power supplied from the medium through the stimulated emission of radiation, hence the acronym laser: Light Amplification by Stimulated Emission of Radiation. As in radio or microwave electronics an amplifier at optical frequencies may be converted into an oscillator by providing positive feedback. This is achieved by enclosing the amplifying medium in an optical cavity formed by two highly-reflecting plane or concave mirrors. Provided that the gain in the medium is sufficient to overcome the losses in the cavity, oscillation will occur and the characteristic laser output beam will be observed. As an electromagnetic wave propagates through the amplifying medium, energy will be given to the wave by stimulated emission while simultaneously it is abstracted from the wave by the reverse process of absorption. In order that there should be a net amplification on a transition connecting non-degenerate states, it is necessary that the number density of atoms in the upper level should exceed that in the lower laser level. This fundamental requirement is known as a *population inversion*.

The first experiments in which a medium with an inverted population was used for the amplification of electromagnetic radiation were performed by Gordon, Zeiger, and Townes (1954). They used a beam of ammonia molecules and obtained oscillation on the 24 GHz microwave inversion transition. The possibility of extending these experiments to the coherent amplification and generation of infrared and visible frequencies was suggested by Schawlow and Townes (1958). Laser oscillation at 6943 Å was first successfully demonstrated by Maiman (1960) using a ruby rod which was optically pumped by a pulsed xenon flash lamp. Soon afterwards Javan *et al.* (1961) reported continuous wave (C.W.) laser oscillation at 1·15 μm and four other infrared wavelengths using a gas discharge in a mixture of helium and neon. Visible output at 6328 Å was later obtained from a similar laser by White and Rigden (1962). Since then laser oscillation has been obtained on many hundreds of radiative transitions in atomic and molecular gases, liquids, and solids under both pulsed and continuously operating conditions. Lasers are now widely used in distance measurement and alignment devices, holography, atomic and molecular physics, and many other areas of pure and applied science. This rapid development is described in a useful collection of Scientific American articles entitled *Lasers and Light* edited by Schawlow (1969). More detailed reviews of the early work in this field have been published by Bennett (1962, 1965).

The main purpose of this chapter is to examine the mechanisms by which population inversions are created in the most important of currently available gas laser systems. First, however, we derive detailed expressions for the gain coefficient in gas discharges and consider the conditions that must be satisfied by the radiative lifetimes and collision rates involved to ensure that gain should exist in steady state systems.

11.2. Population inversion and the atomic gain coefficient

The propagation of light through an excited gas was considered in detail in Chapter 10. If spontaneous emission is neglected, the intensity of radiation of angular frequency ω emerging from a column of gas of length L, $I_\omega(L)$, is related to the intensity of radiation incident at $z=0$ by

$$I_\omega(L) = I_\omega(0) \exp\{-\kappa(\omega)L\} \tag{11.1}$$

where $\kappa(\omega)$ is the absorption coefficient given by equation (10.7). Introducing the relation between the Einstein A- and B-coefficients, equation (9.20), the absorption coefficient becomes

$$\kappa(\omega) = \frac{\pi^2 c^2}{\omega_{ki}^2} g_k A_{ki} \left(\frac{N_i}{g_i} - \frac{N_k}{g_k}\right) g(\omega) \tag{11.2}$$

where $g(\omega)$ describes the atomic lineshape. For systems in thermal equilibrium the population densities per magnetic sub-level are related by the Boltzmann factor

$$\frac{N_k}{g_k} = \frac{N_i}{g_i} \exp\{-(E_k-E_i)/kT\} \tag{11.3}$$

as shown in Fig.11.1(a). Consequently the absorption coefficient is positive at all frequencies, $\kappa(\omega) \geq 0$, and the beam transmitted through the gas is attenuated. This is the situation which exists in most gas discharges. However, there are now many systems in which, under special conditions, a *population inversion*, $N_k/g_k > N_i/g_i$, can be created between two or more levels, as shown in Fig.11.1(b). In this case $\kappa(\omega) < 0$ and a beam of radiation whose angular frequency is close to that of the transition ω_{ki} will be amplified as it passes through the medium. The amplification or gain coefficient at the angular frequency ω may be defined as

$$\alpha^0(\omega) = -\kappa^0(\omega) = \frac{\pi^2 c^2}{\omega_{ki}^2} g_k A_{ki} \left(\frac{N_k^0}{g_k} - \frac{N_i^0}{g_i}\right) g(\omega). \tag{11.4}$$

The gain coefficient, $\alpha^0(\omega)$, given by equation (11.4) is known as the small signal or unsaturated gain coefficient. It determines the gain which exists before amplification of radiation or laser oscillation starts to draw energy from the system thus reducing the population inversion, $(N_k^0/g_k - N_i^0/g_i)$, below the value which exists in the absence of radiation. This topic is treated in detail in sections 13.3-13.5 below.

11.2.1. Inhomogeneous broadening. We shall be concerned mainly with laser lines connecting two stable atomic or molecular levels. Thus in general the lineshape $g(\omega)$ must be represented by a Voigt profile, equation (8.45), and no simple expression for the gain coefficient is possible. However, for laser transitions in the ultraviolet and visible regions of the spectrum, the line profile is dominated by Doppler broadening and in the limit that the Lorentzian

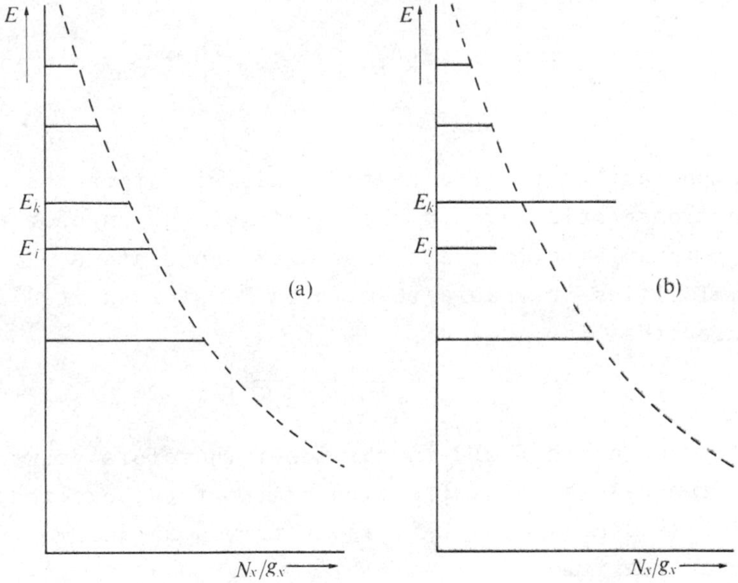

Fig.11.1. Population densities of excited atoms per magnetic sub-level for gas discharges in which (a) thermal equilibrium exists, $N_k/g_k = (N_i/g_i)\exp\{-(E_k-E_i)/kT\}$, and the absorption coefficient of the gas is positive at all frequencies; (b) a population inversion exists between the levels k and i, $N_k/g_k > N_i/g_i$, and light at the frequency $\omega_{ki}/2\pi$ will be amplified by stimulated emission of radiation.

width is much smaller than the Doppler width we have

$$g(\omega) \cong \frac{2}{\Delta\sqrt{\pi}} \exp\{-4(\omega-\omega_{ki})^2/\Delta^2\} \qquad (11.5)$$

where the linewidth parameter Δ is given by

$$\Delta = \frac{2\omega_{ki}}{c} \left(\frac{2kT}{M}\right)^{1/2} . \qquad (11.6)$$

By substituting in equation (11.4), the explicit expression for the gain coefficient in the limit of dominant Doppler broadening becomes (Problem 11.2)

$$\alpha_G^0(\omega) = \frac{\pi^2 c^3}{\omega_{ki}^3} \left(\frac{M}{2\pi kT}\right)^{1/2} g_k A_{ki} \left(\frac{N_k^0}{g_k} - \frac{N_i^0}{g_i}\right) \exp\left\{-\frac{(\omega-\omega_{ki})^2 Mc^2}{2kT\omega_{ki}^2}\right\} \tag{11.7}$$

It is important to realize that in this situation the standing electromagnetic wave in the cavity at the angular frequency ω is able to interact only with those atoms whose axial velocities $\pm v_z$ cause them to be Doppler shifted into resonance, that is

$$\omega = \omega_{ki}\left(1 \pm \frac{v_z}{c}\right). \tag{11.8}$$

The electromagnetic field in the laser therefore draws energy from only a small fraction of the total excited state population and consequently this is termed an *inhomogeneously broadened* transition.

11.2.2. Homogeneous broadening. The Doppler width of spectral lines decreases as we go from the visible into the infrared region of the spectrum and eventually the line profile will be dominated by collision or natural broadening. In this case the gain coefficient has the familiar Lorentzian line profile:

$$g(\omega) \cong \frac{\Gamma_{ki}/2\pi}{(\omega-\omega_{ki})^2+\Gamma_{ki}^2/4} \tag{11.9}$$

and is given explicitly by

$$\alpha_L^0(\omega) = \frac{\pi c^2}{2\omega_{ki}^2} g_k A_{ki}\Gamma_{ki} \left(\frac{N_k^0}{g_k} - \frac{N_i^0}{g_i}\right) \frac{1}{(\omega-\omega_{ki})^2+\Gamma_{ki}^2/4} \tag{11.10}$$

In contrast to the situation discussed in the previous section, all excited atoms in the medium can interact with the wave of angular frequency ω and the transition is said to be *homogeneously broadened*. In the limit that the transition is dominated by natural broadening, $\Gamma_{ki} \to A_{ki}$, the maximum value of the gain coefficient is given by

$$\alpha_L^0(\omega_{ki}) = \frac{2\pi c^2}{\omega_{ki}^2} \Delta N^0 g_k \tag{11.11}$$

where $\Delta N^0 = (N_k^0/g_k - N_i^0/g_i)$ may be termed the population inversion density. Thus for a given inversion density the maximum gain decreases inversely as the square of the transition frequency. This is one of the reasons why laser oscillation is easier to achieve in the infrared than in the visible region of the spectrum and accounts partly for the dearth of laser transitions in the near and vacuum ultraviolet region (Problem 11.3).

11.3. *Transient and steady state population inversion*

Laser emission on some atomic and molecular transitions can be obtained only in a pulsed mode of operation using very rapid discharges. The nitrogen molecular band system at 3371 Å and the copper vapour lines at 5106 Å and 5782 Å are examples of this kind of transition. It is found that in these cases the radiative lifetime of the lower laser level is longer than that of the upper laser level, thus the populations of the two levels concerned can only be transiently inverted. This inversion is rapidly destroyed by the accumulation of atoms in the lower level and the laser pulse, lasting typically 5-10 ns, is self-terminating. Because of the high gain and short inversion times these lasers are frequently operated without feedback mirrors. The output then consists of amplified spontaneous emission and these devices are sometimes misleadingly referred to as 'superradiant lasers'.

Most other laser systems can be operated either on a long pulse, ≥ 1 μs, or on continuous wave output provided that the problems connected with the supply of the necessary electrical power and its dissipation in the discharge tube can be solved. We now wish to derive the conditions which must be satisfied by the lifetimes and collision rates of the levels involved in order that a steady population inversion may be achieved in gas laser systems.

11.3.1. Necessary conditions for steady state inversion. In solid and liquid state lasers the existence of broad absorption bands means that the population inversion can be created by optical excitation using high intensity flash lamps. This process is sometimes called 'optical pumping', a term which we prefer to reserve for the experiments involving free atoms and polarized light which are described in Chapter 17. This method is impracticable for most atomic and molecular gases due to their very narrow absorption lines, and consequently the population inversion is usually created by atom-atom or atom-electron collisions in an electric discharge.†

In order to derive the conditions necessary for steady state population inversion in a gas discharge, let the upper level be populated by collisions and radiative cascade from higher excited levels at the total rate S_k atoms m^{-3} s^{-1}, as shown schematically in Fig.11.2. The corresponding rate for the lower laser level, S_i, specifically excludes the cascade contribution from level k to level i since, as we shall see later, this particular mechanism plays a significantly different role from those of the other contributing processes. The effective lifetimes of the two levels under the conditions existing in the discharge are taken as τ_k and τ_i. These are determined by the sum of the spontaneous transition probabilities to lower levels and by the rate of destruction produced by collisions with electrons and neutral atoms. For instance τ_k is given by

---

†The exceptions to this rule include the cesium vapour laser described by Rabinowitz *et al.* (1962) which is pumped through an accidental coincidence with the intense helium line at 3888 Å, and the atomic iodine laser at 1·315 μm which is pumped by broad-band photodissociation of organic compounds of iodine, such as CF_3I and CH_3I (Kaspar and Pimental 1964).

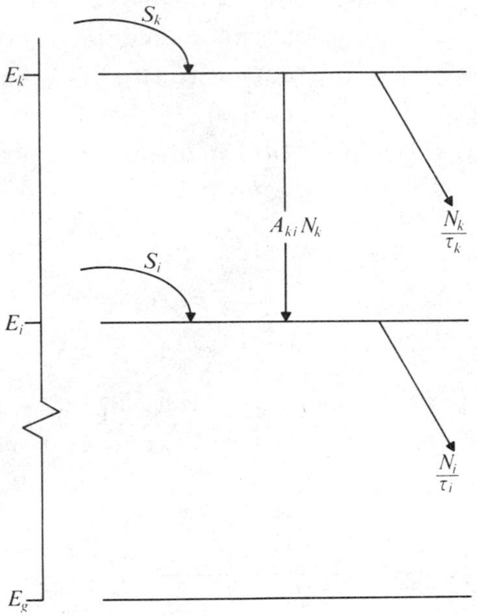

Fig.11.2. Rates of the population and destruction processes for the levels k and i in a gas discharge.

$$1/\tau_k = \sum_i{}' A_{ki} + N_e \sigma_e \overline{v}_e + N\sigma\overline{v} \qquad (11.12)$$

where σ_e and σ are the velocity-averaged cross-sections for destruction by collisions with electrons and neutral atoms whose densities are given by N_e and N respectively. An expression similar to equation (11.12) applies for the total decay rate, $1/\tau_i$, of the lower level. Usually $1/\tau_k$ and $1/\tau_i$ are dominated by the radiative decay rates although atomic collisions often make significant contributions. In the steady state the rates of change of the population densities of the upper and lower level are zero, giving

$$\left.\begin{aligned} \frac{dN_k^0}{dt} &= -\frac{N_k^0}{\tau_k} + S_k = 0 \\ \text{and} \quad \frac{dN_i^0}{dt} &= -\frac{N_i^0}{\tau_i} + S_i + A_{ki}N_k^0 = 0 \end{aligned}\right\} \qquad (11.13)$$

where the term $A_{ki}N_k^0$ accounts for the rate at which atoms are fed into the level i by spontaneous decay from the level k and the superscript zero indicates that the intensity of radiation at the angular frequency ω_{ki} is assumed to be negligible. The equilibrium population densities are given by

$$N_k^0 = S_k\tau_k$$

and (11.14)

$$N_i^0 = (S_i + S_k A_{ki}\tau_k)\tau_i.$$

Thus a steady state population inversion, $N_k^0/g_k > N_i^0/g_i$, will be obtained only if the rates of excitation and decay satisfy the condition

$$\frac{N_k^0 g_i}{N_i^0 g_k} = \frac{S_k\tau_k g_i}{(S_i + S_k A_{ki}\tau_k)\tau_i g_k} > 1. \qquad (11.15)$$

Assuming for the moment that the spontaneous decay rate A_{ki} is negligible, we see that the required population inversion can be achieved most easily if two conditions are satisfied. Firstly, the pumping rate to the upper level, S_k, should be very much larger than the pumping rate to the lower level; that is, a selective population mechanism should exist. Secondly, the lifetime of the upper level, τ_k, should be much longer than that of the lower level, τ_i.

Unfortunately the excitation rate to the lower level can never be zero since even in the absence of collisions with atoms and electrons, the lower laser level is populated by spontaneous emission from the upper level k at the rate $N_k^0 A_{ki}$. A minimum requirement which must be satisfied for a steady state inversion to be possible may be obtained in this case by substituting $S_i=0$ into equation (11.15) giving

$$\frac{1}{A_{ki}} > \frac{g_k}{g_i}\tau_i. \qquad (11.16)$$

Very approximately this condition means that the lifetime of the upper level must exceed that of the lower level and that unless this is satisfied, C.W. laser oscillation is impossible. This is the reason for the inherently pulsed mode of operation of the super-radiant molecular nitrogen transition at 3371 Å referred to in section 11.3. In this laser the lower-level lifetime, $\tau_i \approx 10$ μs, is much longer than that of the upper level, $\tau_k \approx 40$ ns, and the output is self-terminating some 7-10 ns after laser action first commences (Problem 11.4), unless some impurity gas is employed to selectively quench atoms in the lower level.

The condition set by equation (11.16) is, however, only a minimum requirement and even if this is satisfied, C.W. laser oscillation may not be possible for collision processes in the discharge may prevent the attainment of a population inversion by creating lower-level atoms at an excessive rate, $S_i \neq 0$. Many gas laser systems therefore rely on a combination of both a favourable lifetime ratio and a selective population mechanism. To close this discussion we should point out that equations (11.15) and (11.16) illustrate an important application of the lifetime measurements discussed in Chapter 6, particularly for the case of neon and the other noble gases mentioned in section 6.3.5.

11.4. Population inversion mechanisms in gas lasers

11.4.1. Introduction.

By considering the equilibrium populations of excited atoms in a gas discharge, Basov and Krokhin (1962) have derived expressions enabling the conditions under which gain will exist to be specified. However, the collision processes occurring in a gas discharge are so complex and so few of the necessary cross-sections, such as σ_e and σ in equation (11.12), are accurately known that it is impossible to predict theoretically which atomic and molecular species will be useful as gas lasers and on which transitions inversion will be obtained. Most laser transitions have been discovered by empirical investigation of systems which appeared to be promising on some rather general and often speculative arguments. We therefore present a discussion of the five types of population mechanisms which are mainly responsible for inversion in gas lasers without attempting a detailed analysis of any particular system. Although laser oscillation has been obtained on many hundreds of atomic and molecular transitions, we will have space to mention only those systems which seem to be of most importance at the present time. Measurements of the collision cross-sections relevant to some of these systems have already been considered in Chapter 7.

11.4.2. Resonant energy transfer between atoms - the He-Ne laser. Laser oscillation in gases was obtained first using a helium-neon discharge as the active medium. This is still an important system, the red He-Ne output at 6328 Å being probably the only laser line with which most people are familiar. Since the discovery of the He-Ne system in 1961-2 many detailed investigations have been completed and consequently our understanding of it is perhaps the most complete of any laser system.

(a) Population by inelastic atom-atom collisions. When an electric discharge is run in pure helium at pressures of the order of 1 Torr, large populations of atoms accumulate in the two metastable levels $2\ ^1S_0$ and $2\ ^3S_1$. These metastable levels are populated mainly by direct electron excitation from the ground level but also receive some contribution by radiative cascade from higher levels. At current densities of ≈ 200 mA cm^{-2} the steady state population in the metastable level approaches 10^{-5} of that in the ground level. The metastable atoms can be destroyed by collisions with atoms or electrons or by diffusion to the walls of the discharge tube.

If a small amount of neon is added to the helium discharge the mean electron energy is not appreciably altered; a helium-neon ratio of approximately 7:1 has been found to be optimum for discharge tubes of 6 mm bore. Collisions then occur between the metastable helium atoms and ground level neon atoms. As shown in Fig.11.3 there is a close coincidence between the $2\ ^3S_1$ metastable level of helium and the $2s_2$

level of neon, $\Delta E = 0{\cdot}04$ eV. (In Fig.11.3 the Paschen notation has been used for the neon levels. The numerical subscripts in this notation have no spectroscopic significance and it is used simply as a concise method of labelling the levels. The correspondence between the Paschen notation and the j-l coupling notation introduced by Racah is given in tables of atomic energy levels (Moore 1949). However, in neon and the other inert gases most of the levels can be correctly represented only in intermediate coupling.) A similar coincidence exists between the helium $2\ ^1S_0$ metastable level and the $3s_2$ level of neon, $\Delta E = 0{\cdot}05$ eV, and resonant transfer of energy of excitation occurs in atom-atom collisions:

$$\mathrm{He}(2\ ^3S_1) + \mathrm{Ne}(^1S_0) \rightarrow \mathrm{He}(1\ ^1S_0) + \mathrm{Ne}(2s_2) \qquad (11.17)$$

$$\mathrm{He}(2\ ^1S_0) + \mathrm{Ne}(^1S_0) \rightarrow \mathrm{He}(1\ ^1S_0) + \mathrm{Ne}(3s_2). \qquad (11.18)$$

The energy defects involved in these reactions are of the order of kT and the cross-section for excitation transfer may be expected to be quite large, of the order of 10^{-17}-10^{-16} cm^2, as explained in section 6.2.4 (Problem 11.5). The total destructive collision cross-sections of helium metastables by neon atoms have been measured as $\sigma = (3{\cdot}7 \pm 0{\cdot}5) \times 10^{-17}$ cm^2 and $\sigma = (4{\cdot}1 \pm 1{\cdot}0) \times 10^{-16}$ cm^2 for the $2\ ^3S_1$ and $2\ ^1S_0$ levels respectively. Thus the collisions defined by equations (11.17) and (11.18) provide a selective excitation mechanism for the $2s_2$ and $3s_2$ levels of neon which is important for the attainment of a population inversion, as shown by equation (11.15).

(b) The existence of favourable lifetime ratios. In neon the $2s_2$ and $3s_2$ levels are connected to the ground level by strong allowed transitions in the vacuum ultraviolet and at very low pressures they have short radiative lifetimes, ≈ 10-20 ns, as the measurements of Lawrence and Liszt (1969) have shown. However, at neon pressures greater than $0{\cdot}1$ Torr there is almost complete resonance trapping of these lines and the effective lifetimes are increased to

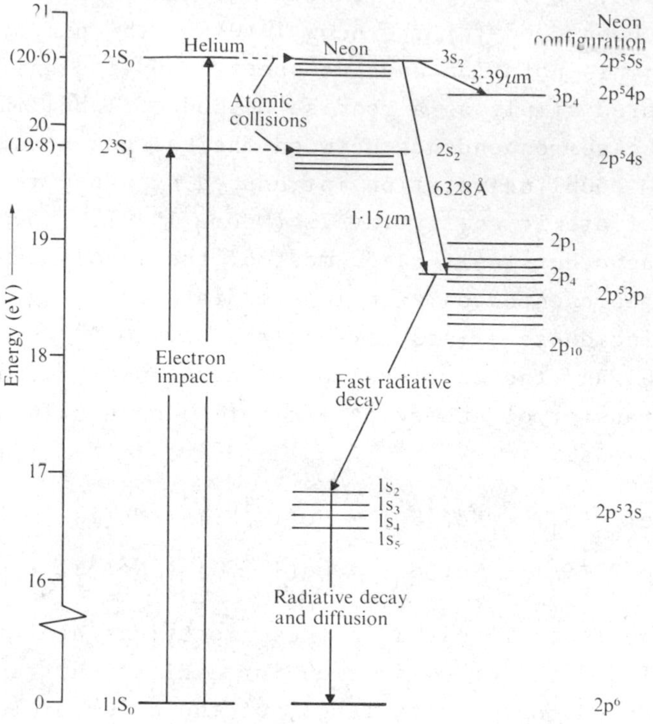

Fig.11.3. Energy levels of helium and neon showing the dominant excitation mechanisms and the neon transitions responsible for laser action in the visible and near infrared.

$\tau(2s_2) = 0\cdot96 \times 10^{-7}$ s and $\tau(3s_2) = 1\cdot1 \times 10^{-7}$ s. In contrast Bennett and Kindlmann (1966) have shown that the lifetimes of the levels of the $2p^53p$ configuration are all of the order of 20 ns, in particular $\tau(2p_4) = 18\cdot5$ ns. Detailed values for the other levels are given in Table 6.1. Consequently in addition to the selective population mechanism, there also exists a favourable lifetime ratio, and a steady state population inversion can be maintained between the $2s_2$ and $2p_4$ levels, producing C.W. laser oscillation at $1\cdot15$ μm. This was the strongest of the five gas laser transitions first reported by Javan *et al.* (1961). Similar inversions exist between the levels ($3s_2$-$3p_4$) and ($3s_2$-$2p_4$) giving strong laser oscillation at 3·39 μm and 6328 Å respectively.

Inelastic collisions with helium atoms in the $2\ ^1S_0$ and $2\ ^3S_1$ metastable levels are only moderately specific and other levels of the 3s and 2s groups are also populated. Since favourable lifetime ratios exist for all the transitions between the levels of the $2p^5 5s$, $2p^5 4s$, and $2p^5 3p$ configurations, many other lines have been made to oscillate on C.W. output; some of these are given in Table 11.1. However, the lines at 6328 Å, 1·15 μm and 3·39 μm remain the strongest and most useful of the He-Ne laser transitions.

(c) Details of laser construction. The amplifying medium of the first gaseous optical maser was excited by a 50 W radio-frequency discharge and the optical cavity was formed by highly reflecting plane mirrors which were located inside the evacuated discharge envelope. This arrangement is experimentally very inconvenient. Moreover, as will be explained in section 12.2 below, it was necessary to maintain the parallelism of the mirrors to within 6 arc seconds to obtain a cavity Q factor which was sufficiently high for oscillation to take place.

The construction of present medium power He-Ne lasers is shown in Fig.11.4. The laser is powered by a d.c. discharge between internal electrodes running typically at 25-50 mA at ≈ 2000 V. Amplification is obtained in the positive column of the discharge but unfortunately this region is extremely sensitive to impurities in the discharge. Contamination of the gas can be prevented by a high temperature bake-out under vacuum and by careful attention to the cleanliness of electrode materials. The optical cavity is formed by concave mirrors mounted externally on a rigid base and provided with fine angle adjustments. The alignment of the concave mirrors is much less critical than that of a plane mirror cavity. The discharge tube is terminated in windows set at the Brewster

TABLE 11.1

A selection of the neon C.W. laser transitions excited in a helium-neon discharge

Transition configurations $2p^5 5s \rightarrow 2p^5 3p$		Transition configurations $2p^5 4s \rightarrow 2p^5 3p$		Transition configurations $2p^5 5s \rightarrow 2p^5 4p$	
Wavelength (μm)	(Paschen notation)	Wavelength (μm)	(Paschen notation)	Wavelength (μm)	(Paschen notation)
0·5939	$3s_2 \rightarrow 2p_8$	1·080	$2s_3 \rightarrow 2p_7$	2·782	$3s_3 \rightarrow 3p_7$
0·6046	$3s_2 \rightarrow 2p_7$	1·084	$2s_2 \rightarrow 2p_6$	2·945	$3s_4 \rightarrow 3p_{10}$
0·6118	$3s_2 \rightarrow 2p_6$	1·114	$2s_4 \rightarrow 2p_8$	3·318	$3s_4 \rightarrow 3p_8$
0·6293	$3s_2 \rightarrow 2p_5$	1·118	$2s_5 \rightarrow 2p_9$	3·333	$3s_2 \rightarrow 3p_5$
0·6328	$3s_2 \rightarrow 2p_4$	1·139	$2s_5 \rightarrow 2p_8$	3·335	$3s_5 \rightarrow 3p_9$
0·6352	$3s_2 \rightarrow 2p_3$	1·141	$2s_2 \rightarrow 2p_5$	3·390	$3s_2 \rightarrow 3p_2$
0·6401	$3s_2 \rightarrow 2p_2$	1·152	$2s_2 \rightarrow 2p_4$	3·391	$3s_2 \rightarrow 3p_4$
0·7305	$3s_2 \rightarrow 2p_1$	1·160	$2s_2 \rightarrow 2p_3$	3·448	$3s_4 \rightarrow 3p_7$
		1·161	$2s_3 \rightarrow 2p_5$	3·583	$3s_5 \rightarrow 3p_6$
		1·177	$2s_2 \rightarrow 2p_2$	3·981	$3s_4 \rightarrow 3p_3$
		1·198	$2s_3 \rightarrow 2p_2$		
		1·207	$2s_5 \rightarrow 2p_6$		
		1·523	$2s_2 \rightarrow 2p_1$		

angle. These eliminate the reflection loss for the component of the electric field linearly polarized in the plane of incidence, but lead to an output beam which is almost completely linearly polarized. Further details of laser cavity design are discussed in section 12.5 below.

An even simpler method of construction is used in low power 6328 Å He-Ne lasers, in which the mirrors are sealed directly on to the ends of the capillary discharge tube. The thick walled capillary provides mechanical stability and no mirror adjustments are required. The output of these devices is typically 0·5 - 1·0 mW and the beam is unpolarized.

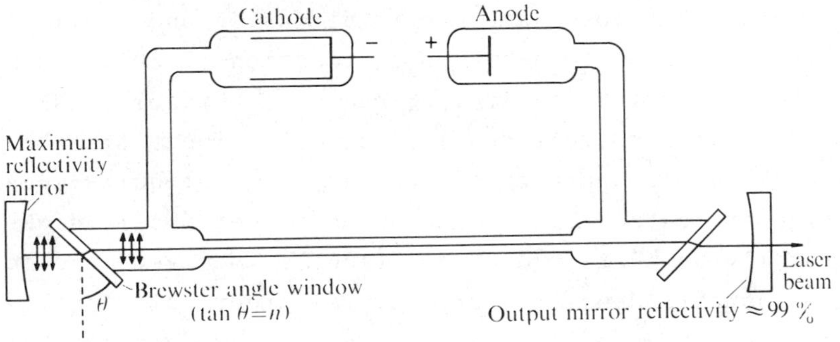

Fig.11.4. Schematic diagram of external mirror He-Ne laser.

(d) Input/output power characteristics of He-Ne lasers. It seems plausible to expect that the density of helium metastable atoms, and hence the output power of the He-Ne laser, could be increased by simply increasing the current through the discharge. However, the results of White and Gordon (1963), shown in Fig.11.5, clearly demonstrate that, for a given pressure and tube diameter, there is a well-defined optimum current. As the discharge current is increased beyond this value the laser output power decreases until eventually oscillation ceases altogether.

To account for this behaviour several different factors

must be considered, the most important of which is the equilibrium helium metastable density, N_1. This is determined by a balance between the electron excitation rate and the destruction rate produced by the combined effects of inelastic electron collisions and diffusion to the walls. Thus in the limit of low neon density, the ratio of the helium metastable density to the helium ground state density, N_0, is given by (Problem 11.6)

$$\frac{N_1}{N_0} = \frac{n_e \sigma_{01} \bar{v}_e}{\Gamma_1' + n_e \sigma_{10} \bar{v}_e} \tag{11.19}$$

where σ_{01} and σ_{10} are the velocity-averaged electron excitation and destruction cross-sections respectively, Γ_1' is the metastable loss rate due to diffusion and collisional excitation transfer to neon atoms, equation (7.30), and n_e is the mean electron density. Since n_e is proportional to the discharge current density we would expect that the ratio N_1/N_0 would approach a constant value at quite moderate discharge currents. This behaviour has been observed experimentally by White and Gordon (1963). However, the measured metastable densities are several orders of magnitude less than the theoretical value predicted by detailed balancing:

$$\left(\frac{N_1}{N_0}\right)_{Sat} = \frac{\sigma_{01}}{\sigma_{10}} = \frac{g_1}{g_0} \exp\{-(E_1-E_0)/kT_e\} \tag{11.20}$$

since in equation (11.19) we have neglected the metastable loss rate produced by ionizing collisions. Nevertheless it is generally correct to say that the maximum output power of lasers, such as the He-Ne and He-Cd systems which are pumped by collisions with metastable species, are restricted to ≤ 100 mW by the saturation of the metastable density.

The density of atoms in the lower laser level is determined mainly by the equilibrium between direct electron excitation from the ground level and radiative decay, giving

$$N_i^0 = N\,\tau_i n_e \sigma_{0i} \bar{v}_e \tag{11.21}$$

where N is the neon ground state density and τ_i is the effective lifetime of the level determined by equation (11.12). From equations (11.19) and (11.21) we find that the gain coefficient at the line centre, $\alpha^0(\omega_{ki})$, which is proportional to $(N_k^0/g_k - N_i^0/g_i)$, is expected to vary with discharge current, I, as

$$\alpha^0(\omega_{ki}) = \frac{K_1 N_0 N I}{\Gamma_1' + K_2 I} - K_3 N I \tag{11.22}$$

where K_1, K_2, and K_3 are constants involving the electron collision cross-sections. As Fig.11.5 shows, this simple expression fits the measurements of the unsaturated gain on the 3·39 μm line remarkably well. It has been found experimentally for a range of discharge tube diameters, d, extending from 1-15 mm that the optimum gain occurred at a constant value of the product pd ≈ 3·6 Torr mm where p is the total pressure. This is consistent with equation (11.22) and arises from the fact that as the diameter or pressure is changed the balance of creation and destruction processes is shifted permitting the use of higher currents before metastable saturation is reached, a point which is discussed in more detail by Gordon and White (1963).

At very high current densities a large population builds up in the neon 1s resonance and metastable levels. This also causes a reduction in the observed gain since the population of the lower laser level is increased by the combined effects of stepwise electron excitation and radiation trapping of the 2p → 1s neon transitions. However, this neon 'metastable bottleneck' is certainly not the main effect limiting the output power of He-Ne lasers.

11.4.3. Resonant energy transfer between molecules - the N_2-CO_2 laser. The quantum efficiency of the He-Ne laser, defined as the ratio of the energy of the emitted photon to the energy of excitation of the upper laser level, is quite low, being less than 5 per cent for most transitions.

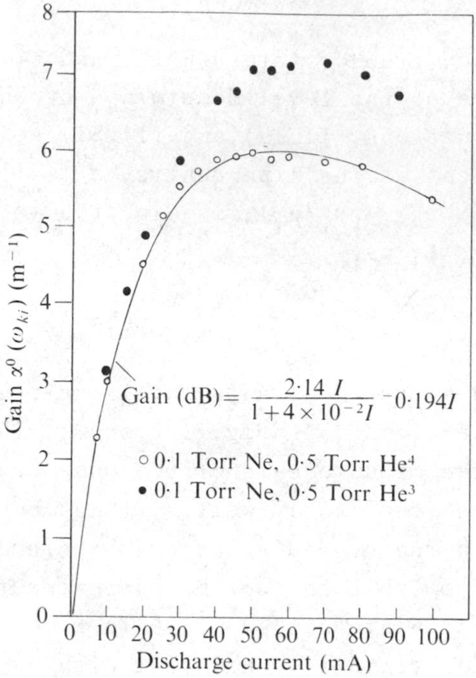

Fig.11.5. Unsaturated gain on the neon transition $3p_4$ - $3s_2$ at 3·39 μm as a function of discharge current. Gas conditions are 5:1 He:Ne mixture; pd = 3·6 Torr mm, d = 6 mm. Experimental results are given for both He^4-Ne and He^3-Ne gas mixtures. (After White and Gordon (1963).)

The actual operating efficiency, defined as the ratio of optical output power to electrical input power, is smaller by a factor of approximately one hundred since only a small fraction of the collisions in the discharge produce atoms in the upper laser levels. In an attempt to overcome this low efficiency Patel (1964a,b) investigated the possibility of laser oscillation on vibrational-rotational transitions in the carbon dioxide molecule. In the initial experiments laser oscillation at several wavelengths around 10·6 μm and 9·6 μm was achieved, but the output power was quite low. Subsequently Patel (1964c,1965) obtained increased output power by mixing the carbon dioxide with molecular nitrogen, but more significantly operating efficiencies of 20-30 per cent were obtained. Carbon dioxide lasers with C.W. output powers of several kW are now widely used for machining, welding, and refining metals, while pulsed systems with peak powers of 10 GW or more are important in current research in the field of thermonuclear fusion in laboratory-generated plasmas. We shall therefore now consider some of the physics of this important laser system.

The molecular vibrational levels pertinent to the CO_2-N_2 laser are shown in Fig.11.6. Each vibrational level is associated with a set of closely spaced rotational levels which are not included in the diagram for the sake of clarity. Nitrogen is a homonuclear diatomic molecule and hence, by symmetry arguments, it cannot possess a permanent electric dipole moment. Thus radiative decay of the excited vibrational levels of the lowest electronic state of the nitrogen molecule is strictly forbidden and the lifetimes of these levels are therefore quite long, $\approx 0{\cdot}1$ s at 1 Torr, being determined by collisional destruction and the diffusion loss to the walls. It is found that between 10-30 per cent of the nitrogen molecules in the laser discharge accumulate in the v=1 level.

The internal vibrations of the triatomic carbon dioxide molecule can be represented approximately by linear combinations of three orthogonal normal modes of vibration The state of the vibrating molecule is given by a set of three vibrational quantum numbers $(v_1 v_2 v_3)$ corresponding to the degree of excitation of each of the three normal modes, namely the symmetric stretching mode represented by the states $(v_1 00)$, the symmetric bending mode represented by $(0 v_2 0)$, and the asymmetric stretching mode represented by $(00 v_3)$. From Fig.11.6 it is clear that the excited vibrational level 001 of the carbon dioxide molecule is in near-perfect coincidence (ΔE=18 cm^{-1}) with the highly populated v=1 vibrational level of molecular nitrogen. Consequently a very efficient and selective population of the 001 level

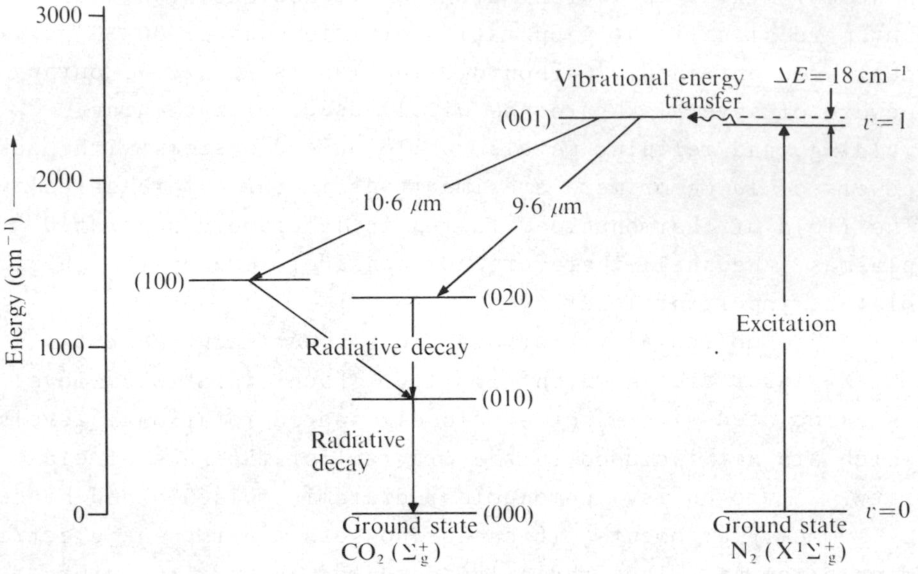

Fig.11.6. Partial energy-level diagram showing low lying vibrational levels of nitrogen and carbon dioxide together with the CO_2 infrared laser transitions pumped by vibrational energy transfer.

occurs by resonant transfer of energy between nitrogen and carbon dioxide molecules.

In addition to this selective excitation process, the 001 vibrational level also has a favourable lifetime ratio with respect to the 100 and 020 levels. Consequently in a wide bore N_2-CO_2 discharge tube at ≈ 10 torr pressure, strong laser oscillation occurs at 10·6 μm and 9·6 μm on the closely spaced rotational transitions of the 001 → 100 and 001 → 020 bands of carbon dioxide. It has been found that the operating efficiency can be increased to 30 per cent by the addition of helium to the discharge. This high overall efficiency is the result of a combination of the following factors:

(a) the levels involved all belong to the electronic ground state of the molecule resulting in a quantum efficiency of approximately 45 per cent;

(b) a very large fraction of the excited nitrogen

molecules accumulate in the v=1 level;

(c) a large fraction of the carbon dioxide molecules excited by electron impact rather than resonant energy transfer still accumulate in the 001 level as a result of energy sharing collisions with molecules in the 000 level;

(d) there is an efficient depopulation of the lower laser levels which is increased still further by the addition of helium to the gas mixture.

Having briefly considered examples of lasers using gases of neutral atoms and molecules, we must now discuss the possibilities which exist for laser action in ionized species.

<u>11.4.4. Penning ionization collisions with He($^3S_1$).</u> Helium atoms in the 2 $^3S_1$ metastable level possess a large amount of internal energy, as shown in Fig.11.3. When these metastable atoms collide with atoms of another species of low ionization potential, for instance atoms of a metallic element M, it is possible for this energy to be used in ionizing the metal atoms. This is an example of a process known as Penning ionization. In the case of the metastable helium atom enough energy may be available for the metallic ion to be simultaneously raised to an excited level:

$$\mathrm{He}(2\ ^3S_1) + \mathrm{M} \rightarrow \mathrm{He}(1\ ^1S_0) + (\mathrm{M}^+)^* + e. \qquad (11.23)$$

Penning ionization collisions of the type represented by equation (11.23) were proposed as the main excitation mechanisms for certain laser transitions discovered in the spectra of zinc and cadmium ions which are listed in Table 11.2. Time-resolved and flowing afterglow studies which have subsequently confirmed this suggestion have been reviewed by Collins (1973) and are described in detail in Chapter 7. The zinc and cadmium ion lasers are just two examples of the class of metal vapour lasers of which a very readable account has been given by Silfvast (1973).

TABLE 11.2

Laser transitions excited by Penning ionization collisions in He-Zn *and* He-Cd *discharges*

Laser wavelength (Å)	Upper laser level	Lower laser level	Active species
5894	$3d^9 4s^2\ ^2D_{3/2}$	$3d^{10} 4p\ ^2P^o_{1/2}$	Zn^+
7478	$3d^9 4s^2\ ^2D_{5/2}$	$3d^{10} 4p\ ^2P^o_{3/2}$	Zn^+
3250	$4d^9 5s^2\ ^2D_{3/2}$	$4d^{10} 5p\ ^2P^o_{1/2}$	Cd^+
4416	$4d^9 5s^2\ ^2D_{5/2}$	$4d^{10} 5p\ ^2P^o_{3/2}$	Cd^+

In a discharge in helium containing a trace of cadmium vapour the cadmium ions are selectively excited to a group of energy levels, shown in Fig.11.7, which are nearly resonant with the energy of the metastable helium atom. The afterglow experiments show that the main excitation process is to levels of the $4d^{10}6s$ and $4d^9 5s^2$ configurations and that the total cross-section for the destruction of He($^3S_1$) by cadmium atoms is large, $\sigma = 6{\cdot}5 \times 10^{-15}$ cm^2. It has also been shown that the rate of production of ions in the $4d^9 5s^2$ configuration is three times as great as that for ions in the $4d^{10}5p$ configuration, thus demonstrating that a selective population mechanism exists. In addition there exists a very favourable lifetime ratio, since $\tau(4d^9 5s^2) = 700$ ns while $\tau(4d^{10}5p) = 1\text{-}2$ ns. Consequently C.W. laser oscillation at 4416 Å and 3250 Å has been obtained in helium discharges when cadmium vapour is introduced from a small reservoir heated by an oven. Typical operating parameters are: helium pressure ≈ 1 Torr; cadmium pressure ≈ 5×10^{-3} Torr; discharge current 100 mA at 6 kV. A similar mechanism is responsible for the excitation of the levels of the $3d^9 4s^2$ configuration of the zinc ion in helium-zinc discharges which produce C.W. laser oscillation at 7478 Å and 5894 Å.

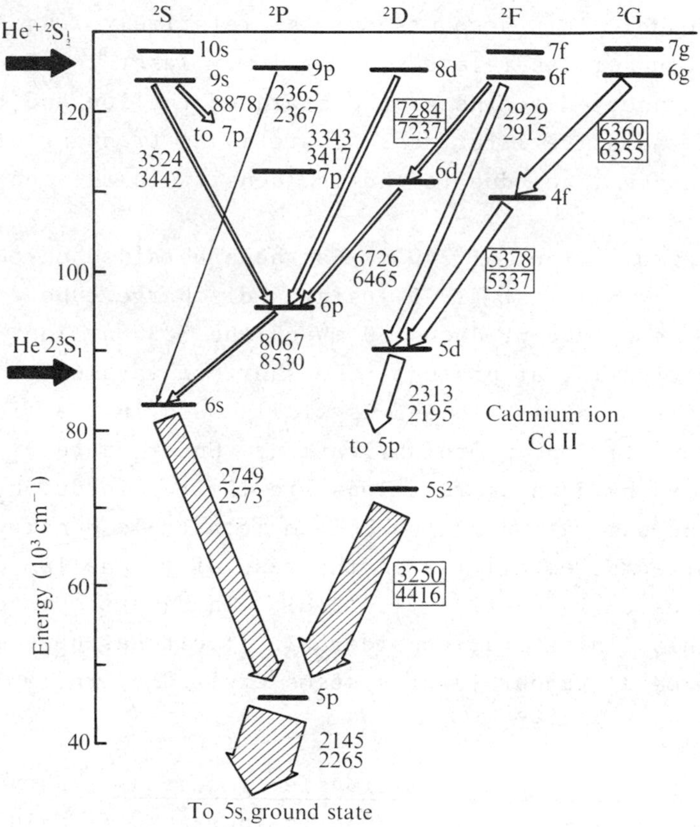

Fig.11.7. Energy-level diagram for Cd^+ showing known laser transitions pumped either by Penning ionization collisions with metastable helium atoms, He($2\ ^3S_1$), or by charge transfer reactions with helium ground state ions, $He^+(1\ ^2S_{1/2})$. The wavelengths of laser transitions are enclosed by rectangles. (After Webb *et al.* (1970).)

The 3250 Å and 4416 Å transitions connect the $4d^9 5s^2\ ^2D$ and the $4d^{10} 5p\ ^2P^o$ terms of CdII and would be strictly forbidden in the central field approximation since they are both two-electron transitions. However, mixing of the $4d^9 5s^2$ and the $4d^{10} 5d$ configurations, both of which give rise to ^2D terms of the same parity and total angular momentum quantum number, means that these laser transitions are weakly allowed. This accounts for the relatively long lifetime of the upper laser levels. The lower laser levels are optically connected to the ground state of the ion and thus have short radiative lifetimes. There is no trapping of these resonance lines due to the low density in the ion ground state.

A maximum output of 300 mW has been obtained on the 4416 Å line by Silfvast (1973) using a discharge tube 2m long. The same tube produced 50 mW on the 3250 Å line. This transition is, at present, the shortest wavelength C.W. laser transition available and is widely used in the fields of photochemistry and photofabrication. Unfortunately, lasers pumped by Penning reactions are limited in output power by the saturation of the helium metastable density at high currents, exactly as in the case of the helium-neon system discussed in section 11.4.2(d). In an attempt to overcome this limitation, considerable effort has been made to develop metal vapour laser systems excited by thermal-energy charge transfer.

<u>11.4.5. Thermal-energy charge transfer reactions.</u> When the current in a helium discharge tube is increased from 100 mA to 10 A the density of helium ions, He^+, increases proportionally in contrast to the helium metastable density which has already reached its maximum value. The helium ion ground state, $He^+(1\,^2S_{1/2})$, lies 24·6 eV above the neutral atom ground state and when this ion collides with a metal atom, this energy is available for the simultaneous ionization and excitation of the metal atom:

$$He^+(1\,^2S_{1/2}) + M \rightarrow He(1\,^1S_0) + (M^+)^* + \Delta E. \quad (11.24)$$

This process is known as thermal-energy charge transfer.

The metal ion is selectively excited to those levels which are in close resonance with the energy of $He^{+}(1\ ^{2}S_{1/2})$, as shown in Figs. 11.7 and 11.9. The process is even more specific than the Penning ionization excitation of equation (11.23) since in the final state there is now no third particle available to share the kinetic energy and momentum released.

TABLE 11.3.

Cadmium ion laser transitions excited by thermal energy charge transfer reactions in a He-Cd *discharge*

Laser wavelength (Å)	Upper laser level	Lower laser level	Charge-transfer cross-section (cm^2)
5337	$4f\ ^{2}F^{o}_{5/2}$	$5d\ ^{2}D_{3/2}$	$< 10^{-18}$
5378	$4f\ ^{2}F^{o}_{7/2}$	$5d\ ^{2}D_{5/2}$	
6355	$6g\ ^{2}G_{7/2}$	$4f\ ^{2}F^{o}_{5/2}$	$7{\cdot}7 \times 10^{-16}$
6360	$6g\ ^{2}G_{9/2}$	$4f\ ^{2}F^{o}_{7/2}$	
7237	$6f\ ^{2}F^{o}_{5/2}$	$6d\ ^{2}D_{3/2}$	$8{\cdot}2 \times 10^{-16}$
7284	$6f\ ^{2}F^{o}_{7/2}$	$6d\ ^{2}D_{5/2}$	
8067	$6p\ ^{2}P^{o}_{3/2}$	$6s\ ^{2}S_{1/2}$	$< 10^{-18}$
8531	$6p\ ^{2}P^{o}_{1/2}$	$6s\ ^{2}S_{1/2}$	
8878	$9s\ ^{2}S_{1/2}$	$7p\ ^{2}P^{o}_{3/2}$	$8{\cdot}2 \times 10^{-16}$

In a helium-cadmium discharge, selective excitation by charge-transfer reactions produces steady-state population inversions between the levels of the $(4d^{10}6g - 4d^{10}4f)$ and $(4d^{10}6f - 4d^{10}6d)$ configurations, producing C.W. laser oscillation on the lines listed in Table 11.3 and shown diagrammatically in Fig.11.7. Radiative cascade from these upper levels is then responsible for laser oscillation on transitions connecting the ($4d^{10}4f$ - $4d^{10}5d$) configurations. Laser oscillation occurs on similar transitions in the ZnII and SeII spectra excited by charge transfer reactions in a helium discharge. Selenium is particularly interesting since Silfvast and Klein (1970) were able to obtain C.W. laser oscillation on 24 visible transitions of SeII ranging in wavelength from 4604 Å to 6535 Å. All the transitions originate from the configuration $4s^24p^25p$ which is close in energy to $He^+(1\ ^2S_{1/2})$. Many new laser transitions in metal vapours excited by thermal-energy charge transfer have been discovered by Piper and Webb (1973) using the hollow-cathode discharge tube shown in Fig.11.8. In this configuration the large potential fall at the cathode produces highly energetic electrons and thus leads to an efficient production of He^+ ions.

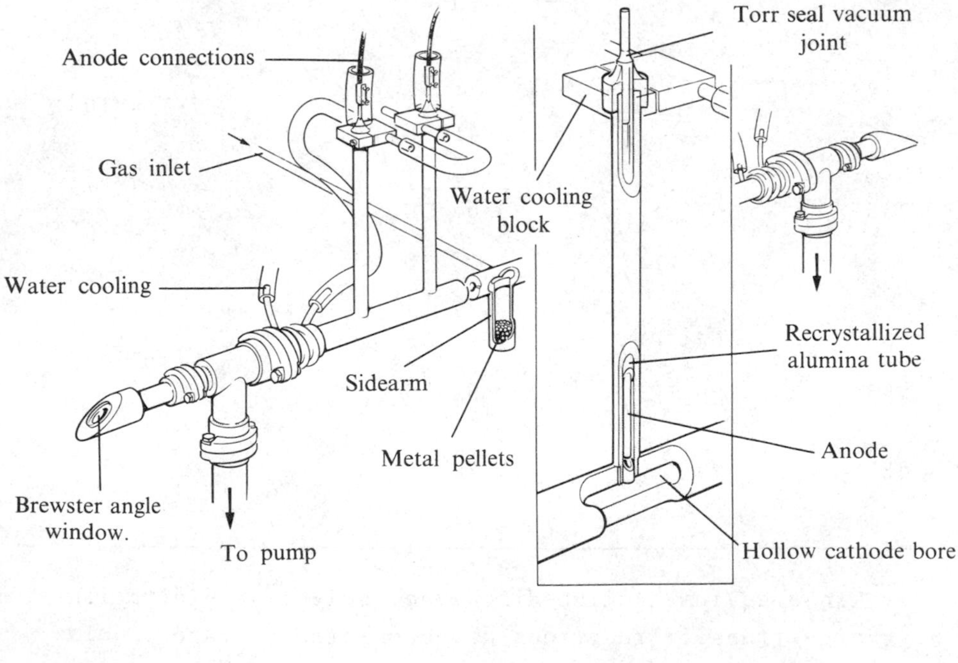

Fig.11.8. Hollow cathode laser discharge tube. (After Piper and Webb (1973).)

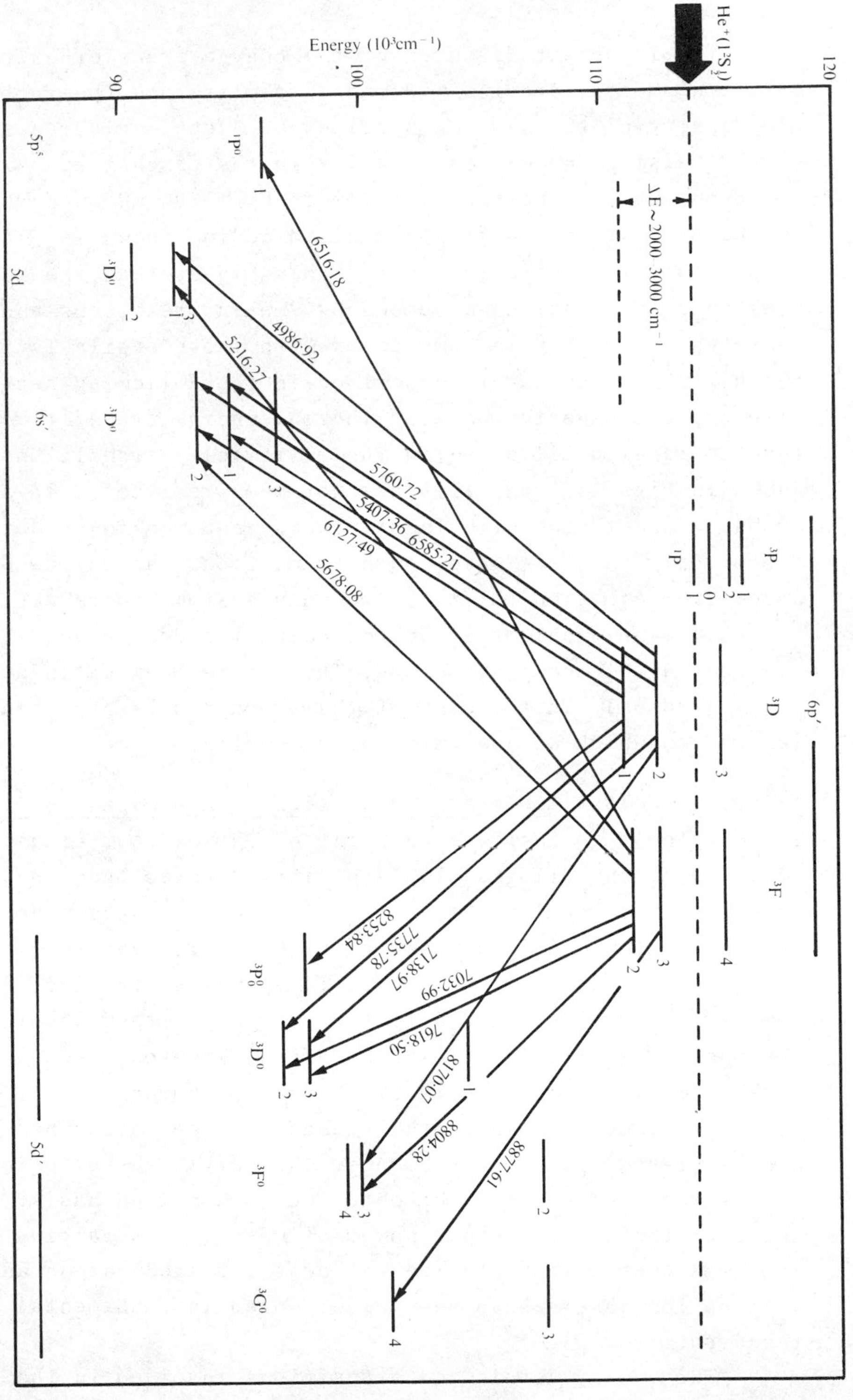

Fig.11.9. Energy level diagram for ionized iodine showing known laser transitions pumped by thermal energy charge transfer reactions with ground state helium ions, He$^+$ ($1\ ^2S_{1/2}$).

The important features of the charge-transfer excitation process are its wide applicability and the fact that the density of the pumping species, He^+, can be increased by simply raising the current density in the discharge. Piper and Webb (1976) report output powers of $\approx$ 100 mW on several of the transitions in singly ionized iodine shown in Fig. 11.9. This is generally considerably higher than the output available on transitions pumped by Penning reactions and competes favourably in some cases with those available from the noble gas ion lasers described in the following section. However, the expectation that thermal-energy transfer would lead to laser oscillation on the ultraviolet transitions shown in Figs.11.7 and 11.9 has not been realized. This is probably due to the high input powers required to produce useful gain on short-wavelength transitions, as may be deduced from equation (11.11). In such systems there are inevitably many problems associated with the design and engineering of the plasma tube. These have been satisfactorily solved only in the case of the argon and krypton ion lasers, to which we now turn our attention.

11.4.6. Inelastic electron collisions — noble gas ion lasers. In an attempt to increase the list of known laser transitions Bell (1964) and Bridges (1964a,b) investigated the behaviour of high current pulsed discharges in the noble gases neon, argon, krypton, and xenon and also in atomic vapours of mercury, carbon, silicon, etc. Within a very short time, 200 new wavelengths were added to the growing list of laser transitions. The strongest lines in argon, krypton, and xenon were made to oscillate on a C.W. basis by Gordon *et al.* (1964), although in most other cases C.W. operation has not been achieved, presumably because the very high current densities required can only be obtained on a pulsed basis. Because of their high output power at wavelengths ranging from the near ultraviolet to the near infrared, argon and krypton ion lasers have been widely used in fundamental research.

The energy levels and transitions involved in the argon

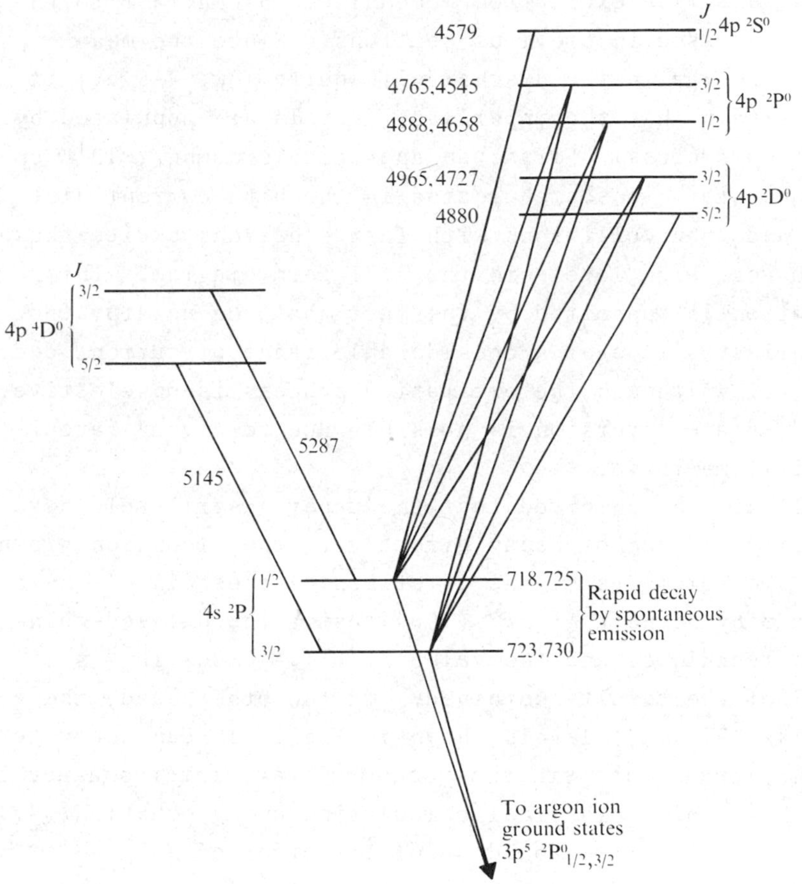

Fig.11.10. Schematic diagram showing ten of the visible laser lines emitted by excited states of Ar^+. Wavelengths are given in Å units; the diagram is not drawn to scale. The levels involved in the laser transitions lie ≈ 20 eV above the argon ion ground state, which is itself 15·6 eV above the ground state of the neutral atom.

ion laser are shown in Fig.11.10. They are typical of the laser transitions which occur in the singly-ionized spectra of all the noble gases. When these devices are operated in the C.W. mode using a positive column discharge there is no obvious selective excitation mechanism, in contrast to the lasers discussed in previous sections. Since the mean electron energy in the discharge is quite low, 4-5 eV, it seems certain that the upper laser levels are populated by a two-step process. First, an appreciable number, 10^{13} cm^{-3}, of ground-state ions are created in the high current discharge and then collisions with fast electrons excite these ions to very high levels in the ArII term diagram. This explanation is supported by the fact that the unsaturated gain varies as I^2 over a considerable range of current densities, I. Although the excitation process is unselective, a steady-state inversion is possible due to a very favourable lifetime ratio.

It will be observed that the upper laser levels have odd parity and cannot decay directly to the argon ion ground state. Measurements of the lifetimes of these levels have been made by Bennett *et al.* (1964) using the delayed-coincidence technique, and the value $\tau(^4D^o) = 7 \cdot 5 \times 10^{-9}$ s is typical of the results obtained. On the other hand, the even parity 4s $^2P_{1/2,3/2}$ levels shown in Fig.11.10 can decay to the ion ground state via the vacuum ultraviolet resonance lines. On the basis of a classical calculation using equation (4.11) the expected lifetime would be of the order of $\tau(^2P) = 3 \cdot 0 \times 10^{-10}$ s. The density of Ar^+ ions in the ground level is sufficiently small that these resonance lines are not affected by resonance trapping and, just as important, the lower laser levels are sufficiently high above the neutral atom ground or metastable levels that the thermal population is negligible in spite of the plasma temperature of 3000 K. Thus there is a very favourable lifetime ratio $\tau_k/\tau_i \approx 20$ and strong C.W. laser oscillation is possible on all the transitions shown in Fig.11.10.

Output powers in the range 1-10 W on the strongest lines at 4880 Å and 5145 Å is standard on commercially available argon ion lasers. A typical argon laser plasma tube would have a diameter of 3 mm, gas pressure of 0·4 Torr and would be operated with a current of 30 amps at 240 V d.c. Thus 7·5 kW of power is dissipated in the small volume of the discharge capillary, most of which is transferred directly to the walls of the tube by ion bombardment. A great deal of research and development has gone into solving the problems connected with this energy dissipation. There are very few materials of the necessary purity which are able to withstand the intense ion bombardment and high wall temperature. They must also possess high thermal conductivities, thus permitting efficient heat transfer to water flowing through a cooling jacket surrounding the discharge. At present graphite or beryllia (BeO) is used for the bore of the discharge tubes. It has been found that the output power of argon ion lasers can be increased by a factor of two or three by applying an axial magnetic field of approximately 0·1 Tesla (1000 gauss) to the discharge tube. The reason for this behaviour is not properly understood. It may be due to an increase in the electron density in the plasma at a given value of the current, owing to the fact that the electrons are now constrained to move in helical orbits along the tube and the loss to the walls is reduced. In other respects, however, the argon ion laser construction resembles that of the He-Ne laser shown in Fig.11.4.

Problems

11.1. A helium-neon laser operating on the 6328 Å line has an output power of 1 mW. The laser beam can be focussed to a small, diffraction-limited, spot in which the intensity is 250 MW m^{-2}. Show that the same intensity would be obtained from a source of black-body radiation having a temperature of $2{\cdot}28 \times 10^{19}$ K.

11.2. Calculate the gain coefficient for the neon lines at 6328 Å and 3·39 μm attained in a helium-neon discharge assuming that an inversion density

$$\frac{N_k^0}{g_k} - \frac{N_i^0}{g_i}$$

of 10^8 atoms cm^{-3} exists for both of the lines. The spontaneous transition probabilities for emission at 6328 Å and 3·39 μm are $1{\cdot}38 \times 10^6 s^{-1}$ and $9{\cdot}66 \times 10^5 s^{-1}$ respectively, and the profiles of both lines may be assumed to be dominated by Doppler broadening at the effective gas temperature of 400 K.

11.3. The proposal is made to construct an X-ray laser operating at a wavelength of 10 Å. Assuming that the width of the X-ray gain profile is dominated by natural broadening, show that to obtain a gain coefficient of 10^{-2} m^{-1}, an inversion density of $6{\cdot}28 \times 10^{16}$ m^{-3} is required.

Assuming that the inverted population radiates spontaneously at 10 Å at the classical rate, estimate the minimum power which must be delivered to the system in order to attain the required inversion density. (Ans: $2{\cdot}77 \times 10^{14}$ W.)

11.4. It is proposed to construct a gas laser operating on a transition between an upper level $|k\rangle$ having a statistical weight $g_k=5$ and a lower level $|i\rangle$ having a statistical weight $g_i=3$. Under the operating conditions in the discharge, the lower level has an effective lifetime of 50 ns and the spontaneous transition probability of the laser line is $A_{ki} = 1{\cdot}35 \times 10^7$ s^{-1}. Is C.W. laser oscillation possible in this system? (Ans: No.)

11.5. The energy separation between the $2\ ^3S_1$ metastable level of helium and the $2s_2$ level of neon is $\Delta E = 0{\cdot}04$ eV. Compare this with the mean kinetic energy of the atoms in a gas at 450 K and make an estimate of the cross-section for excitation transfer.

11.6. In a helium-neon gas discharge it may be assumed that the densities N_0 and N_1 of the helium ground and metastable levels respectively are determined only by collisions with electrons and by the diffusion loss of metastable atoms to the walls of the discharge. Neglecting the effect of other excited levels, show that the ratio of metastable to ground-state helium atom densities is given by equation (11.19).

References

Basov, N.G. and Krokhin, C.N. (1962). *Appl.Opt.* 1, 213.

Bell, W.E. (1964). *Appl.Phys.Lett.* 4, 34.

Bennett, W.R.Jr, Kindlmann, P.J., Mercer, G.N., and Sunderland, J. (1964). *Appl.Phys.Lett.* 5, 158.

_____ and Kindlmann, P.J. (1966). *Phys.Rev.* 149, 38.

Bloom, A.L. (1968). *Gas lasers*, Wiley, New York.

Bridges, W.B. (1964*a*). *Appl.Phys.Lett.* 4, 128.

_____ (1964*b*). *Proc.Inst.elec. and electron.Eng.* 52, 843.

Collins, G.J. (1973). *J.appl.Phys.* 44, 4633.

Gordon, E.I. and White, A.D. (1963). *Appl.Phys.Lett.* 3, 199.

_____, Labuda, E.F., and Bridges, W.B. (1964). *Appl.Phys. Lett.* 4, 178.

Gordon, J.P., Zeiger, H.J., and Townes, C.H. (1954). *Phys. Rev.* 95, 282.

Javan, A., Bennett, W.B. Jr., and Herriott, D.R. (1961). *Phys.Rev.Lett.* 6, 106.

Kaspar, J.V.V. and Pimental, G.C. (1964). *Appl.Phys.Lett.* 5, 231.

Lawrence, G.M. and Liszt, H.S. (1969). *Phys.Rev.* 178, 122.

Maiman, T.H. (1960). *Nature, Lond.* 187, 493.

Moore, C.E.(1949). *Atomic energy levels*, Vol.I. National Bureau of Standards, Circular 467, U.S.Govt.Printing Office, Washington, D.C.

Patel, C.K.N. (1964*a*). *Phys.Rev.Lett.* 12, 588.

_____ (1964*b*). *Phys.Rev.* 136, A1187.

Patel, C.K.N. (1964e). *Phys.Rev.Lett.* 13, 617.

_____ (1965). *Appl.Phys.Lett.* 7, 15.

Piper, J.A. and Webb, C.E. (1973). *J.Phys.D.* 6, 400.

_____ and Webb, C.E. (1976). *I.E.E.E. J.Quantum Electron.* QE-12, 21.

Rabinowitz, P., Jacobs, S., and Gould, G. (1962). *Appl.Opt.* 1, 513.

Schawlow, A.L. and Townes, C.H. (1958). *Phys.Rev.* 112, 1940.

Silfvast, W.T. (1973). *Scient.Am.* 228, No.2, 88.

_____ and Klein, M.B. (1970). *Appl.Phys.Lett.* 17, 400.

Webb, C.E., Turner-Smith, A.R. and Green, J.M. (1970). *J.Phys. B.* 3, L135.

White, A.D. and Gordon, E.I. (1963). *Appl.Phys.Lett.* 3, 197.

_____ and Rigden, J.D. (1962). *Proc.Inst.Radio Engrs.* 50, 2366.

General references and further reading

Bennett, W.R. Jr. (1962). *Appl.Opt.Suppl.* 1, 24.

_____ (1965). *Appl.Opt.Suppl. on chemical lasers*, p.3.

Bloom, A.L. (1968). *Gas lasers*, Wiley, New York.

Schawlow, A.L. (1969). *Lasers and light*. W.H. Freeman & Co., San Francisco.

Sinclair, D.C. and Bell, W.E. (1969). *Gas laser technology*. Holt, Rinehart and Winston, New York.

Willett, C.S. (1974). *Introduction to gas lasers; population inversion mechanisms*. Pergamon, Oxford.

Yariv, A. (1967). *Quantum electronics*. Wiley, New York.

_____ (1971). *Introduction to optical electronics*. Holt, Rinehart & Winston, New York.

12

Resonant modes of optical cavities

In gas discharges at pressures below 10 Torr, the low density of atoms means that the gain of many laser transitions is only a few per cent per metre, making these devices quite unsuitable as optical-frequency amplifiers. However, these systems can usually be made into oscillators if sufficient positive feedback is applied by enclosing the gain medium in an optical cavity formed by highly-reflecting plane or concave mirrors. The temporal and spatial coherence of the laser output is then very largely determined by the properties of the optical cavity. In this chapter we therefore examine the resonant frequencies and spatial distributions of the electromagnetic fields which form the laser cavity modes. We shall discover that only certain combinations of mirror curvatures and spacings form cavities with stable, low-loss modes. For these cavities detailed calculations, which we can only present in outline, have enabled the diffraction losses to be investigated. We shall show finally how the total cavity loss determines the cavity Q-value and hence the intrinsic spectral purity of the laser output.

12.1. Introduction

The optical cavities of many gas lasers must necessarily be rather long in order that amplification in a low-gain medium can offset the unavoidable cavity losses which can seldom be reduced below 1-2 per cent. It is not immediately obvious then that such open-walled structures will indeed possess a set of low-loss modes. The cavity will certainly have very substantial losses for any radiation which propagates in a direction making a large angle with the cavity axis.

Therefore, in order that such a resonator should possess low-loss, high-Q modes, two criteria must be satisfied:

(a) Firstly, on the basis of geometrical optics, there must exist a family of rays which after suffering repeated reflections from the two mirrors, do not miss either mirror until they have made between 20-100 complete transits through the cavity.

(b) Secondly, to avoid excessive losses due to diffraction the second mirror should lie in the near-field diffraction pattern of the first. This condition is satisfied if the Fresnel number, F, of the optical system obeys the inequality

$$F = \frac{a^2}{\lambda L} \geq 1 \tag{12.1}$$

where a is the radius of the mirror aperture, which is assumed to be circular, and L is the mirror separation. The Fresnel number is approximately equal to the number of Fresnel zones seen in the aperture of one mirror from the centre of the other (Problems 12.1 and 12.2).

12.2. Numerical solution of cavity mode problem

A rigorous solution for the eigenmodes of an open resonator has not so far been obtained. However, if the dimensions of the resonator are large compared with the wavelength of light and if the electric field is assumed to be a linearly polarized transverse electromagnetic (TEM) wave, then the Fresnel-Kirchhoff formulation of Huygens' principle can be used. This enables the field distribution on mirror 2, $E_2(u_2,v_2)$, to be related to the field distribution existing on mirror 1 through the surface integral

$$E_2(u_2,v_2) = \frac{ik}{4\pi} \iint_{S_1} E_1(u_1,v_1) \frac{(1+\cos\theta)}{R} \exp(-ikR)\, du_1 dv_1 \tag{12.2}$$

where R is the distance between the points (u_1,v_1) and (u_2,v_2) on the mirrors 1 and 2 respectively, and θ is the angle which the direction of R makes with the normal to the mirrors, as shown in Fig.12.1. The variables (u,v) specifying positions on the mirror surfaces may be expressed in any

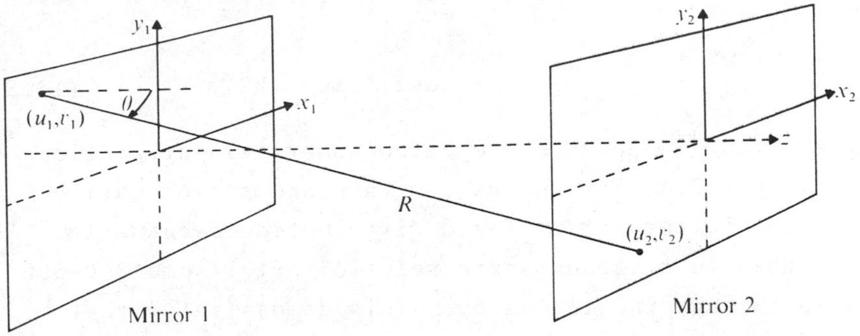

Fig.12.1. Geometry for diffraction calculations in an optical resonator.

convenient orthogonal coordinate system.

By assuming an arbitrary initial field distribution on the first mirror, the field produced at the second mirror may be computed using equation (12.2). This process can then be repeated over and over again for successive transits through the cavity. If there is a steady state solution, i.e. if an eigenmode ists, the field distributions across the mirrors after many transits will be identical except for repeated multiplication by a complex constant $\gamma = \gamma_0 \exp(i\beta)$. This represents the change in amplitude and phase, over and above the normal geometrical phase change given by exp(-ikL), suffered by the wave in travelling from one mirror to the other. In a cavity composed of identical mirrors the mode must have the same distribution over each mirror, thus after p single-pass transits between the mirrors we have

$$E_1^p(u_1,v_1) = \gamma^p E_1^0(u_1,v_1) \qquad (12.3)$$

Substituting equation (12.3) into equation (12.2) we obtain an integral equation for the spatial distribution and propagation constant γ of the eigenmode $E^0(u_1,v_1)$:

$$\gamma E^0(u_2,v_2) = \iint_{S_1} K\, E^0(u_1,v_1)\, du_1 dv_1 \qquad (12.4)$$

where K is known as the kernel of the equation and is given by

$$K = \frac{ik}{4\pi R}(1 + \cos\theta)\exp(-ikR). \qquad (12.5)$$

Fox and Li (1961) showed by repeated numerical integration of equation (12.4) for the case of a plane mirror cavity that an initial arbitrary field distribution eventually settled down to a steady-state solution after some 200-300 transits through the resonator. This demonstrates that eigenmodes do exist in open-sided resonators having finite rectangular or circular apertures. The eigenmodes of rectangular symmetry can be classified as TEM_{mnq} where m,n, and q-1 are the number of nodes in the x, y and z-directions respectively. The z-direction is usually chosen to coincide with the axis of the cavity. For mirrors having circular apertures the modes are designated by TEM_{plq} where p and l give the number of circular and radial nodal lines respectively.

For a plane-parallel Fabry-Perot type cavity for which $F \geq 10$, resonance occurs when the cavity length, L, is equal to an integral number of half wavelengths,

$$L = \frac{q\lambda_q}{2}. \qquad (12.6)$$

For a typical cavity of length L=100 cm we have $q \approx 3 \times 10^6$ at λ=6328 Å. Thus the integer q is always very large, while, as we shall see, the low-loss modes are those for which (m,n) or (p,l) are confined to combinations of small integers by the diffraction losses. Consequently the field distribution of a mode in the (x,y) plane is nearly independent of the longitudinal mode number q. It is convenient to refer to these field configurations, some of which are shown in Fig.12.2, as the transverse or spatial mode distributions.

The fractional power loss per transit, $1 - \gamma_0^2$, obtained by Fox and Li (1961) for the lowest-order transverse modes of a plane mirror cavity having circular apertures, is shown in Fig.12.3 as a function of the Fresnel number

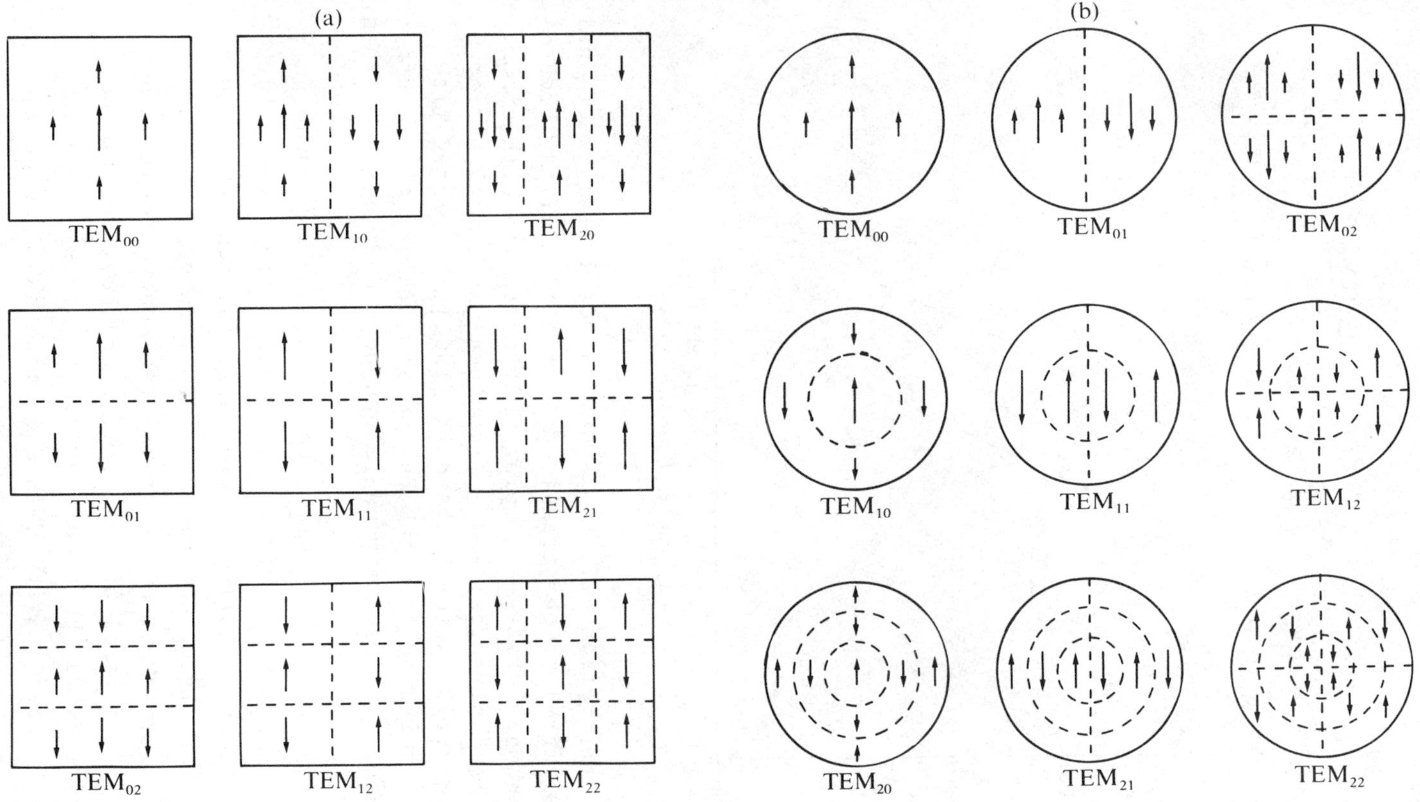

Fig.12.2. Diagrammatic representation of the linearly-polarized field configurations of transverse modes of optical resonators for (a) square, and (b) circular mirror apertures. (After Kogelnik and Li (1966).)

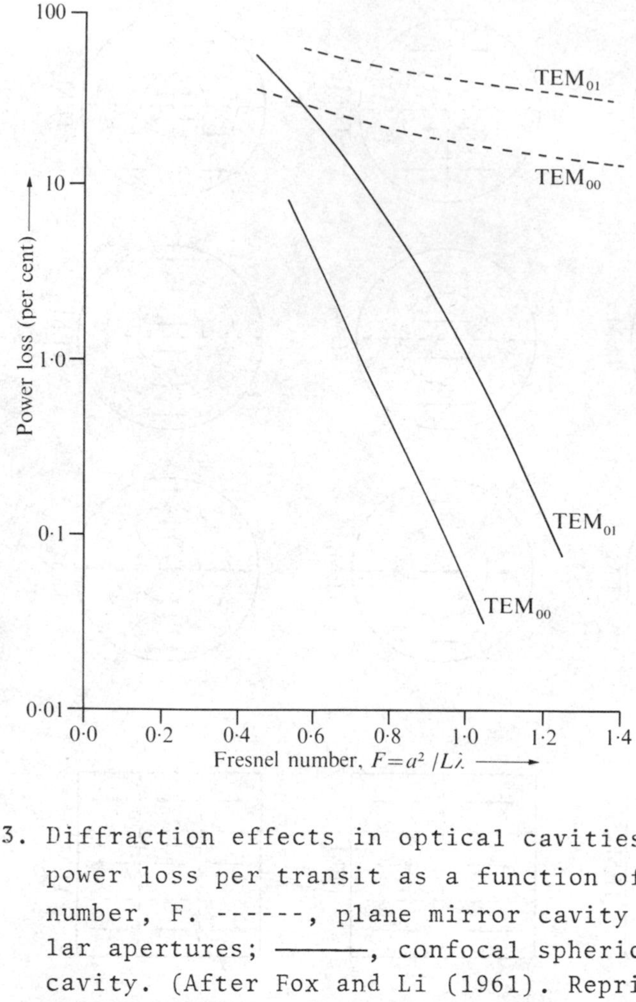

Fig.12.3. Diffraction effects in optical cavities: fractional power loss per transit as a function of the Fresnel number, F. ------, plane mirror cavity with circular apertures; ———, confocal spherical mirror cavity. (After Fox and Li (1961). Reprinted with permission from the *Bell System Technical Journal*, © 1961, The American Telephone and Telegraph Company.)

of the cavity, $F=a^2/\lambda L$. Similar calculations have been performed by Fox and Li (1961) and Boyd and Gordon (1961) for the confocal cavity formed by two identical concave mirrors separated by their common radius of curvature. Their results for the lowest-order transverse modes are also shown in Fig.12.3. It is clear that in the confocal cavity the diffraction loss may be orders of magnitude less than that of the plane mirror cavity. This is reasonable since the curved mirrors tend to reflect light back towards the axis of the resonator. It is also important to notice that the ratio of the diffraction losses of the TEM_{01} and TEM_{00} modes of the plane-parallel resonator is approximately two over a wide range of cavity dimensions. Thus a small additional loss, such as that caused by a speck of dust or a window of poor optical quality, will cause a laser with a plane-parallel cavity to operate in higher-order transverse modes. In contrast, the ratio of the diffraction losses for the confocal configuration is much higher, of the order of 25 for resonators with $F \geq 1$ and it is much easier to ensure operation in the uniphase TEM_{00} mode.

The effects of non-parallelism of the mirrors in a nominally plane-parallel cavity have been studied by Fox and Li (1963). The diffraction loss is found to increase rapidly for only slight deviations from perfect alignment or from perfect optical quality of the windows and mirrors.

For instance, in order to maintain the diffraction loss below 1 per cent per pass at 6328 Å in a 1 m long cavity, having an effective aperture of 3 mm diameter, it is necessary to maintain the parallelism of the mirrors to within 1 arc sec. This is an extremely stringent requirement for a mechanical structure of this size. Fortunately cavities using spherical mirrors are between 10 and 100 times less sensitive to mirror misalignment and consequently the plane-parallel resonator is now seldom used.

12.3. Approximate analytic solutions for transverse modes

Boyd and Gordon (1961) showed that explicit analytical expressions for the electric field distribution in the transverse modes may be obtained by allowing the limits of integration in equation (12.4) to tend to infinity. This is a valid approximation for stable resonators having Fresnel numbers which satisfy the condition $F \gg 1$, i.e. the size of the mirror aperture is large compared with the size of the laser cavity mode. The condition that stable solutions exist for radiation which is propagated back and forth within the resonator is then equivalent to requiring that the field distribution in the far-field diffraction pattern must be identical to that of the near-field distribution. In this case equation (12.4) reduces to a Fourier integral which in cartesian coordinates has the form

$$E^0(x_2,y_2) = A \iint_{-\infty}^{\infty} E^0(x_1,y_1)\, \exp\{-ik(x_1x_2+y_1y_2)/R\}\, dx_1dy_1 \tag{12.7}$$

where A is a constant and now $R \gg L$. Thus the spatial mode distributions are approximately described by functions which are their own Fourier transforms. The solutions of equation (12.7) have the form (Problem 12.3)

$$E^0_{mn}(x,y) = E_0\, H_m\left(\frac{\sqrt{2}x}{w}\right) H_n\left(\frac{\sqrt{2}y}{w}\right) \exp\left\{-\frac{(x^2+y^2)}{w^2}\right\} \tag{12.8}$$

where H_m and H_n are Hermite polynomials of order m and n respectively. Some of the Hermite polynomials of low order are given in Table 12.1. The parameter w has dimensions of length and is often called the beam radius or the spot size. It is determined by the geometry and dimensions of the resonant cavity, as will be explained in section 12.4 below. The field distributions given by equation (12.8) should be familiar since they are identical with the wave-functions of the two-dimensional harmonic oscillator. It is then apparent that the mode distributions corresponding to different values of (m,n) form a complete orthogonal set of functions which may be used for the expansion of any arbitrary field distribution E(x,y).

Frequently the symmetry of the transverse modes is determined by the circular cross-section of the laser discharge tube. In this case the two-dimensional Fourier transform is expressed in polar coordinates (r,θ) as

$$E^0(r_2,\theta_2) = A \int_0^{\infty}\int_0^{2\pi} E^0(r_1,\theta_1)\, \exp\{-ikr_1r_2\cos(\theta_1-\theta_2)/R\}\, r_1dr_1d\theta_1. \tag{12.9}$$

The solutions of equation (12.9) are of the form

Table 12.1.

Hermite polynomials and generalized Laguerre polynomials of low order

Hermite polynomial $H_m(x)$	Laguerre polynomial $L_p^l(x)$
$H_0(x) = 1$	$L_0^l(x) = 1$
$H_1(x) = 2x$	$L_1^l(x) = (l+1) - x$
$H_2(x) = 4x^2 - 2$	$L_2^l(x) = \frac{(l+1)(l+2)}{2!} - (l+2)x + \frac{x^2}{2!}$
$H_3(x) = 8x^3 - 12x$	$L_3^l(x) = \frac{(l+1)(l+2)(l+3)}{3!} - \frac{(l+2)(l+3)x}{2!} + \frac{(l+3)x^2}{2!} - \frac{x^3}{3!}$

$$E_{pl}^0(r,\theta) = E_0\left(\frac{\sqrt{2}r}{w}\right)^l L_p^l\left(\frac{2r^2}{w^2}\right) \exp\left(-\frac{r^2}{w^2}\right) \left\{\begin{matrix}\sin\\ \cos\end{matrix} l\theta\right\} \quad (12.10)$$

where the functions L_p^l are known as the generalized Laguerre polynomials, some of which are given in Table 12.1.

The mode patterns of a laser oscillating in some of the lowest-order modes of rectangular symmetry are shown in Fig.12.4. These should be compared with the diagrammatic representation of electric fields shown in Fig.12.2. The higher-order modes clearly have more extended field distributions and consequently higher diffraction losses than the lower-order modes, as we have already remarked. The wavefronts of these modes contain one or more phase reversals of π arising from the particular form of the Hermite or Laguerre polynomial which determines the field distribution.

The lowest-order mode has identical form in both rectangular and cylindrical symmetry:

$$E_{00}(x,y) = E_{00}^0(r,\theta) = E_0\exp\left(-\frac{r^2}{w^2}\right) \quad (12.11)$$

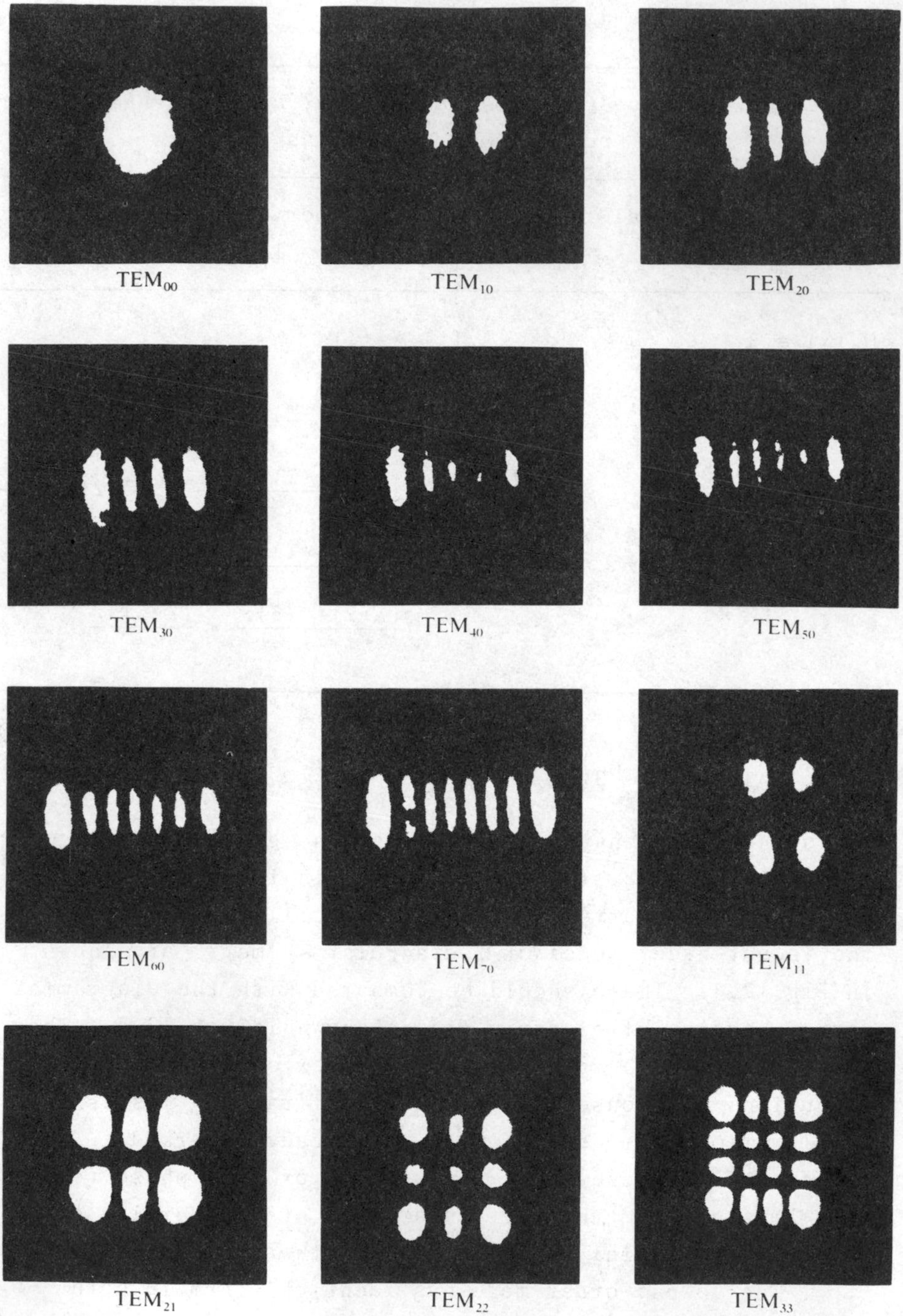

Fig.12.4. Field patterns of a gas laser oscillating in single-low-order transverse modes of rectangular symmetry.

where $r^2=x^2+y^2$. This mode has the lowest diffraction loss and is often referred to as the dominant mode. It is particularly important in laser applications because the electric field has a constant phase across the entire wavefront. In addition the beam from a laser oscillating in the TEM_{00} mode has the smallest angular divergence outside the cavity and it can be focussed down to give the smallest spot size consistent with the theory of diffraction. To ensure operation in the TEM_{00} mode, an adjustable aperture may be placed inside the laser cavity and the diffraction losses of higher-order modes increased until they cease to oscillate. Alternatively, once the bore of the discharge tube is fixed, then the spacing and radii of the mirrors may be chosen to achieve the same result. In this case the numerical calculations by Li (1965) of the diffraction losses in resonators with circular apertures provide a useful guide.

When the laser is oscillating on the single TEM_{00} transverse mode, the phase is constant across the wavefront and the beam may be said to possess complete spatial coherence. This spatial coherence should be regarded as a consequence of the properties of the optical resonator rather than the stimulated emission process which is responsible for the laser oscillation.

12.4. Mode size and cavity stability

The relationship between the characteristic length w appearing in equations (12.8) and (12.10) and the dimensions of the optical cavity was first derived for confocal resonators by Boyd and Gordon (1961) and then extended to more general resonator configurations by Boyd and Kogelnik (1962). Consider a resonator consisting of mirrors 1 and 2 having radii R_1 and R_2 separated by a distance L as shown in Fig. 12.5. The mirror radii are defined as positive if the mirrors are concave and facing inwards to form the laser cavity. Then the radius w_1 of the laser mode at the first mirror is defined as the distance from the axis at which the field drops to 1/e of its maximum amplitude and is given

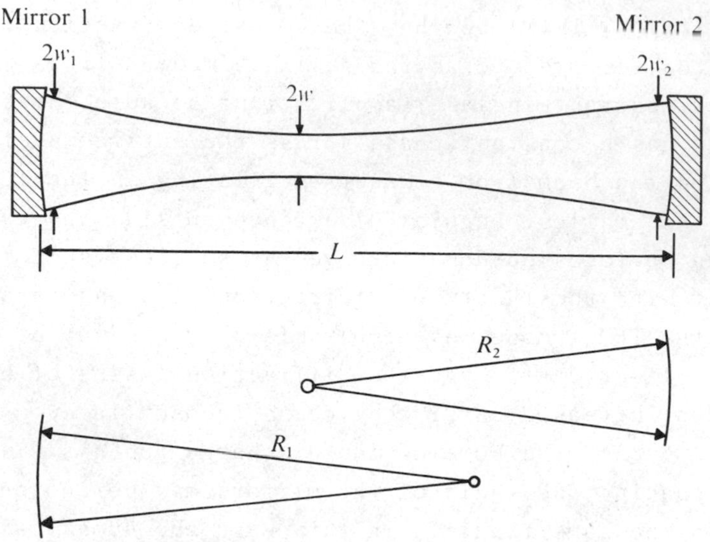

Fig.12.5. Definitions of the parameters R_1, R_2, and L for an optical cavity and the radii characterizing the Gaussian field distribution of the laser transverse modes.

by

$$w_1^4 = \left(\frac{\lambda}{\pi}\right)^2 \left(\frac{R_1^2 L}{R_1+R_2-L}\right) \left(\frac{R_2-L}{R_1-L}\right) . \tag{12.12}$$

The beam radius at the second mirror, w_2, may be obtained by interchanging the indices 1 and 2 throughout equation (12.12). For the symmetrical confocal resonator we have $R_1=R_2=L$ and the mode radius at the mirrors is given by the simple expression (Problem 12.4)

$$w_1 = w_2 = \left(\frac{\lambda L}{\pi}\right)^{1/2} \tag{12.13}$$

Within a concave mirror resonator, the beam contracts to a minimum diameter where the equiphase surface is a plane, as shown in Fig.12.5. The diameter, $2w_0$, at this beam waist is given by

$$w_0^4 = \left(\frac{\lambda}{\pi}\right)^2 \left\{ \frac{L(R_1-L)(R_2-L)(R_1+R_2-L)}{(R_1+R_2-2L)^2} \right\}. \qquad (12.14)$$

Outside the resonator the angular divergence of a transverse mode is measured by the far-field diffraction angle θ, which is defined as the ratio of the spot size at a distance z divided by z in the limit $z \to \infty$. It may be shown that $\theta = \lambda/\pi w_0$, (Problem 12.5).

It can be demonstrated by simple ray tracing that not all combinations of mirror radii and mirror separations represent stable cavities since a beam, initially launched parallel to the cavity axis, may after several reflections start to diverge. The cavity then has a very high diffraction loss and is said to be unstable. The conditions which must be satisfied for the resonator to be stable and so possess low-loss, high-Q modes can be deduced from equation (12.14), since w_0^4 must, in this case, be positive. We therefore require that

$$(R_1+R_2-L)(R_1-L)(R_2-L) \geq 0. \qquad (12.15)$$

For concave mirrors this cavity stability condition may be stated in the form: the cavity will be stable if either both centres of curvature lie inside the cavity and overlap or both centres of curvature lie outside the cavity. In more general cases it is convenient to recast the stability criterion in the form (Problem 12.6):

$$0 \leq \left(1 - \frac{L}{R_1}\right)\left(1 - \frac{L}{R_2}\right) \leq 1. \qquad (12.16)$$

The various resonator geometries can then be represented diagrammatically as in Fig.12.6 where the boundaries between the stable and unstable (shaded regions) are determined by equation (12.16). We find that the symmetric confocal, ($R_1=R_2=L$), the symmetric concentric, ($R_1=R_2=L/2$), and the plane-parallel resonators are all on the verge of instability and may, by accidental misalignment, become extremely lossy. Thus it is virtually impossible to obtain laser oscillation in these cavities with low-gain systems. Fortunately such singularly undesirable configurations of symmetric mirrors may be easily avoided by slightly increasing or decreasing the mirror separation. On the other hand, in high-gain systems such as the pulsed CO_2 lasers, an unstable cavity may not prevent laser action and may offer a convenient way of extracting power from the oscillator, as Sinclair and Cottrell (1966) have demonstrated.

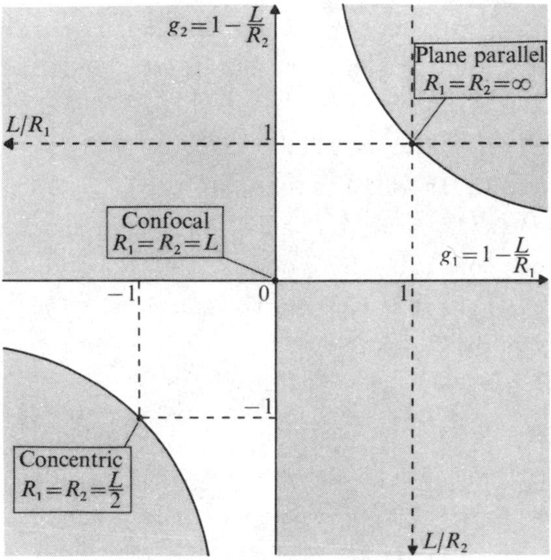

Fig.12.6. Stability diagram for optical resonators. Shaded areas are those for which the stability criterion $0 \leq (1-L/R_1)(1-L/R_2) \leq 1$ is violated. In these regions the cavity has a high loss.

12.5. Design considerations for practical systems

To illustrate some of the considerations involved in designing a simple optical cavity, we consider the low gain He-Ne system operating at 6328 Å. We assume that the design aim is to obtain the maximum output power on the TEM_{00} mode. We arbitrarily choose a discharge tube of length 80 cm, and in this case a mirror separation of 100 cm would be convenient. The diameter of the plasma tube is then determined by making a compromise between achieving the highest gain, which varies inversely with d as explained in section 11.4.2(d), and attaining maximum output power, which increases as d^2 due to the increased volume of active material. We assume that this fixes the bore of the tube at d=4·0 mm. The optimum cavity is now one for which the spot size fills the plasma tube. However, in order that the diffraction loss of the dominant mode should not be too high, the mode radius everywhere within the cavity should be smaller than the bore of the discharge tube by a factor of between 3·5 and 4·0. Hence in this case (Problem 12.7), 1·0 mm $\leq w_{max} \leq$ 1·14 mm. Using equation (12.12), we find that this spot size can be achieved in a symmetrical mirror cavity using either an almost concentric system with $R_1 \approx$ 50·5 cm or a long-radius cavity with R ≈ 48 m. An alternative arrangement would involve an approximately hemispherical system having one plane mirror and one concave mirror with R ≈ 104 cm. The main differences between these cavities is the volume filled by the TEM_{00} mode and the ease of alignment of the cavity mirrors. Although the mode volume of the hemispherical cavity is only one third of the total available volume (since the spot size on the plane mirror approaches zero in the hemispherical limit), it is larger than that of the concentric system and is likely to be the preferred resonator due to the difficulty of aligning mirrors with large radii of curvature.

To ensure that the dominant transverse mode is the TEM_{00} mode, the surface figure of the laser mirrors should be accurate to ≤ λ/20 over the size of the laser beam. Similarly the Brewster angle windows of the laser discharge should have surfaces flat to λ/4 and parallel to one minute of arc, although the tolerance on the actual Brewster angle is less stringent being ≈ 1° (Problem 12.8). The surface scatter from the windows and mirrors should be held to an absolute minimum by a lengthy final state in the optical polishing and by careful cleaning before use. The mirrors are usually coated with alternate $\lambda/4$ layers of high and low refractive index materials. Hard dielectrics like TiO_2 and SiO_2 are now used in place of the softer materials ZnS and cryolite. Reflectances of up to 99·8 per cent with absorption and scattering losses of 0·2 per cent can be routinely obtained. The laser mirrors are held in mounts capable of fine adjustments to within a few seconds of arc and these mounts are rigidly fixed to a quartz or invar base to obtain cavity length stability. Further parameters requiring optimization, such as the transmission of the output mirror, the gas pressure and composition, and the discharge current density are discussed by Smith (1966).

12.6. Cavity Q-factor and resonance linewidth

We have seen in preceding sections that the diffraction losses in optical cavities can be reduced to negligible values by careful design. However, all practical resonators have finite losses associated with the output transmission at the mirrors and so the Q of the cavity, defined by

$$Q = \omega \frac{\text{Energy stored}}{\text{Energy lost per second}} \tag{12.17}$$

is finite. This quality factor Q determines the frequency uncertainty or linewidth of the cavity resonance $\Delta\omega_c$ as we can see by considering the decay of the energy $W(t)$ stored in the cavity when the source of excitation is removed at $t=0$. From equation (12.17) we have

$$-\frac{dW}{dt} = \frac{W\omega}{Q} \tag{12.18}$$

which has solutions of the form $W(t) = W(0)\exp\{-t/(Q/\omega)\}$. The time dependence of the electric field in the cavity is exactly the same as that emitted by the damped harmonic electron oscillator considered in section 8.1, and the width of a resonant mode of the passive cavity is therefore given in angular frequency units by

$$\Delta\omega_c = (\text{Decay time})^{-1} = \omega/Q. \qquad (12.19)$$

To relate $\Delta\omega_c$ to the cavity losses, we consider a collimated beam of radiation of intensity I_0 launched inside the cavity parallel to the cavity axis. After making one complete trip the beam returns to its starting point with an intensity $(I_0-\delta_c I_0)$ where δ_c is the fractional round-trip loss of the cavity. This fractional loss will include the diffraction loss discussed in section 12.2, but will usually be dominated by the losses produced by absorption, scattering, and output transmission. Since a round trip in the cavity takes a time $\Delta t=2L/c$, we have

$$\frac{dI_0}{dt} = \lim_{\Delta t\to 0} \frac{\Delta I_0}{\Delta t} = -I_0 \frac{c\delta_c}{2L}. \qquad (12.20)$$

Thus the cavity decay time is given by $2L/\delta_c c$ and an alternative expression for the cavity resonance width is given by (Problem 12.9)

$$\Delta\omega_c = \frac{c\delta_c}{2L}. \qquad (12.21)$$

For a cavity in which diffraction and scattering losses are negligible, we have $\delta_c \approx 1-\mathcal{R}_1\mathcal{R}_2$ where $\mathcal{R}_1$ and $\mathcal{R}_2$ are the power reflection coefficients of the cavity mirrors at the laser frequency. A typical value of δ_c would be $0\cdot 02$, giving $\Delta\omega_c = 3\times 10^6\ s^{-1} \equiv 0\cdot 5$ MHz for a cavity 1 m long. From equations (12.19) and (12.21) we have

$$Q = \frac{2\omega L}{c\delta_c} = \frac{4\pi L}{\lambda\delta_c}. \qquad (12.22)$$

For the example considered above, the Q-value of the optical cavity is of the order of 10^9, which is very high compared with the maximum Q of the order of 10^4 which is achievable in the microwave region.

The cavity losses determine not only the minimum inversion density required to sustain oscillation, which we shall discuss in Chapter 13, but also the theoretical spectral purity of the laser output. By considering the noise power produced by spontaneous emission into a cavity mode, which cannot be separated from the power derived by stimulated emission, it can be shown that the line width of the laser output is limited to a value (Schawlow and Townes 1958)

$$\Delta\omega_L \approx \frac{\hbar\omega_0(\Delta\omega_c)^2}{2P} \qquad (12.23)$$

where P is the laser output power. Using the value of $\Delta\omega_c$ derived above, we obtain $\Delta\omega_L = 2\pi \times 10^{-3}\ s^{-1}$ for a typical He-Ne laser emitting 1 mW of power at 6328 Å. In many lasers this theoretical limit is never achieved (Problem 12.10) because of mechanical vibrations and temperature fluctuations in the laser resonator. However, by careful design and the use of electronic frequency stabilization schemes it is now possible to approach this theoretical limit very closely, as we shall see in Chapter 13.

Problems

12.1. The cavity of a helium-neon laser has a length of 60 cm and an effective circular aperture of radius 2 mm at the mirrors. Calculate the Fresnel number of this cavity when used for laser oscillation at the wavelengths of 0·633, 1·15 and 3·39 μm respectively. (Ans: 10·5, 5·80, 1·97.)

12.2. The cavity described in Problem 12.1 is equipped with plane mirrors which are initially aligned perfectly parallel to one another. Show that if one of the mirrors is accidentally tilted through a small angle θ, a ray of light which was initially coincident with the axis of the cavity will, after (2N+1) transits between the mirrors, have been displaced off-axis through a distance y where

$$y \cong 2L\theta N(N+1).$$

For successful laser oscillation the beam must complete not less than 61 transits before leaving the cavity. Hence show that the parallelism of the mirrors must be maintained to better than 0·37 sec of arc.

12.3. By expressing the field distribution $E^0(x_1,y_1)$ in a laser cavity mode as a product $E^0(x_1)E^0(y_1)$, show that the double integral in equation (12.7) may be separated into two independent integrals over x_1 and y_1 respectively. Hence, using the Hermite polynomials $H_0(\xi)$ and $H_1(\xi)$ given in Table 12.1, prove that the functions

$$H_0\left(\frac{\sqrt{2}x}{w}\right)\exp\left(-\frac{x^2}{w^2}\right) \quad \text{and} \quad H_1\left(\frac{\sqrt{2}x}{w}\right)\exp\left(-\frac{x^2}{w^2}\right)$$

are their own Fourier transforms.

12.4. In an optical resonator having mirrors of equal curvature and fixed separation L, show that the minimum mode spot size at the mirrors is obtained in the confocal system, $R_1=R_2=L$.

Plot the ratio of the mode spot size at the mirrors of a symmetrical resonator to the value it has in the confocal system as a function of L/R and show that the spot size becomes infinite in the plane-parallel and concentric mirror cavities.

12.5. Show that the 6328 Å output beam from a helium-neon laser having a 1 m long confocal cavity has a minimum angular divergence of $8{\cdot}98 \times 10^{-4}$ rad. The laser beam is expanded by projection through a telescope with a magnification of 100, and directed towards the Moon which is $3{\cdot}78 \times 10^5$ km from the Earth. Calculate the radius of the laser spot on the Moon's surface. (Ans: 3·39 km.)

12.6. Show that

$$\frac{1}{(1-L/R_1)(1-L/R_2)} - \frac{L(R_1+R_2-L)}{(R_1-L)(R_2-L)} = 1$$

and hence or otherwise prove that the cavity stability

criterion, equation (12.15), may be rearranged in the form given in equation (12.16).

12.7. Using the explicit expressions for the electric field distributions given by equation (12.10) and Table 12.1, calculate the fraction of the power in the TEM_{00} and TEM_{01} cylindrically-symmetric modes which is lost when an aperture of diameter 3·5 w is inserted close to one of the mirrors of a symmetric mirror laser cavity. (Ans: $2 \cdot 19 \times 10^{-3}$ and $8 \cdot 89 \times 10^{-3}$.)

12.8. A plane electromagnetic wave falls on the interface between two transparent dielectrics having refractive indices n_1 and n_2. When the direction of propagation makes an angle θ with the normal to boundary and the wave is linearly polarized in the plane of incidence, the reflectivity, r, is given by

$$r = \left(\frac{\sin 2\theta'' - \sin 2\theta}{\sin 2\theta'' + \sin 2\theta}\right)^2$$

where θ'' is the angle of refraction determined by the condition $\sin\theta/\sin\theta'' = n_2/n_1$. Show that the reflectivity is zero at the Brewster angle, $\theta = \tan^{-1}(n_2/n_1)$, and calculate the tolerance on the setting of the Brewster angle window of a laser discharge tube if the reflectivity loss is to be less than $2 \cdot 5 \times 10^{-3}$ per interface. The refractive index of fused silica at 6328Å may be taken as 1·4572. (Ans: 30'.)

12.9. An air-spaced Fabry-Perot étalon is formed of two plane-parallel plates, each having reflectivity $\mathcal{R}$, separated by a distance L. When the étalon is illuminated at normal incidence with monochromatic light of wavelength λ, the ratio of the transmitted to the incident intensity is given by

$$\frac{I_t}{I_0} = \left\{1 + \frac{4\mathcal{R}}{(1-\mathcal{R})^2} \sin^2(\delta/2)\right\}^{-1}$$

where $\delta = 4\pi L/\lambda$. Plot I_t/I_0 as a function of δ for plates of 90 per cent reflectivity and show that the half-width of the transmission maxima is given by

$$\Delta\omega_{1/2} = c(1-\mathcal{R})/L\sqrt{\mathcal{R}}.$$

Prove that this expression reduces to equation (12.21) when $\mathcal{R}$ is close to unity.

This problem demonstrates the link between the widths of the cavity response and the transmission peaks of this well-known interferometer. However, the fringe pattern observed with a Fabry-Perot étalon should not be confused with the mode pattern of a laser having a plane-parallel resonator. The Fabry-Perot étalon is normally used with the plates so close together that all the transverse modes of the corresponding optical cavity are virtually degenerate in frequency. The plane wavefronts assumed in the discussion of the theory of the étalon are composed of an infinite sum over the transverse modes of the cavity.

12.10. An optical resonator designed for use with the argon ion laser transition at 5145 Å consists of two mirrors having reflectivities of 99·8 and 87·8 per cent respectively which are spaced 60 cm apart. Calculate the width of the resonant modes of the passive cavity and the theoretical laser linewidth when the single mode output power is 0·5 W.

Then, using equation (12.6), estimate the change in the resonator length which would result in a shift of the frequency of oscillation equal to the theoretical linewidth of the laser output.

References

Boyd, G.D. and Gordon, J.P. (1961). *Bell Syst.Tech.J.* 40, 489.

_________ and Kogelnik, H. (1962). *Bell Syst.Tech.J.* 41, 1347.

Fox, A.G. and Li, T. (1961). *Bell Syst.Tech.J.* 40, 453.

Fox, A.G. and Li, T. (1963). *Proc.Inst.elec.and electron Eng.* 51, 80.

Li, T. (1965). *Bell Syst.Tech.J.* 44, 917.

Schawlow, A.L. and Townes, C.H. (1958). *Phys.Rev.* 112, 1940.

Sinclair, D.C. and Cottrell, T.H.E. (1966). *Appl.Opt.* 6, 845.

Smith, P.W. (1966). *I.E.E.E. J.Quantum Electron.* 2, 77.

General references and further reading

Kogelnik, H. and Li, T. (1966). *Appl.Opt.* 5, 1550. (This article gives a clear and concise account of the properties of laser cavities.)

Yariv, A. (1971). *Introduction to optical electronics.* Holt, Rinehart and Winston, New York.

13
Saturation characteristics and single-frequency operation of gas lasers

In Chapter 11 we showed how, in some gas discharges under rather special conditions, a population inversion can be created which leads to amplification at visible wavelengths. By enclosing such an amplifier in one of the optical resonators whose properties were discussed in Chapter 12, oscillation at optical frequencies can be obtained. We now consider in more detail the intensity and frequency distribution of the radiation emitted by these laser oscillators.

In this chapter we derive detailed expressions for the gain necessary to achieve oscillation threshold in a cavity whose losses are known. These enable the inversion density required for successful laser operation to be estimated in most cases. We next show how the growth of the radiation density inside the laser cavity reduces the inversion density through the effect of stimulated emission, leading eventually to the establishment of steady-state oscillation at a well-defined output power. This phenomenon is known as gain saturation and is treated in detail for both homogeneously- and inhomogeneously-broadened transitions. Its effect on the measurements of gain coefficients is also briefly discussed.

In lasers with inhomogeneously-broadened transitions it is found that oscillation usually occurs simultaneously on a number of longitudinal cavity modes. The reasons for this behaviour are explained and we examine one method for stabilizing the intensities and inter-mode frequency differences in multi-mode operation. Next we consider several different techniques which have been used to obtain oscillation on a single longitudinal mode and this leads on naturally to a discussion of the output power versus oscillation frequency of single-frequency gas lasers. The experimental observation and theoretical interpretation of the Lamb dip is the main topic of section 13.8.

The Lamb dip is an important manifestation of the saturation of the gain which occurs in inhomogeneously-broadened transitions when the oscillation frequency coincides with the centre of the laser line. A similar phenomenon occurs if the laser frequency is tuned close to the centre of an absorption line of a sample of atoms or molecules interacting with the standing wave field of the laser. This saturated absorption has become an important new technique in atomic and molecular spectroscopy since it removes the limit on the attainable resolution which was formerly imposed by the Doppler broadening of absorption lines.

Finally we discuss the frequency stabilization of gas lasers and show how the techniques of saturated absorption and frequency stabilization have been combined in a recent measurement of the velocity of light which attained a hundred-fold increase in precision. However, since this chapter is very largely concerned with the frequency distribution of the laser output, we first discuss in more detail the frequencies of the modes of the passive optical cavity.

13.1. Frequencies of the resonant cavity modes

In a typical gas laser the frequencies of the longitudinal modes of the optical resonator are separated by only one tenth of the width of the Doppler-broadened gain curve. Consequently most gas lasers oscillate simultaneously on several longitudinal modes and the time dependence of the total laser output depends on the amplitudes, frequencies, and relative phases of these oscillating modes. We therefore now consider in more detail the resonant frequencies of the modes of the passive cavity. These are determined by the phase of the propagation constant $\gamma = \gamma_0 \exp(i\beta)$ introduced in section 12.2. We recall that the propagation constant may be obtained by solving equations (12.4) or (12.7) using either numerical or analytical methods respectively. The cavity will be resonant at the angular frequency ω=kc if, after one round trip involving reflections off both mirrors, the change in phase of $2(kL-\beta)$ is equal to a multiple of 2π. Thus the cavity resonant frequencies are given by

$$\nu_{mnq} = \frac{c}{2L}\left(q + \frac{\beta}{\pi}\right) \tag{13.1}$$

where (q-1) is an integer giving the number of nodes in the electric field along the z axis. Using the approximation that the aperture of the laser mirrors is very large compared with the mode radius, Boyd and Kogelnik (1962) obtained the frequencies of the stable modes of a general two-mirror cavity in the form

$$\nu_{mnq} = \frac{\omega_{mnq}}{2\pi} = \frac{c}{2L}\left[q + \left(\frac{m+n+1}{\pi}\right)\cos^{-1}\left\{\left(1-\frac{L}{R_1}\right)\left(1-\frac{L}{R_2}\right)\right\}^{1/2}\right] \tag{13.2}$$

where R_1 and R_2 are the radii of curvature of the two mirrors which are separated by a distance L. For cavities with circular rather than rectangular symmetry, the factor (m+n+1) in equation (13.2) should be replaced by $(2p+l+1)$.

From equation (13.2) we see that transverse modes belonging to a given longitudinal mode number q generally differ in frequency. This is a consequence of the fact that the phase velocity of the wave in the cavity increases with increasing transverse mode number. However, in a plane-mirror resonator we have $R_1=R_2=\infty$ and the transverse modes belonging to a given longitudinal mode number q are degenerate in frequency:

$$\text{Plane-parallel} \qquad \nu_{mnq} = \frac{cq}{2L}. \tag{13.3}$$

This is in agreement with the result of an earlier, less rigorous derivation given in equation (12.6). Considerable frequency degeneracies also occur in the case of the confocal cavity, $R_1=R_2=L$, where

$$\text{Confocal} \qquad \nu_{mnq} = \frac{c}{4L}(2q+m+n+1). \tag{13.4}$$

In practice, however, even very slight deviations from the

exact plane-parallel or exact confocal geometries cause the transverse modes (m,n) of the same longitudinal mode number q to have frequencies which differ from one another by a few MHz (Problem 13.1).

In contrast, the frequency difference between adjacent longitudinal modes having the same transverse field distribution is the same for all cavity geometries:

$$\Delta\nu = \nu_{mnq+1} - \nu_{mnq} = \frac{c}{2L} . \qquad (13.5)$$

Longitudinal mode spacings for some commonly-used cavity geometries are given in Table 13.1. It is evident that for the longer cavities many longitudinal modes will lie within a Doppler width of the line centre and may oscillate simultaneously provided that the gain at these frequencies is sufficiently high. This is indicated schematically in Fig.13.1.

TABLE 13.1.

Longitudinal mode spacings for typical cavity lengths

Cavity length (cm)	10	30	60	100	180
Mode spacing (MHz)	1500	500	250	150	83

It is important to recognize that this discussion applies only to the resonant frequencies of the empty, or passive, cavity. When the cavity is filled with an amplifying medium, anomalous dispersion causes the refractive index to vary with frequency close to the laser linecentre. As a result the frequencies of the longitudinal modes are no longer exactly equally spaced. However, in gas lasers these frequency shifts are only of the order of one part in a thousand and for the moment we neglect them, returning for

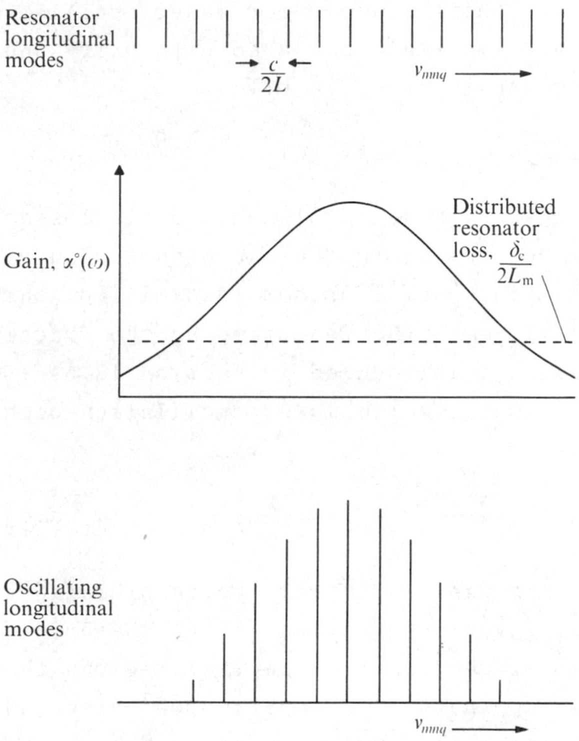

Fig.13.1. Longitudinal modes and the Doppler-broadened gain curve. Laser oscillation is possible on all modes for which $\alpha^0(\omega) \geq \delta_c/2L_m$.

a more detailed discussion of this point in section 13.6.

13.2. Gain required for oscillation

For laser oscillation to occur at the angular frequency ω two conditions must be satisfied. First, ω should equal or be very close to the angular frequency ω_{mnq} of a resonant cavity mode, and second, the gain in the medium due to stimulated emission must exceed the loss per round trip in the cavity. For a medium exhibiting gain at the frequency ω we have from equations (11.1) and (11.4)

$$I_\omega(2L_m) = I_\omega(0)\exp\{2\alpha^0(\omega)L_m\} \qquad (13.6)$$

where L_m is the length of the amplifying medium within the laser cavity. At the threshold for laser oscillation, $2\alpha^0(\omega)L_m$ is generally small compared with unity and the fractional increase in power is given by

$$\{I_\omega(2L_m) - I_\omega(0)\}/I_\omega(0) \approx 2\alpha^0(\omega)L_m. \tag{13.7}$$

The resonator losses, made up of unavoidable losses due to absorption and scattering in the end windows and mirrors, diffraction, and the useful output transmission through one or both mirrors, may be described by the fractional round-trip loss, δ_c, introduced in section 12.6. The threshold condition at which laser oscillation occurs is then given by:

$$2\alpha^0(\omega)L_m = \frac{2\pi^2 c^2}{\omega_{ki}^2} g_k A_{ki} \left(\frac{N_k^0}{g_k} - \frac{N_i^0}{g_i}\right) L_m g(\omega) \geq \delta_c \tag{13.8}$$

where the explicit expression for the unsaturated gain coefficient $\alpha^0(\omega)$ given by equation (11.4) has been introduced. In the ultraviolet and visible spectral regions the line-profile $g(\omega)$ is dominated by Doppler broadening and, using equation (11.7), the threshold condition becomes

$$2\alpha_G^0(\omega)L_m = \frac{2\pi^2 c^3}{\omega_{ki}^3}\left(\frac{M}{2\pi kT}\right)^{1/2} L_m g_k A_{ki}\left(\frac{N_k^0}{g_k} - \frac{N_i^0}{g_i}\right) \exp\left\{-\frac{(\omega-\omega_{ki})^2 Mc^2}{2kT\omega_{ki}^2}\right\}$$
$$\geq \delta_c. \tag{13.9}$$

where M is the atomic mass. In this case the gain profile has the Gaussian lineshape shown in Fig.13.1 and is said to be inhomogeneously broadened. The case of a homogeneously-broadened line is considered in detail in section 13.3 below.

Using equation (13.9), it may be shown (Problem 13.2) that inversion densities of the order of 10^7 - 10^8 cm^{-3} are required to reach the oscillation threshold in visible and near infrared lasers. With the gain just slightly above threshold, laser oscillation will occur only if one of the axial modes, ω_{mnq}, is close to the line centre, ω_{ki}. For a long cavity, $L \geq 50$ cm, reference to Table 13.1 shows that the modes are closely spaced and this condition is therefore satisfied automatically. However, for a short cavity, L $L \leq 15$ cm, the mode spacing is comparable with the Doppler width in many cases. Thus it may be necessary to tune the mode frequency towards line centre by slightly altering the length of the cavity before laser oscillation will commence.

13.3. Gain saturation - homogeneously-broadened transitions

From equation (13.6) it would appear that the intensity of radiation bouncing to and fro inside the laser cavity would increase indefinitely. This is an unphysical situation and occurs because we have so far neglected the effect of stimulated emission on the populations of the laser levels. As soon as oscillation commences, stimulated transitions start to reduce the inversion density below the value that it had in the absence of oscillation. In a time of the order of the inverse of the cavity linewidth, $\Delta\omega_c$, a steady state is established in which the gain at the oscillation frequency is reduced to a value equal to that required to replace the cavity losses. This process is known as *gain saturation*. The changes which occur in the population inversion density, $(N_k/g_k - N_i/g_i)$, between the laser on/laser off equilibrium situations may be quite considerable. They can be detected by measuring the variations in the intensity of spontaneous emission from either the upper or lower laser levels which occur when a shutter inside the laser cavity is alternately opened and closed. If the signal is to be displayed on an oscilloscope, a toothed disk spinning with its axis parallel to that of the laser forms a convenient shutter.

We now proceed to consider gain saturation in detail for the case of a homogeneously-broadened transition. Although experimentally this is not the most common situation the analysis required is somewhat less involved than that necessary in the case of the inhomogeneously-broadened lines, discussed in section 13.4. We need to derive the population densities of atoms in the upper and lower laser levels, N_k and N_i respectively, existing in equilibrium with the laser radiation. These population densities are determined by the rates of depletion and re-population of the levels. Including only the most significant processes and using the notation defined in section 11.3.1 we have:

$$\frac{dN_k}{dt} = S_k - \frac{N_k}{\tau_k} - (N_k B^I_{ki} - N_i B^I_{ik}) \frac{2I(\omega)g(\omega)}{4\pi} \qquad (13.10)$$

$$\frac{dN_i}{dt} = S_i - \frac{N_i}{\tau_i} + N_k A_{ki} + (N_k B^I_{ki} - N_i B^I_{ik}) \frac{2I(\omega)g(\omega)}{4\pi} \qquad (13.11)$$

where S_k, S_i are the rates at which the levels k and i are populated by collisions and radiative transfer from other levels; τ_k, τ_i are the effective lifetimes of the levels under the conditions existing in the discharge, as given by equation (11.12); A_{ki} and B^I_{ki} are the Einstein coefficients for spontaneous and stimulated emission respectively; $2I(\omega) = I_+(\omega)+I_-(\omega)$ is the sum of the intensities of the forward and backward propagating laser beams inside the cavity, and $g(\omega)$ in this case has the Lorentzian line profile given by equation (11.9). It is important to note that in the case of homogeneous broadening all atoms in the sample are able to interact with both components of the electromagnetic field of angular frequency ω. The lineshape factor $g(\omega)$ simply determines the strength of this interaction relative to that at the resonance frequency ω_{ki}. The steady-state solution of equations (13.10) and (13.11) is obtained by setting $\dot{N}_k = \dot{N}_i = 0$. Then, making use of the relation $g_i B^I_{ik} = g_k B^I_{ki}$ we have, after some straightforward manipulation,

$$\frac{N_k}{g_k} - \frac{N_i}{g_i} = \frac{\left\{ \dfrac{S_k \tau_k}{g_k} - (S_i + \alpha S_k) \dfrac{\tau_i}{g_i} \right\}}{\left\{ 1 + \beta \dfrac{g_i B^I_{ik}}{g_k A_{ki}} \dfrac{2I(\omega)g(\omega)}{4\pi} \right\}} \qquad (13.12)$$

where α and β are dimensionless constants given by

$$\alpha = \tau_k A_{ki} \tag{13.13}$$

and

$$\beta = \alpha \left\{ 1 + (1-\alpha) \frac{\tau_i g_k}{\tau_k g_i} \right\}. \tag{13.14}$$

The unsaturated inversion density, $(N_k^0/g_k - N_i^0/g_i)$, which exists under small signal conditions is obtained from equation (13.12) by allowing $I(\omega)\to 0$, giving

$$\frac{N_k^0}{g_k} - \frac{N_i^0}{g_i} = \frac{S_k \tau_k}{g_k} - (S_i + \alpha S_k)\frac{\tau_i}{g_i} . \tag{13.15}$$

Thus the connection between the saturated and unsaturated inversion densities is given by the simple expression

$$\frac{N_k}{g_k} - \frac{N_i}{g_i} = \frac{N_k^0/g_k - N_i^0/g_i}{\{1 + 2I(\omega)/I_S(\omega)\}} \tag{13.16}$$

where the relation $(g_i B_{ik}^I)/(g_k A_{ki}) = (4\pi^3 c^2)/\hbar\omega_{ki}^3$ derived from equation (9.20) has been used to introduce the *saturation parameter* defined by

$$I_S(\omega) = \frac{\hbar\omega_{ki}^3}{\pi^2 c^2} \frac{1}{\beta g(\omega)} = I_S(\omega_{ki}) \left\{ \frac{4(\omega-\omega_{ki})^2 + \Gamma_{ki}^2}{\Gamma_{ki}^2} \right\} \tag{13.17}$$

The saturation parameter has a minimum value on resonance given by

$$I_S(\omega_{ki}) = \frac{\hbar\omega_{ki}^3 \Gamma_{ki}}{2\pi c^2 \beta} \tag{13.18}$$

From equation (13.16) we see that the inversion density which exists in equilibrium with an intracavity radiation intensity of $I(\omega)$ is necessarily less than the unsaturated inversion density, as we deduced earlier by qualitative arguments. We may now combine equations (11.10) and (13.16) to obtain a general expression for the saturated gain coefficient of a homogeneously-broadened transition, giving

$$\alpha_L^S(\omega) = \frac{\alpha_L^0(\omega)}{\{1 + 2I(\omega)/I_S(\omega)\}} \qquad (13.19)$$

Using the explicit expression for $\alpha_L^0(\omega)$ given by equation (11.10), it is a straightforward exercise (Problem 13.3) to show that $\alpha_L^S(\omega)$ is also described by a Lorentzian distribution with a maximum amplitude given by

$$\alpha_L^0(\omega_{ki})/\{1 + 2I(\omega)/I_S(\omega_{ki})\}$$

and a full width at half maximum given by

$$\Delta\omega_{\frac{1}{2}}^S = \Gamma_{ki}\,\{1 + 2I(\omega)/I_S(\omega_{ki})\}^{1/2} \qquad (13.20)$$

This increase in linewidth may be interpreted as due to the effect of the reduction of the effective lifetimes of the upper and lower laser levels caused by stimulated transitions from these levels.

The effect of gain saturation on the behaviour of a homogeneously-broadened gas laser is best explained by assuming that an additional optical component with a variable loss is introduced into the optical cavity of a conventional laser. When the inserted loss is high, as indicated by level (i) in Fig.13.2, laser oscillation is impossible. The intensity in all cavity modes remains negligibly small and the gain has its maximum unsaturated value shown by curve (a), Fig.13.2. As the additional loss is slowly reduced, the point is reached at which the unsaturated gain, $\alpha_L^0(\omega)$, for the longitudinal mode closest to linecentre is sufficient to overcome the distributed cavity losses, $\delta_c/2L_m$. This is indicated by level (ii) in Fig.13.2. Laser oscillation is then established on this mode with an intensity which is obtained from equation (13.19) by equating $\alpha_L^S(\omega)$ to the distributed cavity loss. Further reduction of the inserted loss to level (iii) allows the intensity $I(\omega)$ in this mode to increase. However, at the same time the gain saturation produced by the oscillating mode is sufficient to maintain the gain at all other mode frequencies below the threshold value, as shown by curve (b) in Fig.13.2. Thus we expect lasers with homogeneously-broadened transitions to oscillate preferentially on a single cavity mode.

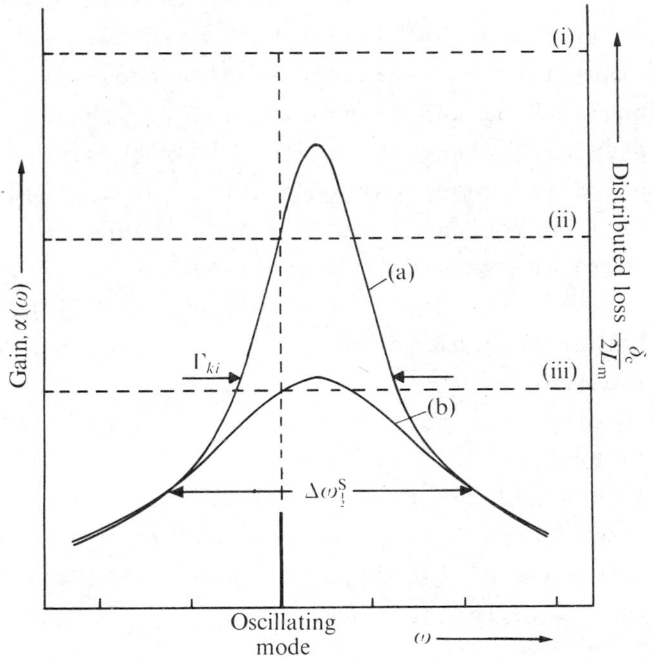

Fig.13.2. Gain curves for homogeneously-broadened laser transition. (a) Cavity loss is so great that oscillation is prevented; (b) cavity loss reduced well below threshold.

This single-frequency operation has been observed by Smith (1971) in a He-Ne waveguide laser oscillating on the 6328 Å line. The narrow bore of the waveguide, 0·43 mm, enabled the laser to be operated at the relatively high pressure of 7 Torr. The pressure-broadened homogeneous linewidth $\Gamma_{ki}/2\pi = 1000$ MHz under these conditions is comparable with the Doppler width of 1700 MHz, and leads to strong mode competition. The spectrum of the laser was analysed using a high finesse, piezoelectrically scanned interferometer and is shown in Fig.13.3(a). In this laser it was possible to tune the frequency continuously up to c/4L from line centre by altering the length of the cavity before the laser jumped to oscillate in the next mode closer to line centre.

However, when the excitation level is well above threshold, it is found that lasers with homogeneously-broadened transitions do oscillate simultaneously in several different longitudinal and transverse modes. This is attributed to a phenomenon known as spatial hole burning. Since the field inside the laser cavity consists of a standing wave there will be regions round the nodal planes of a given mode where the gain is not saturated by that particular mode. This gives other modes with suitable standing field spatial distributions sufficient gain to enable them to oscillate simultaneously.

We have discussed gain saturation in detail for a homogeneously-broadened line since simple explicit expressions can be obtained for the gain coefficient and the intensity of laser radiation. Unfortunately this is not possible in the case of inhomogeneous broadening which we examine in the following section.

13.4. Gain saturation - inhomogeneously-broadened transitions

13.4.1. Theoretical analysis and hole burning. The spectral profile of the spontaneous emission from an excited gas at low pressure is usually a Voigt profile determined by a folding integral of Gaussian and Lorentzian functions, as discussed in section 8.5. An analysis of gain saturation in this situation is considerably more complex than in the case of pure homogeneous broadening and for the sake of simplicity we consider first a sample of excited atoms through which an electromagnetic wave of intensity $I_+(\omega)$ is propagating towards $+\infty$. The electromagnetic wave can now interact only with a restricted class of atoms whose axial velocities, $v \approx (\omega-\omega_{ki})/k$, are such that they are Doppler shifted into resonance with the angular frequency of the wave. Thus when rate equations analogous to equations (13.10) and

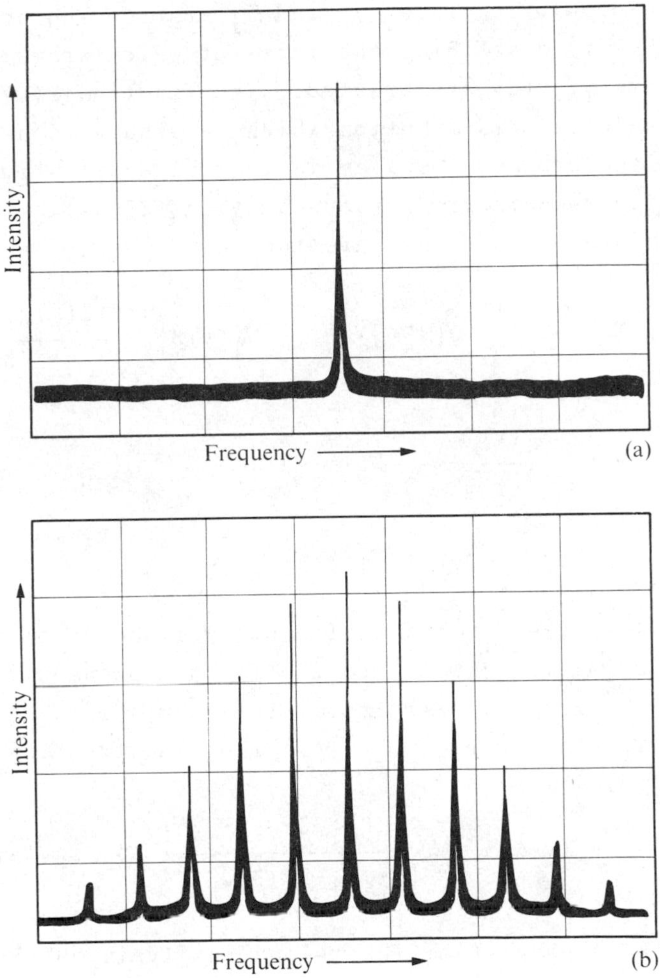

Fig.13.3. Output power spectra of He-Ne lasers obtained using high resolution scanning Fabry-Perot étalons. (a) Single-mode oscillation on the 6328 Å line; high pressure waveguide laser. (b) Multimode oscillation on the 6328 Å line; low pressure gain tube in conventional cavity.

(13.11) are written down they must refer to the population densities, $N_k(v)$ and $N_i(v)$, of atoms moving with a particular velocity v parallel to the direction of propagation. Similarly the re-population rates S_k and S_i must be replaced by $S_kP(v)$ and $S_iP(v)$ describing the rates at which atoms in the velocity class v, v+dv are created. The function P(v) is a normalized velocity distribution which we assume to have the usual Gaussian form for a system in thermal equilibrium at some effective temperature T (see equation (8.40)). In this situation the rate equations have the form

$$\dot{N}_k(v) = S_kP(v) - \frac{N_k(v)}{\tau_k} - \{N_k(v)B^I_{ki} - N_i(v)B^I_{ik}\}\frac{I_+(\omega)g_+(v,\omega)}{4\pi} \tag{13.21}$$

$$\dot{N}_i(v) = S_iP(v) - \frac{N_i(v)}{\tau_i} + A_{ki}N_k(v) + $$
$$+ \{N_k(v)B^I_{ki} - N_i(v)B^I_{ik}\}\frac{I_+(\omega)g_+(v,\omega)}{4\pi} \tag{13.22}$$

where $g_+(v,\omega)$ is the Lorentzian frequency response of atoms, moving with axial velocity v, to the forward propagating beam $I_+(\omega)$. In its own rest frame, the atom sees a wave of angular frequency $\omega(1 - \underline{v}.\underline{k}/c)$. Thus substituting into equation (11.9) we have

$$g_+(v,\omega) = \frac{\Gamma_{ki}/2\pi}{(\omega-\omega_{ki}-kv)^2+\Gamma^2_{ki}/4} . \tag{13.23}$$

As before, the steady-state inversion density is obtained by requiring that $\dot{N}_k(v)=\dot{N}_i(v)=0$, giving

$$\frac{N_k(v)}{g_k} - \frac{N_i(v)}{g_i} = \frac{\left\{\frac{S_k\tau_k}{g_k} - (S_i+\alpha S_k)\frac{\tau_i}{g_i}\right\}P(v)}{\left\{1 + \beta\frac{g_iB^I_{ik}}{g_kA_{ki}}\frac{I_+(\omega)g_+(v,\omega)}{4\pi}\right\}} \tag{13.24}$$

where α and β are defined by equations (13.13) and (13.14) respectively. The total inversion density existing under small signal conditions may be obtained by integrating over the total range of axial velocities, giving

$$\frac{N_k^0}{g_k} - \frac{N_i^0}{g_i} = \int_{-\infty}^{+\infty} \left\{ \frac{N_k(v)}{g_k} - \frac{N_i(v)}{g_i} \right\} dv$$

$$= \frac{S_k \tau_k}{g_k} - (S_i + \alpha S_k) \frac{\tau_i}{g_i} \qquad (13.25)$$

since the velocity distribution function, $P(v)$, is normalized. This enables the relation between the saturated and unsaturated inversion densities to be given in a simple form:

$$\frac{N_k(v)}{g_k} - \frac{N_i(v)}{g_i} = \frac{\left(\frac{N_k^0}{g_k} - \frac{N_i^0}{g_i} \right) P(v)}{\left\{ 1 + \frac{I_+(\omega)}{I_S^+(v,\omega)} \right\}} . \qquad (13.26)$$

The saturation parameter, $I_S^+(v,\omega)$, in this case depends on both v and ω and is given by

$$I_S^+(v,\omega) = \frac{\hbar \omega_{ki}^3}{\pi^2 c^2} \frac{1}{B g_+(v,\omega)}$$

$$= I_S(\omega_{ki}) \left\{ \frac{4(\omega - \omega_{ki} - kv)^2 + \Gamma_{ki}^2}{\Gamma_{ki}^2} \right\} . \qquad (13.27)$$

In the small-signal limit, $I_+(\omega) \to 0$, and equation (13.26) predicts that the inversion density has a Gaussian distribution as a function of the velocity v, as shown by curve (a) in Fig.13.4. However, at finite values of $I_+(\omega)$, a localized reduction of the inversion density occurs around the value $v = (\omega - \omega_{ki})/k$ since at this point $I_S^+(v,\omega)$ has its minimum value. This situation is illustrated by curve (b) in Fig.13.4. The beam of radiation of angular frequency ω propagating through the medium is said to *burn a hole* in the velocity distribution. The width of the hole burned in the inversion density velocity distribution may be determined by assuming that $P(v)$ is constant over the range of velocities considered. Taking the width at half the maximum depth of the hole gives (Problem 13.4)

$$\Delta\omega_{1/2}(\text{hole}) = \Gamma_{ki} \left\{1 + \frac{I_+(\omega)}{I_S(\omega_{ki})}\right\}^{1/2} . \qquad (13.28)$$

As explained in section 13.3, the increase in the hole width at high intensities is the result of an effective reduction in the lifetimes of the upper and lower laser levels caused by stimulated emission and absorption.

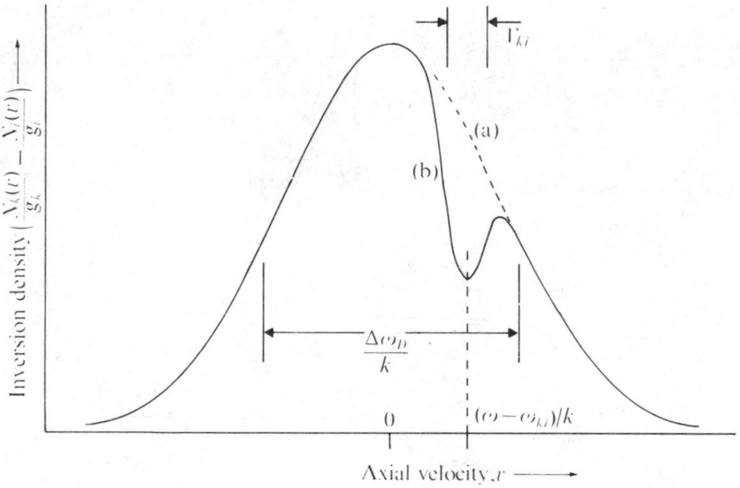

Fig.13.4. Inversion density as a function of atomic axial velocity for a gas discharge interacting with an electromagnetic wave propagating to +∞.
(a) Gaussian distribution of thermal velocities.
(b) 'Hole' burned in normal velocity distribution.

13.4.2. The number of simultaneously oscillating modes. The hole burning phenomenon provides a simple explanation of the operation of lasers with transitions dominated by inhomogeneous broadening. In such a laser, as the inversion density is increased by raising the current through the discharge, oscillation occurs first on the longitudinal mode whose frequency is closest to line centre. The standing wave of the laser field can be separated into two travelling waves. Each of these two counter-propagating waves interacts with atoms in a well-defined velocity class and so burns a separate hole in the velocity distribution. The inversion density as a function of velocity is shown in Fig.13.5 for a laser oscillating on a single mode whose frequency is well separated from the line centre, $\omega-\omega_{ki} > \Gamma_{ki}$. The two holes are symmetrically located about the line centre.

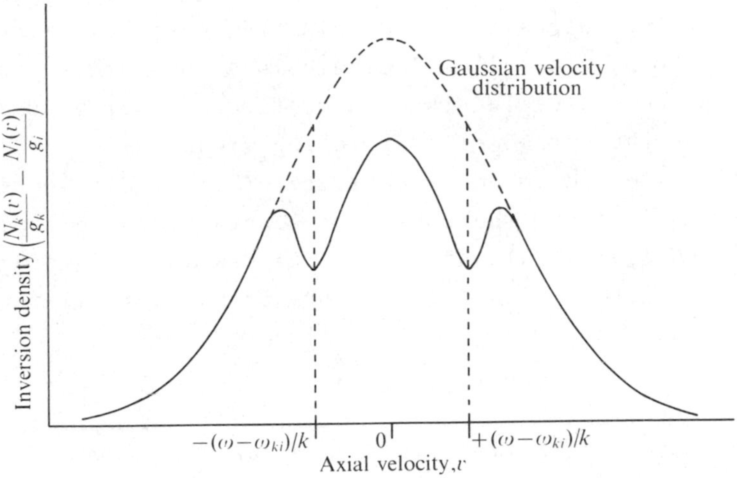

Fig.13.5. Inversion density as a function of atomic axial velocity in a gas laser oscillating on a single axial mode.

In general the frequencies of the longitudinal cavity modes are separated by intervals which are large compared with the linewidth Γ_{ki} and each axial mode is able to draw energy from a different part of the atomic velocity distribution. Thus, as the inversion density is increased still further, the gain at other mode frequencies becomes sufficiently large to overcome the cavity losses and the laser is able to oscillate simultaneously on several longitudinal modes. In this situation a large number of holes are burned in the atomic velocity distribution and therefore no attempt has been made to show this in Fig.13.5. The multimode output spectrum of a typical inhomogeneously-broadened laser transition is shown in Fig.13.3(b). This was obtained using a high finesse piezoelectrically scanned interferometer.

The actual number of longitudinal modes which oscillate simultaneously in an inhomogeneously-broadened gas laser is determined mainly by the Dopper width and the intermode frequency difference, equation (13.5), although the maximum inversion density and the distributed cavity loss $\delta_c/2L_m$ will also have to be taken into account in specific cases (Problem 13.5). Since the intensities and relative phases of these oscillating modes generally fluctuate in a completely random fashion, the coherence length of a multimode laser is usually determined by the spread of oscillating frequencies in spite of the almost perfect monochromaticity of the radiation in each individual mode. The coherence lengths of some typical multimode gas lasers calculated on this basis are given in Table 13.2, together with the Doppler widths appropriate to the various discharge temperatures.

TABLE 13.2.

Doppler widths and effective coherence lengths of typical gas lasers.

Laser	Wavelength (Å)	Effective temperature (K)	Doppler width (MHz)	Coherence length (cm)
He-Ne	6328	400	1700	30
	11 523	400	920	60
	33 912	400	310	150
He-Cd	4416	600	1130	20
Ar^+	4880	3000	3500	10

13.4.3. The saturated gain coefficient. Before we can discuss in detail the propagation of a beam of laser radiation through an inhomogeneously-broadened medium, it is necessary to complete the analysis of saturation which was started in section 13.4.1. The saturated gain coefficient, $\alpha_+^S(v,\omega)$, for atoms in the specified velocity class v,v+dv may be obtained by replacing the unsaturated inversion density in equation (11.10) by the saturated value given in equation (13.26). In addition the lineshape factor, $g(\omega)$, must be replaced by the frequency response, $g_+(v,\omega)$, appropriate for atoms with velocity v interacting with an electromagnetic wave propagating to $+\infty$. The total gain coefficient at the angular frequency ω may then be obtained by summing the contributions from atoms in all possible velocity classes, giving

$$\begin{aligned}
\alpha_+^S(\omega) &= \int_{-\infty}^{+\infty} \alpha_+^S(v,\omega)\, dv \\
&= \frac{\pi^2 c^2}{\omega_{ki}^2} g_k A_{ki} \left(\frac{N_k^0}{g_k} - \frac{N_i^0}{g_i}\right) \times \\
&\quad \times \int_{-\infty}^{+\infty} \frac{\Gamma_{ki}/2\pi}{\{(\omega-\omega_{ki}-kv)^2+\Gamma_{ki}^2/4\}} \frac{P(v)\, dv}{\left\{1 + \dfrac{I_+(\omega)}{I_S^+(v,\omega)}\right\}} . \quad (13.29)
\end{aligned}$$

By some further manipulation this expression may be recast in the form of a folding integral of Lorentzian and Gaussian distributions whose widths are specified by the linewidth parameters $\Gamma_{ki}\{1 + I_+(\omega)/I_S(\omega_{ki})\}^{1/2}$ and Δ respectively. In the general case this means that it is impossible to derive a simple expression for the intensity of the laser beam, unlike the case of homogeneous broadening discussed in section 13.3. However, in the limit that $\Gamma_{ki} \ll \Delta$, the velocity distribution function, $P(v)$, is essentially constant over the region about $v=(\omega-\omega_{ki})/k$ whence the main contribution to the integral in equation (13.29) originates. Taking $P((\omega-\omega_{ki})/k)$ outside the integral sign, the remaining expression can be manipulated to the form

$$\int_{-\infty}^{+\infty} \frac{dx}{x^2+a^2} = \frac{\pi}{a} .$$

Then, using equation (8.40) for $P((\omega-\omega_{ki})/k)$, the saturated gain coefficient in the limit of dominant inhomogeneous broadening becomes

$$\alpha_G^S(\omega) = \frac{\alpha_G^0(\omega)}{\left\{1 + \frac{I_+(\omega)}{I_S(\omega_{ki})}\right\}^{1/2}} \tag{13.30}$$

where

$$\alpha_G^0(\omega) = \frac{\pi^2 c^3}{\omega_{ki}^3}\left(\frac{M}{2\pi kT}\right)^{1/2} g_k A_{ki}\left(\frac{N_k^0}{g_k} - \frac{N_i^0}{g_i}\right) \times$$
$$\times \exp\left\{-\frac{(\omega-\omega_{ki})^2 Mc^2}{2kT\omega_{ki}^2}\right\}. \tag{13.31}$$

In the small-signal limit we see that equation (13.30) reduces to equation (11.7) as expected. Comparing equations (13.19) and (13.30) we find that the gain of a Doppler-broadened transition saturates more slowly than that of a homogeneously-broadened line. This is a consequence of the fact that in the inhomogeneous case only a small fraction of the total number of atoms are on 'speaking terms' with radiation of angular frequency ω.

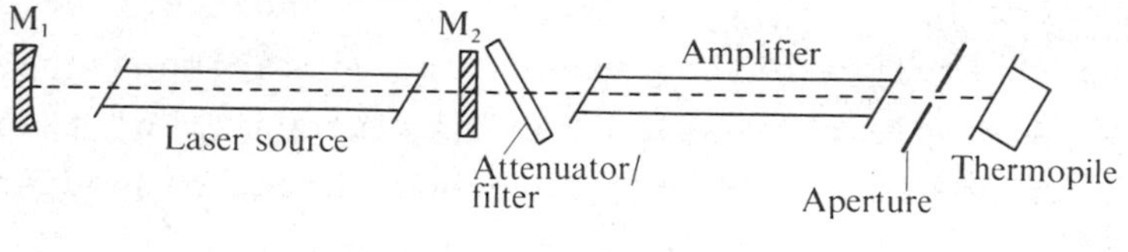

Fig.13.6. Schematic diagram of apparatus used for measurement of gain coefficients in laser amplifiers. (After Bridges (1963).)

13.5. Measurement of gain coefficients

Although positive gain is essential for successful laser oscillation it is, in many cases, quite difficult to perform accurate measurements of the gain coefficient. This is especially true for low-gain transitions or for transitions on which successful laser oscillation has not yet been attained. However, when a laser operating on the transition of interest is available the experimental arrangement may be made quite simple. That used by Bridges (1963) is shown in Fig.13.6. In this experiment the collimated beam from a xenon laser oscillating at 3·508 μm is used as a source for gain measurements in a separate amplifier. The 30 cm long amplifier tube has a bore of 7 mm and is filled with pure Xe or xenon-helium mixtures at various pressures. The discharge in the amplifier is excited by connecting the internal electrodes to the tank circuit of an r.f. oscillator, while that in the laser is a hot cathode d.c. discharge. Before entering the amplifier the laser output power of 0·5 mW is attenuated by Corning glass filters. These filters and the detector are tilted with respect to the optical axis to ensure that the levels of laser oscillation and of gain in the amplifier are not altered by light at 3·51 μm reflected back into their respective discharge tubes. An iris in front of the thermopile is used to eliminate stray radiation from the hot filament of the laser discharge.

If I_1 and I_2 are the intensities of radiation at 3·51 μm entering and leaving the amplifier tube, the gain in decibels in the amplifier is defined as

$$G = 10 \log_{10}(I_2/I_1). \tag{13.32}$$

The ratio (I_2/I_1) is equal to the ratio of the thermopile readings after these have been corrected for the spontaneous emission from the amplifier tube. The experimental results for two different gas fills are shown in Fig.13.7 as a function of the input intensity I_1. In the absence of saturation the dB gain would have a constant value given by

$$G = (10 \log_{10} e)\alpha^0(\omega)L = 4\cdot343\ \alpha^0(\omega)L. \tag{13.33}$$

From Fig.13.7 we see that even at the lowest input signal of 7·8 μW there is clear evidence of gain saturation. For a homogeneously-broadened line we have from equation (13.19)

$$\frac{1}{I(\omega)}\frac{dI(\omega)}{dz} = \alpha_L^S(\omega) = \frac{\alpha_L^0(\omega)}{\left\{1 + \frac{I(\omega)}{I_S(\omega)}\right\}}. \qquad (13.34)$$

Integrating over the amplifier pathlength L we have

$$\alpha_L^0(\omega)L = \ln\left(\frac{I_2(\omega)}{I_1(\omega)}\right) + \left\{\frac{I_2(\omega)}{I_S(\omega)} - \frac{I_1(\omega)}{I_S(\omega)}\right\}. \qquad (13.35)$$

This curve is also plotted in Fig.13.7 and the agreement between the experimental results and the theoretical prediction is seen to be good, considering that the approximation of pure homogeneous line broadening is unlikely to be completely correct in this case. The results shown in Fig.13.7 clearly illustrate the importance of working at the lowest input levels I_1 if measurements of the unsaturated gain $\alpha^0(\omega)$ are required.

When an unambiguous interpretation of the measurements is required it is also necessary to have an accurate knowledge of the spectral distribution of the probe laser for exactly the same reasons as in the absorption measurements described in section 10.6.3.

In the experimental arrangement shown in Fig.13.6 the laser oscillator may be replaced by a conventional discharge tube emitting lines free from self-reversal. However this will necessitate the use of a spectrometer or narrow-band interference filters to isolate the line of interest and the signal levels will be much lower, perhaps requiring phase-sensitive detection. In addition, considerably more care is required to obtain correct collimation of the light from a conventional source.

Gain measurements on pulsed systems introduce further difficulties in signal detection. Moreover the light levels from the amplifier may now be sufficient to drive the detector into a non-linear response regime. This difficulty occurs even with photomultipliers when the anode current exceeds 1/1000 of the current flowing in the dynode voltage resistor chain. Care should be taken to correct for this non-linearity or to reduce the light levels to more suitable values by inserting additional attenuators before the detector.

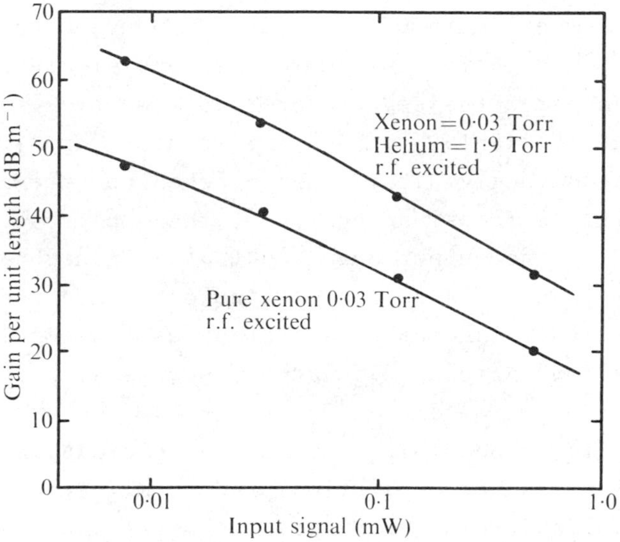

Fig.13.7. Laser amplifier gain as a function of input signal intensity. •, experimental results obtained on the 3·51 μm line of Xe. Solid plot is of the theoretical prediction, equation (13.35), for a homogeneously-broadened line. (After Bridges (1963).)

13.6. Mode-locking of gas lasers

As we explained in section 13.4.2 laser transitions which are dominated by inhomogeneous broadening usually oscillate simultaneously on a number of longitudinal modes. The time-dependent electric field of the optical output of the laser may therefore be written in the form

$$E(t) = \sum_q E_q \exp\{i(k_q z - \omega_q t + \phi_q)\}. \tag{13.36}$$

Thus the total field depends on the amplitudes, frequencies, and relative phases of all the oscillating modes. Generally the mode frequencies are not equally spaced owing to the modification of the passive cavity resonances by anomalous dispersion in the gain medium, as detailed studies by Bennett (1962) and McFarlane (1964) have shown. In addition, random fluctuations in the length of the cavity produce changes in the positions of the holes burned in the Doppler-broadened gain curve. Whenever these holes overlap, two or more modes compete for the available gain and the relative amplitudes of these modes may suddenly alter. Hence, when no attempt is made to stabilize a free-running laser, the amplitude of the electric field in the output beam fluctuates in a random and uncontrollable fashion. As a result the coherence length of a multimode laser is of the same order as that of light emitted from a conventional discharge lamp, as we see from Table 13.2.

However, it is possible to force the oscillating modes to adopt equal frequency separations and to maintain constant amplitudes and relative phases. This is known as *mode-locking* or *phase-locking*. Many different techniques for achieving mode-locking have been developed and these are clearly described in the review articles by Harris (1966) and Smith (1970). In one of the simplest of these methods, described by Hargrove *et al.* (1964), a polished quartz block is placed inside the laser cavity and is driven into acoustic resonance by a piezoelectric crystal. In the quartz block the standing acoustic wave induces periodic refractive index variations which act as a diffraction grating and lead to a cavity loss which is a periodic function of time. This internal modulation produces side bands on each longitudinal mode which have well-defined phase relations with one another. If the frequency of modulation equals the resonator mode spacing of c/2L, and if the depth of loss modulation is sufficiently great, these sidebands act as the main driving terms for the longitudinal modes. Consequently the modes are locked with a constant phase difference in this case of $\phi_{q+1}-\phi_q=\pi$. Using $\omega_q=\omega_0+q\,\Delta\omega$ and $\phi_q=\phi_0+q\pi$, we find that equation (13.36) becomes

$$E(t) = \exp\{i(k_0z-\omega_0t+\phi_0)\}\sum_q(-)^qE_q\exp\{i(\Delta kz-q\Delta\omega t)\}. \tag{13.37}$$

This represents a carrier wave at the centre frequency ω_0 whose envelope is determined by the number and relative amplitudes of the oscillating modes. In the approximation that all the modes have equal amplitudes, it is easily shown (Problem 13.6) that the envelope has the form of a pulse whose width is given approximately by the inverse of the total frequency spread of the oscillating modes. The output of a multimode laser in which many modes are locked together by internal loss modulation therefore consists of a train of very narrow pulses with a pulse repetition frequency of c/2L. Other mode-locking schemes, such as that using an internal phase modulation, produce light which is frequency modulated, but unfortunately space does not permit a description of these techniques.

An alternative picture of how this pulse train is generated can be obtained by considering the radiation propagating back and forth inside the laser. Since the modulator period of 2L/c is equal to the time required for light to make one round trip in the cavity, radiation which sees a loss at the modulator in one cycle will suffer attenuation in repeated passes through the modulator. Only light which arrives when the modulator loss is close to zero will be sufficiently amplified in the gain tube to offset these losses, and thus the radiation inside the cavity builds up into a narrow pulse. In the limit of very narrow pulses the average output power of the mode-locked laser will equal that of the same device when free-running, since the average modulator loss is then effectively zero. Consequently the intensity in the pulses from mode-locked lasers is extremely high and they may be used in experiments on the non-linear optical properties of solids, liquids, and gases.

Unfortunately a lower limit on the width of the pulses from mode-locked gas lasers is imposed by the relatively narrow bandwidth of the Doppler-broadened gain curve. Thus using the He-Ne laser oscillating at 6328 Å, pulse widths of 330 ps are the narrowest obtainable so far. In the cadmium ion laser, however, it is possible to create an artificially broad oscillation bandwidth of approximately 8000 MHz by using the overlapping lines of a carefully prepared mixture of different even isotopes. In this way a continuous train of pulses with pulse widths as narrow as 100 ps was obtained on the 4416 Å and 3250 Å laser lines. Very much shorter pulse lengths may be obtained by mode-locking a C.W. dye laser; the present limit of 1·5 ps is imposed by dispersion in the optical components over the very wide oscillation bandwidth. These pulses are being used to investigate very rapid relaxation processes in liquids and solids, the initiation of thermonuclear fusion by laser-induced compression, and as a carrier for an optical pulse-code modulation system.

13.7. Single-frequency operation of gas lasers

<u>13.7.1. Wavelength selection in gas lasers.</u> Many laser gas discharges create population inversions on several different atomic transitions, ω_{ki}. If the gain coefficients are small, oscillation will usually occur on only one of these lines due to the selective reflectivity of the cavity mirrors. However, in high gain systems, like the krypton ion and argon ion lasers, several transitions will generally oscillate simultaneously, especially if the mirrors have been coated to give a high reflectivity over a broad spectral range. When these lasers are used as light sources for fluorescence or scattering experiments it is first necessary to obtain oscillation on a single wavelength. This is ensured by replacing one mirror of the optical cavity by a Brewster angle prism assembly which has a highly-reflective

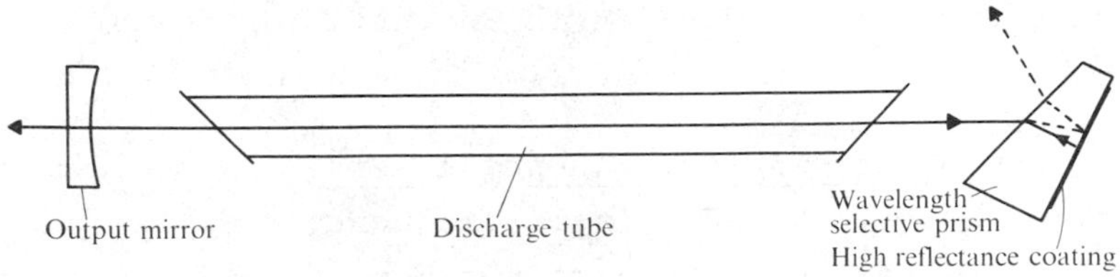

Fig.13.8. Optical cavity with internal Brewster angle prism used to select one of several possible transitions for laser oscillation.

coating on the rear face, as shown in Fig.13.8. Laser oscillation then occurs only on that transition for which refraction and reflection combine to send the light back down the axis of the cavity. By rotating the prism through a small angle, the laser can be tuned in turn to the different possible oscillating wavelengths. This wavelength selection has the added advantage of eliminating the competition for the available gain which occurs when two laser transitions start or terminate on the same energy level. For example, in He-Ne lasers the 6328 Å and 3·39 μm transitions both originate on the $3s_2$ level (Fig.11.3). In medium power lasers the gain saturation produced by the 3·39 μm transition is sufficient to prevent oscillation at 6328 Å unless the infrared line is suppressed in this way.

The resolution obtainable in laser spectroscopy experiments is limited ultimately by the bandwidth of the multimode laser output. This is not usually important in laser-induced fluorescence or Raman scattering experiments. However, for Brillouin scattering, for saturated absorption, and for two-photon absorption experiments, single-frequency lasers are essential. These are also necessary in interferometry and holography whenever the optical path difference exceeds 30 cm. Many different schemes for obtaining single-frequency output from lasers have been reported. We shall consider only a few of the most commonly used techniques. We assume throughout that the laser has been constrained to oscillate on a single wavelength in the TEM_{00} transverse mode.

13.7.2. Low power single-frequency lasers. As the discussion of section 13.4.2 showed, single-frequency oscillation is possible in lasers with long gain tubes, $L_m \geq 50$ cm, if the level of excitation is so low that only the longitudinal mode nearest to the line centre has sufficient gain to overcome the cavity losses. This method is not used very often for the available output power is low and it is difficult to adjust the discharge current to the required level. Moreover drifts in the cavity length lead to large variations in the intensity of the laser output.

An alternative method of securing oscillation on a single longitudinal mode is to choose the length of the cavity so that only one axial mode lies within a Doppler width of the line centre. In the case of the 10·6 μm transition in CO_2, the narrow Doppler-width means that the cavity can be up to 100 cm long and single-frequency output powers of several watts are possible. However, in the visible region the Doppler width is much larger and, referring to Tables 13.1 and 13.2, we see that the cavity length for a single frequency He-Ne laser oscillating on the 6328 Å transition should lie between 10-15 cm (Problem 13.7). Details of the construction of short, single-frequency 6328 Å lasers have been given by Collinson (1965), Baird *et al.* (1965), and Bruce (1971). In these designs the frequency of the oscillating mode is tuned to the line centre by varying the temperature of the laser and relying on the thermal expansion of the resonator. The frequency stability is therefore determined directly by the resolution and stability of the temperature controller.

The disadvantage of these short lasers is that the output power is usually less than 100 μW owing to the short length of the gain tube. Moreover the technique is not applicable to lasers such as the argon ion system, where the high plasma temperature produces a very wide oscillating bandwidth. In this case a cavity length as short as 3 cm would be required for single-frequency oscillation. Consequently interferometric mode selection techniques have been developed for use with high or medium power lasers and we now proceed to consider the two most commonly used devices.

13.7.3. Single-frequency operation using a Fox-Smith interferometer. In most of the interferometric methods of longitudinal mode selection, an arrangement of three or more mirrors is used to form a complex laser resonator which has only one high Q, low-loss, mode lying within a Doppler width of the line centre. For lasers with medium output powers, ≤ 100 mW, one of the most useful of these devices is the Fox-Smith interferometer illustrated in Fig. 13.9(a). It consists of two high-reflectivity mirrors, M_1 and M_2, and a beam splitter, making a two mirror interferometer which replaces the usual output mirror of the cavity. The resonances of this interferometer are spaced by $c/2L_1$ and the reflectivity versus wavelength characteristic is identical to the transmittance versus wavelength characteristic of a Fabry-Perot interferometer, as shown in Fig.13.9(b). If the frequency of a longitudinal mode of the long cavity L_2 coincides with one of these resonances, all the light entering the interferometer is reflected back on itself regardless of the reflectance of the beam splitter, provided only that the mirrors M_1 and M_2 have reflectivities of unity. The reflectance of the beam splitter is important, however, since it determines the Q-value or bandwidth of the interferometer resonance.

The use of this interferometer in He-Ne and argon ion lasers has been described by Smith (1965) and Gorog and Spong (1969) respectively. To ensure that the laser oscillates in a single longitudinal mode, the interferometer length L_1 should be chosen so that $c/2L_1$ is greater than the amplification bandwidth of the laser, and the width of the interferometer resonance should be as narrow as possible. Unfortunately it is not feasible to use a high-reflectivity beam splitter because scattering losses in both the beam splitter and the interferometer mirrors reduce the maximum

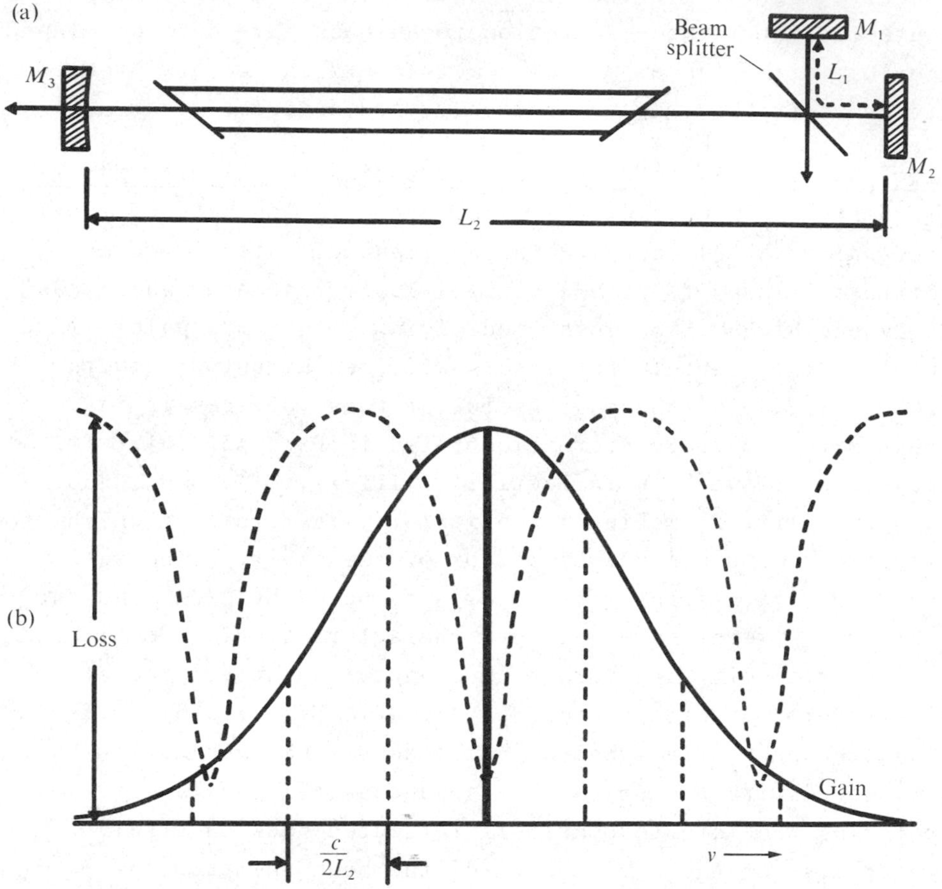

Fig.13.9. Laser cavity using Fox-Smith interferometer for selection of a single longitudinal mode. (a) Mirror geometry; (b) gain/loss curves.

reflectance below the ideal value of unity. In practice beam splitter reflectances of only 30-60 per cent are possible and this results in a rather low finesse for the interferometer. Consequently longitudinal modes of the cavity L_2 lying on either side of the resonant mode also have a moderately high reflectivity at the interferometer and may oscillate in high-gain systems. This technique is therefore not as suitable for use in high power lasers as the internal Fabry-Perot étalon described in the following section.

13.7.4. Single-frequency operation using a Fabry-Perot étalon. In high power systems, such as the argon ion laser, the gain on many of the transitions is so high that additional optical components can be inserted in the optical cavity without laser oscillation being prevented by the unavoidable losses. A single longitudinal mode of the long cavity can then be selected by inserting a Fabry-Perot étalon as shown in Fig.13.10(a). When the étalon is aligned with its axis

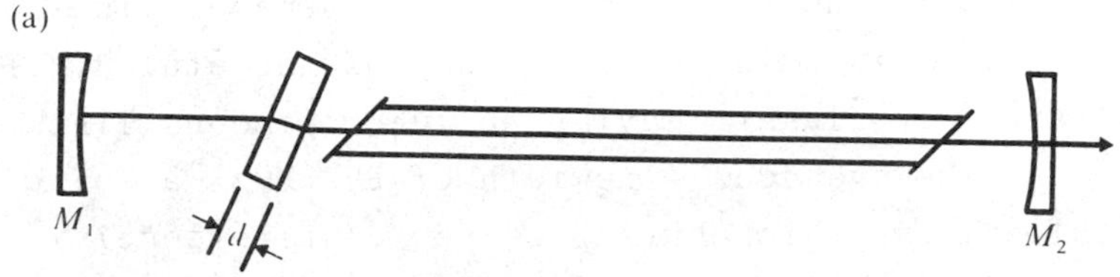

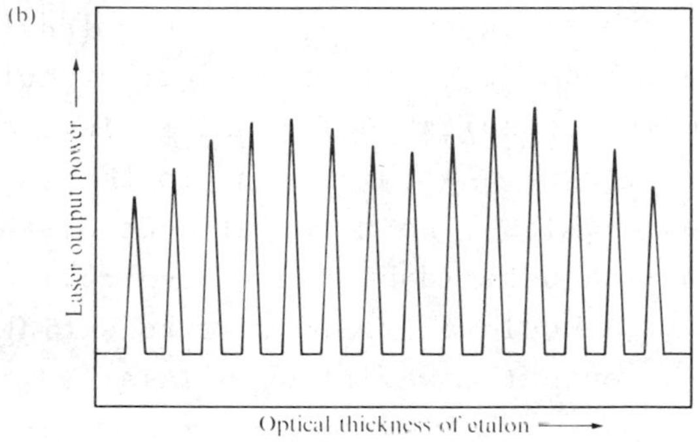

Fig.13.10. Laser cavity using internal Fabry-Perot étalon for mode selection. (a) Mirror geometry; (b) experimental display of laser output as a function of the thickness of the internal Fabry-Perot étalon.

parallel to that of the laser it becomes an integral part of the cavity. The normal cavity resonance frequencies are modified and become very sensitive to the position of the étalon relative to the cavity mirrors. However, if the étalon is tilted so that the angle it makes with the cavity axis is greater than the angular radius of the lowest-order transverse mode, $\theta \geq (\lambda/w)^{1/2}$, it then acts simply as a bandpass transmission filter.

To ensure oscillation on only a single longitudinal mode, the free spectral range of the étalon, c/2nd, should equal or exceed the gain bandwidth of the laser and the reflection finesse should be sufficient to discriminate against frequencies located approximately one homogeneous linewidth from the maximum transmission frequency of the étalon. For a typical argon ion laser, having an inhomogeneous linewidth of 5 GHz and a homogeneous linewidth of 500 MHz, a solid quartz étalon 10 mm thick having 20 per cent reflectivity multi-layer dielectric coatings on each face would be suitable. The operation of such a single-frequency argon ion laser is discussed in detail by Hercher (1969). Approximately 50 per cent of the total multimode power was available in single-frequency operation and the coherence length of the laser was extended from 10 cm to 10 m. Unfortunately this simple technique is not suitable for use in low-gain systems, for the tilt of the étalon with respect to the laser axis eventually causes the laser beam to walk out of the cavity. There is therefore an unavoidable loss, typically of the order of 5 per cent, which would be sufficient to extinguish or greatly reduce the output power of such lasers.

13.7.5. Tuning characteristics of single-frequency lasers with interferometric mode selection. In the single-frequency lasers discussed in the two preceding sections, oscillation will only occur when an interferometer resonance is close to the frequency of one of the longitudinal modes of the main cavity. This adjustment can be made in the Fox-Smith interferometer by mounting the mirror M_1 on a piezoelectrically driven shaft, while in the case of the internal Fabry-Perot it may be accomplished either by varying the tilt of the étalon with respect to the cavity axis or by enclosing the solid étalon in a thermostatically controlled oven and using thermal expansion to vary the thickness d. In both cases, when the length of the main cavity is held fixed and the interferometer resonance frequency is continuously varied, the laser frequency changes discontinuously as the laser hops from one longitudinal mode to the next. The peak amplitudes of the oscillating longitudinal modes in this case trace out the envelope of the single-frequency saturated gain curve, as shown in Fig.13.10(b). The dip which is apparent close to line centre is explained in detail in the following section. To obtain single-frequency output which is continuously tunable over the gain bandwidth of these long lasers, it is necessary to alter the length of the main cavity in synchronism with the frequency scan of the interferometer resonance. Although this is a difficult adjustment, the experiments described in section 13.9.2 below show that it can be performed successfully. However, in our discussion of the output power of single-frequency gas lasers which now follows we assume, for the sake of simplicity, that we are dealing with one of the short single-frequency lasers described in section 13.7.2.

13.8. *Output power versus tuning curves for single-frequency gas lasers*

13.8.1. Experimental observations of the Lamb dip. When the frequency of a single mode laser is tuned over the gain bandwidth of the atomic transition, by altering the length of the cavity for instance, one expects intuitively that the output power should increase to a definite maximum at the line centre. However, the observed output versus frequency curves usually have noticeably flattened peaks and in favourable situations a clearly defined dip occurs in the power output at line centre. This dip was first predicted theoretically by Lamb in summer school lectures he gave during 1962 and 1963 and is one of the main results in an important paper (Lamb 1964), which gives the first reasonably complete account of the theory of gas laser oscillators. The power dip predicted by Lamb was soon confirmed experimentally by the observations of McFarlane *et al.* (1963) and Szöke and Javan (1963).

This Lamb dip, as it is now called, may be explained simply in terms of the hole burning picture introduced by Bennett (1962). When the laser oscillation frequency approaches the line-centre frequency ω_{ki}, the Doppler shift required to bring atoms into resonance with the two travelling waves of the standing wave pattern becomes very small. Consequently these waves are no longer drawing their energy from two completely independent velocity classes and the holes burned in the population distribution of Fig.13.5 must overlap as $\omega \to \omega_{ki}$. Since now both travelling waves are interacting with atoms in the same velocity class, the population inversion density saturates at a value lower than that for the case of a single travelling wave. Hence the laser output power is reduced at the line centre. The width of the dip is determined by the widths of the holes burned in the gain curve and is of the order of the homogeneous linewidth of the laser transition, Γ_{ki}.

Output power versus frequency curves are shown in Fig.13.11 for several different values of the discharge current in a single mode He-Ne laser oscillating on the 1·15 μm transition. The laser discharge tube was filled with isotopically enriched ^{20}Ne and helium since the Lamb dip in the naturally occurring mixture of 91 per cent ^{20}Ne and 9 per cent ^{22}Ne is obscured by the isotope shift of the 1·15 μm line. The laser was tuned by magnetostriction of the invar cavity spacer, and the frequency axis of Fig.13.11 was calibrated by tuning three successive cavity modes, separated by c/2L, across the line centre. At power levels well above threshold the dip in the output power versus tuning curve is clearly visible. The Lamb dip is also apparent in the envelope of the single-frequency power output of the laser shown in Fig.13.10(b), although in this case oscillation occurs only at discrete frequencies across the gain bandwidth for the reasons discussed in section 13.7.5.

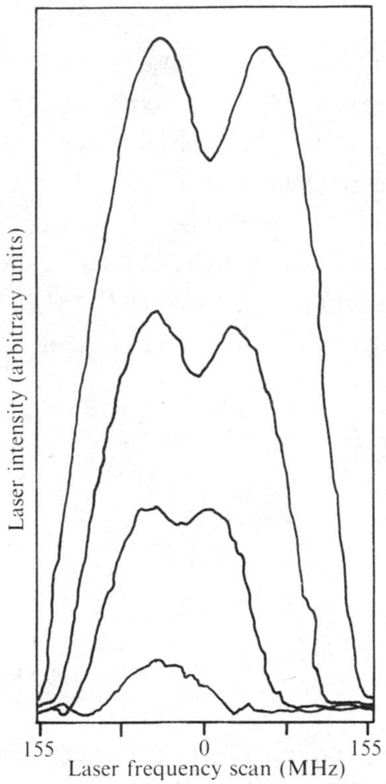

Fig.13.11. Output power versus frequency for single mode helium-neon laser oscillating on the 1·15 μm line. The laser contained helium and isotopically pure ^{20}Ne and results are given for several discharge input power levels. (After Szöke and Javan (1963).)

13.8.2. Simple theory of the Lamb dip. The Lamb dip is another example of the saturation effects which occur in inhomogeneously-broadened laser transitions. The theoretical basis of these effects was discussed in section 13.4 for the case of a travelling wave laser. However, in the case of a single-frequency laser oscillator it is necessary to modify the last term in both equation (13.21) and equation

(13.22) by making the substitution

$$I_+(\omega)g_+(v,\omega) \rightarrow \{I_+(\omega)g_+(v,\omega) + I_-(\omega)g_-(v,\omega)\}.$$

The additional term represents the stimulated transitions induced by the component of the oscillator standing wave which propagates to $-\infty$. For this wave the lineshape factor, $g_-(v,\omega)$, differs from that given in equation (13.23) by the replacement of k by -k. Assuming now that the intensities of the forward and backward travelling waves in the laser oscillator are identical, $I_+(\omega)=I_-(\omega)=I(\omega)$, we find that the saturated inversion density is given by

$$\frac{N_k(v)}{g_k} - \frac{N_i(v)}{g_i} = \left(\frac{N_k^0}{g_k} - \frac{N_i^0}{g_i}\right) P(v) \times$$
$$\times \left[1 + \frac{I(\omega)}{I_S(\omega_{ki})}\left\{\frac{\Gamma_{ki}^2}{4(\omega-\omega_{ki}-kv)^2+\Gamma_{ki}^2} + \frac{\Gamma_{ki}^2}{4(\omega-\omega_{ki}+kv)^2+\Gamma_{ki}^2}\right\}\right]^{-1}. \quad (13.38)$$

The effective saturated gain coefficient in the single-frequency gas laser is then obtained by taking the mean of the saturated gain for the positive- and negative- going waves following the procedure outlined in section 13.4.3 for the case of a single travelling wave. Using equations (11.10) and (13.38) and summing the contributions from atoms in all velocity classes, we have

$$\overline{\alpha^S(\omega)} = \frac{\pi^2c^2}{\omega_{ki}^2} g_k A_{ki} \left(\frac{N_k^0}{g_k} - \frac{N_i^0}{g_i}\right) \times$$
$$\int_{-\infty}^{+\infty} \frac{\frac{1}{2}\{\mathcal{L}(\omega-\omega_{ki}-kv,\Gamma_{ki})+\mathcal{L}(\omega-\omega_{ki}+kv,\Gamma_{ki})\}P(v)\,dv}{\left[1 + \frac{I(\omega)}{I_S(\omega_{ki})}\frac{\pi\Gamma_{ki}}{2}\{\mathcal{L}(\omega-\omega_{ki}-kv,\Gamma_{ki})+\mathcal{L}(\omega-\omega_{ki}+kv,\Gamma_{ki})\}\right]}. \quad (13.39)$$

The steady-state intensity inside the laser cavity, $I(\omega)$, is determined by the condition that the effective saturated gain coefficient just equals the effective loss per unit length in the cavity, thus $\overline{\alpha^S(\omega)} = \delta_c/2L_m$. Unfortunately an analytic expression for the integral appearing in equation (13.39) does not exist and the equation for $I(\omega)$ can only be solved by numerical methods. However, an approximate expression describing the output power as a function of frequency for the single mode laser can be obtained by assuming that the Doppler width is large compared with the homogeneous linewidth. The velocity distribution function, $P(v)$, is then constant over the regions around $v = \pm(\omega-\omega_{ki})/k$ where the main contributions to the integrand of equation (13.39) occur and it may be taken outside the integral sign. Then making a binomial expansion of the denominator to the first order in $I(\omega)/I_S(\omega_{ki})$, we find that equation (13.39) reduces to a sum of Lorentzian functions and products of these functions. It can be shown (Problem 13.8) that the saturated gain coefficient in this approximation reduces to

$$\overline{\alpha^S(\omega)} = \alpha_G^0(\omega)\left[1 + \frac{I(\omega)}{2I_S(\omega_{ki})}\left\{1 + \frac{\Gamma_{ki}^2/4}{(\omega-\omega_{ki})^2+\Gamma_{ki}^2/4}\right\}\right]^{-1}. \qquad (13.40)$$

It follows that the intensity of radiation in the cavity of a single-frequency gas laser is given by

$$\frac{I(\omega)}{2I_S(\omega_{ki})} = \frac{\dfrac{2L_m\alpha_G^0(\omega)}{\delta_c} - 1}{\left\{1 + \dfrac{\Gamma_{ki}^2/4}{(\omega-\omega_{ki})^2+\Gamma_{ki}^2/4}\right\}} \qquad (13.41)$$

where the unsaturated gain coefficient in the limit of dominant inhomogeneous broadening, $\alpha_G^0(\omega)$, is given by equation (13.31). The result given in equation (13.41) accurately describes the output power versus frequency curves for low and medium power lasers when the condition $I(\omega)/I_S(\omega_{ki}) < 0{\cdot}1$ is satisfied. In the case of the He-Ne laser oscillating on the 6328 Å line, this requires $I(\omega) < 2{\cdot}4$ W cm^{-2}. The Gaussian shape of the Doppler-broadened unsaturated gain coefficient, $\alpha_G^0(\omega)$, determines the broad envelope of the tuning curve, while the narrow Lorentzian shape of the Lamb dip comes from the second term in the denominator of equation (13.41). This term is important only when the laser frequency, ω, is within a homogeneous linewidth of the line centre, ω_{ki}.

Detailed studies of the Lamb dip saturation phenomenon at different pressures have been made by Szöke and Javan (1966), Smith (1966), and Cordover and Bonczyk (1969). In addition to determining the homogeneous linewidth in the actual laser discharge, these experiments have shown that the shape of the Lamb dip becomes asymmetric and that there is a pressure shift away from the line centre at high pressures. Since the pressure in a gas laser is fixed within certain limits by the need to have sufficient gain to sustain oscillation, a laser stabilized to the centre of the Lamb dip can never become a primary wavelength standard. Finally we note that our theory of saturation effects was developed only in terms of the rates of contributing processes. It is therefore only an approximate description and the neglect of the phases of atomic wavefunctions and laser radiation fields may be important in certain experiments.

13.9. Saturated absorption spectroscopy using tunable gas lasers

13.9.1. Saturated absorption spectroscopy of atoms.

In the previous section we considered the saturation of the inverted population density in the gain tube produced by interaction with the standing field inside the cavity of a single-frequency laser. When an additional cell is placed in the cavity and filled with atoms or molecules which have an inhomogeneously-broadened absorption line close to that of the laser transition, these atoms and molecules will also be subjected to the laser standing field. The two running-wave components of this field will burn two holes in the velocity distribution of the absorbing atoms which we assume to have a normal rather than an inverted population distribution.

The primary effect of this additional absorption loss in the laser cavity will be to reduce the laser output power. However, as the laser oscillation frequency is tuned towards the centre of the absorption line, the two holes burned in the population of absorbing atoms will start to overlap. At the line centre the population difference, and hence the absorption coefficient of the sample, is reduced by the saturation of atoms with zero axial velocity interacting with both components of the laser standing-wave field. This sharp drop in the absorption coefficient, which occurs about the centre of the absorption line, reduces the internal losses of the cavity and gives rise to a sharp increase in the power output of the laser.

This effect is known as *saturated absorption* and was first observed by Lee and Skolnick (1967) on the power output versus tuning curve for a He-Ne laser oscillating on the 6328 Å transition using the apparatus shown in Fig.13.12(a). This laser had a 42 cm long cavity containing two separate Brewster-angled discharge tubes. The first tube, 7 cm long with 1·5 mm bore, was filled with a mixture of ^{3}He and ^{20}Ne at 3·0 Torr pressure to provide the gain necessary for laser oscillation. The second tube, 15 cm long by 3·0 mm bore, was filled with ^{20}Ne at 0·1 Torr. When an electrical discharge was run in this tube a normal population distribution of excited atoms was obtained which produced significant absorption on the 6328 Å, $2p_4$-$3s_2$ transition of neon. When the laser frequency was tuned across the oscillating bandwidth, the power output showed a narrow inverted Lamb dip on top of the broad Gaussian envelope of the curve (Fig.13.12(b)). The peak of this saturated absorption signal occurred at a frequency 60 MHz below the laser line-centre frequency. Since the pressure in the absorption tube was relatively low, it was assumed that this frequency offset represented the shift of the laser line centre produced by the relatively high pressure existing in the gain tube, as mentioned at the end of section 13.8.2. The width

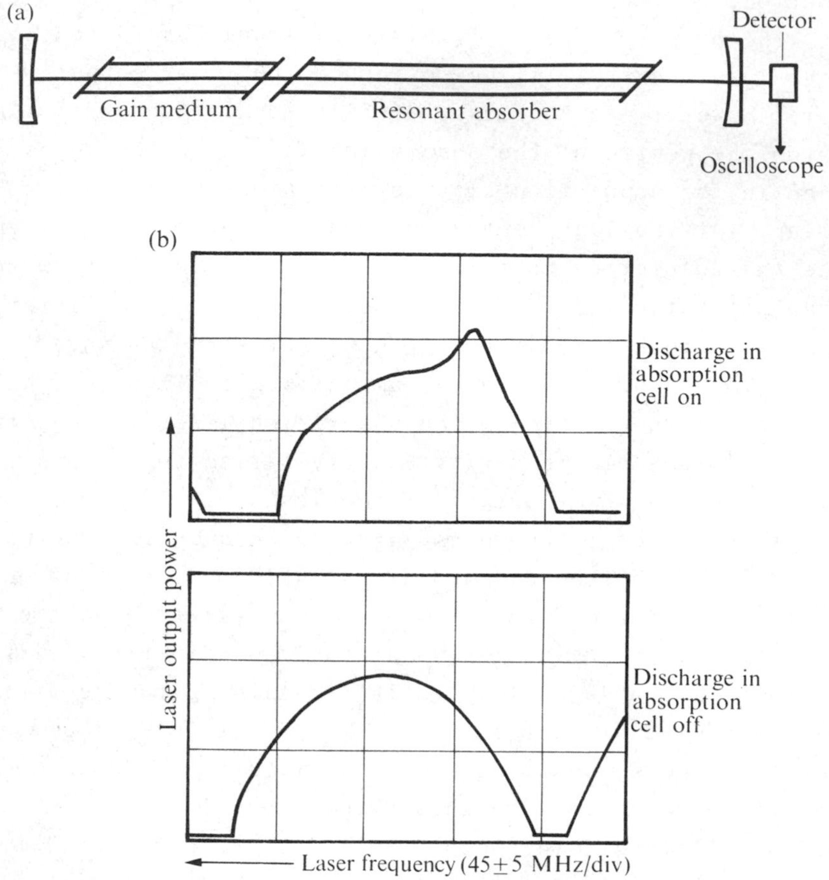

Fig.13.12. Saturated absorption on the 6328 Å transition of a helium-neon laser. (a) Neon absorption tube inside laser resonator. (b) Output power as a function of oscillation frequency showing saturated absorption resonance superimposed on normal Gaussian envelope. (After Lee and Skolnick (1967).)

of the saturated absorption signal was relatively narrow, ≈ 30±5 MHz, and was close to the natural linewidth expected for the 6328 Å transition.

This new technique of saturated absorption spectroscopy

obviously had great possibilities, for here was a method for optical spectroscopy in which the Doppler effect was eliminated and a resolving power of 10^7-10^8 easily attainable. However, saturated spectroscopy is possible only if the absorption feature of interest lies within the tuning range of a single-frequency laser. Unfortunately the tuning range of gas lasers is limited approximately to the Doppler width of the laser line and thus, in the case of atoms, they can usually be applied only to the saturated absorption spectroscopy of the laser transitions themselves.

13.9.2. Saturated absorption spectroscopy of molecules. Fortunately molecular absorption lines are far more numerous than those of atoms and several accidental coincidences exist with known gas laser transitions. The first of these to be studied by saturated absorption spectroscopy was the coincidence between the 6328 Å He-Ne laser and the R(127) rotational absorption line of the 11-5 vibration band in the $X\ {}^1\Sigma^+_{0g} \rightarrow B\ {}^3\Pi^+_{0u}$ electronic transition of molecular iodine, ${}^{127}I_2$. The fluorescence resulting from this absorption process is easily visible in a darkened room if the beam from a 1 mW laser is passed through an evacuated cell containing iodine vapour at room temperature. Hanes and Dahlstrom (1969) were able to resolve 14 hyperfine components in the saturated absorption spectrum of the line using an iodine absorption cell inside the He-Ne laser cavity.

This work was extended by Hänsch *et al.* (1971) using a krypton ion laser, and the apparatus shown in Fig.13.13. In this experiment the iodine absorption cell was placed outside the laser cavity and a standing wave was obtained by arranging two laser beams, each of 10 mW power, to intersect at an angle of less than 2 mrad inside the 20 cm long absorption cell. From equation (13.40) the saturated absorption coefficient when both beams are present in the cell is given approximately by

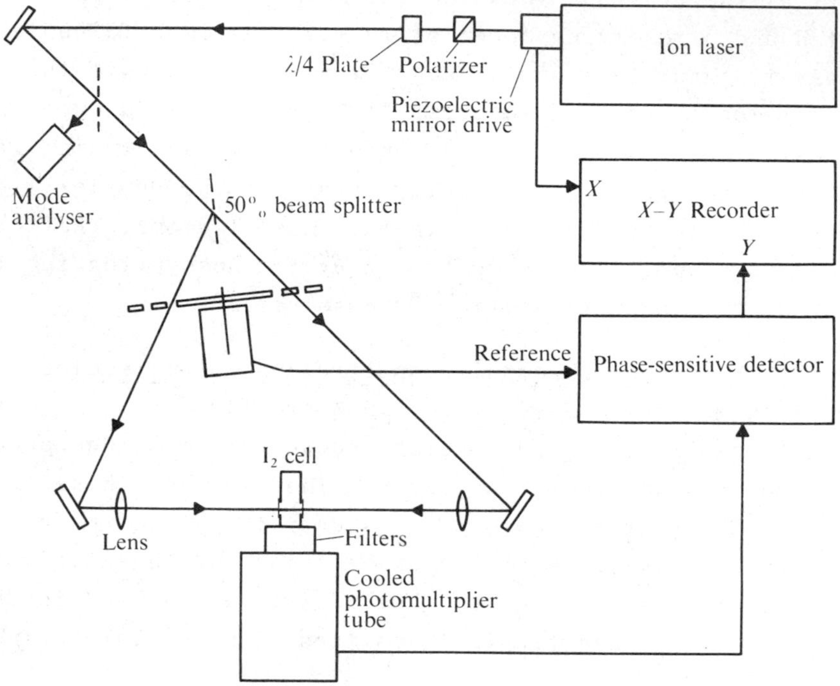

Fig.13.13. Apparatus used by Hänsch *et al.* (1971) for saturated absorption spectroscopy of I_2 vapour.

$$\overline{\kappa^S(\omega)} = \frac{\kappa_G^0(\omega)}{1 + \dfrac{I(\omega)}{2I_S(\omega_{ab})}\left\{1 + \dfrac{\Gamma_{ab}^2/4}{(\omega-\omega_{ab})^2+\Gamma_{ab}^2/4}\right\}} \qquad (13.42)$$

where Γ_{ab} and ω_{ab} are the homogeneous linewidth and angular frequency of the centre of a given molecular absorption line respectively. The direct saturated absorption signal would therefore be observed against a large background signal having the usual Doppler-broadened Gaussian lineshape associated with the unsaturated absorption coefficient, $\kappa_G^0(\omega)$. To remove this background, one of the laser beams is chopped and the modulation produced by saturation of the absorption coefficient is monitored on a probe beam and recovered by phase-sensitive detection. The phase-sensitive detector signal, $S(\omega)$, is thus proportional to the change in the absorption coefficient seen by the positive-going beam when the negative-going beam is removed by the chopper, giving

$$S(\omega) \propto \frac{\overline{\kappa^{S}(\omega)-\kappa_{+}^{S}(\omega)}}{\kappa_{+}^{S}(\omega)} \approx -\frac{I(\omega)}{2I_S(\omega_{ab})}\,\frac{\Gamma_{ab}^{2}/4}{(\omega-\omega_{ab})^2+\Gamma_{ab}^{2}/4}\,. \quad (13.43)$$

In this equation $\kappa_{+}^{S}(\omega)$ has been obtained by modifying equation (13.30).

The saturated absorption signals obtained as the laser frequency is scanned by simultaneously changing the tilt of a mode-selecting internal étalon and adjusting the length of the laser cavity are shown in Fig.13.14. The 5682 Å line

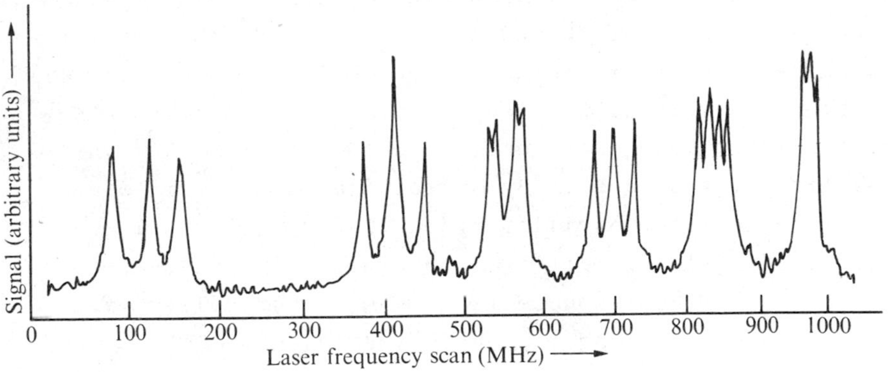

Fig.13.14. Hyperfine structure of the P(117) rotational line of the 21-1 vibrational band of the molecular iodine absorption spectrum at 5682 Å observed by laser saturation spectroscopy. (After Hänsch *et al.* (1971).)

of the krypton ion laser coincides with the P(117) rotational line of the 21-1 vibrational band in the molecular iodine absorption spectrum and the large gain bandwidth of the krypton ion laser allows the complete hyperfine structure of of the P(117) line to be obtained. The saturated absorption peaks have a half width of 10 MHz, corresponding to a resolution of 10^8, and vividly demonstrate the power of this new technique.

13.10. Frequency stabilization of single-frequency gas lasers

13.10.1. Passive frequency stabilization. The intrinsic bandwidth of the output of an idealized single-frequency laser would be of the order of 1 Hz or less, as we showed in section 12.6. In any real laser, however, the output frequency drifts more or less rapidly with time owing to the fact that the instantaneous frequency of oscillation within the bandwidth of the Doppler-gain curve is determined by the optical length of the cavity rather than by the centre frequency of the atomic emission line. Mechanical vibrations, thermal expansion of the cavity spacer, and fluctuations of the refractive index of the air between the cavity mirrors and the Brewster angle windows of the gain tube in an external mirror laser all contribute to the frequency instability of the laser. By surrounding a short internal mirror He-Ne laser oscillating at 6328 Å with an oil bath, Baird *et al.* (1965) initially obtained a frequency stability of 2×10^7 per hour, where the frequency stability of a single-frequency laser may be defined as

$$S(\tau) = \nu_L/\Delta\nu_L(\tau) \qquad (13.44)$$

where ν_L is the average frequency generated by the laser and $\Delta\nu_L(\tau)$ is some measure of the frequency fluctuations during the period of observation τ. Similar passively stabilized lasers have been described by Collinson (1965), Bruce (1971), and others. The laser constructed by Baird *et al.* (1965) was later modified to include temperature control to $\pm 0{\cdot}01\ ^{\circ}C$ and eventually a frequency stability of 2×10^8 per day was attained. This is, however, the upper limit imposed by the oil-bath thermostat and the thermal expansion of the fused quartz spacer (Problem 13.9). To obtain further improvement in the long-term stability, a more sensitive system for controlling the length of the resonator is required.

13.10.2. Frequency stabilization using the Lamb dip. The power output versus frequency characteristic of a single-mode laser provides one way of detecting changes in the oscillation frequency and Fig.13.15 shows how this may be used to

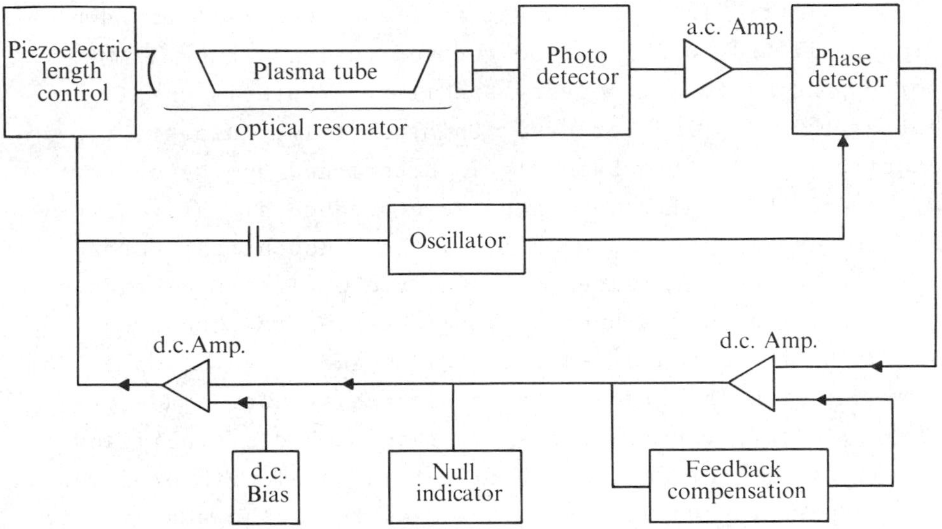

Fig.13.15. Block diagram of laser frequency stabilization scheme using the Lamb dip. (After Bloom (1968).)

stabilize a laser to the centre of the Lamb dip. Such lasers were, until recently, available commercially from Spectra-Physics Ltd. One of the resonator mirrors is mounted on a piezoelectric element which is driven by a.c. voltage, thus sweeping the laser frequency periodically over a small range. The output power from the other end of the laser is detected by a photocell where the a.c. signal is amplified and fed to a phase-sensitive detector. The d.c. output voltage from the phase-sensitive detector is proportional to the first derivative of the power output versus tuning curves shown in Fig.13.11, and thus goes from a negative to a positive value as the laser frequency is tuned across the centre of the Lamb dip. This provides an error signal which is amplified and used to adjust the d.c. bias voltage applied to the piezoelectric element. Thus the laser frequency is servo controlled on to the centre of the Lamb dip.

The width of the Lamb dip is determined by the homogeneous linewidth and is of the order of 110 MHz in a He-Ne laser oscillating at 6328 Å and filled to a total pressure of 1·35 Torr. Thus the long-term frequency drift of Lamb dip stabilized lasers can be reduced to less than ±1 MHz, corresponding to a frequency stability or precision of better than 2×10^{8} per day. Unfortunately, however, the exact profile of the Lamb dip is determined by the effects of interatomic collisions and these produce slightly asymmetric line shapes and a shift in the linecentre frequency. Thus the absolute frequency of individual lasers depends on the filling pressure and has been observed to change by 40 MHz over the range 2-4 Torr (Bloom and Wright 1966). Comparisons with the krypton 86 wavelength standard made by Mielenz *et al.* (1968) have shown that Lamb dip stabilized lasers of similar design may differ in wavelength by more than than 1 part in 10^{7} and this is a measure of the accuracy or reproducibility of the device. Thus these lasers cannot be used as wavelength or frequency standards and it is necessary to consider alternative frequency-stabilization techniques.

13.10.3. Stabilization to molecular resonance frequencies.

The combination of moderately high operating pressures and large pressure shifts effectively rules out any self-stabilized laser oscillator as a primary standard. On the other hand, such oscillators may be used as sources to generate saturated absorption resonances in molecular systems, as discussed in section 13.9.2. Such resonances have several advantages for frequency stabilization schemes:

(a) the absorbing gas can often be at a low enough pressure that collisional broadening and line shifts are negligible;

(b) the width of the saturated absorption line may be extremely narrow if the absorption occurs from the ground state to a long-lived excited level of the molecule;

(c) the design parameters of the laser oscillator and of the absorption cell may be optimized independently of one another.

A large number of these accidental coincidences of molecular absorption lines with gas laser transitions are now known and some of the most important examples are listed by Hall (1973) in a review of saturated absorption spectroscopy.

(a) Stabilization of the 6328 Å helium-neon laser to saturated absorption resonances in molecular iodine. A schematic diagram of a simple molecular iodine stabilized helium-neon laser is shown in Fig.13.16. The operating principle is very similar to that of the Lamb dip stabilization scheme discussed in section 13.10.2; however, the saturated absorption peaks in iodine have a linewidth of only 4 MHz and by locking the laser to one of the hyperfine components of the iodine line, a frequency stability of 5×10^{11} for a 10 s sampling time can be attained.

The spectral distribution and frequency stability of the output of these stabilized lasers cannot be studied by the methods of conventional high-resolution spectroscopy. Instead it is usually necessary to construct two stabilized lasers, one of which may be regarded as a local oscillator, and to investigate the frequency stability using optical heterodyne techniques. As shown in Fig.13.17 the outputs of both lasers are superimposed coherently on the surface of a square-law detector, such as a photodiode or photomultiplier. The detector output signal then contains a component corresponding to the beat note or frequency difference between the two lasers. The power spectrum of this optical heterodyne signal may be displayed directly on an r.f. spectrum analyser and its mean frequency determined pre-

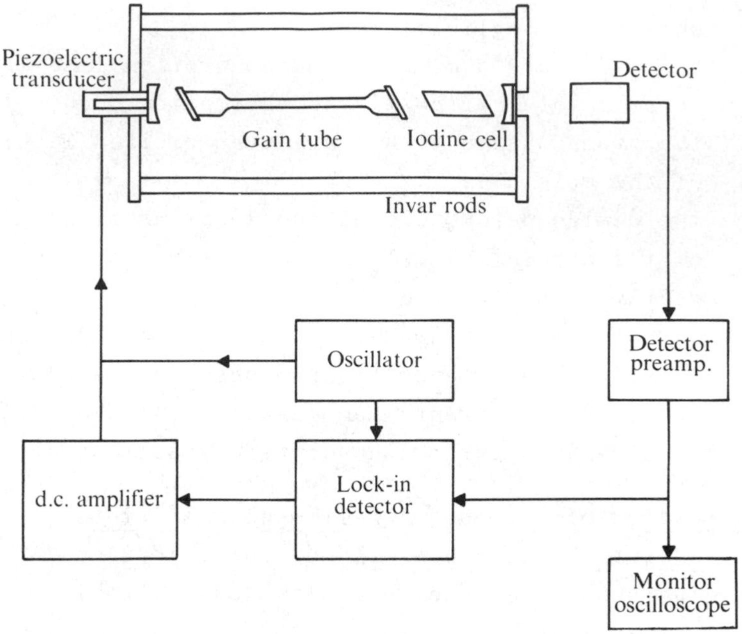

Fig.13.16. Schematic diagram showing stabilization of the frequency of a 6328 Å He-Ne laser to one of the saturated resonances of iodine molecules contained in an intracavity absorption cell.

cisely by means of a frequency counter.

As in heterodyne detection at r.f. frequencies, the optical heterodyne technique eliminates the problems caused by intrinsic or dark current noise in the detector and the effective signal-to-noise level is limited only by the quantum efficiency of the detector. On the other hand because lasers have frequencies of the order of 2×10^{14} Hz, this technique is only applicable if laser frequency stabilities of better than a few parts in 10^8 have been attained; although of course this restriction does not apply to homodyne detection in which one laser acts both as source and local oscillator.

By careful comparison with the ^{86}Kr standard, Schweitzer *et al.* (1973) have shown that the frequency reproducibility of

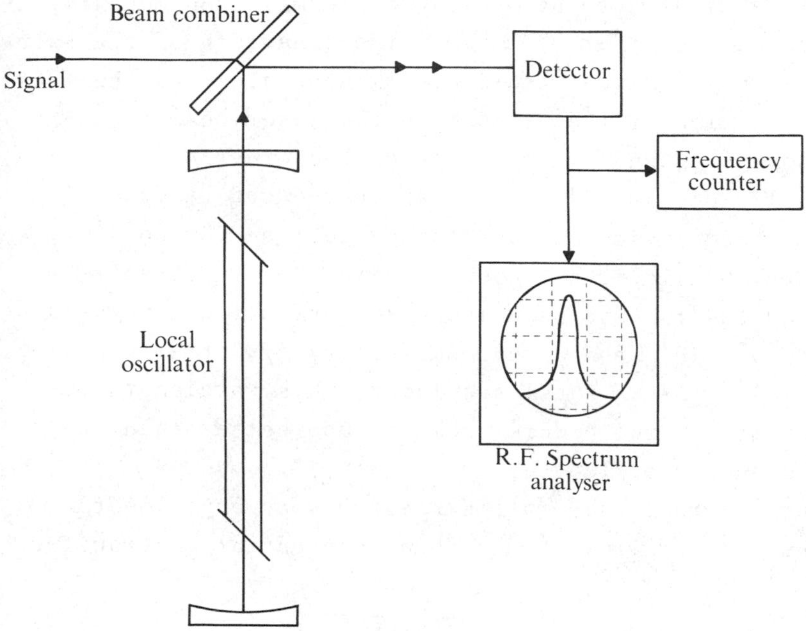

Fig.13.17. Schematic diagram of apparatus for laser spectroscopy using optical heterodyne techniques.

these iodine stabilized helium-neon lasers is about 1 part in 10^{10}. Thus they provide extremely useful secondary wavelength standards for work in the visible region of the spectrum. In spite of this, even greater accuracy and frequency stability have been attained using the infrared helium-neon line at 3·39 μm.

(b) Stabilization of the 3·39 μm helium-neon laser. Some of the most spectacular demonstrations of the power of the saturated absorption technique have been performed by Hall and his co-workers (Barger and Hall 1969, Uzgiris *et al.* 1971, Hall and Bordé 1973) using the helium-neon laser line at 3·3913 μm. By adjusting the helium pressure inside the laser, the centre of the neon gain curve at 3·39 μm can be made to coincide almost exactly with the P(7) line of the ν_3 band of methane, CH_4. In methane the radiative lifetime of the upper molecular level involved in the absorption transition is estimated to be as long as 10 ms. Consequently at pressures of the order of 1 mTorr the linewidth of the saturated absorption resonance is determined mainly by the finite interaction time of the atoms with the laser beam in the absorption cell, and is typically of the order of 100 kHz (Problem 13.10). The linewidth can be reduced by expanding the laser beam inside the absorption cell and by cooling the methane to liquid nitrogen temperatures. Methane-stabilized lasers have been shown to have a frequency stability of 3×10^{13} and a reproducibility of better than 1 part in 10^{11}. As we shall see later, this wavelength reproducibility is two orders of magnitude better than the present standard of length.

For saturated absorption spectroscopy experiments using the 3·39 μm line it was found convenient to transfer

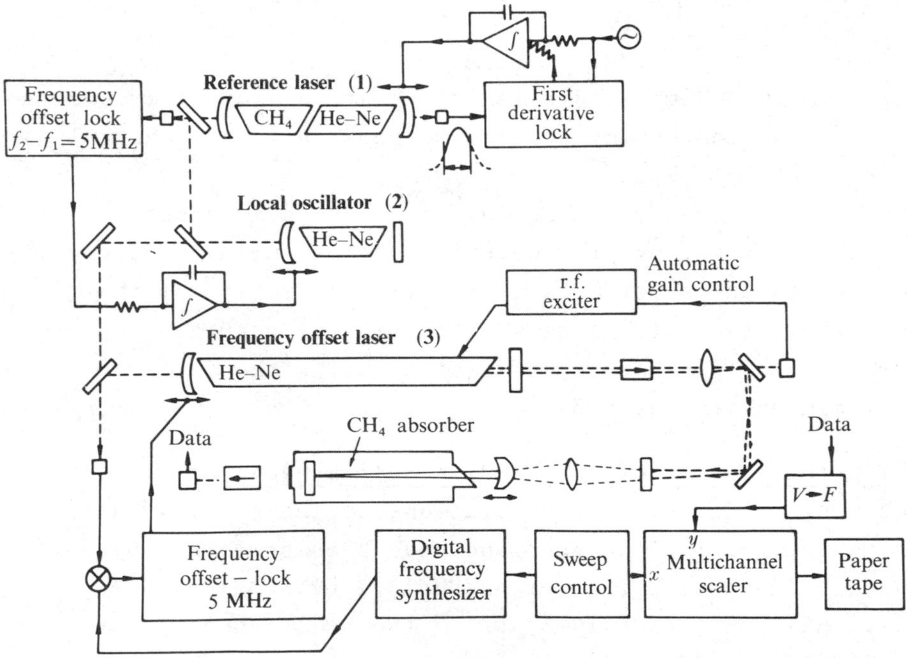

Fig.13.18. Schematic diagram of apparatus used by Barger and Hall (1969) for saturated absorption experiments in methane employing the frequency-offset locking technique. (After Hall (1973).)

the high-frequency stability of the methane-stabilized reference laser to a second more powerful probe laser by a technique known as frequency offset locking. The apparatus involved in these experiments is shown schematically in Fig.13.18. In the first stage the reference laser 1 is stabilized to the line centre of the methane saturated absorption resonance by means of an internal absorption cell in the usual way. Then the outputs of lasers 1 and 2 are mixed together and the frequency of the local oscillator laser is servo-controlled so that the beat note is exactly 5000·0 kHz. This transfers the frequency stability of laser 1 to laser 2 and also locks the local oscillator laser to a frequency offset by 5 MHz from the centre of the methane resonance. In the second stage the output of the powerful probe laser 3 is mixed with that of laser 2 and the beat note is servo-controlled to a frequency which may be set anywhere between 0·25 MHz and 15 MHz. This laser 3 combines great frequency stability with continuously variable output frequency and may be used for ultra-precise saturated absorption spectroscopy. Fig.13.19 shows the first derivative of the methane saturated absorption resonances obtained in this way using an external absorption cell. The curves obtained at various pressures are very accurately described by derivatives of Lorentzian line profiles and the half-width (FWHM) due to collision broadening is given by

$$\Delta\nu_{1/2} = (32{\cdot}6 \pm 1{\cdot}2)\ P \quad \text{kHz}$$

where P is the methane pressure measured in mTorr. The reader will appreciate the remarkable nature of these experiments when he realizes that a spectroscopic resolution of the order of 10^9 has been achieved at optical frequencies.

<u>13.10.4. Measurement of the velocity of light.</u> A knowledge of the velocity of light is now important in many diverse fields such as geophysical distance measurement, studies of the planets using microwave radar and in conversions between electrostatic and electromagnetic units. The extremely

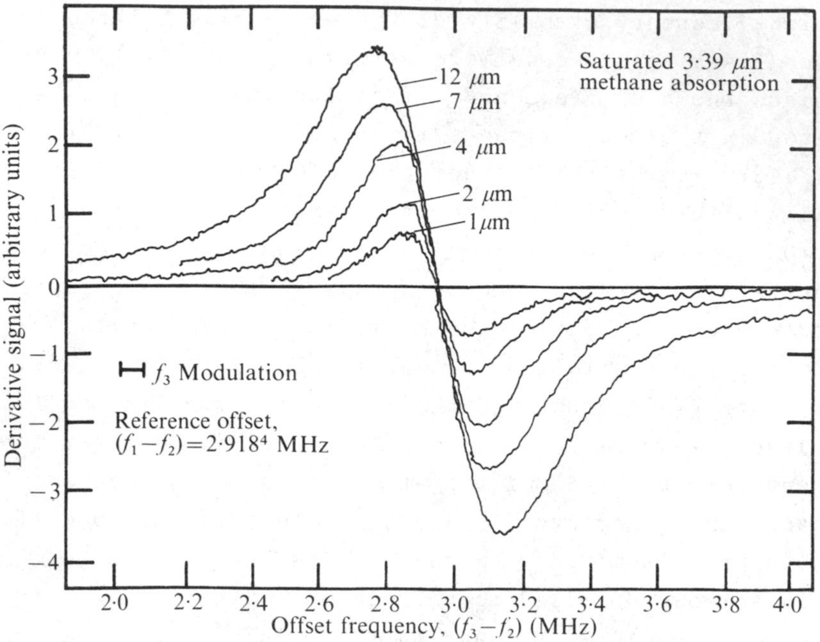

Fig.13.19. First derivative of the output power of laser number 3 versus offset-lock frequency (f_3-f_2). The curves show the saturated absorption resonance on the P(7) line of the ν_3 band of methane at different methane pressures and are very accurately described by derivatives of Lorentzian functions. (After Barger and Hall (1969).)

high accuracy with which the 3·39 μm He-Ne laser can be locked to the methane saturated absorption resonance has led to the possibility of redetermining the velocity of light with an accuracy which is about 100 times better than that of the previously accepted value of 299 792 500 ± 100 m s^{-1} obtained by Froome (1958). The determination involves the measurement of the wavelength and frequency of a stabilized laser giving c = λ ν. The wavelength of the 3·39 μm He-Ne laser stabilized to the methane saturated absorption line was measured in terms of the ^{86}Kr wavelength standard at 6057 Å by Barger and Hall (1973) using a servo-controlled plane-parallel Fabry-Perot interferometer. The accuracy of the wavelength measurement was limited to $\pm 1{\cdot}2 \times 10^{-4}$ Å by the small intrinsic asymmetry of the 6057 Å line emitted by the standard lamp. The Consultative Committee on the Definition of the Metre (CCDM) has recommended that when such asymmetries are observed, the defined wavelength (Table 13.3) should be taken to apply to a point on the krypton line half way between the maximum and the centre of gravity of the observed line profile. Using this definition, the result for the 3·39 μm neon line given in Table 13.3 is the mean of the two values reported by Barger and Hall (1973) in which they assumed that the standard applied either to the maximum or to the centre of gravity of the krypton line.

The frequency of the 3·39 μm methane-stabilized laser has been determined in terms of the cesium beam primary frequency standard (described in section 18.7) by Evenson *et al.* (1973). Since a direct comparison of the frequency of the standard at approximately 9·2 GHz and that of the laser at 88 THz is not possible at present, the experiment involved frequency multiplication using the chain of klystrons and far infrared lasers shown schematically in Fig.13.20. At each step in the multiplication chain one measures the beat frequency between an harmonic of the lower-frequency oscillator and the fundamental of the higher-frequency oscillator. At the low-frequency end of the chain conventional silicon point contact harmonic generator-mixers are used, but in the final stages novel tungsten-nickel diodes must be employed. These consist of a 2 mm long, 25 μm diameter tungsten catwhisker with a sharpened tip which is lightly contacted to the nickel surface. These diodes act as antennae, harmonic generators, and mixers up to the 88 THz frequency of the 3·39 μm laser, but require more than 50 mW of laser power for adequate signal-to-noise. The result of this frequency measurement is given in Table 13.3. The estimated error of 6 parts in 10^{10} does not represent the final limit of accuracy since, given sufficient care, frequency measurements can be made in terms of the cesium primary standard to a few

TABLE 13.3

Measurement of the velocity of light by determination of the frequencies and wavelengths of stabilized lasers

Source	Frequency (kHz)	Wavelength (fm)	Velocity of light ($m\ s^{-1}$)	Reference
^{86}Kr lamp	—	605 780 210·5	—	Defined standard
^{133}Cs atomic beam	9 192 631·770	—	—	Defined standard
He-Ne laser methane stabilized	88 376 181 627 ±50	3 392 231 390 ±12	299 792 457·4 ± 1·1	(ν) Evenson *et al.* (1973) (λ) Barger and Hall (1973) (c) Evenson *et al.* (1972)
CO_2 laser	32 176 079 482 ±28	9 317 246 348 ±24	299 792 459·0 ± 0·8	(ν) Blaney *et al.* (1973) (λ) Jolliffe *et al* (1975) (c) Blaney *et al.* (1975)

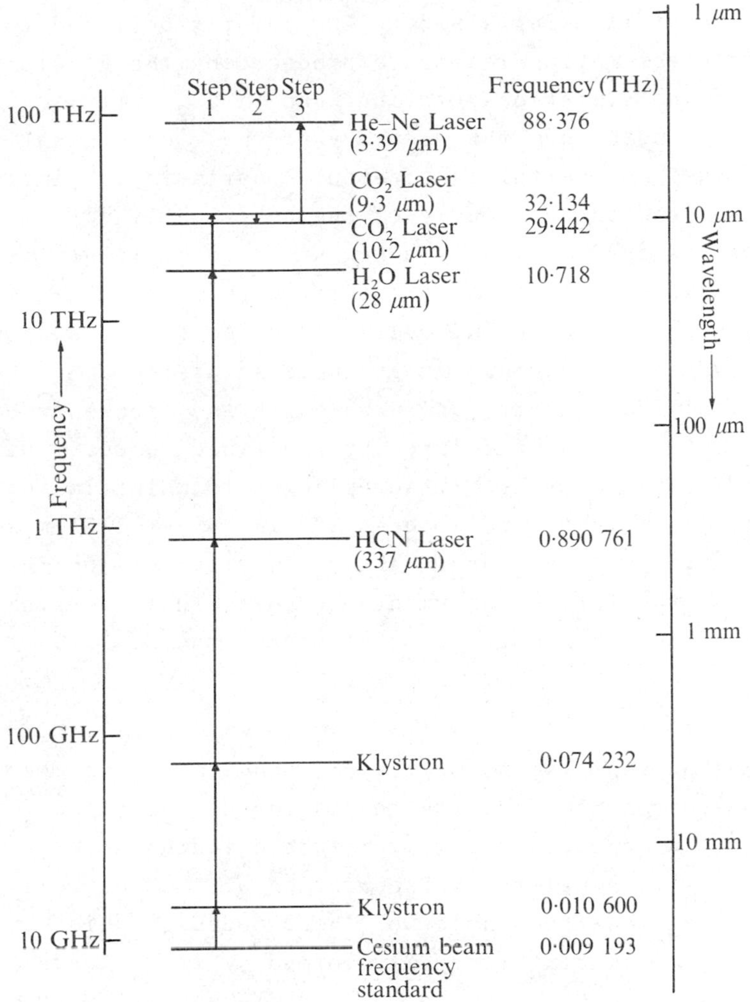

Fig.13.20. Diagrammatic representation of frequency synthesis chain used in determination of the velocity of light.

parts in 10^{13}.

When these wavelength and frequency measurements are combined the result for the velocity of light, given in Table 13.3, is about two orders of magnitude more accurate than the previously accepted value. This result has been confirmed by independent measurements made at the National Physical Laboratory using a 9·3 μm CO_2 laser stabilized to the saturated absorption resonance produced by the R(12) component of the vibration-rotation band in CO_2. It is interesting to note that the accuracy of these determinations is limited by the uncertainty of about 3 parts in 10^9 introduced by the inherent asymmetry of the emission line from the krypton standard. Clearly a new standard of length is required. However, in the near future there is the intriguing possibility of being able to combine the standards of length and time/frequency in a single saturated absorption stabilized laser. The wavelength and frequency of this device would then be related by a defined velocity of light. Indeed the CCDM is already pledged to maintain the velocity of light at the value c = (299 792 458 ± 1·2) m s^{-1} whatever changes may be made in the definitions of length and time units in the future, assuming of course that there are no unsuspected errors in the present measurements.

Problems

13.1. Derive an expression for the frequencies of the modes of a hemispherical cavity consisting of one plane mirror and one concave mirror, whose radius of curvature R is equal to the separation of the mirrors L_1. Show that the frequencies of these modes coincide with those of a concentric cavity formed by two concave mirrors, radius R, separated by a distance $L_2 = 2R$.

Calculate the frequency separation between the transverse modes of an almost plane-parallel cavity in which two concave mirrors of 50 m radius of curvature are separated by a distance of 1 m. (Ans: 9·56 MHz.)

13.2. Calculate the population inversion density per unit volume required to achieve the threshold of oscillation in a helium-neon laser operating at 632·8 nm. The transition probability of the 632·8 nm neon line is estimated to be $A_{ki} \approx 5{\cdot}1 \times 10^6\ s^{-1}$, $g_k = 3$, and the full width at half maximum intensity of the Doppler profile of this transition under the conditions existing in the laser discharge is 1700 MHz. The cavity mirrors each have a reflectivity of 99·9 per cent at the laser wavelength and the active length of the discharge is $L_m = 0{\cdot}60$ m. (Ans: $3{\cdot}7 \times 10^7\ cm^{-3}$.)

13.3. Show that the variation of the saturated gain coefficient $\alpha_L^S(\omega)$ as a function of the angular frequency ω for a homogeneously-broadened laser transition is described by a Lorentzian function whose maximum value is

$$\alpha_L^0(\omega_{ki})/\{1 + \frac{2I(\omega)}{I_S(\omega_{ki})}\}$$

and whose full width at half maximum amplitude is given by

$$\Delta\omega_{1/2}^S = \Gamma_{ki}\{1 + \frac{2I(\omega)}{I_S(\omega_{ki})}\}^{1/2} .$$

The intensity of laser radiation inside the cavity is given by $I(\omega)$, and $I_S(\omega_{ki})$ is the saturation parameter defined by equation (13.18).

A laser with a homogeneously-broadened line is operated in a cavity in which one mirror is perfectly reflecting while the other has an output transmission of t. The total round trip loss of the cavity is therefore given by $\delta = t + \delta'$ where δ' represents the unavoidable fractional cavity losses. Plot the laser output power for several different values of δ' as a function of the mirror transmission t and show that the mirror transmission for maximum output power is given by

$$t_{optimum} = (2L_m\alpha_L^0(\omega_{ki})\delta')^{1/2} - \delta'.$$

13.4. When inhomogeneous broadening is dominant, prove that the width of the hole burned in the velocity distribution of the laser population inversion density is given by

$$\Delta\omega_{1/2}(\text{hole}) = \Gamma_{ki}\{1 + \frac{I_+(\omega)}{I_S(\omega_{ki})}\}^{1/2} = \frac{\omega_0}{c}\ \Delta v_z(\text{hole})$$

where $I_+(\omega)$ is the intensity of laser radiation inside the cavity propagating towards $+\infty$ and $I_S(\omega_{ki})$ is the saturation parameter defined by equation (13.18).

Calculate the saturation parameter for a helium-neon laser oscillating on the 632·8 nm transition, assuming the following values for the necessary parameters:

(a) Statistical weight and effective lifetime of upper laser level $g_k=3$, $\tau_k=12{\cdot}2$ ns.

(b) Statistical weight and effective lifetime of lower laser level $g_i=5$, $\tau_i=5{\cdot}88$ ns.

(c) Spontaneous transition probability of the 632·8 nm line, $A_{ki} = 1{\cdot}38 \times 10^6\ s^{-1}$.

(d) Halfwidth of Lorentzian frequency response, $\Gamma_{ki}/2\pi = 144$ MHz. (Ans: 20·5 W cm^{-2}.)

13.5. A helium-cadmium laser is operated on the 4416 Å transition with a constant length of active discharge in a cavity whose length can be varied continuously. At the linecentre the round-trip gain, $2L_m\alpha^0_G(\omega_{ki})$, is 5 per cent and the round-trip cavity loss δ is 3 per cent. The gain profile varies as a Gaussian distribution whose width is determined by the temperature of 600 K existing in the laser discharge tube. Show that oscillation can only occur within a restricted frequency range on either side of line centre and calculate the maximum length of the cavity for which oscillation is restricted to a single longitudinal mode. (Ans: 31 cm.)

13.6. In a mode-locked laser the phases of adjacent longitudinal modes q are separated by a constant value of $\phi_{q+1} - \phi_q = \pi$. When the laser is oscillating simultaneously on N longitudinal modes all having the same amplitude, show that the laser output consists of a sequence of pulses with a repetition frequency of c/2L and a pulse width given approximately by $\Delta t = 2L/Nc$.

13.7. The gain profile of the 6328 Å line in a helium-neon laser may be assumed to be dominated by Doppler broadening with a full width at half maximum intensity of 1700 MHz. Calculate the optimum cavity length for single frequency oscillation. (Hint: maximize the difference in gain per round trip for adjacent longitudinal modes one of which is tuned to the linecentre) (Ans: 15 cm.)

13.8. Show that equation (13.39) for the saturated gain coefficient may be reduced to equation (13.40) in the approximation that the velocity distribution function P(v), is very much wider than the homogeneous line width Γ_{ki}. (Hint: make use of the properties of the folding integrals of Lorentzian and Gaussian functions which are listed in section 8.6).

13.9. The mirrors of a 6328 Å single-frequency helium-neon laser are separated by a spacer formed from a cylinder of fused quartz 14 cm long. The entire laser is enclosed in an oil-bath whose temperature is maintained constant to within $\pm\ 0{\cdot}01\ ^{o}C$. Calculate the frequency stability expected for this device given that the linear expansion coefficient of fused quartz is $4{\cdot}7 \times 10^{-7}$ per ^{o}C.

The laser output is directed into a two-beam interferometer. Estimate the maximum optical path difference in the interferometer at which two beam interference fringes would still be visible. (Ans: stability = $2{\cdot}1 \times 10^{8}$; path difference = 21 m.)

13.10. Estimate the contributions to the linewidth of the 3·39 μm saturated absorption signal obtained from a sample of methane gas, CH_4, from the following physical processes:

(a) the radiative lifetime of the upper level in methane which is approximately 10 ms;

(b) collision broadening in methane at a pressure of 1 mTorr assuming a collision cross-section of 3×10^{-15} cm^2 at a temperature of 300 K;

(c) the passage of the molecules across the finite radius, r=0·75 mm, of the laser beam in the methane absorption cell. (Ans: (a) $\Delta\nu$=16 Hz; (b) $\Delta\nu$=1·4 kHz; (c) $\Delta\nu$=133 kHz.)

References

Baird, K.M., Smith, D.S., Hanes, G.R., and Tsunekane, S. (1965). *Appl.Opt.* 4, 569.

Barger, R.L. and Hall, J.L. (1969). *Phys.Rev.Lett.* 22, 4.

__________ and __________ (1973). *Appl.Phys.Lett.* 22, 196.

Bennett, W.R. Jr. (1962). *Phys.Rev.* 126, 580.

Blaney, T.G., Bradley, C.C., Edwards, G.T., Knight, D.J.E., Woods, P.T., and Jolliffe, B.W. (1973). *Nature,Lond.* 244, 504.

Blaney, T.G., Bradley, C.C., Edwards, G.T., Jolliffe, B.W., Knight, D.J.E., Rowley, W.R.C., Shotton, K.C., and Woods, P.T. (1975). *Nature,Lond.* 251, 46.

Bloom, A.L. (1968). *Gas lasers.* Wiley, New York.

__________ and Wright, D.L. (1966). *Appl.Opt.* 5, 1528.

Boyd, G.D. and Kogelnik, H. (1962). *Bell Syst.tech.J.* 41, 1347.

Bridges, W.B. (1963). *Appl.Phys.Lett.* 3, 45.

Bruce, C.F. (1971). *Appl.Opt.* 10, 880.

Collinson, J.A. (1965). *Bell Syst.tech.J.* 44, 1511.

Cordover, R.H. and Bonczyk, P.A. (1969). *Phys.Rev.* 188, 696.

Evenson, K.M., Wells, J.S., Petersen, F.R., Danielson, B.L., Day, G.W., Barger, R.L., and Hall, J.L. (1972). *Phys.Rev.Lett.* 29, 1346.

Evenson, K.M., Wells, J.S., Petersen, F.R., Danielson, B.L. and Day, G.W. (1973). *Appl.Phys.Lett.* 22, 192.

Froome, K.D. (1958). *Proc.R.Soc.* 247A, 109.

Gorog, L. and Spong, F.W. (1969). *R.C.A. Rev.* 30, 277.

Hall, J.L. (1973). *Atomic physics*, Vol.3. (Eds. S.J.Smith and G.K. Walters). Plenum, New York.

_________ and Bordé, C. (1973). *Phys.Rev.Lett.* 30, 1101.

Hanes, G.R. and Dahlstrom, C.E. (1969). *Appl.Phys.Lett.* 14, 362.

Hänsch, T.W., Levenson, M.D., and Schawlow, A.L. (1971). *Phys.Rev.Lett.* 26, 946.

Hargrove, L.E., Fork, R.L., and Pollock, M.A. (1964). *Appl.Phys.Lett.* 5, 4.

Harris, S.E. (1966). *Appl.Opt.* 5, 1639.

Hercher, M. (1969). *Appl.Opt.* 8, 1103.

Jolliffe, B.W., Rowley, W.R.C., Shotton, K.C., Wallard, A.J., and Woods, P.T. (1975). *Nature,Lond.* 251, 46.

Lamb, W.E. Jr. (1964). *Phys.Rev.* 134, A1429.

Lee, P.H. and Skolnick, M.L. (1967). *Appl.Phys.Lett.* 10, 303.

McFarlane, R.A. (1964). *Phys.Rev.* 135, A543.

_____________, Bennett, W.R.Jr., and Lamb, W.E. Jr. (1963). *Appl.Phys.Lett.* 2, 189.

Mielenz, K.D., Nefflen, K.F., Rowley, W.R.C., Wilson, D.C. and Engelhard, E. (1968). *Appl.Opt.* 7, 289.

Schweitzer, W.G., Kessler, E.G., Deslattes, R.D., Layer, H.P., and Whetstone, J.R. (1973). *Appl.Opt.* 12, 2927.

Smith, P.W. (1965). *I.E.E.E.J. Quantum Electron.* QE-1, 343.

__________ (1966). *J.Appl.Phys.* 37, 2089.

Smith, P.W. (1970). *Proc.Inst.elec.and electron.Eng.* 58, 1342.

__________ (1971). *Appl.Phys.Lett.* 19, 132.

Szöke, A. and Javan, A. (1963). *Phys.Rev.Lett.* 10, 521.

________ and ________ (1966). *Phys.Rev.* 145, 137.

Uzgiris E.E., Hall, J.L., and Barger, R.L. (1971). *Phys.Rev.Lett.* 26, 289.

General references and further reading

A non-mathematical account of laser saturation spectroscopy is given in the article by

Feld, M. and Letokhov, V.S. (1973). Sci.Amer. 229, No.6, p.69.

A comprehensive review of mode selection techniques in lasers is given by

Smith, P.W. (1972). *Proc.Inst.elec.and electron.Eng.* 60, 422.

14

Tunable dye lasers and atomic spectroscopy

14.1. Introduction

In the previous three chapters we have discussed the properties of gas lasers and have shown how they can be designed for single frequency oscillation, and also how the output frequency may be tuned continuously over the bandwidth of the Doppler-broadened gain curve. Unfortunately this tuning range is relatively narrow and the application of these gas lasers to atomic and molecular spectroscopy is restricted to studies of the laser transitions themselves, or to accidental coincidences with molecular absorption lines. It would therefore seem that the new and powerful technique of saturated absorption spectroscopy was also of relatively limited applicability.

Fortunately, in the past decade, several different types of narrow-bandwidth widely tunable lasers have been developed and of these the organic dye laser has played the most prominent role in atomic and molecular spectroscopy. These dye lasers provide tunable coherent radiation throughout the wavelength range from the near ultraviolet to the near infrared, and moreover, by optical harmonic and laser difference frequency generation, the accessible spectral region can be extended into the vacuum ultraviolet and the far infrared.

This new device has led to a rapid growth in the nonlinear spectroscopy of atoms and molecules. In addition the narrow spectral bandwidth and great intensity per unit spectral range of these dye lasers have made it possible to extend the range of classic spectroscopic techniques such as absorption and fluorescence spectroscopy. Even the more precise spectroscopic methods such as the Hanle effect, the optical double resonance, and the optical pumping techniques have all benefitted from the increasing availability of tunable dye lasers. Selective step-wise excitation using dye lasers now enables atomic energy levels to be studied which have the same parity as the ground state and which consequently cannot be efficiently excited by conventional lamps.

Thus dye lasers have led to an enormous increase in activity in many areas of atomic physics. In the sections which follow we will briefly discuss the principles of dye laser operation and the construction of some typical laser systems. We will then consider a few of the recent applications of these devices to atomic spectroscopy. Further information on dye lasers and their spectroscopic applications is contained in the excellent collections of review articles edited by Schäfer (1973), Walther (1976) and Shimoda (1976).

14.2. Tunable organic dye lasers

14.2.1. Principles of dye laser operation. Many organic compounds that absorb strongly in certain regions of the visible spectrum also fluoresce very efficiently, emitting radiation which covers a large wavelength range. The first descriptions of stimulated emission from these fluorescent organic dyes in liquid solution were reported almost simultaneously by Sorokin and Lankard (1966) and Schäfer *et al.* (1966). It was not long before Soffer and McFarland (1967) had demonstrated that the stimulated emission was also tunable and the rapid development of tunable dye lasers had commenced.

The energy-level scheme of a typical organic dye molecule in dilute solution is shown schematically in Fig. 14.1. It consists of a ground state S_0 and a series of excited singlet levels $S_1, S_2, \ldots\ldots$ together with another series of triplet levels $T_1, T_2, \ldots$ in which the lowest level lies about 15 000 cm^{-1} above the ground state S_0. The energy level separation S_0-S_1 is typically about 20 000 cm^{-1}. In the singlet states the spin of the active electron and that of the remainder of the molecule are antiparallel, while in the triplet states the spins are parallel. Transitions between states of the same multiplicity give rise to the in-

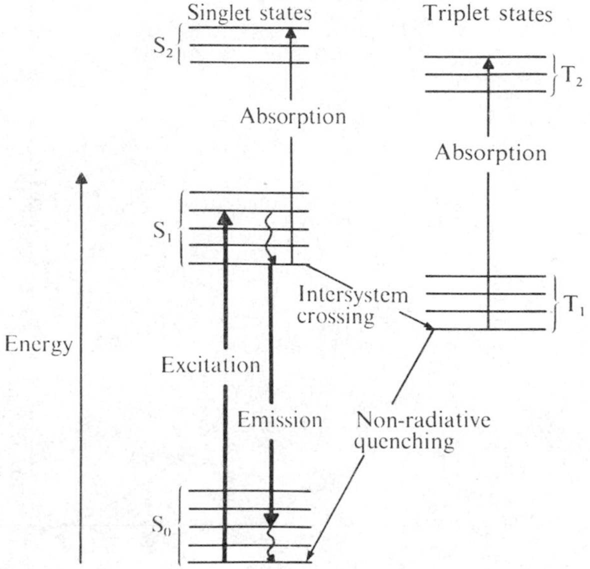

Fig.14.1. Schematic energy level diagram of typical dye molecule. (After Hänsch (1973).)

tense absorption and fluorescence spectra of the dye while singlet ↔ triplet radiative transitions involve a spin flip and are therefore far less probable. Each electronic level is also associated with an array of vibrational and rotational levels. The vibrational levels are spaced by intervals of 1400-1700 cm^{-1} while the spacing of the rotational levels is smaller by a factor of approximately 100 and consequently is too small to be shown in Fig.14.1. Due to rapid relaxation processes the rotational and vibrational levels are smeared out to form broad continuous energy bands. These account for the continuous absorption and emission spectra, examples of which are shown in Fig.14.2 for the case of the well-known laser dye rhodamine 6G in ethyl alcohol solution. The colour of the dye is determined by the broad absorption band $S_0 \rightarrow S_1$ which results from the excitation of an electron in a π-orbital.

When the dye solution is illuminated by light whose wavelength falls in the absorption band, molecules are op-

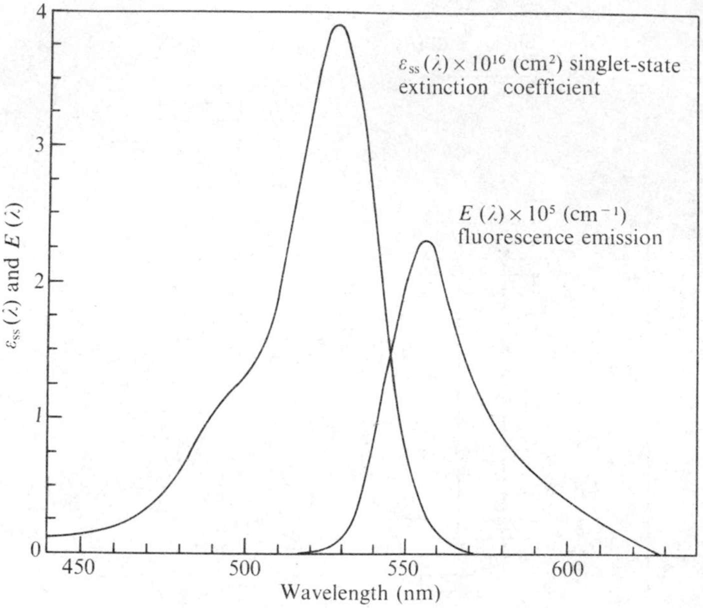

Fig.14.2. Absorption and fluorescence spectra of the laser dye rhodamine 6G in ethanol solution. (After Hänsch (1973).)

tically excited from the level S_0 into some rotational-vibrational level b belonging to the excited singlet state, S_1. Following the excitation, rapid collisions with other molecules dissipate the excess vibrational-rotational energy and the molecule relaxes to the lowest vibrational level of the S_1 state in a time of the order of 10^{-11}-10^{-12} s. From here the molecule can decay by spontaneous emission, with a radiative lifetime $\tau_s \approx 10^{-9}$ s, to any of the rotational-vibrational levels of the ground state. Consequently the emitted light is of longer wavelength than the pumping radiation. Finally non-radiative relaxation processes return the molecule to the v=0 level of the electronic ground state, S_0.

If the intensity of the pumping radiation is very high, exceeding about 100 kW cm^{-2}, a population inversion between S_1 and S_0 may be attained. Light amplification by the stimulated emission of radiation is then possible over almost the entire fluorescence band with the exception of that part which is effectively overlapped by the absorption band of the molecule. In the absence of frequency-selective feedback, the dye laser will oscillate on a band approximately 10-50 Å wide close to the peak of the fluorescence curve. However, due to the rapid thermalization of the vibrational and rotational levels, the spectral profile of the gain curve is essentially homogeneously broadened and it is possible to channel almost the entire available energy into a narrow spectral range by using a laser cavity with wavelength-selective feedback. Thus continuously tunable dye laser oscillation may be obtained.

Unfortunately there are several processes which compete with the fluorescent decay mode of the molecule and therefore reduce the efficiency of the laser. The most important of these are non-radiative relaxation to the ground state $S_1 \rightarrow S_0$, and non-radiative intersystem crossing $S_1 \rightarrow T_1$. This latter process is particularly undesirable since a large population of molecules may build up in the triplet level owing to the relatively long lifetime, $\tau_T = 10^{-3}$-10^{-7} s, of this metastable state. Absorption on the allowed transition such as $T_1 \rightarrow T_2$ can then cause considerable cavity losses if these absorption bands overlap the $S_1 \rightarrow S_0$ fluorescence band, and laser oscillation may be quenched or prevented completely. This triplet loss may be reduced by adding small quantities of chemicals such as cyclooctatetrene to the dye solution, for these favour the non-radiative transitions $T_1 \rightarrow S_0$ and thus shorten the effective lifetime of the T_1 level.

To date laser action has been achieved on several hundred organic dyes, many of which are listed in the survey by Warden and Gough (1971). Examples of some of the most useful dyes together with their respective tuning ranges are shown in Fig.14.3. The tuning range of certain dyes can be extended by adjusting the pH value of the solvent. Thus Shank *et al.* (1970) have been able to cover the entire

No.	Dye	Family	Structure
1	3,3[1] Diethylthia-tricarbocyanine iodide	Polymethine	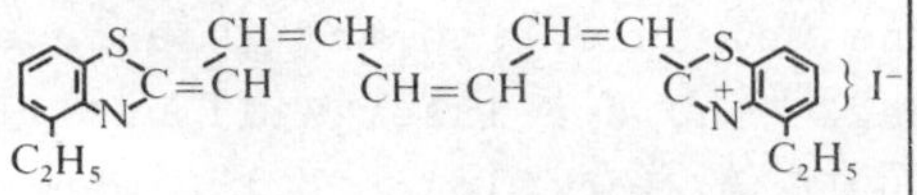
2	Rhodamine 6G	Xanthene	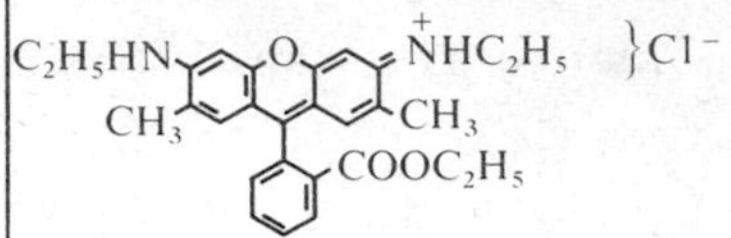
3	4-Methyl-umbelliferone	Coumarin	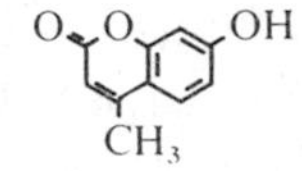
4	2,5 Diphenyl-oxazole	Oxazole	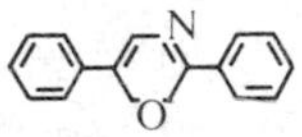

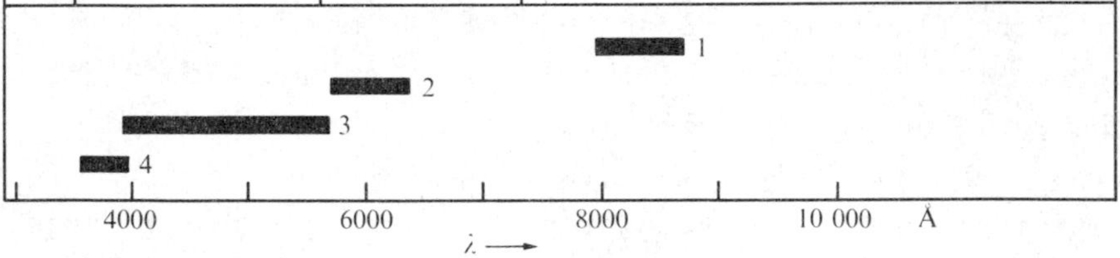

Fig.14.3. Examples of laser dyes, their structure and respective tuning ranges. (After Hänsch (1973).)

range of the spectrum from the near ultraviolet to the yellow, 390-544 nm, using 4-methyl-umbelliferone and its excited acidic complex. Mixtures of different dyes in the same solution have also been used to extend the tuning range or to transfer energy to a dye whose absorption bands did not match the emission of the pump source. However, for use in high-resolution spectroscopy wide tuning range in a single dye is of less importance than high fluorescence efficiency. Happily new dyes which have been especially synthesized for high efficiency over particular spectral regions are becoming increasingly available. The fluorescence efficiency, the triplet losses, and the absorption spectrum of the dye all impose stringent requirements on the properties of the dye laser pump source and we now turn to a consideration of the main types of sources used at present.

14.2.2. Flashlamp - pumped dye lasers. Many laser dyes with emission bands ranging from the near ultraviolet to the infrared have been successfully pumped using flashlamps. These lamps consist of quartz tubes filled with xenon or some other gas at a relatively high pressure and they are excited by a pulsed high-current discharge from a storage capacitor. Flashlamps with short risetimes, of the order of 100 ns or less, are the most suitable for use with dye lasers. The flashlamp and dye cell may be coaxial or alternatively the lamp and cell may be held at the foci of an elliptical cylindrical reflector as indicated schematically in Fig.14.4. Difficulties are encountered in these lasers

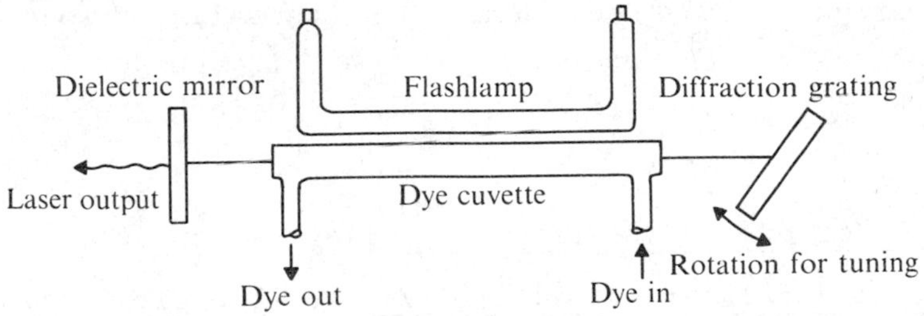

Fig.14.4. Schematic diagram of flashlamp-pumped dye laser. (After Hänsch (1973).)

because of the large amount of thermal energy transferred to the dye solution. This causes refractive index variations throughout the dye cell and leads to very large diffraction losses in the optical cavity. In order to overcome these thermal schlieren effects, the solution is usually circulated rapidly through the dye cuvette by means of a small pump.

In order to achieve a narrow bandwidth tunable output one mirror of the normal laser cavity is usually replaced by a diffraction grating as shown in Fig.14.4. The grating normal makes an angle θ with the axis of the cavity and in this Littrow arrangement the condition

$$2\,d\sin\theta = m\lambda \qquad m = 1, 2, \ldots\ldots \qquad (14.1)$$

must be satisfied for radiation to be reflected back along the cavity axis. In this equation λ is the oscillating wavelength of the laser and d is the grating spacing. Light of other wavelengths is not reflected back along the cavity axis and consequently this radiation sees a very lossy resonator and oscillation is prevented. Thus narrow bandwidth laser output is obtained and wavelength tuning may be accomplished simply by rotating the grating. Prisms, Fabry-Perot étalons, and combinations of these elements with diffraction gratings have all been used as tuning elements in flashlamp-pumped dye lasers and the bandwidths obtained range typically from 0·3-3·0 Å. Flashlamp-pumped dye lasers can be very cheap and simple to make and output energies of up to several Joules per pulse have been obtained. Generally, however, the pulse repetition rate is too low and the output spectral bandwidth is too large to permit their use in high-resolution atomic spectroscopy.

14.2.3. Laser-pumped dye lasers - pulsed systems.

(a) Solid state laser-pumped dye lasers. The first reported dye lasers were pumped by giant pulse (Q-switched) solid state lasers such as the ruby laser at 694·3 nm and the frequency-doubled neodymium laser at 530·0 nm. These pump sources continue to play an important part in applications requiring very high peak powers. The inherently short pulse length of the Q-switched solid state lasers, 5-100 ns, effectively eliminates the problems of triplet state absorption and dye laser efficiencies of up to 50 per cent have been reported in certain cases.

These laser-pumped dyes exhibit very large gain coefficients, $\approx 10^3\ mm^{-1}$, and low-Q optical cavities containing several rather lossy tuning elements may be used. Thus by combining a Littrow mounted échelle grating and one or more Fabry-Perot étalons, single mode operation with bandwidths of less than 0·01 Å can be achieved. Ruby laser pumping is essential for most of the infrared dye lasers since the absorption bands of these molecules lie towards the red end of the spectrum and moreover these dyes are easily destroyed by photodissociation when sources of shorter wavelength are used. However, solid state laser-pumped dyes are not restricted to the long wavelength end of the spectrum, for efficient second and fourth harmonic generation makes the ruby (347·2 nm) and neodymium (265·0 nm) lasers very effective pump sources for dye lasers in the near ultraviolet.

Laser pumping certainly provides the easiest route to tunable dye laser radiation once the pump laser exists. Unfortunately the very low repetition rate of many of the Q-switched solid state lasers is a great disadvantage in experiments in atomic and molecular spectroscopy.

(b) Molecular nitrogen laser-pumped dye lasers. A particularly reliable and convenient pump source is the pulsed nitrogen laser operating at 337·1 nm. The short wavelength of this laser radiation excites many dyes to high-lying singlet levels, but in all cases the molecules relax very quickly to the bottom edge of the lowest excited singlet level, dissipating the excess energy in the solvent, and dye laser oscillation occurs on the $S_1 \rightarrow S_0$ transition. Since most dyes have a strong absorption band in the ultraviolet region the nitrogen laser provides an almost universal pump source. The short pulse length and high repetition frequency of this laser provide a convenience similar to that of C.W. operation and it is one of the most widely used systems in atomic spectroscopy.

A schematic diagram of the nitrogen laser-pumped dye laser system developed by Hänsch (1972) is shown in Fig. 14.5. The nitrogen laser consists of a rectangular channel ≈ 1 m long through which a rapid discharge is passed from a triggered high-voltage capacitor system. Nitrogen molecules are excited to the $C\ ^3\Pi_u$ state by collisions with fast electrons and a transient inversion is created on the $B\ ^3\Pi_g \leftrightarrow C\ ^3\Pi_u$ ultraviolet emission band. The radiation emitted by the laser is self-terminating because the lower level has a longer lifetime than that of the upper level and in most of these devices the output consists of a pulse of amplified spontaneous emission

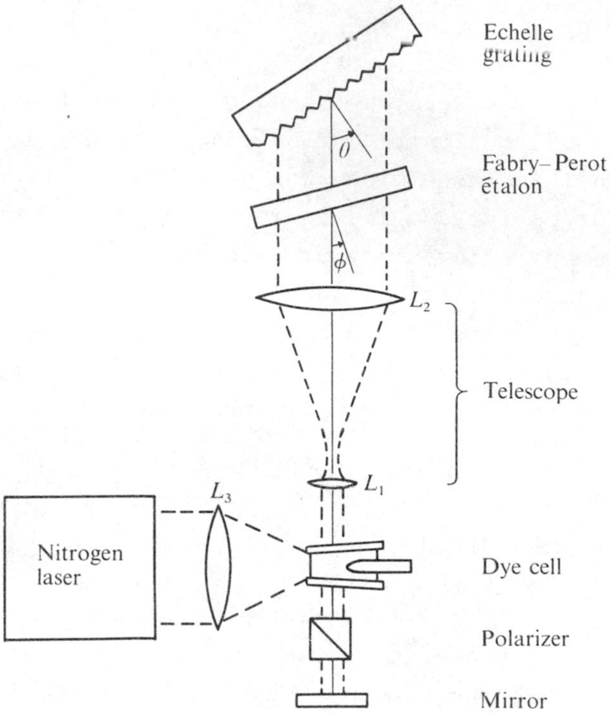

Fig.14.5. Dye laser with narrow bandwidth output pumped by pulsed nitrogen laser. (After Hänsch (1973).)

at 337·1 nm lasting 7-10 ns and having a peak power of ≈ 300 kW

The radiation from the nitrogen laser emerges in the form of a beam of rectangular cross-section, approximately 5 mm × 40 mm, which is focussed by a spherical quartz lens into a line near the inner wall of the dye cell. The active volume of the dye forms a cylindrical filament about 0·2 mm diameter and 10 mm long having a single pass gain which approaches 10^3 mm^{-1} under these conditions. The optical cavity of the dye laser is about 40 cm long and consists of a plane dielectrically-coated mirror at one end and a Littrow-mounted diffraction grating at the other. Owing to diffraction, the rather small active cross-section in the dye results in a substantial angular spread of the emerging radiation and this would normally limit the bandwidth obtainable with angle-dependent wavelength selectors such as gratings and étalons. However, this problem may be overcome by using a telescope as a beam expander within the cavity so illuminating the whole aperture of the grating. In this system a bandwidth of 0·03-0·05 Å can be obtained using the grating and telescope alone and the output bandwidth can be reduced by a further factor of ten by inserting a tilted Fabry-Perot étalon into the cavity. As explained in section 13.7.4 this acts simply as a narrow bandpass filter with the transmission maxima determined by the condition

$$2\,\mu\,t\,\cos\phi = n\lambda \qquad (14.2)$$

where ϕ is the tilt angle, and μ and t are the refractive index and thickness of the étalon respectively. The laser wavelength can be tuned continuously over a range of several Ångstroms by altering the tilt angles of the étalon and grating simultaneously. The wavelength stability is limited by temperature changes in the étalon, but can be as good as 0·01 Å over several hours.

In dye lasers of this type the efficiency can be as high as 20 per cent. At low repetition rates thermal schlieren effects are absent because the pulse length is so short, 5-10 ns, and the output beam is nearly diffraction limited. However, for repetition rates above 10 Hz it is necessary to circulate the dye transversely through the cell. Peak optical output powers of several kW can then be generated at repetition rates approaching 100 Hz.

14.2.4. Argon laser-pumped dye lasers - C.W. systems. The very high intensity required to pump dye lasers on a C.W. basis can so far be obtained only by using the tightly focussed beam of an argon or krypton ion laser, as shown in Fig.14.6. In this particular design, developed by Kogelnik *et al.* (1972), the argon laser beam enters the dye laser cavity through a Brewster angle prism and is focussed to a spot of approximately 10 μm diameter in the tilted dye cell

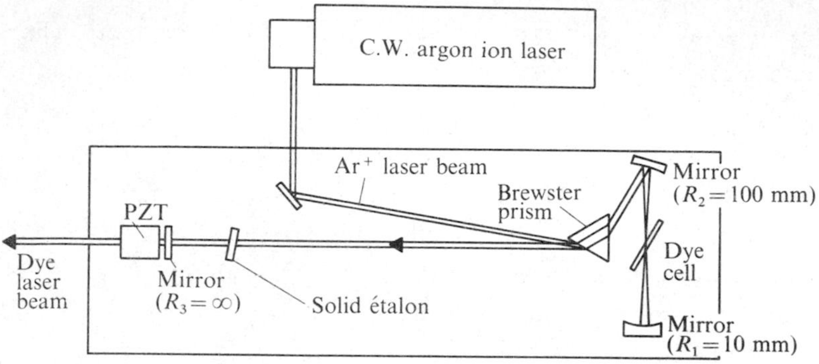

Fig.14.6. Argon ion laser-pumped C.W. dye laser using three mirror cavity. (After Walther (1974).)

by an off-axis, short radius of curvature mirror. The folded arrangement of three mirrors and the Brewster angle prism then form the dye laser cavity and astigmatism is compensated by careful choice of the radii of curvature of the mirrors. Unlike the pumping schemes discussed previously, the single pass gain in C.W. systems seldom exceeds a few per cent and a low-loss optical cavity is necessary for successful dye laser oscillation. This means that diffraction gratings cannot be used as tuning elements; however, even the low dispersion provided by a single prism is sufficient to reduce the output bandwidth to ≈ 0·25 Å, for the amplified radiation now makes a much larger number of transits before leaving the cavity.

In order to reduce the bandwidth still further it is first essential to provide high thermal and mechanical stability, for instance by mounting all the optical components on a massive invar base. Oscillation on a single cavity mode can then be achieved by employing a tilted intracavity étalon and bandwidths as low as 1·5-5·0 MHz have been attained. With this étalon fixed, the dye laser output can be tuned over the free spectral range, c/2L ≈ 500 MHz, of the main cavity by mounting one of the mirrors on a piezoelectric drive and thus varying the length of the cavity. Continuous tuning over the bandwidth determined by the prism is considerably more difficult and requires synchronous tuning of the lengths of both the cavity and the mode-selecting étalon.

For many experiments in atomic and molecular spectroscopy C.W. dye lasers would seem to be the natural choice were it not for the fact that, of all the systems we have discussed, they are the most difficult to operate successfully. Moreover their output is often limited to the 5400-6500 Å region by the fact that only a few dyes such as rhodamine 6G, rhodamine B and coumarin 6 can be made to oscillate on a C.W. basis when the strong blue-green argon lines at 4880 Å and 5145 Å are used as the pump. It is possible to extend the operating range to 4000-7000 Å by pumping with the ultraviolet lines of the krypton or argon ion lasers, but here pump powers of 10-15 W or more are usually necessary.

<u>14.2.5. Comparison of different dye lasers.</u> Representative values of the output power, pulse duration, and tuning range of dye lasers pumped by the four schemes we have just considered are given in Table 14.1. The table also includes values of the minimum output bandwidth but these should be treated with some caution since results in individual cases are very largely determined by the alignment and stability of the dye laser cavity. Moreover this figure may be increased by several orders of magnitude should the number of frequency selective elements in the cavity be reduced.

The range of tunable dye laser radiation may be extended down to 230 nm by frequency doubling or optical sum-frequency generation using non-linear optical materials. Similarly coherent infrared radiation can be generated using difference frequency mixing in non-linear crystals or by stimulated Raman scattering in alkali metal vapours, as indicated in Table 14.2. However, the efficiency of these conversion processes is often rather low, especially in the infrared region, and experiments outside the normal

TABLE 14.1.

Operating parameters of dye lasers

Pump source	Output power (peak value)	Pulse duration (ns)	Repetition rate (Hz)	Approximate minimum bandwidth (MHz)	Wavelength range (nm)
Flashlamp	20 - 50 kW	300 - 500	10 - 100	4	350 - 850
Ruby or neodymium laser	0·5 - 5·0 MW	5 - 20	≈ 1	50	350 - 1200
Molecular nitrogen laser	2·0 - 200 kW	2 - 8	30 - 500	500	350 - 1200
Argon ion laser	50 - 500 mW	—	C.W.	0·5	400 - 700

TABLE 14.2

Wavelength range of ultraviolet and infrared radiation generated by dye laser sum and difference frequency mixing in non-linear media

Non-linear optical material	Wavelength range	Input power (peak)	Conversion efficiency (per cent)
Ammonium dihydrogen phosphate (ADP)	280 - 290 nm	20 MW	10
Cooled ADP	250 - 325 nm	60 kW	10
Lithium formate monohydrate	230 - 300 nm	50 kW	2
Proustite (Ag_3AsS_3)	3·2 - 5·6 μm	900 kW	2
Lithium iodate	4·1 - 5·2 μm	4 MW	10^{-2}
Proustite	10·1 - 12·7 μm	290 kW	10^{-6}
Sodium/potassium vapour	2·21 - 23·4 μm	20 kW	10^{-5}

tuning range of dye lasers are still rather difficult.

When shorter pulse lengths are necessary these can be generated by mode locking a dye laser as discussed in section 13.6. The very wide gain bandwidth of dye lasers should theoretically permit the generation of sub-picosecond pulses, but dispersion in the cavity optics and other problems have so far imposed a lower limit of about 1·5 ps on the attainable pulse length.

14.3. Saturated absorption spectroscopy using tunable dye lasers

14.3.1. Saturation spectroscopy. The resolution attainable in conventional emission or absorption spectroscopy of gases in the visible region of the spectrum is limited by the large Doppler broadening produced by random thermal motion of the atoms or molecules. As explained in section 13.9 laser saturated absorption spectroscopy virtually eliminates Doppler broadening since the beam of monochromatic light interacts only with atoms or molecules in a narrow interval of the total axial velocity distribution. If the laser intensity is sufficiently strong, the velocity-selective population changes which it induces can be detected by a second counter-propagating laser beam and the saturated absorption resonance which occurs when the laser frequency is tuned to line centre may be located very precisely. Until recently these experiments were restricted to gas laser transitions or to molecular absorption lines with which they happened to coincide. Now, however, this restriction has been eliminated by the development of stable, narrow-bandwidth tunable dye lasers.

The first application of tunable dye lasers to the high-resolution saturated absorption spectroscopy of atoms was reported by Hänsch, Shahin, and Schawlow (1971). They studied the yellow resonance lines of sodium, $3\ {}^2S_{1/2} - 3\ {}^2P_{1/2,3/2}$ at 5896 Å and 5890 Å using a rhodamine 6G dye laser pumped by a pulsed nitrogen laser. The bandwidth of the tunable laser output was 300 MHz, but this was reduced still further by passing the beam through an external confocal interferometer approximately 1 m long. This interferometer serves as a very narrow bandpass filter and an output linewidth of approximately 7 MHz was obtained. Using saturated absorption spectroscopy, the hyperfine splittings of both the ground state and the excited levels of sodium were resolved, corresponding to an optical resolution of about 1 part in 10^7. Although these splittings have in fact been measured more accurately using the atomic beam and optical double resonance techniques described in the following chapters, this experiment demonstrated for the first time that it was possible to apply the methods of saturation spectroscopy to virtually any atomic or molecular absorption line throughout the visible region of the spectrum. Since this pioneering experiment several other transitions have been studied by saturation spectroscopy, but perhaps the most important experiment reported so far concerns the high resolution study of the Balmer α line of atomic hydrogen.

14.3.2. Fine structure of atomic hydrogen. The red Balmer α line of atomic hydrogen, $n=3 \rightarrow n=2$ at 6563 Å, is perhaps the most extensively studied of all spectral lines. In spite of this effort, precise measurements of the structure and wavelength of this important line have always been extremely difficult because of its very large Doppler width. In fact, owing to the fine-structure splitting of the $n=2$ and $n=3$ levels, the line consists of seven closely spaced components, as shown in Fig.14.7(a). According to the Dirac theory, the energy of a hydrogenic ion in the state $|nljm\rangle$ is given by

$$\frac{E_{n,j}}{hc} = -\frac{R_M Z^2}{n^2}\left\{1 + \left(\frac{\alpha^2 Z^2}{n}\right)\left(\frac{1}{j+\frac{1}{2}} - \frac{3}{4n}\right) + \right.$$
$$\left. + \text{(terms of higher order in } \alpha^2 Z^2)\right\} \qquad (14.3)$$

where $\alpha = e^2/4\pi\varepsilon_0\hbar c \approx 1/137$ is called the fine structure constant and R_M is the Rydberg constant for the isotope of mass M under consideration. This theory predicts that levels of the same j but with different values of l will be degenerate. The Dirac theory, however, is not quite in agreement with experiment, for Lamb and Retherford (1947) were able to show by means of an atomic beam radio-frequency resonance experiment that the $^2S_{1/2}$ and $^2P_{1/2}$ levels be-

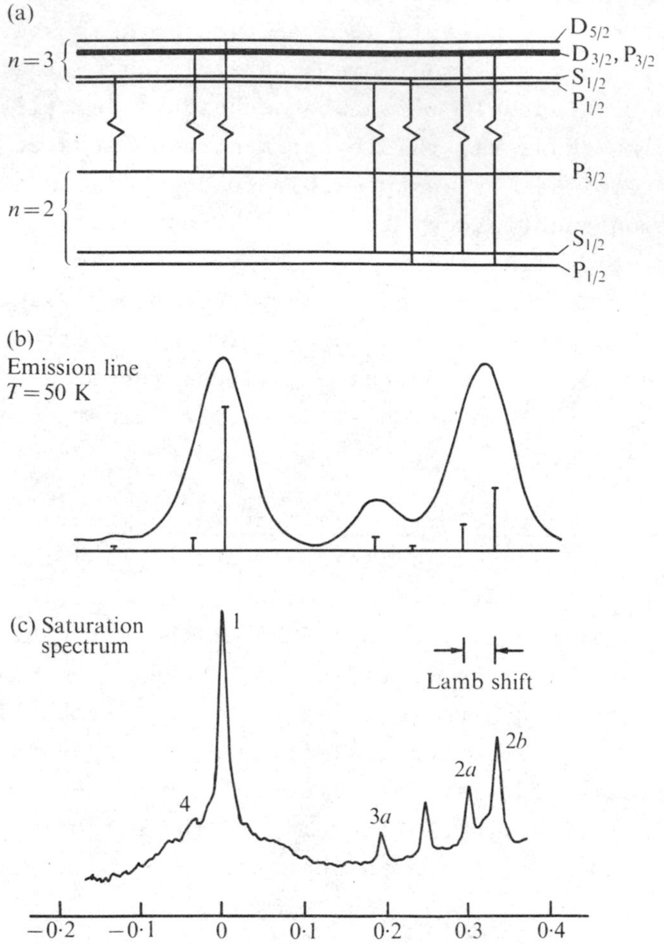

Fig.14.7. Fine structure in hydrogenic systems - the Balmer α line of deuterium. (a) Energy levels and allowed fine-structure transitions. (b) Profile of emission line from cooled deuterium gas discharge and fine structure components with theoretical relative transition probabilities. (c) Saturation spectrum of D_α line showing optically resolved Lamb shift. (After Hänsch (1975).)

longing to the n=2 state in atomic hydrogen were in fact separated by an interval of approximately 1058 MHz, as shown in Fig.14.7(a). The level degeneracy in l is similarly removed in other hydrogenic states, but the shifts are considerably smaller than those of the ^2S levels.

This effect may be explained in terms of the quantum theory of electrodynamics. According to this theory each mode of the quantized radiation field possesses a zero-point energy of $\hbar\omega/2$. This implies that, even in the absence of external radiation, the mean square value of the time-dependent electric field is finite and that a hydrogen atom will experience a perturbation produced by the fluctuations in this field. These zero-point fluctuations cause the electron to wobble randomly in its orbit and so smear the charge over a greater volume of space. Since the electron is bound to the nucleus by a non-uniform electric field, the reduction in electron density causes a shift in the atomic energy levels. This Lamb shift, as it is now called, is greatest for those states in which $|\psi(0)|^2$ is finite, i.e. the n $^2S_{1/2}$ states.

In hydrogen the fine-structure components of the Balmer α line always overlap considerably when observed by conventional spectroscopic techniques. Although the heavier isotopes deuterium and tritium may be used and the discharge tube may be cooled with liquid hydrogen or helium, the problem of Doppler broadening is far from being eliminated, as Fig.14.7 (b) demonstrates. A complete analysis of this blend of unresolved components is impossible, even though the fine-structure intervals have been measured very precisely using the radio-frequency spectroscopic techniques described in Chapters 16 and 18, for the relative intensities of the fine-structure components observed in most sources do not obey any simple rule. Thus an accurate determination of the wavelengths of the different components is very difficult. On the other hand these measurements are very important since the value of the Rydberg constant, defined by

$$R_\infty = \left(\frac{\mu_0 c^2}{4\pi}\right)^2 \left(\frac{me^4}{4\pi\hbar^3 c}\right), \tag{14.4}$$

is determined from the wavelengths of these fine-structure components after small corrections have been made for the Dirac fine structure, the Lamb shifts and the finite mass of the nucleus. These experimental difficulties led Series (1971) to suggest that the accuracy of 1 part in 10^7 claimed for the value of the Rydberg constant in the assessment of fundamental constants by Taylor, Parker, and Langenberg (1969) may well be over-optimistic. Nevertheless it has become apparent over the last few years that any adjustments and consistency tests applied to the values of the fundamental constants will in future require a value for R_∞ which is accurate to a few parts in 10^8. This stimulated Kibble *et al.* (1973) and Kessler (1973) to remeasure the Rydberg using conventional high-resolution techniques. But before their experiments were completed it became apparent that a dramatic improvement in accuracy was feasible, for in 1972 Hänsch and his co-workers (Hänsch, Shahin, and Schawlow 1972) succeeded in resolving the fine structure components of H_α using tunable dye laser saturated absorption spectroscopy.

14.3.3. Determination of the Rydberg constant by laser saturation spectroscopy. A simplified diagram of the apparatus used for saturated absorption spectroscopy of atomic hydrogen is shown in Fig.14.8. In order to be able to observe absorption on the Balmer α line it is first necessary to prepare hydrogen atoms in the n=2 level. Thus the absorption cell consists of a section 15 cm long in the centre of the positive column of a pulsed discharge. The tube is operated at room temperature and electrolytically generated hydrogen or deuterium gas is continuously pumped through it at 0·1-1·0 Torr pressure. The saturated absorption on the Balmer α transition is measured during the discharge afterglow since this almost completely eliminates

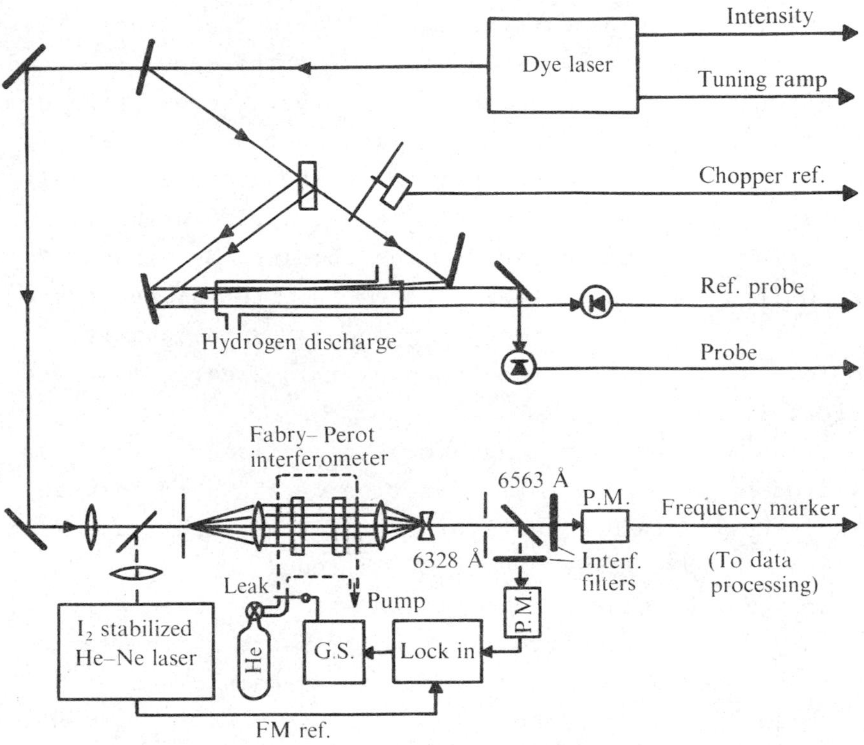

Fig.14.8. Schematic diagram of laser saturated absorption spectrometer and wavelength comparator for measurement of the Rydberg constant. (After Hänsch (1975).)

the effects due to Stark broadening and level mixing induced by the charged particles and electric fields which are present in the active discharge.

The nitrogen laser-pumped dye laser used in these experiments has already been described in section 14.2.3(b). It was again necessary to reduce the dye laser bandwidth to about 30 MHz by passing the beam through an external confocal étalon. For the absorption measurements the dye laser output was split into two weak probe beams about 1 mm diameter and one strong counter-propagating saturating beam. This beam is alternately chopped and overlaps just one of the probe beams. When the dye laser is tuned to one of the fine-structure components of the Balmer α line, these two counter-propagating waves interact with the same atoms and the strong saturating beam bleaches a path through the sample for the overlapping probe beam. The second probe beam enables a compensation to be made to the observed signals for the random fluctuations of the dye laser intensity. The saturation spectrum of the D_α line obtained by Hänsch, Nayfeh, Lee, Curry, and Shahin (1974) using this technique is shown in Fig.14.7(c). Four strong components can be observed and for the first time the Lamb shift is clearly resolved at optical frequencies.

The absolute wavelength of the dye laser output was measured by sending a small fraction of the beam through a Fabry-Perot étalon and recording the observed transmission peaks as the dye laser frequency is tuned across the Balmer α line. The spacing of the étalon was determined very accurately by manually controlling the pressure of the surrounding gas so that the interferometer was maintained on a transmission peak for the 6328 Å line of an auxiliary helium-neon laser. The wavelength of this laser is precisely known, for the output frequency was electronically stabilized to one hyperfine component of the $^{129}I_2$ absorption line, as described in section 13.10.3(a).

For the determination of the Rydberg constant, the strong component $2\ ^2P_{3/2} - 3\ ^2D_{5/2}$ of the Balmer α line was chosen since it possesses the smallest hyperfine structure. The saturated absorption profile of this component observed in deuterium is shown in Fig.14.9, together with the dye laser Fabry-Perot transmission peaks which were recorded simultaneously. The absolute wavelength of the fine-structure component may thus be determined by measuring its separation from the nearest interferometer resonance. The results for the Balmer α lines of hydrogen and deuterium are given in Table 14.3. Using theoretical calculations, corrections have been made to these wavelengths for the effects of fine structure, Lamb shifts, and the finite

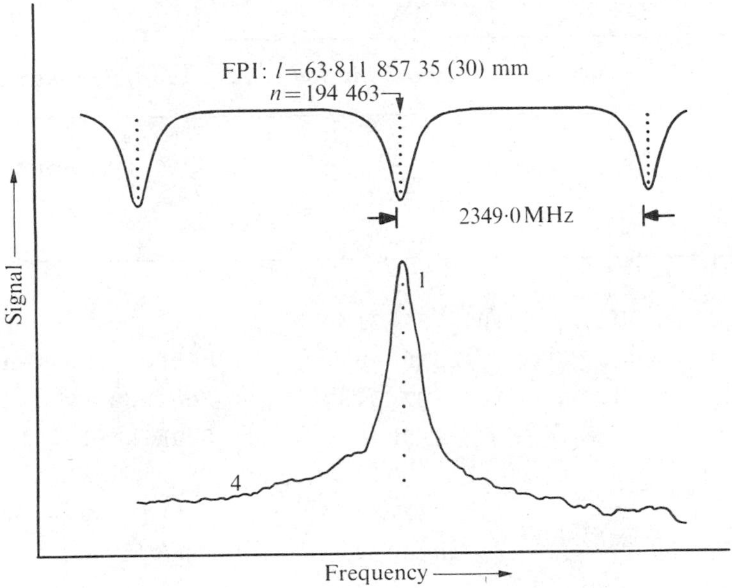

Fig.14.9. Section of saturation spectrum of Balmer α line of deuterium together with transmission maxima of Fabry-Perot interferometer. The wavelength of the component 2 $^2P_{3/2}$ - 3 $^2D_{5/2}$, labelled (1) in the diagram, is determined by measuring its separation from the transmission peak corresponding to the 194 463th order. (After Hänsch (1975).)

nuclear mass, and new values of the Rydberg constant obtained which are accurate to about 1 part in 10^8. It is possible that even this high precision will be exceeded in future measurements, which may then confirm the slight systematic difference between these most recent measurements and the less accurate values obtained by Kibble *et al.* (1973) and Kessler (1973) using conventional high-resolution spectroscopy.

Our interest in the spectrum of hydrogen, which at first glance seems so simple and yet is the most rewarding

TABLE 14.3.

Results of recent determinations of the Rydberg constant

Transition	R_∞ (cm^{-1})	Reference
$H(2\ {}^2P_{3/2} - 3\ {}^2D_{5/2})$	109 737·3130 ±0·0006	Hänsch, Nayfeh, Lee, Curry and Shahin (1974)
$D(2\ {}^2P_{3/2} - 3\ {}^2D_{5/2})$	109 737·3150 ±0·0006	
D(n=2 - n=3)	109 737·326 ±0·008	Kibble *et al.* (1973)
He^+(n=3 - n=4)	109 737·3208 ±0·0085	Kessler (1973)

from a theoretical point of view, is not yet exhausted, for atomic hydrogen is also being probed by the newly developed technique of two-photon absorption spectroscopy.

14.4. Two-photon absorption spectroscopy

<u>14.4.1. Principles of the technique.</u> In section 7.4 we discussed in detail the spontaneous decay of excited atoms by the simultaneous emission of two electric dipole photons. Consideration of arguments similar to those used in Chapter 9 in the derivation of the Einstein relations show that, corresponding to this spontaneous decay process, there must also exist stimulated transitions involving the simultaneous emission or absorption of two photons. Such multiple-quantum transitions were observed experimentally in radio-frequency resonance experiments by Brossel *et al.* (1954) and Kusch (1954), and a theory of this effect for magnetic dipole

transitions has been developed by Winter (1959). However, it was not until the development of lasers that sufficiently intense sources of radiation were available to demonstrate the existence of these effects at optical frequencies. From the spectroscopic point of view, the phenomenon remained something of a curiosity until the development of narrow-bandwidth tunable dye lasers. The growing interest recently in the two-photon absorption technique arises from the fact that, like saturated absorption spectroscopy, it allows the Doppler broadening of an absorption line to be completely eliminated, as we explain below.

We consider an atom in the ground state E_i moving with velocity $\underline{v}$ through the standing wave radiation field of a tunable dye laser whose output frequency is $\omega/2\pi$. In its own rest frame the atom sees two electromagnetic waves travelling in opposite directions with angular frequencies of $\omega(1-v_x/c)$ and $\omega(1+v_x/c)$ respectively. If the dye laser frequency is such that the atom can now reach an excited state of energy E_k by simultaneously absorbing one photon from each of these waves, then the following resonance condition must be satisfied:

$$E_k - E_i = \hbar\omega(1 - \frac{v_x}{c}) + \hbar\omega(1 + \frac{v_x}{c}) = 2\hbar\omega. \quad (14.5)$$

The terms depending on the velocity of the atom cancel out, since the absorption of a photon from each of the counter-propagating beams leaves the linear momentum of the atom unchanged. Consequently the Doppler broadening of the two-photon absorption line is completely eliminated and the width of the absorption resonance should theoretically be of the same order of magnitude as the natural linewidth determined by the homogeneous broadening of the upper and lower levels involved. Two-photon absorption experiments require rather more laser power than saturated absorption experiments but have the advantage that at resonance all atoms in the sample contribute to the signal instead of only those atoms which have zero axial velocity. In one sense, however, the two techniques are complementary, for saturated absorption can be applied only to transitions between levels of different parity while two-photon absorption requires that the parities of the levels involved should be the same.

14.4.2. Two-photon spectroscopy of sodium atoms. The elimination of Doppler broadening in multiphoton optical transitions was studied theoretically by Cagnac *et al.* (1973) and soon verified experimentally in sodium vapour by three independent research groups: Biraben *et al.* (1974a), Levenson and Bloembergen (1974), and Hänsch, Harvey, Meisel, and Schawlow (1974). The apparatus used in improved experiments by Biraben *et al.*(1974b) is shown in Fig.14.10. The output

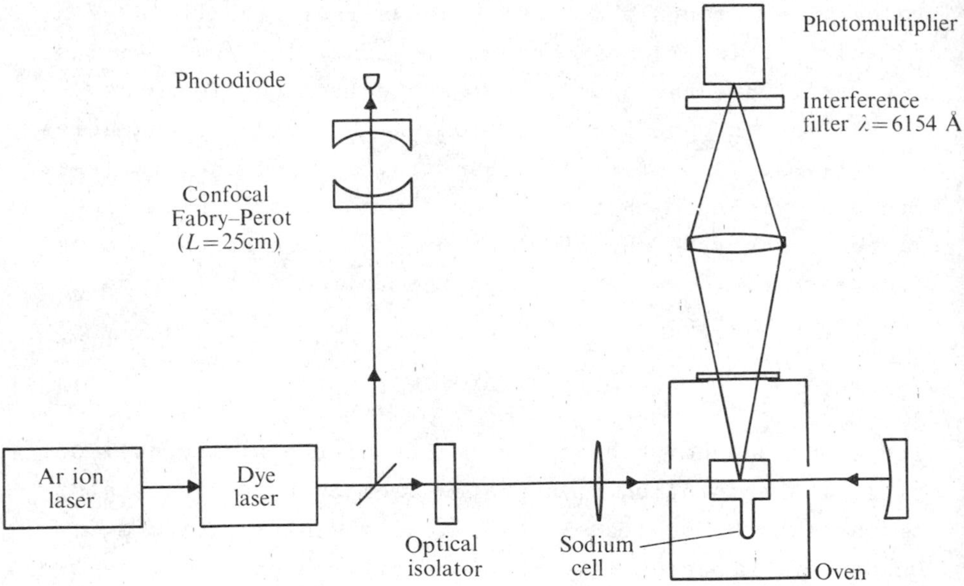

Fig.14.10. Apparatus for two-photon absorption spectroscopy of sodium vapour using a C.W. dye laser. (After Biraben *et al.*(1974b).)

of a 1 W C.W. argon ion laser is used to pump a C.W. dye laser system producing about 10 mW of tunable radiation with an output bandwidth of ≈ 10 MHz. The dye laser radiation is focussed into a cell containing sodium vapour and the transmitted light is then refocussed into the cell using a concave mirror to produce the laser standing wave field necessary for the elimination of the Doppler effect. When the dye laser output is close to 6022 Å, sodium atoms in the 3s ground state can simultaneously absorb two photons and make a transition to the excited 5s state, as shown in Fig.14.11 (a). This resonant two-photon absorption is monitored using the fluorescent light emitted at 6154 Å and 6160 Å as the atoms in the 5s state decay spontaneously by electric dipole transitions to the 3 $^2P_{1/2,3/2}$ levels. A recording of the photomultiplier current obtained as the dye laser frequency is tuned over the two-photon absorption resonance is shown in Fig.14.11(b).

Two narrow peaks of 24 MHz halfwidth are observed superimposed on a broad Gaussian background due to single photon absorption by Na_2 molecules in the vapour. These peaks correspond to the two-photon transitions permitted between the hyperfine levels of the 3s and 5s states and the separation of the peaks can be determined by simultaneously recording the transmission fringes of a confocal Fabry-Perot étalon illuminated by the dye laser output. This measurement permits the hyperfine structure constant for the 5s state to be determined, since that of the 3s ground state is already accurately known from atomic beam resonance experiments.

This experiment clearly demonstrates the power of the new two-photon absorption technique: the structure of high-lying atomic levels having the same parity as the ground state can now be studied at a resolution which is limited only by the natural widths of the levels involved.

<u>14.4.3. Two-photon spectroscopy of atomic hydrogen.</u> Two-photon spectroscopy allows one to detect transitions between levels having the same parity, and obviously it would be very desirable to observe the 1 $^2S_{1/2}$ - 2 $^2S_{1/2}$ transition in atomic hydrogen by Doppler-free two-photon absorption spectroscopy. In this case the upper level is

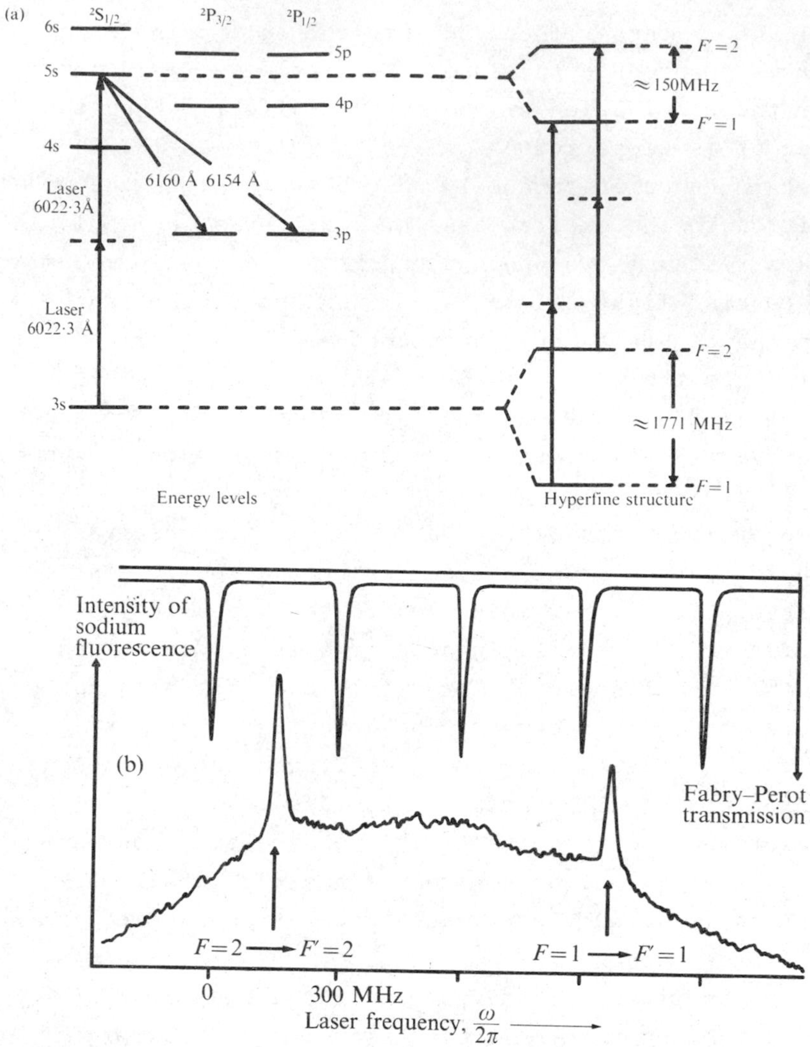

Fig.14.11. (a) Partial energy-level diagram of sodium showing two-photon transitions and hyperfine structure of the 3s and 5s levels. (b) Intensity of sodium fluorescence at 6154 Å and 6160 Å as a function of dye laser frequency showing two-photon resonance signals. (After Biraben *et al.* (1974b).)

metastable, as discussed in Chapter 7, and the ultimate width of the resonance should be extremely narrow. A precise determination of the frequency of this transition would therefore permit an even more accurate measurement of the Rydberg constant than that described in section 14.3.3.

Unfortunately, however, the $1\ {}^2S_{1/2} - 2\ {}^2P_{1/2,3/2}$ Lyman α absorption line, which is almost coincident with the $1\ {}^2S_{1/2} - 2\ {}^2S_{1/2}$ transition, lies in the vacuum ultraviolet at 1215 Å. Even the wavelength of 2430 Å required for two-photon absorption lies outside the range of present narrow-bandwidth dye laser systems. In spite of these difficulties Hänsch *et al.* (1975) have recently succeeded in making the first measurements on this interesting transition using the frequency-doubled output of a pulsed dye laser.

The apparatus used in these experiments is shown schematically in Fig.14.12. The nitrogen laser-pumped dye laser consists of a pressure-tuned oscillator-amplifier sys-

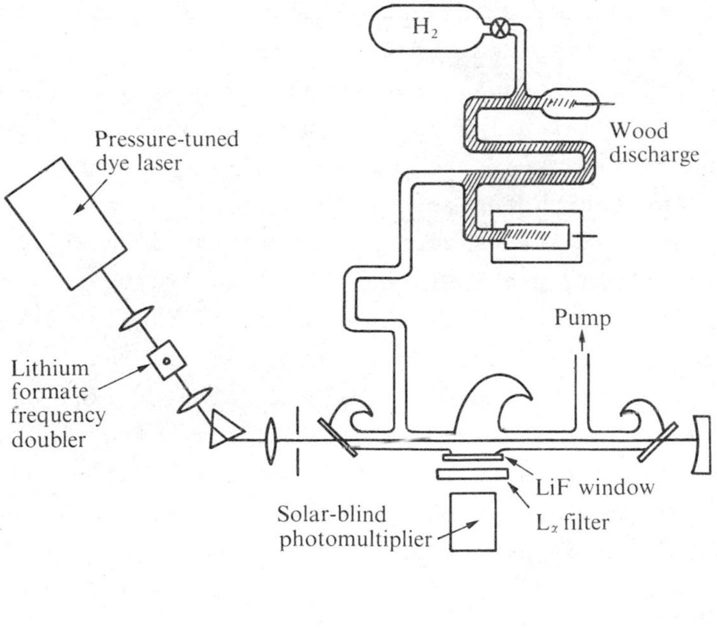

Fig.14.12. Apparatus used for two-photon absorption spectroscopy of atomic hydrogen. (After Hänsch *et al.* (1975).)

tem which has been described in detail by Wallenstein and Hänsch (1974, 1975). This laser generates pulses 10 ns long with a peak optical power at 4860 Å of 30-50 kW and an output bandwidth of 120 MHz. The blue laser output is frequency doubled in a 10 mm long crystal of lithium formate monohydrate, and 2430 Å radiation is generated with a peak power of about 600 W. This radiation is then focussed into the Pyrex absorption chamber through a quartz Brewster-angle end window. Atomic hydrogen or deuterium is pumped through the cell from a discharge source and the recombination of atoms is inhibited by applying a thin coating of syrupy phosphoric acid to the Pyrex walls of the cell.

As before, the standing wave laser field necessary for two-photon absorption is obtained by refocussing the transmitted light back into the cell using an external spherical mirror. The absorption resonance is monitored by observing the intensity of fluorescent light in the Lyman α line which is generated by collisional excitation transfer of hydrogen atoms from the $2\ ^2S_{1/2}$ to the $2\ ^2P_{1/2,3/2}$ levels. The radiation is transmitted through a MgF_2 window and detected by a solar-blind photomultiplier; typical Lyman α counting rates are of the order of 10-20 photons per laser pulse. The two-photon absorption signal recorded in this way on the 1s-2s transition in deuterium is shown in Fig. 14.13(b). Using the Bohr theory of the hydrogen atom it is a simple exercise to show that the Balmer β line coincides with the dye laser fundamental wavelength. Thus the conventional absorption profile of the Balmer β line may be recorded simultaneously by sending a small fraction of the blue dye laser light through a 15 cm long section of the positive column of a Wood's discharge tube, as shown in Fig.14.13(a). A comparison of these two spectra enables the Lamb shift of the ground state $1\ ^2S_{1/2}$ to be determined:

$$\mathscr{S}(1\ ^2S_{1/2}, H) = 8{\cdot}6 \pm 0{\cdot}8 \text{ GHz}$$

and

$$\mathscr{S}(1\ ^2S_{1/2}, D) = 8{\cdot}3 \pm 0{\cdot}3 \text{ GHz}.$$

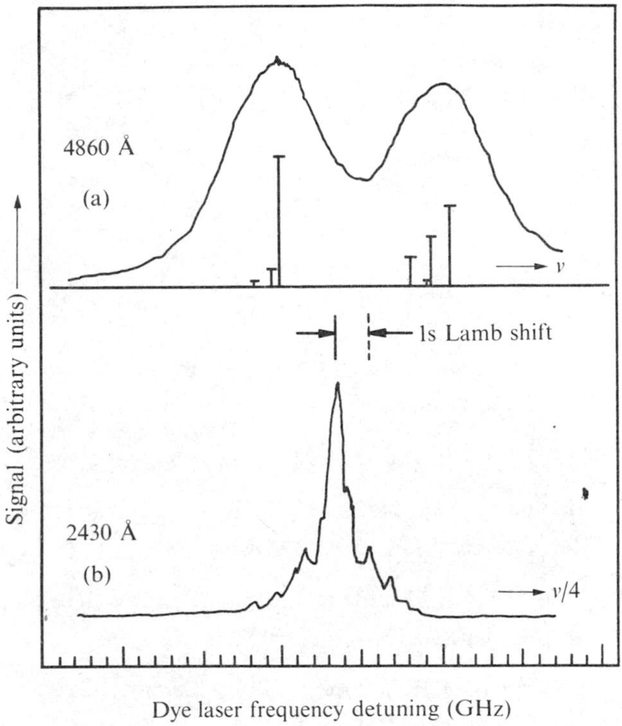

Fig.14.13. (a) Profile of the deuterium Balmer β line observed in absorption together with theoretical fine structure. (b) Lyman α fluorescence signal recorded simultaneously on the 1 $^2S_{1/2}$ - 2 $^2S_{1/2}$ two-photon resonance in deuterium. (After Hänsch *et al.* (1975).)

These results are in good agreement with theory and are rather more accurate than previous measurements made by conventional vacuum ultraviolet spectroscopic techniques. Lee *et al.* (1975), however, have already substantially improved the precision of these experiments by resolving the fine-structure components of the Balmer β line using the saturated absorption technique. It appears, therefore, that the development of atomic and molecular spectroscopy using tunable dye lasers will continue to be very rapid and the reader will of necessity have to consult the current literature to learn the present state of the art.

References

Biraben, F., Cagnac, B., and Grynberg, G. (1974a). *Phys.Rev. Lett.* 32, 643.

__________, __________, and __________ (1974b). *Phys.Lett.* 49A, 71.

Brossel, B., Cagnac, B., and Kastler, A. (1954). *J.Phys. Radium, Paris*, 15, 6.

Cagnac, B., Grynberg, G., and Biraben, F. (1973). *J.Phys., Paris*, 34, 845.

Hänsch, T.W. (1972). *Appl.Opt.* 11, 895.

__________, Harvey, K., Meisil, G., and Schawlow, A.L. (1974). *Opt.Comm.* 11, 50.

__________, Lee, S.A., Wallenstein, R., and Wieman, C. (1975). *Phys.Rev.Lett.* 34, 307.

__________, Nayfeh, M.H., Lee, S.A., Curry, S.M., and Shahin, I.S. (1974). *Phys.Rev.Lett.* 32, 1336.

__________, Shahin, I.S., and Schawlow, A.L. (1971). *Phys. Rev.Lett.* 27, 707.

__________, __________, and __________ (1972). *Nature (Physical Science), Lond.* 235, 63.

Kessler, E.G. (1973). *Phys.Rev.* 7A, 408.

Kibble, B.P., Rowley, W.R.C., Shawyer, R.E., and Series, G.W. (1973). *J.Phys.B.* 6, 1079.

Kogelnik, H., Dienes, A., Ippen, E.P., and Shank, C.V. (1972). *I.E.E.E. J.Quantum Electron.* QE-8, 373.

Kusch, P. (1954). *Phys.Rev.* 93, 1022.

Lamb, W.E. Jr, and Retherford R.C. (1947). *Phys.Rev.* 72, 241.

Lee, S.A., Wallenstein, R., and Hänsch, T.W. (1975). *Phys.Rev. Lett.* 35, 1262.

Levenson, M.D. and Bloembergen, N. (1974). *Phys.Rev.Lett.* 32, 645.

Schäfer, F.P., Schmidt, W., and Volze, J. (1966). *Appl.Phys. Lett.* 9, 306.

Series, G.W. (1971). *Proceedings of the international conference on precision measurements and fundamental constants.* NBS Special Pub. 343. U.S.Government Printing Office, Washington, D.C.

Shank, C.V., Dienes, A., Trozzolo, A.M., and Myer, J.A. (1970). *Appl.Phys.Lett.* 16, 405.

Soffer, B.H. and McFarland, B.B. (1967). *Appl.Phys.Lett.* 10, 266.

Sorokin, P.P. and Lankard, J.P. (1966). *IBM J.Res.Dev.* 10, 162.

Taylor, B.N., Parker, W.H. and Langenberg, D.N. (1969). *Rev.mod.Phys.* 41, 375.

Wallenstein, R. and Hänsch, T.W. (1974). *Appl.Opt.* 13, 1625.

__________ and __________ (1975). *Opt.Comm.* 14, 353.

Warden, J.T. and Gough, L. (1971). *Appl.Phys.Lett.* 19, 345.

Winter, J.M. (1959). *Annls.Phys.* 4, 745.

General references and further reading

The principles and construction of dye lasers have been reviewed by:

Schäfer, F.P. (Ed.) (1973). *Dye lasers.* Springer-Verlag, Berlin.

Snaveley, B.B. (1969). *Proc.Inst.elec.and electron.Eng.* 57, 1374.

Sorokin, P. (1969). *Sci.Amer.* 220, No.2, 30.

The applications of tunable dye lasers to atomic spectroscopy have been reviewed by:

Hänsch, T.W. (1973). *Atomic physics*, Vol.3. (Eds. S.J. Smith and G.K. Walters) Plenum Press, New York.

Hänsch, T.W. (1975). *Atomic physics*, Vol.4. (Eds. G. zu Pulitz, E.W. Weber and A. Winnacker), Plenum Press, New York.

Lange, W., Luther, J., and Steudel, A. (1974). *Adv.atom and molec.Phys.* 10, 173.

Walther, H. (1974). *Physica Scripta* 9, 297.

__________ (Ed.) (1976). *Laser spectroscopy of atoms and molecules*. Springer-Verlag, Berlin.

Further information of interest is contained in the review articles by Bjorkholm, J.E. and by Hänsch, T.W. in:

Proceedings of the international school of physics 'Enrico Fermi'; Non-linear spectroscopy, Course LVI, Varenna 1975. Academic Press, New York.

and in a recent publication:

Shimoda, K. (Ed.) (1976). *High-resolution laser spectroscopy*. Springer-Verlag, Berlin.

15
The Hanle effect and the theory of resonance fluorescence experiments

From the beginning of this century experiments using resonance radiation and resonance fluorescence have been largely responsible for our increasing understanding of excited atoms and their interaction with radiation. In the 1920's the polarization of resonance fluorescence from atoms subjected to external magnetic fields was studied in detail by Hanle and used to measure radiative atomic lifetimes. More recently Brossel and Kastler (1949) and Kastler (1950) pointed out that polarized resonance radiation could be used to produce and detect differences in the populations of the Zeeman sub-states of both excited and ground state atoms. Following these suggestions, the techniques of magnetic resonance were widely applied to a study of bulk samples of free atoms. These experiments enabled detailed information about the Zeeman and hyperfine structure of excited and ground levels of atoms to be obtained, together with measurements of radiative lifetimes and interatomic collisional relaxation rates. The techniques of magnetic depolarization of resonance fluorescence, magnetic-optical double-resonance, and optical pumping of metastable and ground-state atoms now cover an important area in the field of atomic physics and will be disucssed in this and the two subsequent chapters.

This chapter opens with an account of resonance fluorescence and its depolarization by external magnetic fields, a phenomenon now known as the Hanle effect. Experiments of this type in mercury vapour are described and we develop a classical theory to explain the shape of the observed signals. This is followed by a discussion of the applications of this technique to the accurate measurement of atomic lifetimes. For the sake of simplicity the effects of interatomic collisions and of trapping or reabsorption of resonance radiation in these experiments are not considered until sections 16.4 and 16.5.

Next we proceed to develop the theory of resonance fluorescence experiments using the ensemble density matrix to describe the system of atoms. The important concepts of optical and radio-frequency coherence and of the interference of atomic states are discussed in detail. As an illustration of this theory general expressions describing the Hanle effect experiments are obtained. These are evaluated in detail for the frequently employed example of atoms whose angular momentum quantum numbers in the ground and excited levels are $J_g=0$ and $J_e=1$ respectively. Finally resonance fluorescence experiments using pulsed or modulated excitation are described.

We stress the fact that in this chapter we are concerned only with the low field Zeeman effect of the even isotopes of an element. This simplification is not fundamental and is made purely for the sake of clarity of exposition. The effects of hyperfine structure in the odd isotopes and of the decoupling of the electronic and nuclear spins which occurs in large magnetic fields will be considered in Chapter 18.

15.1. Resonance radiation and resonance fluorescence

The simple spectra of many elements are dominated by one or two lines of enormous intensity, the most familiar example being the intense yellow emission from the sodium lamps used in street lighting. These transitions, which in sodium occur at 5896 Å and 5890 Å, are known as the *resonance lines* of the given element and often have absorption oscillator strengths close to unity. Generally they are the spectral lines of longest wavelength connecting the excited levels with the ground state by means of allowed electric dipole transitions.

When the yellow light from a sodium lamp is focussed into an evacuated cell containing sodium vapour in equilibrium with the metal, only a little stray light due to reflection will be seen when the cell is cold. However, at temperatures of the order of 100°C a faint cone of scattered light becomes visible in the body of the cell, especially if the cell is viewed at right-angles to the incident beam. The scattered light observed in this experiment is known as *resonance fluorescence* and was first studied in detail by Wood (1913). As the cell temperature is increased still further the resonance fluorescence rapidly becomes stronger but the edges of the cone of light become increasingly diffuse owing to multiple scattering, until at 200°C the whole bulb begins to glow with resonance radiation. At very high vapour pressures the resonance fluorescence is concentrated in a thin layer close to the front face of the cell and at temperatures above 500°C, specular reflection of the incident light occurs.

This phenomenon can be easily explained in terms of the classical theory: the incident radiation sets up dipole oscillations in the medium which re-radiate electromagnetic waves of the same frequency. Resonance fluorescence is thus a special case of the scattering of light in which the frequency of the incident electromagnetic wave coincidences with the natural frequency of the internal vibrations of the atomic electrons. Both the classical and quantum theory of light scattering are described in detail by Loudon (1973). However, for many of the experiments discussed in this and the two subsequent chapters, the scattering process can be considered to consist of the two separate events of excitation and radiative decay. In our present example we would say that sodium atoms in the ground level $3\,^2S_{1/2}$ absorb photons from the beam of resonance radiation and are raised to either of the $3\,^2P_{1/2,3/2}$ levels. The excited atoms in these levels have a mean lifetime of $\approx 1{\cdot}6 \times 10^{-8}$ s and then decay spontaneously, re-radiating the yellow fluorescent light in all directions.

Resonance fluorescence can be excited in the vapour of many other elements provided that suitable light sources and resonance cells are available. The equipment required in the case of mercury vapour is particularly simple and is shown schematically in Fig.15.1. Mercury has the advantage that a suitable vapour pressure, $\approx 1{\cdot}2 \times 10^{-3}$ Torr, is obtained at room temperature and, in contrast to sodium, it

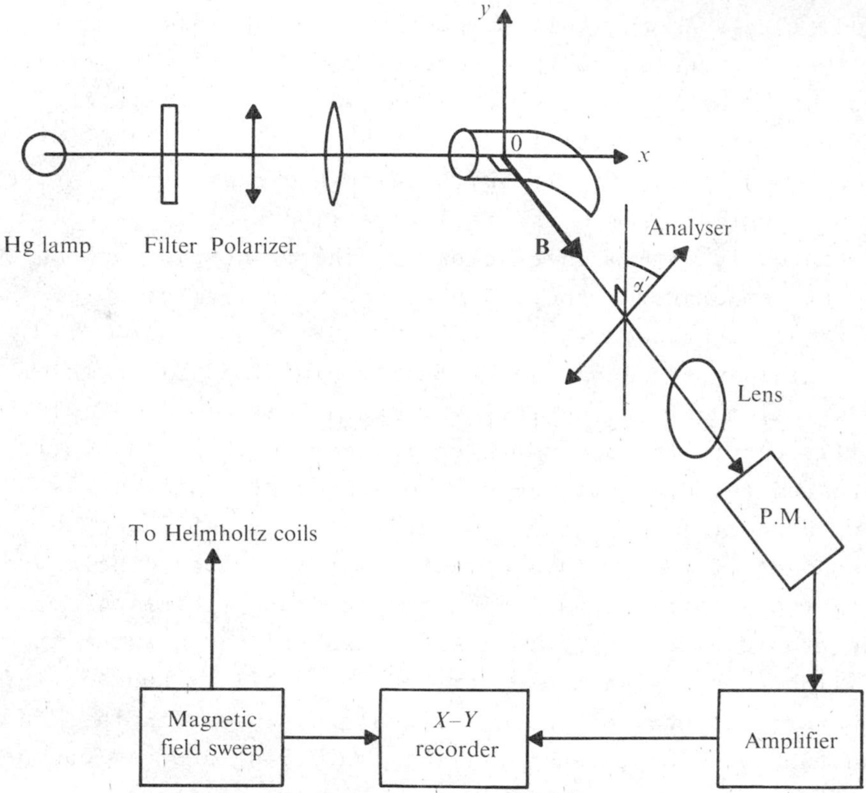

Fig.15.1. Schematic diagram of the apparatus used for resonance fluorescence and Hanle effect experiments on the $^3P_1$ level of mercury.

does not react with Pyrex or quartz. Of more fundamental importance is the fact that natural mercury consists of 70 per cent of even isotopes whereas ^{23}Na has a complex hyperfine structure. The only disadvantage is that the resonance lines of mercury are in the ultraviolet, as Fig.15.2 shows. The transition at 1850 Å is the true resonance line but as it lies in a region where molecular oxygen is strongly absorbing, all the optical paths in this case would have to be in vacuum or filled with nitrogen. Resonance fluorescence experiments using the intercombination line at 2537 Å avoid this difficulty and are relatively easy to carry out. The progressive break down of L-S coupling in the heavier elements which makes this one of the most intense lines in the mercury spectrum was discussed in section 5.5.5.

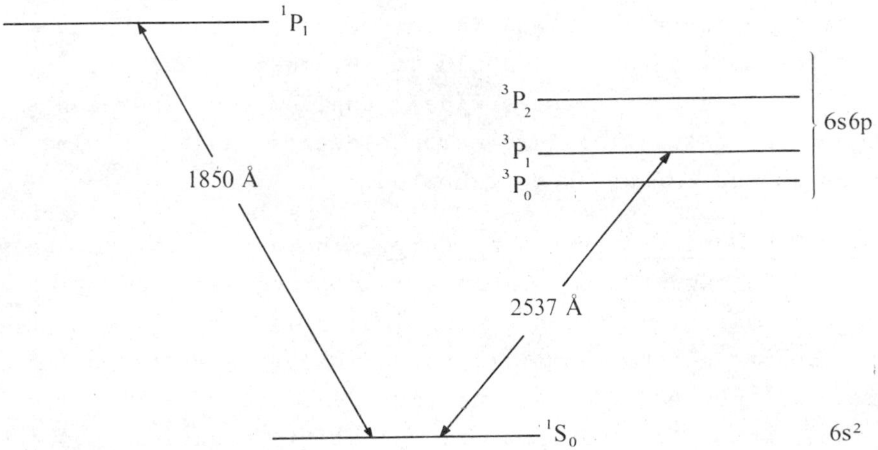

Fig.15.2. Resonance lines and first few energy levels of mercury.

15.2. Magnetic depolarization of resonance radiation - the Hanle effect

15.2.1. Historical introduction. In one of the earliest investigations of resonance fluorescence, Lord Rayleigh (1922) showed that the radiation scattered from mercury vapour was polarized when the atoms were excited by polarized light. Soon afterwards Wood and Ellet (1924) showed that the polarization of the fluorescent light was destroyed by applying small magnetic fields to the resonance cell. Further experimental studies of this effect were made by Hanle (1924), who also worked out a classical theory describing the influence of the magnetic field on the polarization of the resonantly scattered light. He showed that the effect, which now bears his name, could be used to measure the lifetimes of excited atoms. Interest in the field of atomic physics in general and resonance fluorescence in particular declined over the next two decades, but recently the Hanle effect has been developed into one of the most reliable methods for measuring the lifetimes of excited levels of atoms and molecules. Although we shall concentrate our attention on the alkalis and the elements of group IIB, namely Zn, Cd, and Hg, it should be remembered that this technique has been much more widely applied and experiments on the noble gases, the rare earth elements, and molecules such as NO and OH have been reported.

15.2.2. The Hanle effect in mercury. The apparatus necessary to study the magnetic depolarization of resonance radiation in mercury vapour is shown schematically in Fig.15.1. Light from a quartz mercury lamp is passed through a filter which removes all but the 2537 Å line and then through a linear polarizer whose transmission axis is at right angles to the direction of the magnetic field $\underline{B}$. This field is produced in a Helmholtz coil pair which is not shown in the diagram. The resonance radiation is focussed by a quartz lens into a quartz cell placed at the centre of the Helmholtz coils. This resonance cell is prepared by evacuation to pressures below 10^{-7} Torr and is then sealed off after a small quantity of mercury has been distilled into it. For the present we shall assume that the cell contains only even isotopes of mercury. The atoms of the attenuated vapour in the resonance cell are excited by absorption of the linearly polarized light. As we will see in section 15.5, these excited atoms must be described by wavefunctions which are a coherent superposition of the $m_J=\pm 1$ states belonging to the $6\ ^3P_1$ level. Nevertheless after a mean lifetime of $\tau \approx 10^{-7}$ s, the excited atoms decay spontaneously back to the ground level, re-emitting the 2537 Å radiation.

The fluorescent light emitted in a direction at right-angles to the direction of the incident radiation is collected by a fused silica lens, passed through a linear polarizer, and is detected by a photomultiplier. As the magnetic field is slowly varied from -5 G to +5 G, the intensity of fluorescent light observed undergoes pronounced variations, displaying either a Lorentzian or dispersion shape centred at the zero-field position as shown in Fig.15.3. The experimental signal is produced by a change in the state of polarization of the fluorescent light from almost perfect polarization at zero magnetic field to complete depolarization at

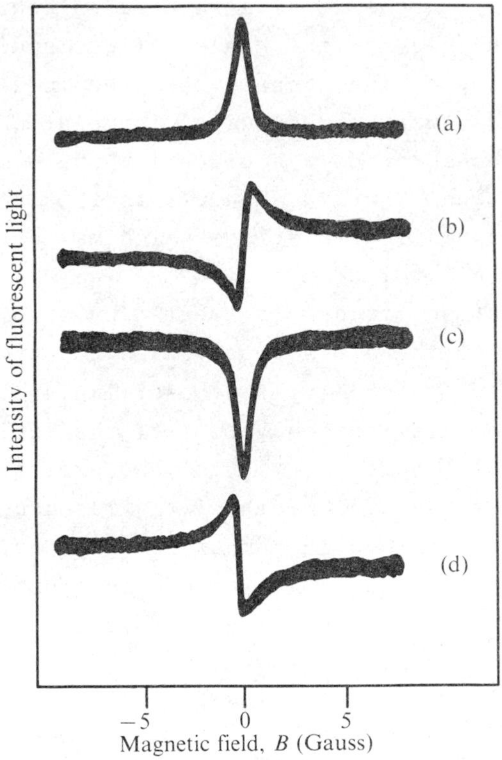

Fig.15.3. Hanle effect or zero-field level-crossing signals for the 6 $^3P_1$ level of mercury. The four curves correspond to rotation of the polarizer in successive steps of $\pi/4$. (After Kibble and Series (1961).)

fields of only 5-10 G. This constitutes the Hanle effect or the magnetic depolarization of resonance radiation. These experiments are also sometimes referred to as zero-field level-crossing experiments for reasons which will be explained in section 15.6.1.

15.2.3. Classical theory of the Hanle effect. The detailed dependence of the intensity of scattered light as a function of the magnetic field strength may be derived by applying the classical model in which an excited atom is represented by a single, harmonically oscillating electron. This approach is useful because it gives a clear picture of the processes responsible for the Hanle effect and correctly predicts the shape of the signals observed in all experiments. The more rigorous quantum-mechanical formulation of the theory will be considered in section 15.5 below, where it will become apparent that the Hanle effect is analagous to the perturbed angular correlations which may be observed in the γ-ray decay of excited nuclei.

The excitation process is treated by assuming that the electron in one of the atoms receives an impulse at the moment of excitation, t_0, which starts it oscillating in a direction specified by the polarization vector of the incident radiation. This simple representation of the excitation process is valid provided that the width of the resonance line emitted by the lamp is very broad in comparison with both the natural linewidth and the Zeeman splitting of the atoms in the resonance cell. The excited electron, oscillating at the angular frequency, ω_0, now radiates in the usual dipole distribution pattern producing an electric field at a point on the axis of observation given by

$$\underline{E}(t) = E(0)\exp\{-i(\omega_0 - i\Gamma/2)(t-t_0)\}\hat{\underline{j}}. \qquad (15.1)$$

For the present we assume that the damping constant is determined only by the radiative lifetime of the excited atom, $\Gamma=1/\tau$. From equation (15.1) we see that at zero magnetic field the fluorescent light is linearly polarized in a direction parallel to the electric vector of the incident radiation.

However, when the magnetic field is finite, the oscillating electron also experiences the Lorentz force of $-e\underline{v}\wedge\underline{B}$ which causes the plane of oscillation to precess about the field direction with the Larmor angular frequency given by

$$\omega_L = g_J \frac{e}{2m} B = (g_J \mu_B B)/\hbar \qquad (15.2)$$

where g_J is the Landé factor for the excited $6\ ^3P_1$ level and $\mu_B = e\hbar/2m$ is the Bohr magneton. In the absence of a polarizer, a hypothetical observer looking towards the resonance cell from the position of the detector would see the electron tracing out the paths shown in Fig.15.4. As the magnetic field strength is progressively increased, the excited electron is able to complete more and more of the rosette before its energy has been radiated away and consequently the electric field emitted suffers progressive depolarization. This precession of the classical electron is associated in the quantum-mechanical theory with the time development of the transverse magnetic moment created by the coherent excitation of the $m_J = \pm 1$ levels.

At time t, the plane of oscillation of the electron makes an angle $\{\omega_L(t-t_0)-\alpha'\}$ with the transmission axis of the polarizer, and the intensity of light recorded by the detector is given by

$$I(B,t_0) = I_0 \exp\{-\Gamma(t-t_0)\}\cos^2\{\omega_L(t-t_0)-\alpha'\}. \qquad (15.3)$$

This damped modulation of the intensity at twice the Larmor frequency has been observed in time-resolved experiments which are described in section 15.8 below. For the moment we are concerned with steady-state experiments in which the intensity of fluorescent light is measured for a sample of atoms that were excited at a constant rate, R, for all times from $t_0 = -\infty$ to the time of observation t. In this case we have

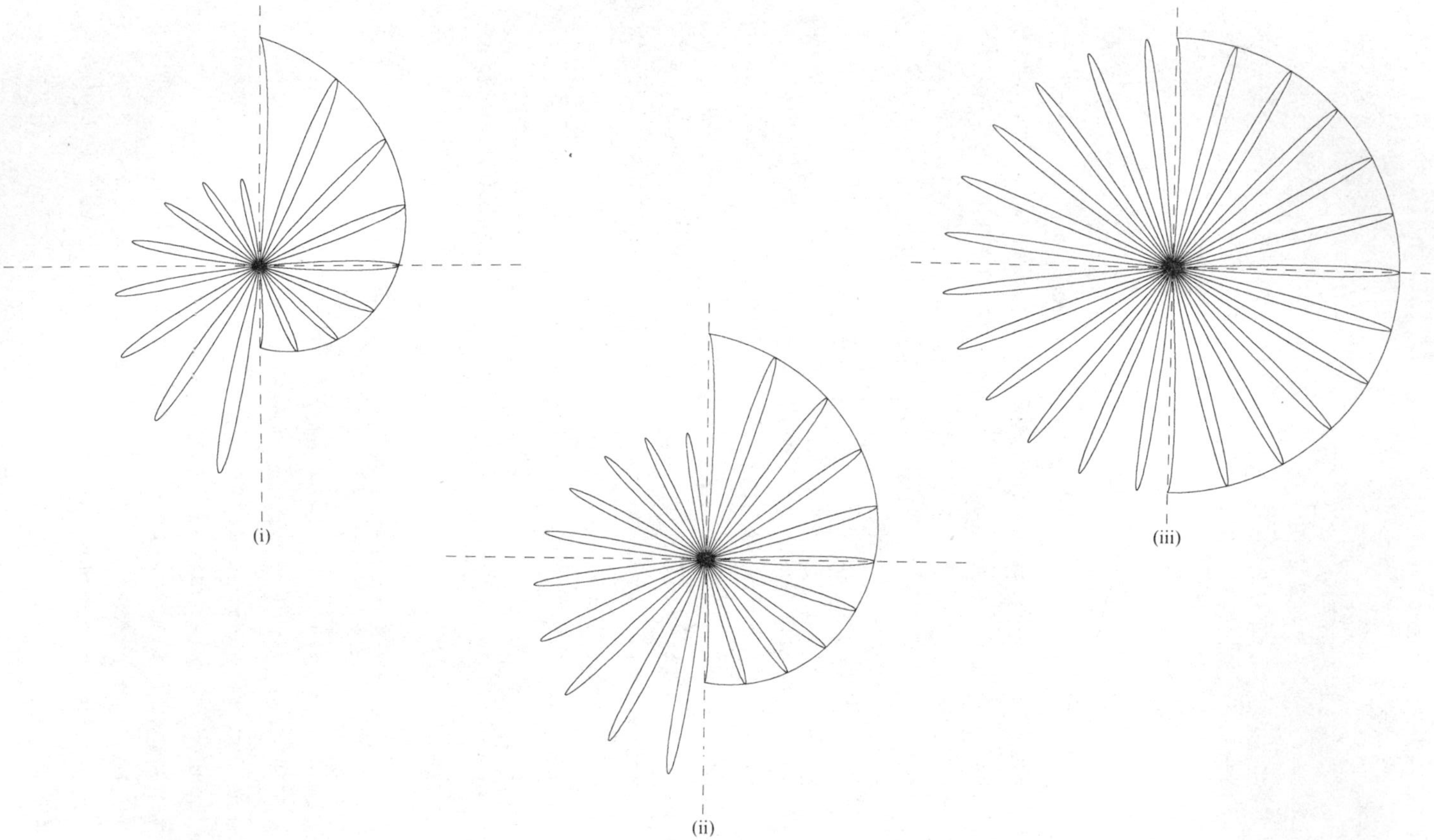

Fig.15.4. Classical explanation of the Hanle effect. The atoms behave like damped harmonic oscillators. The orbits of the oscillating electron for weak, medium, and strong magnetic fields are shown from left to right in the figure.

$$
\begin{aligned}
I(B) &= R\int_{-\infty}^{t} I(B,t_0)\,dt_0 \\
&= \frac{I_0R}{2}\int_{-\infty}^{t} \exp\{-\Gamma(t-t_0)\}[1+\cos 2\{\omega_L(t-t_0)-\alpha'\}]\,dt_0 \\
&= \frac{I_0R}{2}\left\{\frac{1}{\Gamma}+\frac{\Gamma\cos 2\alpha'}{\Gamma^2+4\omega_L^2}+\frac{2\omega_L\sin 2\alpha'}{\Gamma^2+4\omega_L^2}\right\} \quad . \qquad (15.4)
\end{aligned}
$$

Thus in this experimental geometry, the shape of the Hanle effect signal depends on the orientation of the polarizer in the detection beam. The signals have a Lorentzian shape for $\alpha'=0$ or $\pi/2$ and a dispersion shape for $\alpha'=\pi/4$ or $3\pi/4$ when plotted as a function of the magnetic field dependent variable ω_L, as shown in Fig.15.3. It is a remarkable fact that the field-dependent terms of equation (15.4) correctly describe the shape of the Hanle signals for any atomic system, regardless of the angular momentum quantum numbers that characterize the levels. A quantum-mechanical analysis is necessary only for levels with hyperfine structure or for nearly degenerate fine-structure multiplets.

15.2.4. Hanle effect width and determination of lifetimes. Hanle effect experiments can in fact be performed in a variety of geometrical arrangements differing from that shown in Fig.15.1 (Problem 15.1), and in some of these the use of polarized light is unnecessary. However, in most cases it is arranged that the observed signal has the Lorentzian form. In these experiments the field-dependent term falls to half its maximum value at magnetic fields $B_{\pm}$ given by

$$\pm\, 2\omega_L = 2g_J\mu_B B_{\pm}/\hbar = \Gamma,$$

as shown in Fig.15.5. The full width of the signal, ΔB, at half the maximum intensity is therefore connected with the radiative lifetime through the relation

$$g_J\mu_B\,\Delta B/\hbar = \Gamma = 1/\tau. \qquad (15.5)$$

Thus the lifetime of the excited level may be determined

directly from the width of the measured magnetic depolarization curve provided, of course, that the Landé factor g_J of the level is known, (Problem 15.2). This width is usually obtained by a detailed fit of the theoretical lineshape to the observed signal rather than by a single measurement of ΔB at the half-intensity level.

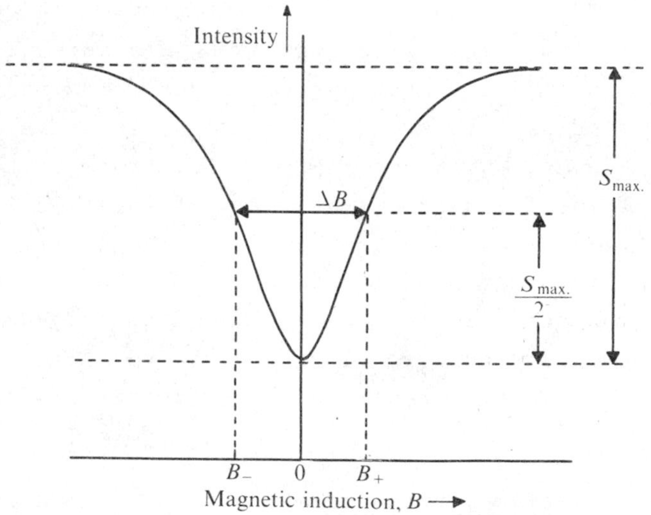

Fig.15.5. Schematic Hanle signal with the definition of ΔB, the full width at half maximum intensity.

It is important to note that although optical radiation is being used in these experiments both for excitation and detection, the width of the depolarization signal is determined by the natural linewidth of the excited level rather than by the Doppler width of the optical line. The detection system is in fact equally sensitive to all frequencies within the line profile and the shape of the optical line is not resolved. The depolarization signal appears effectively at zero (d.c.) frequency and hence we may say that Doppler broadening is theoretically impossible.

Over the last fifteen years this simple and powerful technique has been used to obtain accurate measurements of the radiative lifetimes of a large number of excited levels, some of which are given in Tables 15.1 and 15.2. Representative experiments in this field are described in the papers referred to in these tables. In a number of cases the recent measurements differ significantly from the results obtained by the early workers in this field and tabulated in Mitchell and Zemansky (1966). These discrepancies are generally due to the effects of radiation trapping and collisional broadening which were not thoroughly understood until about 1965. Since these effects also occur in experiments involving optical double resonance, we defer a discussion of them until Chapter 16. We now consider how the range of applicability of Hanle effect experiments has been extended by the use of electron bombardment and excitation from metastable levels.

15.3. Excitation by electron impact

It is difficult to apply the Hanle effect to levels above the resonance level using optical excitation from the ground state because of the low oscillator strength and short wavelength of many of the absorption lines.† Thus in an effort to extend the number of accessible levels several investigators have used electron impact excitation. It is well known that when atoms are excited by a collimated beam of electrons whose energy is at or just above the excitation threshold, the light emitted is strongly polarized. In order to create the atomic polarization necessary for Hanle effect signals using electron bombardment excitation, it is necessary that the electron velocity vector should be perpendicular to the applied magnetic field $\underline{B}$.

In these experiments the experimental chamber usually consists of a small triode valve structure in a glass envelope sealed on to a vacuum and gas handling system, as

---

†The second members of the principal series in the alkalis Na, K, Rb, and Cs form an exception to this rule.

TABLE 15.1.

Lifetimes of the n $^2P_{3/2}$ level of alkali atoms and their isoelectronic ions obtained by resonance fluorescence experiments

Atom or ion	Resonance line wavelength D$_1$ (Å)	D$_2$ (Å)	Lifetime (ns)	Absorption f-value of D$_2$ line	Reference
Li(n=2)	6708	6708	27·8 ± 0·8	0·50	
Na(n=3)	5896	5890	16·0 ± 0·5	0.650	(a)
Mg^+(n=3)	2803	2796	3·67± 0·18	0·64	(b)
K(n=4)	7699	7665	26·0 ± 0·5	0·678	(c)
Ca^+(n=4)	3968	3934	6·72± 0·20	0·66	(b)
Rb(n=5)	7948	7800	25·5 ± 0·5	0·715	(a)
Sr^+(n=5)	4216	4078	6·53± 0·2	0·71	(b)
Cs(n=6)	8944	8521	32·7 ± 1·5	0·666	(a)
Ba^+(n=6)	4934	4554	6·27± 0·25	0·74	(b)

(a) Schmieder *et al.*(1970)
(b) Gallagher (1967)
(c) Schmieder *et al.*(1968).

shown in Fig.15.6. The electrons are produced by a heated oxide-coated cathode and are accelerated into the grid-anode space where atomic excitation takes place. The emitted light is detected in a manner similar to that described in section 15.2, the only difference here being that a monochromator or narrow-band interference filter must be used to isolate the spectral line of interest. This technique has been used, for instance, in Hanle effect lifetime measurements in helium and neon (Faure *et al.* 1963).

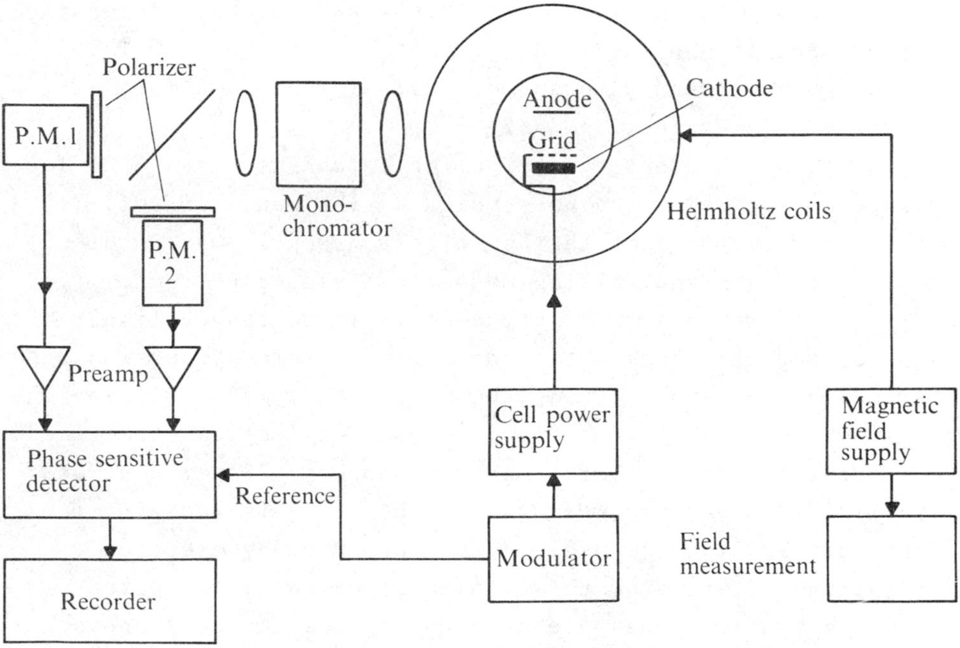

Fig.15.6. Sample cell and electrode structure for Hanle effect and optical double-resonance experiments using electron impact excitation. (After Pebay-Peyroula (1969).)

However, electron impact excitation in hot-cathode triodes is not suitable for elements that poison the cathode or require high temperatures for vapourization. To overcome these difficulties Lombardi and Pebay-Peyroula (1965) developed a method using electron excitation in an intense r.f. discharge. The experimental chamber in this case consists of a thin cylindrical cell placed between two flat condenser plates which form the termination of a $\lambda/4$ length of r.f.

line. The principle of the technique is that at low enough pressures and with high r.f. electric field strengths between the plates of the condenser, the electrons in the discharge oscillate in a direction parallel to the applied electric field. Typical operating parameters are 0·1 Torr pressure in the cell with r.f. electric field strengths of 250 V/cm at 250 MHz. The electron mean free path is of the order of 1 mm and the mean energy is around 20 eV. This technique has been used successfully for Hanle effect experiments in Ca, Cd, and He.

Although electron excitation extends the range of levels and elements which can be studied by the Hanle effect, it suffers from the fact that an electron, moving at right-angles to the depolarizing magnetic field, experiences the Lorentz force $-e\underline{v}\wedge\underline{B}$. Its trajectory is no longer linear but curved and this causes the admixture of some dispersion shaped signal on the wings of an otherwise Lorentzian profile. The correction which must be made for this effect can sometimes change the measured lifetime by as much as 20 per cent and so limits the accuracy of this technique. In addition to this difficulty, the Hanle effect produced by electron excitation suffers from the problem of cascade. It often happens that the mean electron energy must be well above threshold to obtain sufficient signal. In this situation levels above that of interest are excited and the atomic polarization created in these is carried down to lower levels by radiative cascade. This situation has been treated theoretically by Nedelec (1966), and the results show that the observed signal will then be the product of the Hanle signals expected on the cascading and observed transitions, as shown in Fig.15.8. If the lifetimes of the two levels are sufficiently different the two contributions to the experimental curve may be resolved, but this is often not the case.

Hanle effect signals have also been observed in the light emitted by d.c. rather than r.f. gas discharges. Using the apparatus shown in Fig.15.7, Carrington and Corney (1971) were able to show conclusively that, in the neon discharge at pressures in the range 1-10 Torr, the signals were created by optical excitation from highly populated metastable levels by absorption of the light produced within the discharge itself. Although the exciting radiation is necessarily unpolarized, the geometrical anisotropy of the excitation, which in this case is due to the use of a long, narrow discharge tube, and a suitable orientation of the magnetic field allow useful Hanle effect signals to be obtained. However, at pressures below 0.1 Torr, the mean free path of electrons within the discharge becomes sufficiently large to

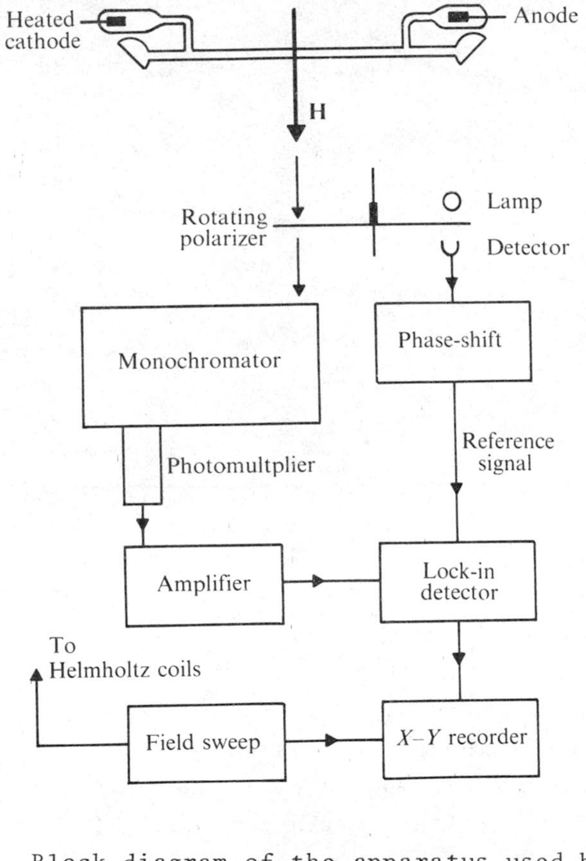

Fig.15.7. Block diagram of the apparatus used by Carrington and Corney (1971) in studies of the Hanle effect in noble gases excited by d.c. discharges.

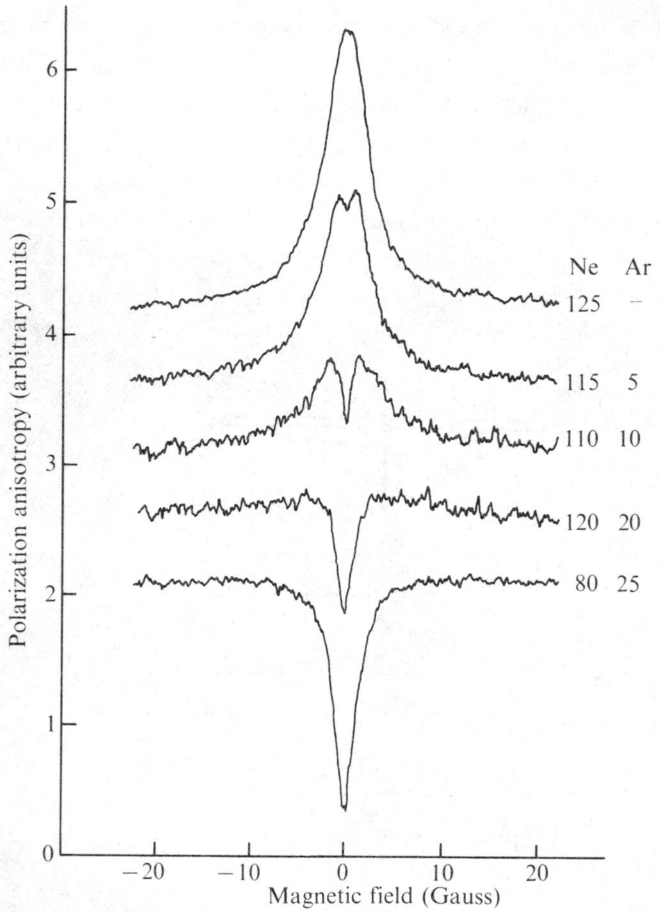

Fig.15.8. Hanle effect on the $1s_5$ - $2p_9$ transition of neon at 6402 Å showing inversion of the signal at low pressure due to cascade of alignment from higher excited levels. Mixtures of argon and neon were used in the gas discharge and the partial pressures of the two components are shown in the diagram in mTorr. (After Carrington (1972).)

produce direct alignment, often in levels above those studied. Radiative cascade then feeds this alignment into the lower levels, just as in the case of electron excitation, and the resulting signals have a complex shape as shown in Fig.15.8.

Thus although the use of discharges and electron bombardment has widened the range of applicability of Hanle effect experiments, the signals are often weak and their interpretation is complicated by the large number of competing processes which occur in these systems.

15.4. Range and accuracy of lifetime measurements

The atomic lifetimes measured by the resonance fluorescence techniques range from $(2{\cdot}0 \pm 0{\cdot}2) \times 10^{-5}$ s for the 4s4p $^3P_1$ level of zinc to $(5{\cdot}69 \pm 0{\cdot}23) \times 10^{-10}$ s for the 1s2p $^1P_1$ level of helium, although the majority of measurements would fall in the range 10^{-6}-10^{-9} s. The range is limited at long lifetimes by the low resonance scattering cross-section of levels with small f-values and by the effects of wall collisions in the resonance cell. Magnetic field inhomogeneities over the scattering cell are also a problem, thus for the $^3P_1$ level of zinc a field inhomogeneity of 1 mG over the cell would broaden the signal by approximately 10 per cent in the absence of wall collisions. Although it is possible to construct Helmholtz coils of the required homogeneity, it is difficult to eliminate the effects of stray a.c. and d.c. magnetic fields at levels below 1 mG.

At the short lifetime end of the range, the application of the Hanle effect becomes more difficult because the fields required for appreciable depolarization also produce Zeeman splittings of the states of the absorbing atom which are a substantial fraction of the Doppler width of the source. The approximation that the spectral profile of the exciting radiation is essentially flat over the absorption profile of the cell is no longer valid and it is found that the intensity of fluorescent radiation decreases on the wings of the Hanle signal. This effect has been observed in Hanle effect experiments in Hg, He, and Xe. It can be reduced to some extent by artificially broadening the lamp profile by applying a magnetic field of a few hundred gauss to the source of resonance radiation.

The relatively simple apparatus, high sensitivity, and the selective nature of the optical excitation used in many resonance fluorescence experiments combine to make this method the most accurate method for measuring atomic lifetimes. In many cases the experimental measurements may be made at densities which are so low that the effect of resonance trapping and collision broadening are completely absent. In these cases the experimental results are usually quoted with errors in the range 3-5 per cent. The method is suitable for precision lifetime measurements and may for this reason allow a set of relative oscillator strengths obtained by the absorption or emission methods to be placed on a reliable absolute basis.

The extension of the Hanle effect measurements to non-resonance levels has usually resulted in increased experimental difficulties and a consequent increase in the uncertainty of the lifetime measurements, although the results are still usually accurate to better than 10 per cent.

15.5. Theory of resonance fluorescence experiments

15.5.1. Introduction of the density matrix. We now wish to prepare the basis of the theoretical analysis which will be used to describe the resonance fluorescence experiments discussed in this and the two subsequent chapters. We are concerned with experiments in which a sample of atoms is illuminated with resonance radiation and a signal is obtained by monitoring either the intensity or the polarization of the fluorescent light. Alternatively the amount of light absorbed by the sample can be measured as in the optical pumping experiments described in Chapter 17. From the survey of quantum mechanics given in Chapter 3, we know that the most general description of the ith atom in the sample involves a linear superposition of eigenfunctions with time-dependent coefficients:

$$\Psi_i(t) = \sum_m a_m^{(i)}(t)\,|m\rangle. \qquad (15.6)$$

TABLE 15.2.

Lifetimes of the nsnp $^1P_1$ *and* $^3P_1$ *levels of Group IIB elements and the absorption f-values of the corresponding resonance lines*

Atom	Multiplicity of transition	Resonance line wavelength (Å)	Lifetime (ns)	Absorption f-value	Reference
Zn(n=4)	Singlet	2139	1·41 ± 0·04	1·46	Lurio *et al.* (1964)
	Triplet	3076	$(2{\cdot}0 \pm 0{\cdot}2) \times 10^4$	$2{\cdot}13 \times 10^{-4}$	Byron *et al.* (1964b)
Cd(n=5)	Singlet	2288	1·66 ± 0·05	1·42	Lurio and Novick (1964)
	Triplet	3261	$(2{\cdot}39 \pm 0{\cdot}04) \times 10^3$	$2{\cdot}00 \times 10^{-3}$	Byron *et al.* (1964a)
Hg(n=6)	Singlet	1850	1·36 ± 0·05	1·18	Lurio (1965)
	Triplet	2537	$(1{\cdot}18 \pm 0{\cdot}02) \times 10^2$	$2{\cdot}45 \times 10^{-2}$	Barrat (1959)

The expectation value of any physical observable, represented for instance by the operator $\mathcal{M}$, is then given by

$$\langle \Psi_i | \mathcal{M} | \Psi_i \rangle = \sum_{m,n} \rho_{mn}^{(i)} \langle n | \mathcal{M} | m \rangle \tag{15.7}$$

where

$$\rho_{mn}^{(i)} = a_m^{(i)}(t)\, a_n^{(i)}(t)^*. \tag{15.8}$$

We see that the observable properties of this atom are determined not so much by the individual coefficients $a_m^{(i)}(t)$ but by the products of the amplitudes represented by $\rho_{mn}^{(i)}$. These may conveniently be arranged in the form of a matrix. If the atomic wavefunction is changed, then the expectation value of $\mathcal{M}$ is also changed because the matrix of the coefficients $\rho_{mn}^{(i)}$ has been altered. However, the matrix elements of the operator $\langle n | \mathcal{M} | m \rangle$ will remain the same. Alternatively if we wish to calculate the expectation value of a different operator, the matrix $(\rho_{mn}^{(i)})$ is unaltered but the operator matrix elements appearing in equation (15.7) will be changed. Thus the matrix $(\rho_{mn}^{(i)})$, called the *atomic density matrix*, would seem to be a more useful description of the system than the original atomic wavefunction.

In resonance fluorescence experiments, however, we are never able to study just a single atom, rather we are forced to investigate the properties of a sample containing N atoms. The measurable properties of this sample are then given in terms of the average values, $\langle \mathcal{M} \rangle$, of a set of physical observables taken over the ensemble of independent atoms where

$$\langle \mathcal{M} \rangle = \frac{1}{N} \sum_{i=1}^{N} \langle \Psi_i | \mathcal{M} | \Psi_i \rangle$$

$$= \frac{1}{N} \sum_{i=1}^{N} \sum_{m,n} \rho_{mn}^{(i)} \langle n | \mathcal{M} | m \rangle. \tag{15.9}$$

For this reason it is now convenient to use the average values of the products of the probability amplitudes to define the elements of the *density matrix* ρ of the *ensemble*:

$$\rho_{mn} = \langle m|\rho|n\rangle = \frac{1}{N}\sum_{i=1}^{N}\rho_{mn}^{(i)} . \tag{15.10}$$

From equations (15.9) and (15.10) we see that the mean value of the observable $\mathscr{M}$ can be expressed in terms of the ensemble density matrix by

$$\langle\mathscr{M}\rangle = \sum_{n,m}\langle m|\rho|n\rangle\langle n|\mathscr{M}|m\rangle = \mathrm{Tr}(\rho\mathscr{M}). \tag{15.11}$$

where Tr indicates that the trace of the product matrix is to be taken.

Although the density matrix is less familiar than the description in terms of atomic wavefunctions, it has the advantage that its elements have an immediate physical significance. For instance the probability of finding an atom of the vapour in the state $|m\rangle$ is given by ρ_{mm}. These diagonal elements of the density matrix also determine the magnetization of the sample when it is placed in an external magnetic field. The description of the ensemble in terms of the density matrix means that it is not necessary to know the wavefunctions of individual atoms, and indeed it may be quite impossible to write down a wavefunction for each atom owing to the effects of interatomic forces or of interactions between the atoms and the radiation fields (Problem 15.3).

<u>15.5.2. The Liouville equation.</u> We now derive the differential equation which controls the time development of the density matrix operator ρ. From equations (15.8) and (15.10) the rate of change of the matrix element ρ_{mn} is given by

$$\frac{\partial\rho_{mn}}{\partial t} = \frac{1}{N}\sum_{i=1}^{N}\{\dot{a}_m^{(i)}a_n^{(i)*} + a_m^{(i)}\dot{a}_n^{(i)*}\} . \tag{15.12}$$

Since the time dependence of the probability amplitude $a_m^{(i)}(t)$ is given approximately by $\exp(-iE_m t/\hbar)$, we see that off-diagonal elements of the density matrix are periodic functions of time with characteristic angular frequencies given by $(E_m - E_n)t/\hbar$. The detailed development of the atomic wavefunction is controlled by the Schrödinger equation

$$i\hbar \frac{\partial \Psi}{\partial t} = \mathcal{H} \Psi$$

once the Hamiltonian operator is specified. Substituting equation (15.6) into this equation gives

$$\dot{a}_m^{(i)} = \frac{1}{i\hbar} \sum_k \langle m|\mathcal{H}|k\rangle a_k^{(i)} \tag{15.13}$$

Using this result in equation (15.12) together with the definition of the density matrix, equation (15.10), and the Hermitian property of $\mathcal{H}$ leads to the required differential equation:

$$\frac{\partial \rho_{mn}}{\partial t} = \frac{1}{i\hbar} \sum_k (\mathcal{H}_{mk}\rho_{kn} - \rho_{mk}\mathcal{H}_{kn}). \tag{15.14}$$

In operator form, equation (15.14) becomes

$$\frac{\partial \rho}{\partial t} = \frac{1}{i\hbar} [\mathcal{H},\rho] \tag{15.15}$$

where the square bracket denotes the commutator product $(\mathcal{H}\rho - \rho\mathcal{H})$. This result is known as the *Liouville equation*. The theory of resonance fluorescence experiments therefore reduces to a study of the solutions of the Liouville equation for appropriate forms of the Hamiltonian operator $\mathcal{H}$.

15.5.3. Application of the density matrix to the theory of resonance fluorescence. Much of the theoretical development of the density matrix which follows applies equally well to all types of resonance fluorescence experiment. We shall, however, use the Hanle effect as a simple example of the general treatment. We start by assuming that the Hamiltonian $\mathcal{H}$, is the sum of a large time-independent operator $\mathcal{H}_0'$ and smaller perturbations $\mathcal{H}_1$ and $\mathcal{H}_2$ which represent the effects of optical excitation and radiative decay respectively. In this situation the Liouville equation becomes

$$\frac{\partial\rho}{\partial t} = \frac{1}{i\hbar}[\mathcal{H}_0',\rho] + \frac{1}{i\hbar}[\mathcal{H}_1,\rho] + \frac{1}{i\hbar}[\mathcal{H}_2,\rho]$$

$$= \frac{1}{i\hbar}[\mathcal{H}_0',\rho] + \frac{d}{dt}^{(1)}\rho + \frac{d}{dt}^{(2)}\rho\ . \qquad (15.16)$$

The time-independent Hamiltonian $\mathcal{H}_0'$ is itself the sum of an operator $\mathcal{H}_0$, which determines the unperturbed eigenvalues and eigenfunctions of the atomic electrons, and an operator $\mathcal{H}_{pert}$, which describes the interaction of the atom with static external magnetic fields. In Chapters 16 and 17 additional terms will be introduced into equation (15.16) to account for the effects of magnetic resonance and optical pumping respectively.

In resonance fluorescence experiments we are usually interested in just two electronic levels, one of which is normally the electronic ground state. The wavefunctions required for the expansion of this restricted density matrix consist of the ground-state basis functions $|\mu\rangle$ and those of the excited state represented by $|m\rangle$. These wavefunctions are eigenfunctions of the angular momentum operators $\underline{J}^2$ and J_z with angular momentum quantum numbers given by (J_g,μ) and (J_e,m) respectively.

<u>15.5.4. The excitation process.</u> The perturbation, $\mathcal{H}_1$, represents the interaction between the atoms and the incident flux of resonance radiation. If, for the sake of simplicity the radiation field is treated classically then the perturbation has the form $\mathcal{H}_1 = -\underline{p}.\underline{E}(t)$. Most of the experiments with which we shall be concerned have been performed with conventional resonance lamps. The radiation from these sources has a wide spectral bandwidth, implying very limited temporal coherence, and the spatial coherence is also very small. Thus in resonance fluorescence experiments an ensemble electric dipole moment which oscillates at the optical frequency of the incident radiation cannot be created in the sample. All off-diagonal elements of the density matrix of the form $\rho_{\mu m}$ are therefore identically zero and we say that

the density matrix possesses no *optical coherence.*† In these experiments the density matrix therefore reduces to the sum of a ground-state part ρ_g and an excited state part ρ_e. Since at present we are chiefly interested in the time development of ρ_e, we will assume for the sake of simplicity that ρ_g has only diagonal matrix elements, i.e. $\rho_{\mu\mu'}=0$ unless $\mu=\mu'$.

The wide spectral bandwidth of the source of resonance radiation also means that the transition of the atom from the ground to the excited level is effectively instantaneous on the time scale during which the excited atoms evolve significantly. Thus the effect of the perturbation $\mathcal{H}_1$ can be treated in terms of a rate process, as discussed in detail in section 9.3. When the incident radiation is in a pure state of polarization, i.e. either σ^+, σ^- or π polarization, the atoms of the sample are excited to a pure Zeeman sub-state of the excited level. The rate at which atoms are generated in the sub-state $|m\rangle$ by optical excitation is then

$$\frac{d}{dt}^{(1)} \rho_{mm} = \sum_{\mu} P_{\mu m}\rho_{\mu\mu} \tag{15.17}$$

where $P_{\mu m}$ is the absorption transition probability given by equation (9.40). However, this result does not give a complete description of the excited state density matrix since it determines only the diagonal matrix elements of ρ_e. In Hanle effect experiments these terms are responsible for a background of resonance fluorescence which is field-independent. The magnetic depolarization signal is observed

---

†Finite values of the off-diagonal elements $\rho_{\mu m}$ would imply that the phases of the oscillating electric dipole moments induced in different atoms were coherently related to one another. Such a situation can only be achieved if the atoms are stationary and are interacting with a beam of radiation of high spatial and temporal coherence, as for instance obtained from a laser.

only when the off-diagonal elements of ρ_e are also non-zero. Since the concepts involved in this case are so important they are considered separately in the following section.

<u>15.5.5. Radio-frequency coherence and interference of atomic states.</u> The off-diagonal elements of the excited state atomic density matrix, $\rho^{(i)}_{mm'}$, are complex numbers whose phases ϕ_i are determined by the phase differences of the probability amplitude coefficients $a^{(i)}_m$ and $a^{(i)}_{m'}$, as shown by equation (15.8). When the average is performed over the ensemble of N atoms in equation (15.10) two different physical situations may be distinguished. In the first the ensemble of excited atoms is prepared in such a way that all values of the phase ϕ_i are equally probable. For sufficiently large N this automatically leads to the vanishing of the off-diagonal elements, $\rho_{mm'}$, of the ensemble density matrix.

In the second case the distribution of phases is not isotropic and the off-diagonal matrix elements are consequently finite, $\rho_{mm'} \neq 0$. For this to happen it is necessary that the excited atoms are prepared in such a way that the phase difference between the interfering states $|m\rangle$ and $|m'\rangle$ is the same for all atoms of the ensemble. One method of achieving this is by means of optical excitation using light which is a mixture of different polarization components. If the polarization vector of the incident radiation, $\underline{\hat{e}}$, is expanded in terms of spherical unit vectors in the form

$$\underline{\hat{e}} = \sum_{q=-1}^{+1} (-)^q \, \mathcal{E}_q \, \underline{\hat{\varepsilon}}_{-q} \tag{15.18}$$

where $\hat{\varepsilon}_0$ coincides with the axis of quantization, then we require at least two of the coefficients $\mathcal{E}_q$ to be non-zero. In this case the rate at which the off-diagonal elements of the excited state density matrix are created is obtained by generalizing equations (9.40) and (15.17), giving:

$$\frac{d}{dt}{}^{(1)}\rho_{mm'} = \frac{\pi U(\omega)}{\varepsilon_0 \hbar^2} \sum_{\mu} \langle m|\underline{\hat{e}}.\underline{D}|\mu\rangle \langle \mu|\underline{\hat{e}}^*.\underline{D}|m'\rangle \rho_{\mu\mu} \, . \tag{15.19}$$

To avoid confusion with the density operator, it is now more convenient to represent the energy density of the incident radiation by $U(\omega)$ and for similar reasons the electric dipole moment operator for the atom is taken as $\underline{D} = -\sum_i e\underline{r}_i$. In contrast to the situation considered in section 15.5.4, the radiation now excites the atoms to a superposition of different Zeeman sub-states. In this case we may say that the *polarization* of the beam of radiation is *coherent* and that this coherence is transferred to the atoms.

Moreover, radiation which is polarization coherent possesses a finite component of angular momentum at right-angles to the magnetic field $\underline{B}$ and this angular momentum is also transferred to the atoms at the moment of excitation. Thus finite values of the off-diagonal matrix elements $\rho_{mm'}$ also imply that there exist finite expectation values of the transverse components of the angular momentum operators $\langle J_\pm \rangle, \langle J_\pm^2 \rangle$ etc. and of the transverse components of the magnetic dipole moment, $\langle M_x \rangle$ and $\langle M_y \rangle$. Again we should note that this situation only occurs when the angular momentum components of different atoms in the direction perpendicular to the axis of quantization are coherently phased rather than being oriented at random.

The time-dependence associated with the off-diagonal elements $\rho_{mm'}$ is given by $\exp\{-i(E_m - E_{m'})t/\hbar\}$ and can be regarded as a manifestation of the interference between the atomic states $|m\rangle$ and $|m'\rangle$. When these Zeeman sub-states belong to the same level of the atomic fine or hyperfine structure, the angular frequency $(E_m - E_{m'})/\hbar$ lies in the radio band and we say that the ensemble of atoms possesses *radio-frequency or Hertzian coherence*. Since Hanle effect experiments are performed at zero (d.c.) frequency, this Hertzian coherence manifests itself as a change in the polarization of the fluorescent light rather than as a radio-frequency beat. Radio-frequency modulation of the fluorescent light can be observed if the Hertzian coherence is generated using pulsed or modulated excitation, as described in sections 15.8 and 15.9 below. Light beats can also be observed when the radio-frequency coherence is generated by magnetic resonance in excited levels, a topic which is considered in the following chapter.

<u>15.5.6. The relaxation processes.</u> The second perturbation term in equation (15.16) represents the effect of relaxation processes. At present we are concerned only with the excited-state density matrix and assume that the relaxation is due only to spontaneous emission. Since this is again an essentially random process, being triggered by the zero-point fluctuations of the vacuum radiation fields, the effect of $\mathcal{H}_2$ can be represented as a rate process and we have

$$\frac{d}{dt}^{(2)} \rho_{mm'} = -\Gamma \rho_{mm'} \tag{15.20}$$

where $\Gamma = \sum_i A_{ki}$ is the spontaneous decay rate. Interatomic collisions and resonance trapping of the fluorescent radiation make additional contributions to the relaxation of the excited state but a detailed discussion of these effects is reserved until sections 16.4 and 16.5.

15.6. Theory of the Hanle effect

<u>15.6.1. The excited-state density matrix.</u> We now apply the general formalism developed in the previous paragraphs to the particular case of the Hanle effect. In these experiments the excited atoms are subjected to a static external magnetic field $\underline{B}$ whose direction is chosen as the axis of quantization. In the absence of hyperfine structure the time-independent Hamiltonian for the system becomes

$$\mathcal{H}_0' = \mathcal{H}_0 + g_J\mu_B BJ_z = \mathcal{H}_0 + \hbar\omega_L J_z \tag{15.21}$$

where J_z is the z-component of the dimensionless angular momentum operator introduced in section 3.6 and ω_L is the Larmor angular frequency defined by equation (15.2). The Hamiltonian operator $\mathcal{H}_0$ determines the unperturbed energy levels of the atom. Using equation (15.21) in the Liouville equation (15.16), we discover that the elements of the excited-state density matrix are solutions of the equation

$$\frac{\partial \rho_{mm'}}{\partial t} = -i\omega_L \langle m|[J_z,\rho]|m'\rangle + \frac{d}{dt}^{(1)} \rho_{mm'} + \frac{d}{dt}^{(2)} \rho_{mm'} . \quad (15.22)$$

Substituting the explicit expressions for the excitation and relaxation rate processes from equations (15.19) and (15.20) and using the fact that the states $|m\rangle$ and $|m'\rangle$ are eigenfunctions of the operator J_z, we have

$$\frac{\partial \rho_{mm'}}{\partial t} + i\omega_L(m-m')\rho_{mm'} + \Gamma\rho_{mm'} = \frac{\pi U(\omega)}{\varepsilon_0 \hbar^2} \sum_{\mu} F_{mm'}\, \rho_{\mu\mu} \quad (15.23)$$

where $F_{mm'} = \langle m|\hat{\underline{e}}.\underline{D}|\mu\rangle\langle\mu|\hat{\underline{e}}^*.\underline{D}|m'\rangle$ is often called the excitation matrix. We again note that the off-diagonal elements of the excited-state density matrix are periodic functions of time with an associated angular frequency of $\omega_L(m-m')$. It is interesting to note that the time-development of the transverse components of the excited-state magnetic dipole moment is identical to the Larmor precession of the electron oscillators introduced in the classical model of section 15.2.

In most experiments the excited atoms are created by a lamp whose intensity is independent of time and we can obtain the steady-state density matrix by setting $\dot{\rho}_{mm'}=0$ in equation (15.23), giving

$$\rho_{mm'} = \frac{\pi U(\omega)}{\varepsilon_0 \hbar^2} \sum_{\mu} \frac{F_{mm'}\rho_{\mu\mu}}{\Gamma + i\omega_L(m-m')} . \quad (15.24)$$

We see that appreciable Hertzian coherence, implying large off-diagonal elements of ρ_e, is created only when the angular frequency $\omega_L(m-m')$ is comparable to or less than the relaxation rate Γ. Since Γ determines the width of the excited-state energy levels in angular frequency units, this coherence is destroyed when the magnetic field is large enough that different Zeeman sub-levels no longer overlap. This is the reason why Hanle effect investigations are sometimes known as zero-field level-crossing experiments.

15.6.2. The polarization of the fluorescent light. The Hanle effect signal is usually obtained by measuring the intensity of the fluorescent light with polarization vector $\underline{\hat{e}}'$ emitted in some well-defined direction $\underline{\hat{r}}$. From equations (2.70) and (5.8) it can be shown (Problem 15.4) that the intensity of light with this polarization emitted when an excited atom in the sub-state $|m\rangle$ decays to the ground-state sub-level $|\mu'\rangle$ is

$$\frac{dI}{d\Omega} = \frac{\omega^3}{8\pi^2\varepsilon_0\hbar c^3}\,|\langle m|\underline{\hat{e}}'.\underline{D}|\mu'\rangle|^2 \qquad (15.25)$$

when measured in terms of photons/s-steradian. To include the case of excited atoms described by an arbitrary density matrix, we generalize equation (15.25) by forming the fluorescent light monitoring operator L_F defined by

$$L_F = \frac{\omega^3}{8\pi^2\varepsilon_0\hbar c^3}\sum_{\mu'}\underline{\hat{e}}'.\underline{D}|\mu'\rangle\langle\mu'|\underline{\hat{e}}'^*.\underline{D} \qquad (15.26)$$

Thus the observed intensity of fluorescent light can be obtained using equations (15.11), (15.24), and (15.25) in the form

$$\frac{dI}{d\Omega} = \mathrm{Tr}(\rho\, L_F)$$

$$= \frac{U(\omega)}{8\pi\varepsilon_0^2}\left(\frac{\omega}{\hbar c}\right)^3\sum_{\substack{mm'\\ \mu\mu'}}\frac{F_{mm'}G_{m'm}}{\Gamma+i(m-m')\omega_L}\,\rho_{\mu\mu} \qquad (15.27)$$

where $G_{m'm} = \langle m'|\underline{\hat{e}}'.\underline{D}|\mu'\rangle\langle\mu'|\underline{\hat{e}}'^*.\underline{D}|m\rangle$ is called the emission matrix.

We see that field-dependent terms appear in the denominator of equation (15.27) when $m \neq m'$, i.e. the Hanle effect signal is a direct result of the Hertzian coherence created in the excited state by excitation with coherently polarized light. We can describe the phenomenon as the result of a quantum-mechanical interference between the scattering amplitudes for the two possible routes from the initial ground level sub-state $|\mu\rangle$ to the final sub-state

$|\mu'\rangle$, as indicated diagrammatically in Fig.15.9. This quantum-mechanical interference manifests itself as a change in the polarization and spatial distribution of the scattered resonance radiation as the separation of the Zeeman sub-states of the excited level is varied. The interference effect disappears when these levels are separated by more than their natural width Γ.

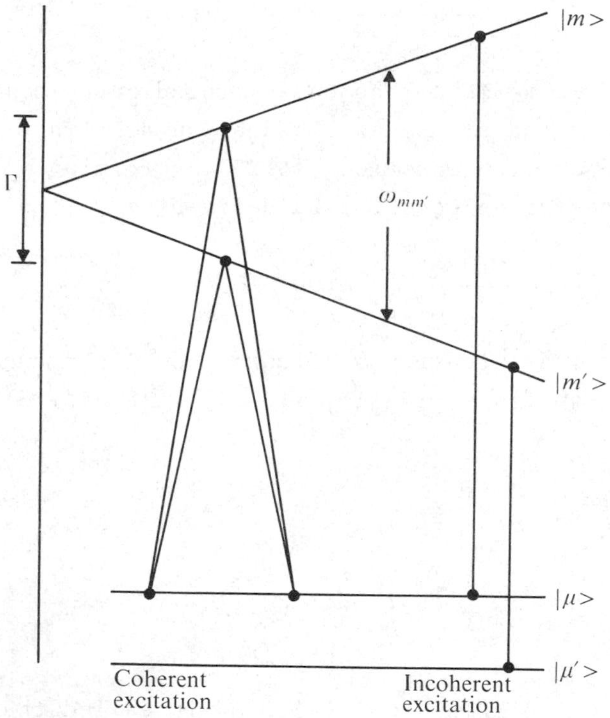

Fig.15.9. Energy level diagram illustrating the quantum theory of the Hanle effect. The excited states $|m\rangle$ and $|m'\rangle$ are degenerate at zero magnetic field. In a finite magnetic field, the excitation of $|m\rangle$ and $|m'\rangle$ from a single ground level sub-state $|\mu\rangle$ by coherently polarized light results in interference effects in the coherently scattered radiation provided that $\omega_{mm'} = (m-m')\omega_L \leq \Gamma$ where Γ is the natural radiative width of the levels.

It should be clear from this discussion that similar effects are to be expected when the Zeeman levels of an atom with hyperfine structure intersect at large magnetic fields. These high-field level-crossing signals are discussed in detail in Chapter 18. Similar changes in the angular distribution of electromagnetic radiation are well known in nuclear physics and the perturbation of the angular correlation of cascading γ-rays produced by a large external field is widely used in the spectroscopy of excited nuclei.

15.6.3. Effect of hyperfine structure on the Hanle signal. For an element having an odd isotope with nuclear spin I, equation (15.27) shows that the principal effect of hyperfine structure in the Hanle effect experiments will be to replace the Landé factor g_J used in the calculation of the Larmor frequency, equation (15.2), by the hyperfine g-factor g_F, where, to a good approximation

$$g_F \cong g_J\left\{\frac{F(F+1) + J(J+1) - I(I+1)}{2F(F+1)}\right\} \qquad (15.28)$$

and F is the total angular momentum quantum number of a given hyperfine component of the excited energy level. Thus, except for the case when J=I, the different hyperfine levels have different values of g_F and so produce Hanle effect signals with different widths. Since it is not usually possible to excite to just a single hyperfine level, the observed signal will depend on the detailed form of the emission and absorption profiles of the lamp and resonance cell respectively. It is therefore difficult to use the odd-isotope Hanle signals for accurate lifetime measurements, and experiments on pure samples of the even isotopes are to be preferred.

These are not possible in the case of sodium and potassium and here an even more complex situation arises, for these elements have an excited state hyperfine structure in zero field which is only slightly larger than the radiation width of the levels. The magnetic fields required for the Hanle effect are sufficiently large to cause appreciable decoupling of the electronic and nuclear spin angular momentum vectors, $\underline{J}$ and $\underline{I}$. In this situation the excitation of different hyperfine levels can no longer be treated as independent processes and the coherence created is responsible for level-crossing signals at finite fields. These level-crossing signals overlap the zero-field Hanle signals and therefore a detailed comparison of the experimental and theoretical signals over a wide range of magnetic fields is required to determine the lifetimes and hyperfine structure constants of the levels. This topic will be discussed again in Chapter 18.

15.7. *Theory of resonance fluorescence in the* $J_e=1 \rightarrow J_g=0$ *case*

15.7.1. Details of the experimental geometry. The excitation and emission matrices appearing in equation (15.27) are determined solely by the geometry of the apparatus and the orientation of the polarizers used. They are therefore a common feature of many different types of resonance fluorescence experiment and we demonstrate their manipulation in some detail. We consider a sample of atoms excited by a beam of resonance radiation whose direction of propagation is defined by the angles (θ,ϕ) in a spherical coordinate system. We assume that the light is linearly polarized with its electric vector making an angle α with the unit vector $\hat{\underline{\theta}}$ as shown in Fig.15.10.

In cartesian coordinates it may be shown (Problem 15.5) that the polarization vector of the incident radiation is given by

$$\hat{\underline{e}} = \hat{\underline{i}}(\cos\theta\cos\phi\cos\alpha - \sin\phi\sin\alpha) + \hat{\underline{j}}(\cos\theta\sin\phi\cos\alpha + \cos\phi\sin\alpha) - \\ - \hat{\underline{k}}\sin\theta\cos\alpha. \tag{15.29}$$

For the evaluation of the atomic matrix elements, however, it is more convenient to expand $\hat{\underline{e}}$ in terms of the spherical unit vectors $\hat{\underline{\varepsilon}}_{\pm 1} = \mp \frac{1}{\sqrt{2}}(\hat{\underline{i}} \pm i\hat{\underline{j}})$; $\hat{\underline{\varepsilon}}_0 = \hat{\underline{k}}$ in the form given by

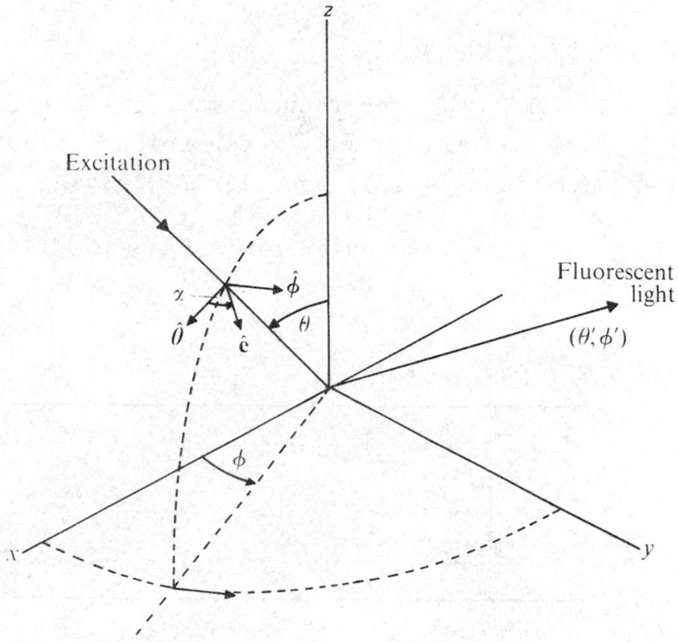

Fig.15.10. Geometry for theoretical description of resonance fluorescence experiments.

equation (15.19). In this system we have

$$\begin{aligned} \mathcal{E}_{\pm 1} &= \mp \frac{1}{\sqrt{2}} (e_x \pm ic_y) = \mp \frac{1}{\sqrt{2}} (\cos\theta\cos\alpha \pm i\sin\alpha)e^{\pm i\phi} \\ \mathcal{E}_0 &= e_z = -\sin\theta\cos\alpha \end{aligned} \tag{15.30}$$

The electric dipole moment operator, $\underline{D}$, may be expanded in the same manner and it can then be shown that

$$\underline{e}.\underline{D} = -(\mathcal{E}_{+1}D_{-1} + \mathcal{E}_{-1}D_{+1}) + \mathcal{E}_0 D_0. \tag{15.31}$$

We must now evaluate the matrix elements of the electric dipole operator. We choose as an example a system in which $J_g=0$ and $J_e=1$ for the numerical values of the matrix elements in this case are particularly simple. Using equation (5.18) we have

$$\begin{aligned} \langle 0|\mathcal{E}_{\pm1}D_{\mp1}|\pm1\rangle &= -\langle D\rangle\mathcal{E}_{\pm1} \\ \langle 0|\mathcal{E}_0 D_0|0\rangle &= \langle D\rangle\mathcal{E}_0 \end{aligned} \qquad (15.32)$$

where $\langle D\rangle = \langle \gamma 0 \| \underline{D} \| \gamma' 1\rangle$ is the reduced matrix element of the operator $\underline{D}$. The excitation matrix can now be calculated with the help of equations (15.31) and (15.32), giving

$$F_{mm'} = \langle m|\hat{\underline{e}}.\underline{D}|\mu\rangle \langle \mu|\hat{\underline{e}}^*.\underline{D}|m'\rangle = \mathscr{F}|\langle D\rangle|^2$$

where

$\mathscr{F} =$ (15.33)

m \ m'	1	0	-1
1	$-\mathcal{E}_{+1}\mathcal{E}_{-1}$	$-\mathcal{E}_0\mathcal{E}_{-1}$	$-\mathcal{E}_{-1}^2$
0	$\mathcal{E}_{+1}\mathcal{E}_0$	$\mathcal{E}_0^2$	$\mathcal{E}_{-1}\mathcal{E}_0$
-1	$-\mathcal{E}_{+1}^2$	$-\mathcal{E}_0\mathcal{E}_{+1}$	$-\mathcal{E}_{-1}\mathcal{E}_{+1}$

The de-excitation matrix for light emitted in the direction (θ',ϕ') and linearly polarized parallel to the electric vector $\hat{\underline{e}}'$ is defined by

$$G_{m'm} = \langle m'|\hat{\underline{e}}'.\underline{D}|\mu'\rangle \langle \mu'|\hat{\underline{e}}'^*.\underline{D}|m\rangle \qquad (15.34)$$

where $\hat{\underline{e}}'$ makes an angle α' with the direction of the unit vector $\hat{\underline{\theta}}'$. In this geometry $G_{m'm}$ is identical to $F_{mm'}$ except that all quantities in equation (15.33) are now distinguished by a prime. The matrices for the case of unpolarized light may be obtained by summing those for light polarized in directions α and $\alpha+\pi/2$ respectively.

15.7.2. Application to Hanle effect experiments. We now use the excitation and emission matrices derived above to obtain expressions for the Hanle effect signals in the $J_g=0 \leftrightarrow J_e=1$ system. These would apply, for instance, to experiments on the even isotopes of Zn, Cd, Hg, and several other elements. With the rather general geometrical arrangement of Fig.15.10 it is convenient to separate the signal,

$dI/d\Omega$, into contributions made by terms in equation (15.27) for which $\Delta m = |m-m'|$ is constant. Using the explicit expressions for $F_{mm'}$ and $G_{m'm}$ given by equation (15.33), we then have

$$\frac{dI}{d\Omega} = \frac{dI}{d\Omega}(\Delta m=0) + \frac{dI}{d\Omega}(\Delta m=1) + \frac{dI}{d\Omega}(\Delta m=2)$$

where

$$\frac{dI}{d\Omega}(\Delta m=0) = \frac{C}{2\Gamma}(1-\mathcal{E}_0^2 - \mathcal{E}_0'^2 + 3\mathcal{E}_0^2\mathcal{E}_0'^2 \tag{15.35}$$

$$\frac{dI}{d\Omega}(\Delta m=1) = -\frac{2C\mathcal{E}_0\mathcal{E}_0'}{\Gamma^2+\omega_L^2}\{\Gamma(\mathcal{E}_{+1}\mathcal{E}_{-1}' + \mathcal{E}_{-1}\mathcal{E}_{+1}') + $$
$$+ i\omega_L(\mathcal{E}_{+1}\mathcal{E}_{-1}' - \mathcal{E}_{-1}\mathcal{E}_{+1}')\} \tag{15.36}$$

$$\frac{dI}{d\Omega}(\Delta m=2) = \frac{C}{\Gamma^2+4\omega_L^2}\{\Gamma(\mathcal{E}_{+1}^2\mathcal{E}_{-1}'^2 + \mathcal{E}_{-1}^2\mathcal{E}_{+1}'^2) + $$
$$+ 2i\omega_L(\mathcal{E}_{+1}^2\mathcal{E}_{-1}'^2 - \mathcal{E}_{-1}^2\mathcal{E}_{+1}'^2)\} \tag{15.37}$$

and

$$C = \frac{U(\omega)}{8\pi\varepsilon_0^2}\left(\frac{\omega}{\hbar c}\right)^3 \langle D\rangle^4 \rho_{00}.$$

In this example ρ_{00} is the density of atoms in the ground state. This separation into three terms also has a physical significance for the density matrix of the ensemble of excited atoms can be rearranged into components which transform under rotations as tensors of rank 0, 1, and 2 respectively. These components are known as the population, orientation, and alignment of the system. In fact the different expressions given by equations (15.35)-(15.37) describe almost exactly the signals arising from the population $dI/d\Omega$ ($\Delta m=0$), orientation $dI/d\Omega$ ($\Delta m=1$), and alignment $dI/d\Omega(\Delta m=2)$ of the sample of excited atoms. We shall return to this topic when we consider the effects of relaxation by collisions and the trapping of resonance radiation in section 16.4.

For the particular geometry shown in Fig.15.1 we have

($\theta=\pi/2$, $\phi=0$, $\alpha=\pi/2$) and ($\theta'=0$, $\phi'=\pi/2$). By substituting these values into equation (15.30) to obtain explicit values for $\mathcal{E}_q$ and $\mathcal{E}_{q'}$, it can be shown (Problem 15.6) that equations (15.35)-(15.37) reduce to

$$\frac{dI}{d\Omega}(\Delta m=0) = \frac{C}{2\Gamma}$$

$$\frac{dI}{d\Omega}(\Delta m=1) = 0$$

$$\frac{dI}{d\Omega}(\Delta m=2) = \frac{C}{\Gamma^2+4\omega_L^2}(\Gamma\cos 2\alpha' + 2\omega_L \sin 2\alpha')$$

a result which is identical to that obtained by the classical theory, equation (15.4). This detailed analysis proves the validity of the statement made in section 15.2 that the shapes of the Hanle effect signals are correctly predicted by the classical theory. However, in systems other than the $J_g=0 \leftrightarrow J_e=1$ we can see by referring to equation (5.18) that the field-dependent term is generally a smaller fraction of the constant background signal arising from the terms with $\Delta m=0$. This reduces the signal-to-noise level and may necessitate the use of phase-sensitive detectors or other signal averaging techniques.

15.7.3. Effect of deviations from perfect geometry. In most resonance fluorescence experiments lenses and light pipes subtending large solid angles are used in both the exciting and detection arms of the apparatus in order to increase the size of the signal. In this situation the observed signal must obviously be obtained by averaging the expressions given in equations (15.35)-(15.37) over the finite solid angle used. The main result of this in a geometry where the pure Lorentzian-shaped $\Delta m=2$ signal is expected is to cause some slight admixtures of the dispersion-shaped $\Delta m=2$ signal, and possibly contributions from the $\Delta m=1$ signals as well. This leads to a slight asymmetry and broadening of the observed signals which can usually be corrected in the analysis of the data.

<u>15.7.4. Estimate of the size of the signal.</u> The size of the resonance fluorescence signal is determined by the rate Q at which photoelectrons are produced at the cathode of the photomultiplier by the scattered light. In equilibrium the rate at which photons are re-emitted is equal to the rate at which they are absorbed, which in turn is determined by the effective absorption cross-section, σ_{eff}. Thus for a resonance cell of length d and cross-sectional area A which is filled with absorbing atoms at a uniform density n, we have

$$Q = F\,A\,d\,n\,\sigma_{eff}\,K \qquad (15.38)$$

where F is the flux of resonance radiation photons incident on the cell, and K is the combined collection and detection efficiency factor. The effective cross-section may be obtained from equation (9.51) by multiplying the total absorption cross-section by the normalization factor for the folding integral of the Gaussian emission and absorption profiles of the resonance lamp and cell, giving

$$\sigma_{eff} = 4\pi\left(\frac{\pi\ \ln 2}{\Delta\omega_s^2+\Delta\omega_c^2}\right)^{1/2} r_0 c f_{ik} \qquad (15.39)$$

where $\Delta\omega_s$ and $\Delta\omega_c$ are the Doppler widths of the light source and absorbing atoms respectively. The large value of the resonance scattering cross-section is one of the main reasons for the great sensitivity of resonance fluorescence experiments. The factor K in equation (15.38) is given by the quantum efficiency of the photomultiplier multiplied by the solid angle subtended by the collection optics expressed as a fraction of 4π.

For typical values of the parameters in equations (15.38) and (15.39), we find (Problem 15.7) that $Q \approx 3 \times 10^7\ s^{-1}$. This is much larger than either the dark current of the photomultiplier ($\approx 10^3\ s^{-1}$) or the shot noise due to random fluctuations in the photon counting statistics ($\approx Q^{1/2}$). Thus in many resonance fluorescence experiments excellent signal-to-noise ratios can be obtained, and experiments at very low densities or using only milligram quantities of mass-separated stable or radioactive isotopes are possible. However, equation (15.38) does emphasize the need for intense light sources in these experiments and the importance of achieving high collection and detection efficiency for the fluorescent light.

15.8. Resonance fluorescence experiments using pulsed excitation

15.8.1. Introduction and experimental techniques. In previous sections we drew attention to the fact that, in both the classical and quantum theories, expressions derived for the intensity of resonance fluorescence from atoms subjected to an external magnetic field, equations (15.3) and (15.23) respectively, contain terms which may lead to a modulation of the intensity at the Larmor frequency or its second harmonic. This radio-frequency modulation has been observed in several different kinds of experiment, the simplest of which makes use of pulsed excitation and time-resolved detection of the fluorescent light.

The first of these experiments were performed simultaneously by Dodd *et al.* (1964) and Aleksandrov (1964). The improved apparatus used by Dodd *et al.* (1967) in a more detailed study of this phenomenon is shown in Fig.15.11. Resonance fluorescence in cadmium vapour excited by the intercombination line $5\ ^1S_0 - 5\ ^3P_1$ at 3261 Å was chosen for this investigation. The cadmium atoms in a heated resonance cell were excited to the $5\ ^3P_1$ level by a pulse of resonance radiation lasting $0{\cdot}2\mu s$. This excitation pulse was obtained by passing the light from a commercial cadmium lamp through a Kerr cell shutter operated by a high voltage pulse unit. The time dependence of the resonance fluorescence emitted in a direction at right-angles to both the magnetic field and the direction of the incident light was

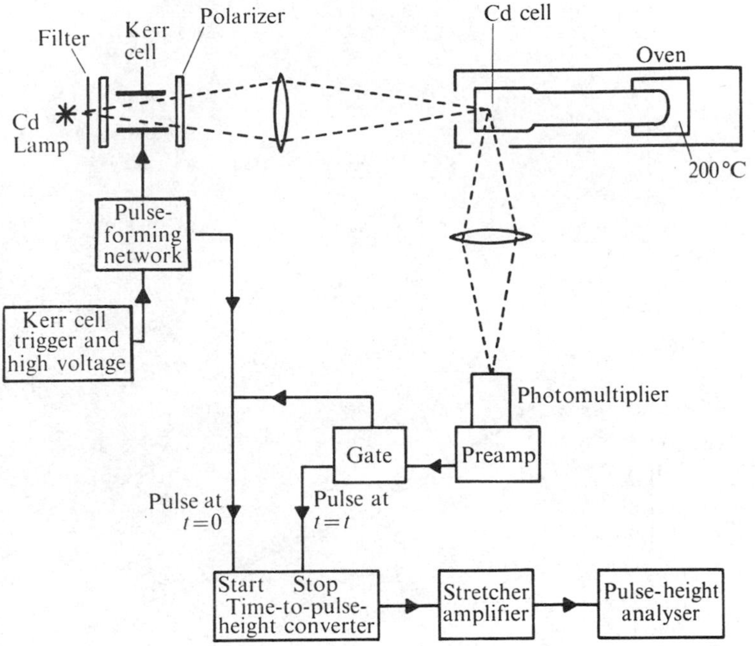

Fig.15.11. Block diagram of the apparatus of Dodd *et al.* (1967) for cadmium resonance fluorescence experiments using pulsed excitation.

studied using the single-photon counting techniques described in section 6.3.3. The result of a seven-hour-long run obtained with the cadmium vapour at 200°C and an applied magnetic field of 345 mG is shown in Fig.15.12(a). This displays the number of counts recorded in each channel of the analyser after a correction has been made for the background produced by photomultiplier dark current pulses and by the light reflected from the cell walls. The experimental trace can clearly be resolved into two distinct contributions consisting of a decaying but unmodulated intensity and a damped intensity modulation as shown in Figs.15.12(b) and (c) respectively. The observed angular frequency of modulation was $(9{\cdot}25 \pm 0{\cdot}30) \times 10^6\ s^{-1}$, in good agreement with that predicted from the known g_J factor for the 5 $^3P_1$ level Problem (15.9).

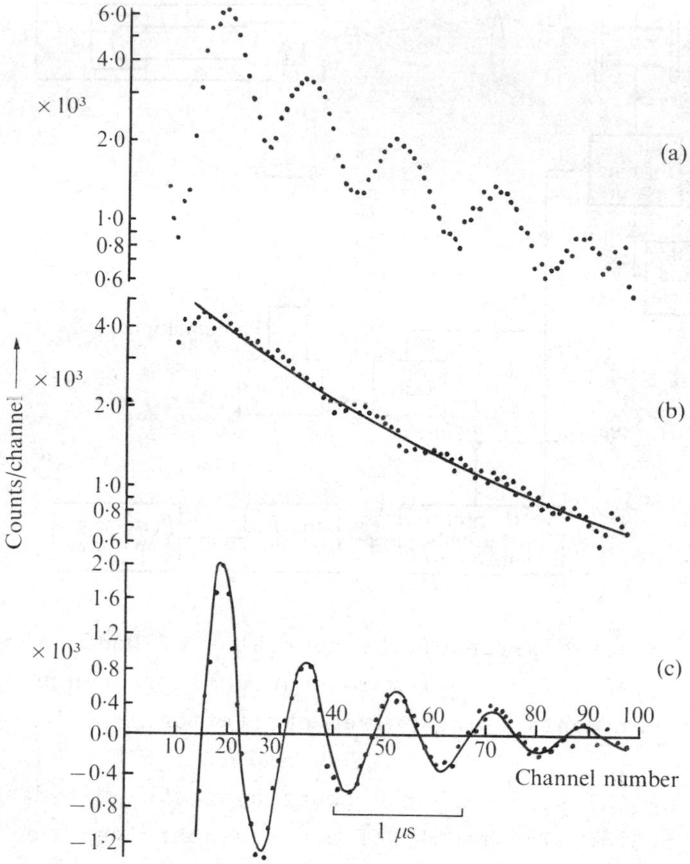

Fig.15.12. Transient decay of cadmium atoms excited by a pulse of polarized resonance radiation. (a) The experimentally observed fluorescence as a function of time. (b) The exponential decay of unmodulated fluorescent light. (c) The modulated component of the fluorescent light. The modulation is produced by Larmor precession in the applied magnetic field of 345 mG. (After Dodd *et al.* (1967).)

15.8.2. Theoretical interpretation. For the moment we concentrate on the interpretation of the signal shown in Fig.15.12(c). The phenomenon can be explained classically by imagining that the exciting pulse created a set of electric dipole oscillators with their axes all initially aligned parallel to the electric vector of the incident radiation (Problem 15.9). Subsequently these dipoles precess at the Larmor angular frequency about the magnetic field direction while simultaneously they suffer damping by re-radiation. The angular distribution of the electric dipole radiation pattern then produces maxima in the observed intensity every time that the axis of the dipoles is at right-angles to the direction of observation.

A general expression for the intensity of light observed in these experiments may be obtained using the density matrix formalism developed in section 15.6. We rewrite equation (15.23) allowing for the fact that the energy density of the incident radiation, $U(\omega,t)$, is now a function of time:

$$\frac{\partial \rho_{mm'}}{\partial t} + (\Gamma + ix)\rho_{mm'} = \frac{\pi U(\omega,t)}{\varepsilon_0 \hbar^2} \sum_{\mu} F_{mm'}\rho_{\mu\mu} \qquad (15.40)$$

where $x = \omega_L(m-m')$. Multiplying by the integrating factor $\exp\{(\Gamma+ix)t\}$ gives

$$\frac{d}{dt}[\rho_{mm'}\exp\{(\Gamma+ix)t\}] = \frac{\pi}{\varepsilon_0 \hbar^2} \sum_{\mu} F_{mm'}\rho_{\mu\mu} U(\omega,t)\exp\{(\Gamma+ix)t\}. \qquad (15.41)$$

For the sake of simplicity we now assume that the pulse of incident radiation has a rectangular shape,

$$\text{i.e. } U(\omega,t) = U(\omega) \quad \text{for } t_0 < t < t_0 + \Delta t_0$$
$$= 0 \quad \text{for all other values of } t.$$

Then integrating equation (15.41) we have for $t > t_0 + \Delta t_0$:

$$\rho_{mm'}(t) = \frac{U(\omega)\pi}{\varepsilon_0\hbar^2}\sum_{\mu} F_{mm'}\rho_{\mu\mu}[\exp\{(\Gamma+ix)(t_0+\Delta t_0)\}-\exp\{(\Gamma+ix)t_o\}]\times$$
$$\times\frac{\exp\{-(\Gamma+ix)t\}}{\Gamma+ix} \tag{15.42}$$

where we have assumed that all the atoms in the sample are in the ground state before the arrival of the excitation pulse, i.e. $\rho_{mm'}(t)=0$ for $t \le t_0$. The terms in equation (15.42) which involve Δt_0 are dependent on the length of the exciting pulse. The simplest situation occurs when the pulse is very much shorter than either the radiative lifetime or the period associated with the Larmor precession, i.e. $\Gamma\Delta t_0 < \omega_L\,\Delta t_0 \ll 1$. In this limit we find that equation (15.42) reduces to

$$\rho_{mm'}(t) = \frac{U(\omega)\pi}{\varepsilon_0\hbar^2}\sum_{\mu} F_{mm'}\rho_{\mu\mu}\exp\{-(\Gamma+ix)(t-t_0)\}. \tag{15.43}$$

Finally, applying the monitoring operator L_F, we obtain a general expression for the intensity of fluorescent light in experiments using pulsed excitation:

$$\frac{dI}{d\Omega} = \frac{U(\omega)}{8\pi\varepsilon_0^2}\left(\frac{\omega}{\hbar c}\right)^3\sum_{\substack{mm'\\ \mu\mu'}} F_{mm'}G_{m'm}\rho_{\mu\mu}\exp[-\{\Gamma+i\omega_L(m-m')\}(t-t_o)]\ . \tag{15.44}$$

This calculation makes it clear that light beats are associated with the time evolution of the off-diagonal elements of the excited-state density matrix. Consequently they can only be observed in pulsed experiments if the light which excites the atoms of the sample is also polarization coherent. Only then is the necessary Hertzian coherence created in the excited-state density matrix. The theory predicts that modulation at the angular frequencies ω_L and $2\omega_L$ will be detectable depending on the geometry and polarization used in the experiment.

<u>15.8.3. Application to the $J_e=1$ case.</u> In the notation of section 15.7.1 the geometrical arrangement used by Dodd *et al.*(1967), and shown in Fig.15.11, is described by the

directions ($\theta=\pi/2$, $\phi=0$, $\alpha=\pi/2$) and ($\theta'=\pi/2$, $\phi'=\pi/2$). The excitation and emission matrices for this geometry may be evaluated using equations (15.30) and (15.33) and it can then be shown that the signal predicted by equation (15.44) is given by

$$\frac{dI}{d\Omega} = \frac{C}{2}[1 - \cos\{2\omega_L(t-t_0)\}]\exp\{-\Gamma(t-t_0)\} \qquad (15.45)$$

where the signals obtained with the analyser at angles of α' and $\alpha'+\pi/2$ have been summed corresponding to the detection of unpolarized light.

The theory therefore predicts an intensity modulation of 100 per cent in this geometry when the excitation pulse is sufficiently short. For pulses which do not satisfy the condition $\omega_L\Delta t_0 \ll 1$ it is found that the depth of modulation is reduced and that there is a phase shift in the modulation term. The reason for this can be easily visualized on the classical model, for the Larmor precession during a long excitation pulse causes the dipoles to be distributed through an angle $\omega_L\Delta t_0$ in the plane perpendicular to $\underline{B}$, rather like a fan. In the limit that $\omega_L\Delta t_0 \gg 1$, the dipoles are isotropically distributed in this plane and no modulation is observed.

In the experiments of Dodd *et al.*(1967) the short pulse criteria is reasonably well satisfied since $\omega_L\Delta t_0 \approx 0{\cdot}14$ and the reduced depth of modulation apparent in Fig.15.12(a) is due to depolarization by collisions with foreign gases contaminating the resonance cell. Fluorescence from the F=1/2 levels of the odd isotopes ^{111}Cd and ^{113}Cd (both I=1/2), which have a total abundance of 25 per cent in the natural cadmium used in the resonance cell, also contribute to this background since in this geometry interference effects from levels separated by $\Delta m=1$ are not observable. The signal from the F=3/2 level of the odd isotopes is modulated, however, at a frequency which is 2/3 of that of the even isotope signal, as evaluation of equation (15.28) will show. This small additional component of the fluorescent light can be clearly identified by detailed analysis of the experimental data.

One of the principal difficulties in these time-resolved experiments is the low intensity of the conventional sources of pulsed resonance radiation, as the seven-hour running time of Dodd *et al.* (1967) emphasizes. Excitation by means of pulsed tunable dye lasers seems likely to overcome this difficulty and we now briefly refer to some recent experiments using this technique.

15.8.4. Pulsed laser excitation. Quantum beats produced by dye laser excitation have been observed by Gornik *et al.* (1972) on the $6\,s^2\ {}^1S_0$-$6s6p\ {}^3P_1$ intercombination line of ytterbium at 5556Å. In this experiment the conventional resonance cell was replaced by an atomic beam of ytterbium and the atoms were excited by radiation from a nitrogen laser-pumped dye laser using a solution of sodium fluorescein and esculin in methyl alcohol. The dye laser radiation was linearly polarized in a direction at right-angles to the magnetic field applied to the atoms in the scattering chamber and the laser output was tuned on to the resonance line by a combination of three tilted solid quartz étalons and an interference filter. At a pulse repetition rate of 30 Hz a peak output power of several hundred watts was obtained with a spectral bandwidth of $< 50 \times 10^{-3}\ cm^{-1}$. The laser pulse length of 7 ns is approximately a hundred times shorter than the radiative lifetime of the ${}^3P_1$ level, thus the analysis of section 15.8.2 may be applied without any modification.

The fluorescent light from the atomic beam was monitored by a photomultiplier and recorded in a Tektronix transient analyser. Quantum beat signals having excellent signal-to-noise ratios and displaying 100 per cent depth of modulation could be obtained with a single pulse of excitation. From a measurement of the modulation frequency (2·99 MHz) and a knowledge of the applied magnetic field (0·714 G), the Landé g-factor for the ${}^3P_1$ level may be obtained to an accuracy of about one per cent. Moreover the exponential envelope of the signal yields a radiative lifetime of

$\tau(6\ ^3P_1) = 860 \pm 43$ ns, in good agreement with previous determinations.

The g-factors and lifetimes of the lowest-lying $^1P_1$ states in barium and calcium have also been measured using the quantum beat technique by Schenck *et al.*(1973). However, a much larger number of lifetime measurements have been made in zero magnetic field by simply observing the exponential decay of the fluorescent light following dye laser excitation on a fast oscilloscope. A representative sample of the available results is given in Table 15.3. As the table indicates, this new technique is applicable to measurements over a wide range of lifetimes. Perhaps equally interesting is the new information on collisional quenching cross-sections which may also be obtained by this selective optical excitation method.

TABLE 15.3.

Radiative lifetimes of atomic and molecular levels obtained using pulsed dye laser excitation

Atom or molecule	Level	Lifetime	Reference
Na	$3p\ ^2P_{1/2}$	$16{\cdot}4 \pm 0{\cdot}6$ ns	Erdmann *et al.*(1972)
Ca	$4s5s\ ^3S_1$	$10{\cdot}7 \pm 1{\cdot}0$ ns	Gornik *et al.*(1973)
Mg	$3s3p\ ^3P_1$	$2{\cdot}2 \pm 0{\cdot}2$ ms	Wright *et al.*(1974)
I_2	$B\ ^3\Pi^+_{0u}$ ($v'=0$ - $v'=25$)	$1{\cdot}6$ - $2{\cdot}1$ μs	Sakurai *et al.*(1971)
Br_2	$B\ ^3\Pi^+_{0u}$ ($v'=1$ - $v'=31$)	$0{\cdot}14$ - $1{\cdot}3$ μs	Capelle *et al.*(1971)
NO_2	Unknown	41 - 43 μs	Sakurai and Capelle (1970).

15.9. Resonance fluorescence experiments using modulated excitation

<u>15.9.1. Introduction and experimental technique.</u> The classical theory of resonance fluorescence, in which the atoms are treated as dipole oscillators precessing at the Larmor frequency, leads one to predict that interesting effects will also occur if the atoms are excited by light whose intensity is periodically modulated. As the external magnetic field is varied in these experiments a point is reached at which the Larmor frequency, ω_L, equals the angular frequency of the modulation, f. Additional atoms will then be excited in phase with the Larmor precession of those already in the upper state and a significant transverse magnetic moment is built up in the ensemble of excited atoms. We would therefore expect a resonant increase in the amplitude of modulation of the fluorescent light when the condition $\omega_L = f$ is satisfied.

The first resonance fluorescence experiments using intensity-modulated excitation were performed by Aleksandrov (1963) and independently by Corney and Series (1964 a,b). The apparatus used by Corney (1968) in a more detailed investigation of the phenomenon is shown in Fig.15.13. The $5\ ^3P_1$ level of cadmium was again chosen for study. The light source was an electrodeless capillary discharge tube made from fused silica and was excited by applying a radio-frequency voltage at 231 kHz to external electrodes of aluminium sheet. It was found that the radiation from the lamp was intensity modulated at 462 kHz owing to a 100 per cent modulation of the mean energy of the electrons in the discharge which occurs at twice the frequency of the applied voltage (Harries and von Engel 1954). However, the depth of modulation of the 3261 Å radiation emitted by the lamp was only of the order of 1 per cent, indicating that direct electron impact excitation accounts for only a small fraction of the total population of the $5\ ^3P_1$ level. The light from the discharge was focussed on to a cylindrical resonance cell in an oven heated by hot air to 200°C. The

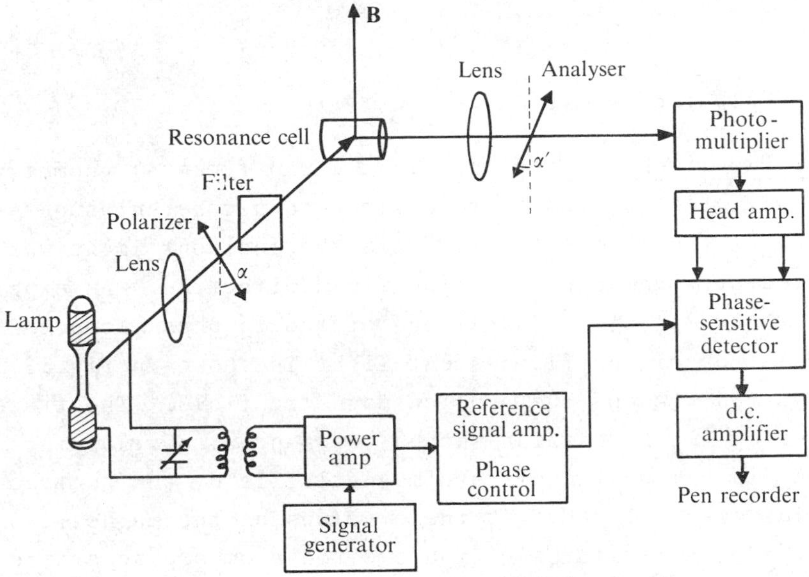

Fig.15.13. Schematic diagram of the apparatus used in resonance fluorescence experiments on cadmium vapour excited by intensity-modulated light.

magnetic field was produced by a Helmholtz coil pair having a mean diameter of 24·9 cm and was uniform to 2 per cent over the volume of the cell. Stray d.c. magnetic fields were reduced to less than 0·5 mG using three mutually orthogonal Helmholtz coil pairs and care was taken to eliminate stray a.c. fields by careful location of electronic equipment.

The intensity of the fluorescent light was monitored by a photomultiplier and the time-dependent signal was isolated by an amplifier tuned to 462 kHz. The radio-frequency modulation of the fluorescent light was readily confirmed by this method, but the small percentage modulation of the incident light made phase-sensitive detection essential for detailed studies. This technique had the advantage that the amplitudes of the in-phase and quadrature components of the fluorescent radiation,

$$I(t) = A \cos ft + B \sin ft,$$

could be studied separately.

15.9.2. Experimental results. In the experimental geometry shown in Fig.15.13 atoms were excited to a coherent superposition of the $m_J = \pm 1$ states when the incident light was polarized at right-angles to the field direction. An experimental recording of the amplitude of the in-phase component of modulation of the fluorescent light is shown in Fig.15.14 (a) as a function of the applied magnetic field. The trace consists of two Lorentzian shaped resonances, overlapping slightly in the region of zero magnetic field, and centred at fields of ± 0·110 G. At these values of the magnetic field the $m_J = \pm 1$ states of the excited atom are separated by a frequency interval $2\omega_L/2\pi = 462$ kHz, as expected. The resonances can thus be regarded as the effect of interference between the σ^+ and σ^- polarized Zeeman components of the fluorescent light. The theory developed in the next section shows that the amplitude of the in-phase component is given by

$$A = \left\{ \frac{2\Gamma}{\Gamma^2 + f^2} - \frac{\Gamma}{\Gamma^2 + (2\omega_L + f)^2} - \frac{\Gamma}{\Gamma^2 + (2\omega_L - f)^2} \right\} I_0 \Gamma \ . \quad (15.46)$$

This expression was evaluated using the experimental lifetime for the 5 $^3P_1$ level of $\tau = 2 \cdot 25 \times 10^{-6}$ s obtained by Barrat and Butaux (1961), and is plotted in Fig.15.14(b). The agreement between the experimental and theoretical curves is very satisfactory.

15.9.3. Theory of resonance fluorescence excited by modulated light. Theoretical expressions for the intensity of light observed in resonance fluorescence experiments using modulated excitation can be obtained by a simple extension of equation (15.41). We now assume that the energy density of the incident radiation, $U(\omega,t)$, is amplitude modulated at

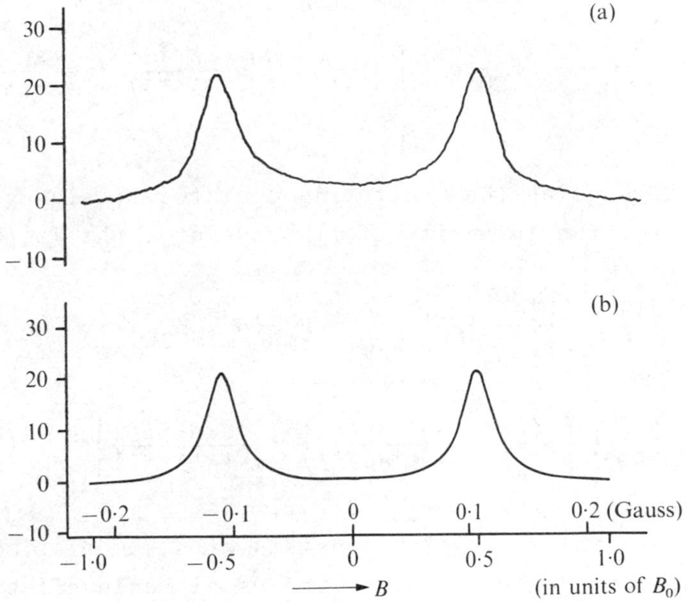

Fig.15.14. Study of the fluorescent radiation from a sample of cadmium atoms excited by light which is intensity modulated at the frequency $f/2\pi$ = 462 kHz. (a) Experimental recording of the in-phase component of the modulated fluorescence as a function of applied field, B. (b) Plot of the theoretical expression, equation (15.46). Resonances occur when $B = \pm B_0/2$ where $B_0 = \hbar f/g_J\mu_B$.

the angular frequency f. Thus $U(\omega,t)$ is given by

$$U(\omega,t) = U(\omega)\,(1 + a\cos ft) \qquad (15.47)$$

where a is the depth of modulation. Substituting into equation (15.41), we have

$$\frac{d}{dt}[\rho_{mm'}\exp\{(\Gamma+ix)t\}] = \frac{\pi U(\omega)}{\varepsilon_0\hbar^2}\sum_{\mu} F_{mm'}\rho_{\mu\mu}(1+a\cos ft)\exp\{(\Gamma+ix)t\}. \qquad (15.48)$$

Integrating over t from $t = -\infty$ to the instant of observation t, and assuming $\rho_{mm'}(-\infty)=0$, we have

$$\rho_{mm'}(t) = \frac{\pi U(\omega)}{\varepsilon_0\hbar^2}\sum_{\mu} F_{mm'}\rho_{\mu\mu}\left[\frac{1}{\Gamma+ix}+\frac{a}{2}\left\{\frac{\exp(ift)}{\Gamma+i(x+f)}+\frac{\exp(-ift)}{\Gamma+i(x-f)}\right\}\right]. \tag{15.49}$$

Finally, introducing the monitoring operator L_F, the general expression for the intensity of fluorescent light is given by

$$\frac{dI}{d\Omega} = \frac{U(\omega)}{8\pi\varepsilon_0^2}\left(\frac{\omega}{\hbar c}\right)^3 \sum_{\substack{mm'\\ \mu\mu'}} F_{mm'}G_{m'm}\rho_{\mu\mu} \times$$

$$\times\left[\frac{1}{\Gamma+ix}+\frac{a}{2}\left\{\frac{\exp(ift)}{\Gamma+i(x+f)}+\frac{\exp(-ift)}{\Gamma+i(x-f)}\right\}\right] \tag{15.50}$$

where $x=\omega_L(m-m')$.

This general expression consists of a sum of time-independent terms which describe the usual Hanle effect signals and terms modulated at the angular frequency f which are resonant when $\omega_L(m-m') = \pm f$. These terms control the amplitude of modulation of the fluorescent light and originate in the off-diagonal elements of the excited state density matrix. When the condition $f \gg \Gamma$ is satisfied, excitation by amplitude-modulated light creates substantial Hertzian coherence at fields which are well separated from the zero-field level-crossing region.

15.9.4. Application to $J_e=1$ case. The geometrical arrangement shown in Fig.15.13 is described by the angles $(\theta=\pi/2,\ \phi=0,\alpha)$ and $(\theta'=\pi/2,\ \phi'=\pi/2,\alpha')$ in the notation of section 15.7.1. Using equations (15.30) and (15.33) to evaluate the excitation and emission matrices in this geometry we obtain, after substitution in equation (15.50),

$$\frac{dI}{d\Omega}(\Delta m=0) = C(\cos^2\alpha\cos^2\alpha'+\tfrac{1}{2}\sin^2\alpha\sin^2\alpha')$$

$$\times\left\{\frac{1}{\Gamma}+\frac{a}{2}\left(\frac{2\Gamma\cos ft}{\Gamma^2+f^2}+\frac{2f\sin ft}{\Gamma^2+f^2}\right)\right\}. \tag{15.51a}$$

$$\frac{dI}{d\Omega}(\Delta m=1) = C\,\frac{\sin 2\alpha \sin 2\alpha'}{2}\Bigg[\frac{\omega_L}{\Gamma^2+\omega_L^2} + \frac{a}{2}\cos ft\left\{\frac{\omega_L-f}{\Gamma^2+(\omega_L-f)^2}+\frac{\omega_L+f}{\Gamma^2+(\omega_L+f)^2}\right\} + \frac{a}{2}\sin ft\left\{\frac{\Gamma}{\Gamma^2+(\omega_L-f)^2}-\frac{\Gamma}{\Gamma^2+(\omega_L+f)^2}\right\}\Bigg] \quad (15.51b)$$

$$\frac{dI}{d\Omega}(\Delta m=2) = C\,\frac{\sin^2\alpha \sin^2\alpha'}{2}\Bigg[\frac{\Gamma}{\Gamma^2+4\omega_L^2} + \frac{a}{2}\cos ft\left\{\frac{\Gamma}{\Gamma^2+(2\omega_L+f)^2}+\frac{\Gamma}{\Gamma^2+(2\omega_L-f)^2}\right\} + \frac{a}{2}\sin ft\left\{\frac{2\omega_L+f}{\Gamma^2+(2\omega_L+f)^2}-\frac{2\omega_L-f}{\Gamma^2+(2\omega_L-f)^2}\right\}\Bigg]. \quad (15.51c)$$

The time-independent terms in equations (15.51a,b,c) describe Hanle effect signals excited by the unmodulated component of the incident radiation density, equation (15.47), while the time-dependent terms in equation (15.51a) describe the phase shift and depth of modulation which is expected when a system having a damping constant Γ is periodically excited (Problem 15.10). Since this is a population effect these terms are independent of the applied magnetic field. By contrast the time-dependent terms in equations (15.51b,c) are resonant at fields given by $\omega_L = \pm f$ and $2\omega_L = \pm f$, corresponding to interferences between states with $\Delta m=1$ and $\Delta m=2$ respectively. These resonances in the amplitude of modulation have the familiar Lorentzian or dispersion lineshape, depending on the phase of the modulation detected.

The signals shown in Fig.15.14a were obtained in this geometry by detecting unpolarized fluorescent light. The required theoretical expression is obtained from equation (15.51) by summing expressions with the polarizer angles set at α' and $\alpha'+\pi/2$. The contributions from the terms with $\Delta m=1$ vanish, corresponding to the fact that the orthogonal states of polarization, σ and π, of the fluorescent light can no longer interfere. In this case equation (15.51) reduces to

$$\frac{dI}{d\Omega}\,(\Delta m=0) = \frac{C}{2}(1+\cos^2\alpha)\left[\frac{1}{\Gamma} + \frac{a}{2}\left(\frac{2\Gamma\cos ft}{\Gamma^2+f^2} + \frac{2f\sin ft}{\Gamma^2+f^2}\right)\right]$$

$$\begin{aligned}\frac{dI}{d\Omega}\,(\Delta m=2) = & -\frac{C}{2}\sin^2\alpha\left[\frac{\Gamma}{\Gamma^2+4\omega_L^2} + \right.\\ & + \frac{a}{2}\cos ft\left\{\frac{\Gamma}{\Gamma^2+(2\omega_L+f)^2} + \frac{\Gamma}{\Gamma^2+(2\omega_L-f)^2}\right\} + \\ & \left. + \frac{a}{2}\sin ft\left\{\frac{2\omega_L+f}{\Gamma^2+(2\omega_L+f)^2} - \frac{2\omega_L-f}{\Gamma^2+(2\omega_L-f)^2}\right\}\right].\end{aligned} \quad (15.52)$$

As the polarization angle α of the incident radiation is varied from 0 to $\pi/2$ the intensity of the resonant terms increases from zero to a maximum value. This corresponds to the change from π polarization when only the state $m=0$ is being excited, to σ polarization when atoms are excited to a coherent superposition of $m=\pm1$ states. For $\alpha=\pi/2$ the expression given in equation (15.46) for the amplitude of the in-phase component of modulation is now easily verified, while the amplitude of the sinft component of modulation is given by

$$B = \left\{\frac{2f}{\Gamma^2+f^2} - \frac{2\omega_L+f}{\Gamma^2+(2\omega_L+f)^2} + \frac{2\omega_L-f}{\Gamma^2+(2\omega_L-f)^2}\right\} I_0\Gamma$$

and displays dispersion-shaped resonances at $2\omega_L=\pm f$. Further detailed studies of the resonance signals excited by modulated light are described by Corney (1968).

Problems

15.1. The Hanle effect is observed using the apparatus shown in Fig.15.1. Prove that the polarization of the fluorescent light, defined by $P=(I_y-I_x)/(I_y+I_x)$, is given by $P=\Gamma^2/(\Gamma^2+4\omega_L^2)$.

The geometry of the apparatus is then changed so that the magnetic field, the direction of the incident light, and the axis of observation are mutually orthogonal; and the atoms are excited by unpolarized resonance radiation. Using the classical description of the Hanle effect show that the intensity of the fluorescent light observed without a polarizer is given by

$$I(B) = \frac{I_0 R}{2}\{\frac{3}{\Gamma} - \frac{\Gamma}{\Gamma^2+4\omega_L^2}\}.$$

15.2. The widths of the Hanle signals, ΔB, observed in mercury vapour at low densities using the 1850 Å and 2537 Å resonance lines are found to have the values 85·1 ± 5·26 G and 0·649 ± 0·011 G respectively. The Landé g-factors for the 6 $^1P_1$ and 6 $^3P_1$ levels have the values 1·02 and 1·486 respectively. Calculate the radiative lifetimes of these levels and the absorption oscillator strengths of the corresponding resonance lines.

(Ans: $\tau(^1P_1) = (1{\cdot}31 \pm 0{\cdot}08) \times 10^{-9}$ s; $f_{ik} = 1{\cdot}18$
$\tau(^3P_1) = (1{\cdot}18 \pm 0{\cdot}02) \times 10^{-7}$ s; $f_{ik} = 0{\cdot}0245$.)

15.3. The density matrix of an ensemble of spin $\frac{1}{2}$ particles in the $|m_s\rangle$ representation is given by

$$\rho = \begin{pmatrix} 1/2 & 0 \\ 0 & 1/2 \end{pmatrix}.$$

Using the spin angular momentum operators

$$s_x = \begin{pmatrix} 0 & 1 \\ 1 & 0 \end{pmatrix}, \quad s_y = \begin{pmatrix} 0 & -1 \\ 1 & 0 \end{pmatrix}, \quad s_z = \begin{pmatrix} 1 & 0 \\ 0 & -1 \end{pmatrix}$$

show that $\langle s_x\rangle = \langle s_y\rangle = \langle s_z\rangle = 0$.

In another experiment every particle of the ensemble is prepared in a coherent superposition of the $m_s = \pm\frac{1}{2}$ states such that

$$\rho = \begin{pmatrix} 3/4 & 1/4 \\ 1/4 & 1/4 \end{pmatrix}.$$

Show now that $\langle s_x\rangle = \langle s_z\rangle = \frac{1}{2}$ and $\langle s_y\rangle = 0$.

15.4. Show that the intensity of light of polarization $\underline{\hat{e}}'$ emitted as an atom in the Zeeman sub-level $|m\rangle$ decays radiatively to the sub-level $|\mu'\rangle$ of the ground state is given by equation (15.25).

15.5. The polarization vector $\underline{\hat{e}}$ shown in Fig.15.10 is expressed in terms of the unit vectors $(\underline{\hat{\theta}},\underline{\hat{\phi}})$ by $\underline{\hat{e}} = \underline{\hat{\theta}}\cos\alpha + \underline{\hat{\phi}}\sin\alpha$. Using the relationships which exist between the unit vectors $(\underline{\hat{i}},\underline{\hat{j}},\underline{\hat{k}})$ and $(\underline{\hat{r}},\underline{\hat{\theta}},\underline{\hat{\phi}})$ show that $\underline{\hat{e}}$ may also be expressed in the form given by equation (15.29). Hence complete the detailed derivation of the excitation matrix, equation (15.33).

15.6. The Hanle effect geometry of Fig.15.1 corresponds to the case $(\theta=\pi/2,\ \phi=0,\ \alpha=\pi/2)$ and $(\theta'=0,\phi'=\pi/2)$. By substitution in equations (15.30) and equations (15.35) -(15.37), show that the intensity of the fluorescent light observed in this experiment for the case $J_e=1 \leftrightarrow J_g=0$ is given by

$$\frac{dI}{d\Omega} = \frac{C}{2\Gamma}\left\{\frac{1}{\Gamma} + \frac{2\Gamma\cos2\alpha'}{\Gamma^2+4\omega_L^2} + \frac{4\omega_L\sin2\alpha'}{\Gamma^2+4\omega_L^2}\right\}.$$

15.7. The quantum theory of the Hanle effect predicts that the intensity of the fluorescent light is determined by the factor C/Γ, where

$$C = \frac{U(\omega)}{8\pi\varepsilon_0^2}\left(\frac{\omega}{\hbar c}\right)^3 \langle D\rangle^4 \rho_{00}.$$

Hence show that when this fluorescent light is detected by a photomultiplier, the rate Q at which photoelectrons are released from the cathode surface is given by equation (15.38).

Calculate Q assuming the following values for the relevant parameters:

Total flux of resonance radiation incident on the cell	$= 10^{15}$ photons $cm^{-2}s^{-1}$
Illuminated area of resonance cell	$= 3\ cm^2$

Length of resonance cell	=	2 cm
Density of ground state atoms in cell	=	10^{10} cm^{-3}
Combined collection and cathode quantum-efficiency factor	=	5×10^{-5}
Absorption oscillator strength of resonance line	=	1×10^{-3}
Doppler width of both source and absorption cell	=	1500 MHz

(Ans: 1.1×10^{8}.)

15.8. In a time-resolved experiment a sample of cadmium atoms is excited to the $5\ ^3P_1$ level by a short pulse of linearly polarized resonance radiation. The atoms are subjected to an external magnetic field of 0·345 G and the exponential decay of the fluorescent light is observed to be intensity modulated. Calculate the modulation frequencies expected for atoms of both the even (I=0) and the odd (I=1/2) isotopes of cadmium. (Ans: $\omega/2\pi$ = 1·45 MHz; 0·724 MHz.)

15.9. The geometry of a pulsed resonance fluorescence experiment is described by the angles ($\theta=\pi/2$, $\phi=0$, $\alpha=\pi/2$) and ($\theta'=\pi/2, \phi'=\pi/2$). Using expressions similar to those developed in section 15.2.3, show that the intensity of the fluorescent light observed is given by

$$\frac{dI}{d\Omega} = \frac{C}{2}[1 - \cos\{2\omega_L(t-t_0)\}]\exp\{-\Gamma(t-t_0)\}.$$

15.10. A sample of gas at low pressure in zero magnetic field is subjected to a periodic excitation process, R(t), which has a fundamental angular frequency ω. By expanding R(t) as a Fourier series show that the intensity of light emitted by atoms or molecules in the excited level k is modulated and that the signal at the fundamental frequency is given by

$$S_\omega = S_0 \cos(\omega t - \phi_1)$$

where $S_0 = CA_{ki}\{(a_1^2 + b_1^2)/(\omega^2 + \Gamma^2)\}^{1/2}$ and $\phi_1 = \tan^{-1}(\omega/\Gamma)$.

The spontaneous transition probability of the observed line is A_{ki} and $\Gamma = 1/\tau_k$ is the effective total decay rate of the excited level k; a_1 and b_1 are the in-phase and quadrature Fourier expansion coefficients of R(t) at the fundamental frequency and C is an arbitrary constant. In such an experiment the phase shift ϕ_1 observed on the $A\ ^1\Pi \rightarrow X\ ^1\Sigma$ transition of NO^+ is 62° at the fundamental frequency $\omega/2\pi = 5{\cdot}5$ MHz. Calculate the effective lifetime of the excited $A\ ^1\Pi$ state.
(Ans: $\tau_k = 5{\cdot}4 \times 10^{-8}$ s.)

References

Aleksandrov, E.B. (1963). *Optics Spectrosc.* 14, 233.

____________ (1964). *Optics Spectrosc.* 17, 522.

Barrat, J.P. (1959). *J.Phys.,Paris* 20, 657.

________ and Butaux, J. (1961). *C.r.hebd.Séanc.Acad.Sci., Paris* 253, 2668.

Brossel, J. and Kastler, A. (1949). *C.r.hebd.Séanc.Acad.Sci., Paris* 229, 1213.

Byron, F.W.Jr., McDermott, M.N., and Novick, R. (1964a). *Phys.Rev.* 134, A615.

________, ________, ________ Perry, B.W., and Saloman, E.B. (1964b). *Phys.Rev.* 134, A47.

Capelle, G., Sakurai, K., and Broida, H.P. (1971). *J.Chem. Phys.* 54, 1728.

Carrington, C.G. (1972). *J.Phys.B.* 5, 1572.

________ and Corney, A. (1971). *J.Phys.B.* 4, 849.

Corney, A. (1968). *J.Phys.B.* 1, 458.

Corney, A. and Series, G.W. (1964a). *Proc.phys.Soc.,Lond.* 83, 207.

________ and __________ (1964b). *Proc.phys.Soc.,Lond.* 83, 213.

Dodd, J.N., Kaul, R.D., and Warrington, D.M. (1964). *Proc. Phys.Soc.,Lond.* 84, 176.

________, Sandle, W.J., and Zisserman, D. (1967). *Proc. Phys.Soc.,Lond.* 92, 497.

Erdmann, T.A., Figger, H., and Walther, H. (1972). *Opt.Comm.* 6, 166.

Faure, A., Nedelec, O., and Pebay-Peyroula, J.C. (1963). *C.r.hebd.Séanc.Acad.Sci.,Paris.* 256, 5088.

Gallagher, A. (1967). *Phys.Rev.* 157, 24.

Gornik, W., Kaiser, D., Lange, W., Luther, J. and Schultz, H.-H. (1972). *Opt.Comm.* 6, 327.

________, ________, ________, Luther, J., Meier, K., Radloff, H.-H., and Schultz, H. -H. (1973). *Phys.Lett.* 45A, 219.

Hanle, W. (1924). *Z. Phys.* 30, 93.

Harries, W.L. and von Engel, A. (1954). *Proc.R.Soc.* A222,490.

Kastler, A. (1950). *J.Phys.,Paris* 11, 255.

Kibble, B.P. and Series, G.W. (1961). *Proc.Phys.Soc.,Lond.*78,70 .

Lombardi, M. and Pebay-Peyroula, J.C. (1965). *C.r.hebd.Séanc. Acad.Sci.,Paris* 261, 1485.

Lurio, A. (1965). *Phys.Rev.* 140, A1505.

Lurio, A. and Novick, R. (1964). *Phys.Rev.* 134, A608.

Lurio, A., de Zafra, R.L., and Goschen, R.J. (1964). *Phys.Rev.* 134, A1198.

Nedelec, O. (1966). *J.Phys.,Paris* 27, 660.

Pebay-Peyroula, J.C. (1969). *Physics of the one- and two-electron atoms*. (Eds. F. Bopp and H. Kleinpoppen) p. 348. North-Holland, Amsterdam.

Rayleigh, Lord (1922). *Proc.R.Soc.* 102, 190.

Sakurai, K. and Capelle, G. (1970). *J.chem.Phys.* 53, 3764.

______ ______ and Broida, H.P. (1971). *J.chem. Phys.* 54, 1220.

Schenck, P., Hilborn, R.C., and Metcalf, H. (1973). *Phys. Rev.Lett.* 31, 189.

Schmieder, R.W., Lurio, A., and Happer, W. (1968). *Phys.Rev.* 173, 76.

______, ______, ______ and Khadjavi, A. (1970). *Phys.Rev.* A2, 1216.

Wood, R.W. (1913). *Researches in physical optics*, Vols. I and II. Columbia University Press, New York.

______ and Ellett, A. (1924). *Phys.Rev.* 24, 243.

Wright, J.J., Dawson, J.F., and Balling, L.C. (1974). *Phys.Rev.* A9, 83.

General references and further reading

A wealth of information about early experiments on resonance fluorescence is contained in two text-books

Mitchell, A.C.G. and Zemansky, M.W. (1966). *Resonance radiation and excited atoms*. Cambridge University Press, London.

Wood, R.W. (1934). *Physical optics*. Macmillan, New York.

A clear account of the theory of the interaction of atoms and radiation has recently been published by

Loudon, R. (1973). *The quantum theory of light*. Clarendon Press, Oxford.

The theory of the density matrix is reviewed in an article by

Fano, U. (1957). *Rev.Mod.Phys.* 29, 74.

and the theoretical description of resonance fluorescence experiments in terms of the density matrix is developed in articles by

Barrat, J.P. and Cohen-Tannoudji, C. (1961). *J.Phys.,Paris.* 22, 329 and 443.

and

Cohen-Tannoudji, C. (1962). *Annls.Phys.* 7, 423 and 469.

The physical phenomena associated with the Hertzian coherence of the density matrix are discussed by

Novikov, L.N., Pokazan'ev, V.G. and Skrotskii, G.V. (1970). *Soviet Phys.Usp.* 13, 384.

____________, Skrotskii, G.V., and Solomakho, G.I. (1974). *Soviet Phys.Usp.* 17, 542.

16

Optical double resonance experiments

16.1. Magnetic resonance and excited atoms

16.1.1. Introduction. In our discussion of the Hanle effect in the previous chapter, we assumed that the g-factors of the excited atomic levels were already well known or could be calculated to sufficient accuracy from the Landé formula

$$g_J = g_l\{J(J+1) + L(L+1) - S(S+1)\}/2J(J+1) + \\ + g_s\{J(J+1) + S(S+1) - L(L+1)\}/2J(J+1).$$

However, in the spectra of mercury and many other elements the the existence of intercombination lines shows that L-S coupling does not hold rigorously and that small deviations of the actual g-factors from the L-S values may be observed. These deviations contain useful information about the atomic wavefunctions and a technique which would permit the accurate measurement of the Zeeman splittings of excited levels is clearly very desirable. It would also be interesting to make accurate measurements of the hyperfine structures of the excited levels of odd isotopes since these give information about the magnetic dipole and electric quadrupole moments of the atomic nucleus.

Unfortunately, however, conventional techniques of high-resolution optical spectroscopy are limited by the large Doppler widths of spectral lines, typically of the order of $0{\cdot}050\ \text{cm}^{-1} \equiv 1500$ MHz. Many of the hyperfine structure intervals that must be measured lie in the range $(3\text{-}30) \times 10^{-3}\ \text{cm}^{-1} \equiv (90\text{-}900)$ MHz and low field Zeeman splittings are generally even smaller. Thus the techniques of optical spectroscopy are incapable of achieving the required resolution or precision. This is mainly due to the fact that a small energy separation is being measured indirectly by taking the difference between two very large optical frequencies. From equation (8.42) we recall that the Doppler width is proportional to the observed frequency.

If, therefore, some means could be devised to measure the splittings between the excited levels directly, then the Doppler width associated with the signal would be reduced to a negligible value.

This desirable result is in fact achieved in the magnetic resonance technique, which is well known in atomic beam experiments and in solid state physics. In investigations of ground-state Zeeman splittings by this method, the atoms are subjected to a static magnetic field $\underline{B}=B\hat{\underline{k}}$ which splits the magnetic sublevels by an amount $\hbar\omega_L$ where $\omega_L=g_J\mu_B B/\hbar$ is the Larmor angular frequency. In addition the atoms experience a time dependent field $\underline{B}_1$ which rotates in a plane perpendicular to $\underline{B}$ at the angular frequency ω_0. The coils which generate this magnetic field may be regarded as a source of radio frequency photons having an energy $\hbar\omega_0$. When the frequency of the r.f. field is such that $\omega_0=\omega_L$, magnetic dipole transitions are induced between the levels with the absorption or stimulated emission of quanta of the radio frequency field. In solid state experiments these transitions are usually detected by means of the sharp increase in the r.f. power absorbed by the sample which occurs at resonance. Since the transition probabilities for stimulated emission and absorption between a given pair of states are equal, it is necessary to establish a population difference between them before the resonance can be detected. In solid state experiments this population inequality is provided by the Boltzmann factor $\sim \exp(-g_J\mu_B B/kT)$ and can be enhanced by working at low temperatures.

For several years it seemed unlikely that the magnetic resonance method could be applied to excited atoms because of the low population of atoms that can be created in excited levels and the high temperatures required to attain appreciable vapour densities. However, it was pointed out by Brossel and Kastler (1949) that the excitation of atoms using polarized resonance radiation allowed very large population differences to be established between excited sub-levels, and that the change in the polarization of the fluorescent light which occurs at resonance provides an extremely sensitive method of detecting the induced magnetic dipole transitions. This method of studying excited atoms in which optical resonance fluorescence and radio frequency magnetic resonance are combined is known as the *optical double-resonance* technique. We shall now describe the pioneering experiments in this field carried out by Brossel and Bitter (1952).

16.1.2. Description of the Brossel-Bitter experiment. The apparatus used by Brossel and Bitter (1952) in the first successful application of the optical double-resonance method is shown in Fig.16.1(a). The excited $6\ {}^3P_1$ level of mercury was again chosen for study, although similar work on many other elements has been reported subsequently. Light from a low pressure mercury lamp is passed through a linear polarizer oriented so that the electric vector of the radiation is parallel to the constant magnetic field $\underline{B}$ applied to the atoms in the resonance cell. This field is produced by Helmholtz coils which are not shown in the diagram. With this polarization the incident 2537 Å radiation can only stimulate electric dipole transitions for which $\Delta m_J=0$, Fig.16.1(b). If we restrict our attention for the moment to the atoms of the even mercury isotopes, only the $m_J=0$ state of the excited level is populated. In the absence of other perturbations these atoms would decay after a mean lifetime $\tau = 1/\Gamma$, re-emitting light which is also polarized parallel to $\underline{B}$ (π polarization).

If, however, the excited atoms are subjected to the influence of a strong radio frequency field $\underline{B}_1$, rotating in a plane at right-angles to the steady field $\underline{B}$, then conditions are right for magnetic resonance to occur. When the Larmor frequency ω_L of the excited atoms in the field $\underline{B}$ approaches the angular frequency ω_0 of the applied r.f. magnetic field, absorption and stimulated emission of magnetic dipole radiation transfers the atoms to the $m_J = \pm 1$

states. Since atoms in the $m_J = \pm 1$ states decay by emitting light which is left and right circularly polarized (σ^+ and σ^- respectively in Fig.16.1(b)), the magnetic dipole transi-

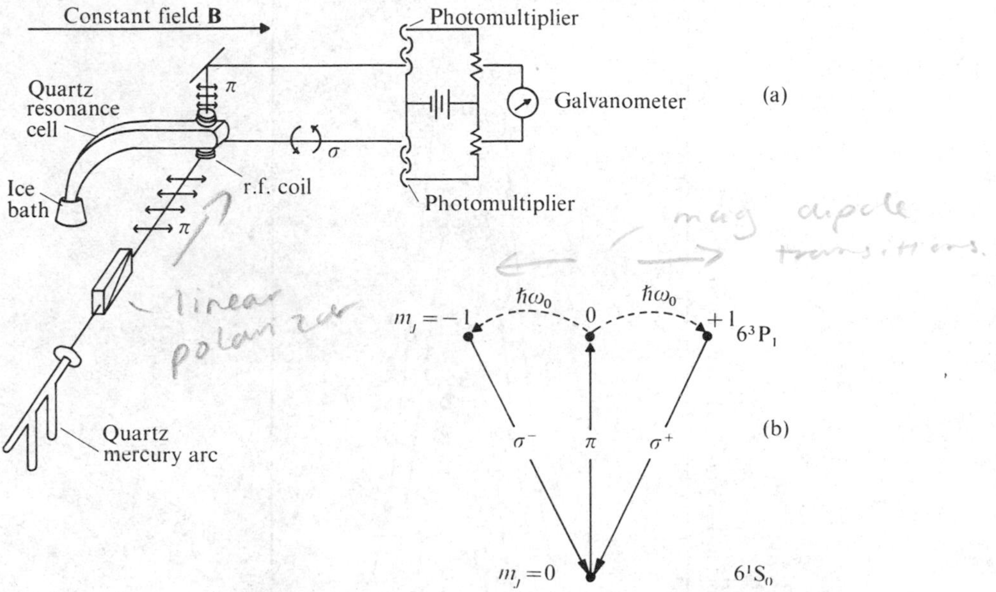

Fig.16.1. (a) Schematic diagram of the apparatus used by Brossel and Bitter (1952) for optical double-resonance experiments on the $6\ ^3P_1$ level of mercury. (b) The optical excitation and fluorescent decay processes involved together with the magnetic dipole transitions induced between the Zeeman sub-levels of the $6\ ^3P_1$ level.

tions will be accompanied by an increase in σ-polarized and a decrease in π-polarized resonance fluorescence. These changes were detected by two photomultipliers which received light emitted in the direction of, and perpendicular to, the field $\underline{B}$. The photomultiplier anode currents were sent in opposition through a bridge network which enabled the noise produced by lamp fluctuations to be eliminated and at the same time doubled the observed signal. The difference current was measured on a galvanometer and when these readings were plotted as a function of the field strength B, a beautiful set of resonance curves was obtained, as shown in Fig.16.2. When the strength of the r.f.

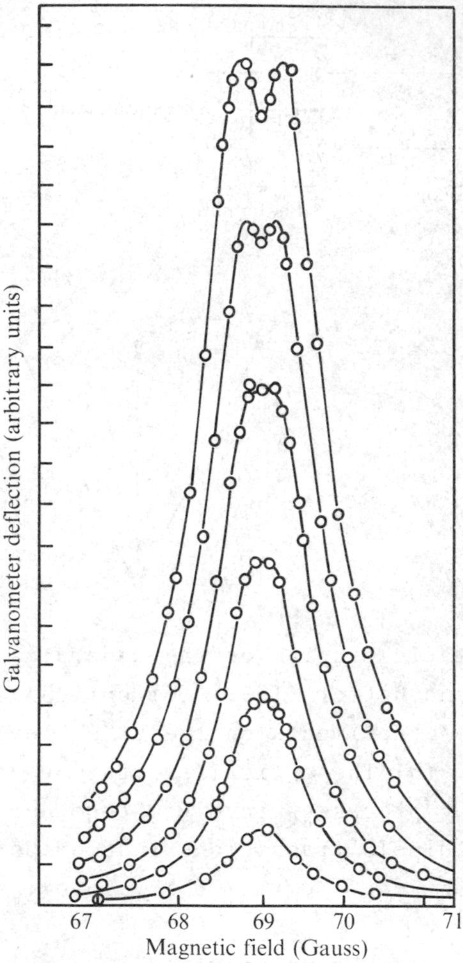

Fig.16.2. Optical double-resonance signals observed on the $6\ {}^1S_0$ - $6\ {}^3P_1$ transition of mercury at 2537 Å as a function of the d.c. magnetic field. The curves illustrate the effect of increasing the magnitude of the r.f. field, $\underline{B}_1$. (After Brossel and Bitter (1952).)

magnetic field was increased the amplitude of the signal grew, corresponding to an increase in the stimulated transition probability. The resonance curves also broadened and eventually became double peaked as the stimulated transitions reduced the population differences between excited states.

By measuring the frequency of the r.f. field and calibrating the static field by means of proton resonance, Brossel and Bitter deduced a value for the Landé g_J-factor of the $6\ ^3P_1$ level of the even mercury isotopes of

$$g_J = 1{\cdot}4838 \pm 0{\cdot}0004.$$

This differs quite significantly from the L-S coupling value (Problem 16.1).

However, the importance of this experiment lies not so much in the result quoted above as in the demonstration of the feasibility of the optical double-resonance method. This pioneering work stimulated renewed interest in resonance fluorescence experiments and opened up a new branch of atomic physics. We shall now derive a detailed expression for the shape of the resonance signal and consider what further information can be obtained by careful studies of the linewidth.

16.2. Theory of the Brossel-Bitter experiment

<u>16.2.1. Classical theory of precession of magnetic dipoles in external magnetic fields.</u> The theory of magnetic resonance can be treated quantum mechanically by introducing into Schrödinger's equation a time-dependent perturbation operator, $-\underline{\mu}.\underline{B}(t)$, representing the combined effect of the magnetic fields applied to the atom. The time-dependent transition probabilities between different magnetic sublevels may then be calculated. It is instructive, however, to approach the calculation first from the classical point of view and to attempt to describe the motion of a classical magnetic dipole moment $\underline{\mu}$ in the combination of static and time-dependent fields. The ideas developed in this discussion of magnetic resonance experiments are useful in many

situations and moreover they lead to expressions for the shape of optical double-resonance signals which are in agreement with observations for the case of an excited level with $J_e=1$ decaying to a ground state with $J_g=0$.

(a) Precession in a static magnetic field. We first examine the behaviour of a magnetic dipole moment $\underline{\mu}$ in a static magnetic field $\underline{B}=B\hat{\underline{k}}$ pointing along the z-axis. The magnetic dipole moment is assumed to be associated with an angular momentum vector $\hbar\underline{J}$ by the relation $\underline{\mu}=-\gamma\hbar\underline{J}$ where $\gamma = g_J\mu_B/\hbar$ is called the magneto-gyric ratio. The magnetic dipole experiences a torque which acts to change the direction of the angular momentum vector, giving the equation of motion

$$\hbar \frac{d\underline{J}}{dt} = \underline{\mu} \wedge \underline{B}. \tag{16.1}$$

Substituting for $\underline{J}$ in this equation we have

$$\frac{d\underline{\mu}}{dt} = -\gamma \; \underline{\mu}\wedge\underline{B}. \tag{16.2}$$

It is not difficult to show (Problem 16.2) that the steady state solution of equation (16.2) may be written in the form

$$\begin{aligned} \mu_x &= \mu \sin\theta \cos\omega_L t \\ \mu_y &= \mu \sin\theta \sin\omega_L t \\ \mu_z &= \mu \cos\theta. \end{aligned} \tag{16.3}$$

This solution corresponds to a precession of the magnetic dipole about the direction of the magnetic field at the Larmor angular frequency $\omega_L = \gamma B$ with $\underline{\mu}$ inclined at some constant but arbitrary angle θ, as shown in Fig.16.3.

(b) Equation of motion in rotating coordinate systems. When we consider a magnetic dipole moment subjected in addition to time-dependent fields, it is first of all appropriate to discuss the motion in a coordinate system which rotates at the angular velocity $\underline{\omega}_0$ with respect to the laboratory frame. The rate of change $\partial\underline{A}/\partial t$ of a vector in the rotating frame is related to the rate of change $d\underline{A}/dt$ in the laboratory frame by

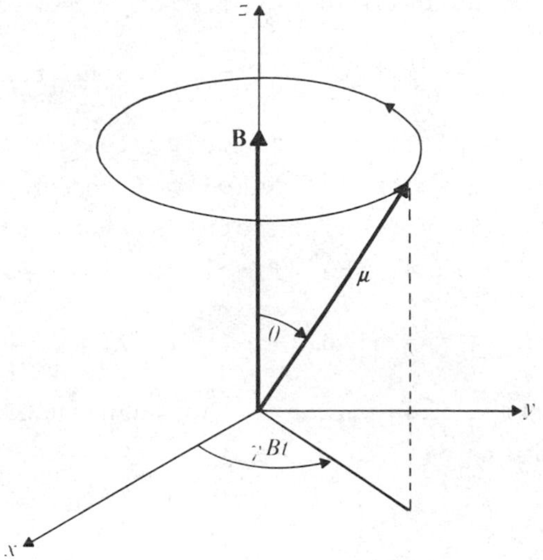

Fig.16.3. Precession of a magnetic dipole moment $\underline{\mu}$ about a steady magnetic field $\underline{B}$.

$$\frac{\partial \underline{A}}{\partial t} = \frac{d\underline{A}}{dt} - \underline{\omega}_0 \wedge \underline{A}. \tag{16.4}$$

Applying equation (16.4) to the rate of change of $\underline{\mu}$ given by equation (16.2), we have

$$\frac{\partial \underline{\mu}}{\partial t} = -\gamma\ \underline{\mu} \wedge \underline{B} - \underline{\omega}_0 \wedge \underline{\mu} = -\gamma\ \underline{\mu} \wedge (\underline{B} - \underline{\omega}_0/\gamma). \tag{16.5}$$

Thus the equation of motion in the rotating frame is identical to that in the laboratory frame when $\underline{B}$ is replaced by the effective field

$$\underline{B}_e = \underline{B} - \underline{\omega}_0/\gamma. \tag{16.6}$$

In particular if $\underline{\omega}_0 = \gamma\underline{B}$ then $\partial\underline{\mu}/\partial t = 0$ and the magnetic moment is at rest in the rotating frame, thus confirming the solution of the equation of motion given by equation (16.3).

(c) Application to time-dependent fields. We now consider

a magnetic moment $\underline{\mu}$ subjected simultaneously to a static field $\underline{B}=B\hat{\underline{k}}$ and a radio frequency field

$$\underline{B}_1 = \hat{\underline{i}}\, B_1\cos\omega_0 t + \hat{\underline{j}}\, B_1\sin\omega_0 t \qquad (16.7)$$

which rotates in a plane perpendicular to the z-axis in the counterclockwise sense at the angular frequency ω_0. In a coordinate system 0x'y'z which rotates about the z-axis at the same rate (Fig.16.4) the effective field is

$$\underline{B}_e = \hat{\underline{i}}'B_1 + \hat{\underline{k}}(B-\omega_0/\gamma) \qquad (16.8)$$

and is clearly time independent. The magnitude of the effective field is

$$B_e = \{B_1^2 + (B-\omega_0/\gamma)^2\}^{1/2} \qquad (16.9)$$

and it is inclined at the angle β to the z-axis, where

$$\tan\beta = \frac{B_1}{B-\omega_0/\gamma}\,. \qquad (16.10)$$

From the discussion of section 16.2.1(a), it follows that in the rotating frame the magnetic dipole precesses at the angular frequency $\omega_e = \gamma B_e$ about the fixed direction of the effective field $\underline{B}_e$ inclined at a constant arbitrary angle. This precession is indicated in Fig.16.4. In the rotating frame the vector $\underline{\mu}$ traces out the surface of a cone whose axis is defined by the direction of the effective field, and the tip of the vector moves on a circle in a plane perpendicular to $\underline{B}_e$. In this plane a magnetic dipole $\underline{\mu}$, originally created parallel to the z-axis at some time t_0, will have precessed through the angle ψ by time t where

$$\psi = \gamma B_e(t-t_0). \qquad (16.11)$$

At this later time t, the angle α that the dipole moment makes with the z-axis is obtained by applying the cosine law to the two triangles indicated in Fig.16.4, giving

$$\cos\alpha = \cos^2\beta + \sin^2\beta\,\cos\{\gamma B_e(t-t_0)\}. \qquad (16.12)$$

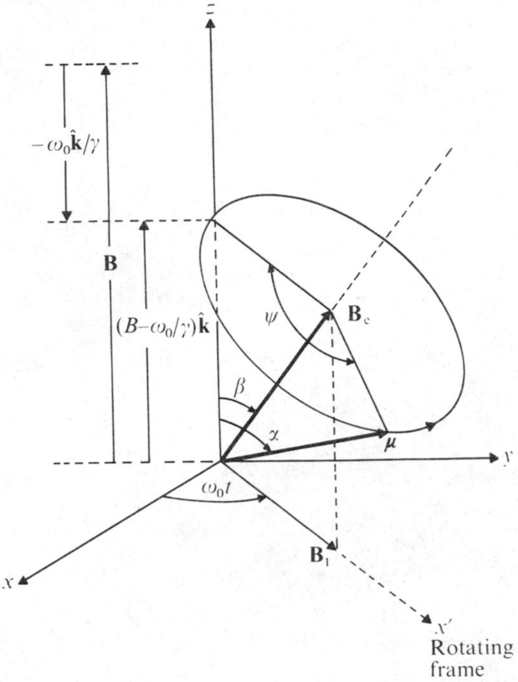

Fig.16.4. The components of the effective magnetic induction $\underline{B}_e$ in a coordinate system rotating with the angular velocity $\omega_0\hat{\underline{k}}$. The field in the laboratory frame is given by $\hat{\underline{i}}\ B_1 \cos\omega_0 t + \hat{\underline{j}}\ B_1 \sin\omega_0 t + \hat{\underline{k}}\ B$.

In a combination of static and rotating fields the z-component of the magnetic dipole moment,

$$\mu_z = \mu \cos\alpha, \qquad (16.13)$$

is therefore a periodic function of time. On resonance, when $\omega_L=\omega_0$, we have $\beta=\pi/2$ and μ_z oscillates sinusoidally between $\pm\mu$ at a rate determined by the field strength B_1.

16.2.2. Interpretation in terms of transition probabilities. This result, derived on the basis of classical mechanics, can now be interpreted quantum mechanically by assuming that the magnetic moment $\underline{\mu}$ is associated with the spin of an electron. In the laboratory frame the electron wavefunction may be written as a linear superposition of spin up, $\chi(+1/2)$, and spin down, $\chi(-1/2)$, eigenfunctions:

$$\psi(t) = a_+(t)\chi(+1/2) + a_-(t)\chi(-1/2). \tag{16.14}$$

The expectation value of the magnetic moment of the system is

$$\langle\mu_z\rangle = \langle\psi|-2\mu_B S_z|\psi\rangle = -\mu_B(|a_+|^2 - |a_-|^2). \tag{16.15}$$

Equating equations (16.13) and (16.15), we have

$$|a_+|^2 - |a_-|^2 = -\cos\alpha. \tag{16.16}$$

Finally combining this equation with the normalization condition $|a_+|^2 + |a_-|^2 = 1$ we have

$$\begin{aligned} |a_+|^2 &= (1-\cos\alpha)/2 = \sin^2(\alpha/2) \\ |a_-|^2 &= (1+\cos\alpha)/2 = \cos^2(\alpha/2). \end{aligned} \tag{16.17}$$

These equations describe the motion of an electron which at time t_0 was prepared in the spin down state corresponding to a classical dipole moment pointing along the z-axis. Thus the probability of a transition occurring to the spin up state is given by

$$\begin{aligned} P(t,t_0) = |a_+|^2 &= \sin^2\beta \sin^2\{\gamma B_e(t-t_0)/2\} \\ &= \frac{\omega_1^2}{\omega_1^2 + (\omega_L - \omega_0)^2} \sin^2\{\omega_e(t-t_0)/2\} \end{aligned} \tag{16.18}$$

where $\omega_1 = \gamma B_1$ and $\omega_e = \gamma B_e$. On resonance the electron oscillates periodically between the spin-down and the spin-up states, as shown in Fig.16.5. The rate at which the transition occurs is determined by the strength of the rotating field B_1. The classical precession of the magnetic dipole moment in the combined magnetic fields can therefore be interpreted quantum mechanically in terms of magnetic dipole transitions stimulated by the applied radio-frequency field $\underline{B}_1$ (Problem 16.3).

16.2.3. Magnetic dipole transitions between isolated excited states. We now consider magnetic dipole transitions between two isolated excited atomic states which have the same

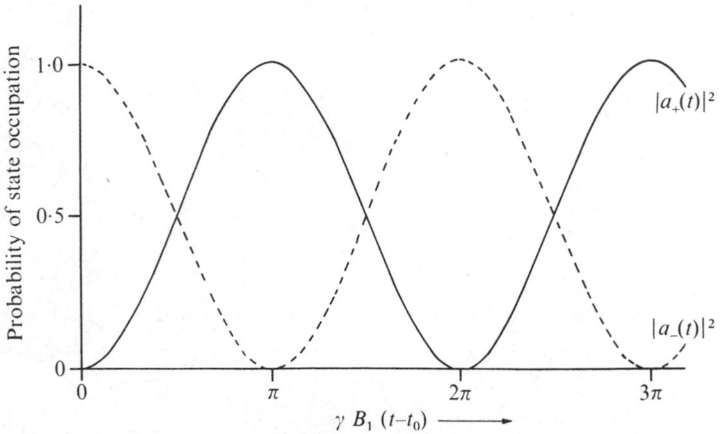

Fig.16.5. A plot of the time-dependent probabilities, $|a_+(t)|^2$ and $|a_-(t)|^2$, of finding an electron in the $m_s = +\frac{1}{2}$ and $m_s = -\frac{1}{2}$ states respectively. The electron was prepared in the $m_s = -\frac{1}{2}$ state at time $t=t_0$ and the magnetic resonance condition $\omega_0 = \gamma B = \omega_L$, is assumed to be satisfied exactly.

radiative decay rate Γ. This calculation would apply to the Brossel-Bitter experiment, for example, if the fluorescent light was transmitted through a circular polarizer, thus allowing the increase in the σ^+ polarized radiation alone to be measured. We are then concerned only with transitions from the $m_J=0$ to the $m_J=+1$ states. The intensity of light emitted at time t by atoms which were prepared in the state $m_J=0$ at time t_0 is given by

$$I(t,t_0) = I_0P(t,t_0)\ \exp\{-\Gamma(t-t_0)\} \qquad (16.19)$$

where $P(t,t_0)$ is the magnetic resonance transition probability and the exponential factor takes account of the radiative decay of the excited atoms. Using equation (16.18) the resonance fluorescence signal obtained from a sample of atoms which are excited at a uniform rate R from $t_0 = -\infty$ is

$$I = R\int_{-\infty}^{t} I(t,t_0)dt_0$$

$$= I_0R\left\{\frac{\omega_1^2}{\omega_1^2+(\omega_L-\omega_0)^2}\right\}\int_{-\infty}^{t}\sin^2\{\omega_e(t-t_0)/2\}\exp\{-\Gamma(t-t_0)\}dt_0$$

$$= \frac{RI_0}{\Gamma}\,\frac{\omega_1^2}{\Gamma^2+\omega_1^2+(\omega_L-\omega_0)^2}\,. \qquad (16.20)$$

Not unexpectedly this is a Lorentzian-shaped signal centred at the magnetic field $B_0=\omega_0/\gamma$. The full width at half the maximum intensity,

$$\Delta\omega_{1/2} = 2(\Gamma^2+\gamma^2B_1^2)^{1/2}, \qquad (16.21)$$

is determined, at low r.f. powers, by the radiative widths of the two energy levels involved. The magnetic dipole transitions stimulated by the external r.f. field artificially shorten the lifetime of the states and lead to the power broadening of the resonance signals which is evident in Fig.16.2. This power broadening is expressed quantitatively in equation (16.21).

16.2.4. Derivation of the Brossel-Bitter formula. In a static magnetic field the 6 $^3P_1$ level of mercury splits into three magnetic sub-states which are essentially equally spaced for fields less than 10^5 G. Thus the simple result presented above does not apply for the detection scheme used in the Brossel-Bitter experiment and equation (16.20) must therefore be modified. The magnetic dipole transition probabilities for a system with a multiplicity of equally spaced levels have been calculated quantum mechanically by Majorana (1932) and Rabi (1932). However, we prefer to return to our classical model in which the excited atoms are represented as electric dipole oscillators. When π-polarized excitation is used, as in the Brossel-Bitter experiment, the axes of the dipoles at the instant of excitation are parallel to the z-direction. Under the combined influence of the static and rotating magnetic fields these dipoles precess until after a time interval $(t-t_0)$ they are inclined at the angle α to the z-axis, where $\cos\alpha$ is given by equation (16.12). The σ-polarized resonance fluorescence signal observed in the direction of the magnetic field (see Fig.16.1) is proportional to the square of the component of the electric dipole moment perpendicular to $\underline{B}$, giving

$$\begin{aligned} I(t,t_0) &= I_0\sin^2\alpha\exp\{-\Gamma(t-t_0)\} \\ &= 2I_0\sin^2\beta[(1-\tfrac{3}{4}\sin^2\beta) - \cos^2\beta\cos\{\omega_e(t-t_0)\} - \\ &\quad - \tfrac{1}{4}\sin^2\beta\cos\{2\omega_e(t-t_0)\}]\exp\{-\Gamma(t-t_0)\}. \end{aligned} \qquad (16.22)$$

The signal observed in a steady state experiment is obtained by integrating equation (16.22) from $t_0 = -\infty$ to t; after some manipulation we obtain (Problem 16.4)

$$\begin{aligned} I &= \frac{RI_0}{4\Gamma}\sin^2\beta\left\{(1-\tfrac{3}{4}\sin^2\beta) - \frac{\Gamma^2\cos^2\beta}{\Gamma^2+\omega_e^2} - \frac{\Gamma^2\sin^2\beta}{4(\Gamma^2+4\omega_e^2)}\right\} \\ &= \frac{RI_0}{4\Gamma}\frac{\omega_1^2}{\omega_1^2+(\omega_L-\omega_0)^2}\left\{\frac{\omega_1^2}{\Gamma^2+4\omega_1^2+4(\omega_L-\omega_0)^2} + \frac{(\omega_L-\omega_0)^2}{\Gamma^2+\omega_1^2+(\omega_L-\omega_0)^2}\right\}. \end{aligned} \qquad (16.23)$$

In the original experiments of Brossel and Bitter (1952) this expression was checked by determining the values of RI_0 and ω_1 which fitted the experimental data at resonance and then plotting equation (16.23) using a value for Γ which was obtained by a method described below. As may be seen from Fig.16.2 there is excellent agreement between the theoretical curves and the experimental points over the entire range of values of both static and r.f. field strengths. When $\omega_1 = \gamma B_1 \ll \Gamma$ the curves are bell shaped with a maximum at $B_0=\omega_0/\gamma$. As the strength of the rotating field is increased, the amplitude of the resonance signal grows rapidly, as we see in Fig.16.2. Eventually, when the condition $\omega_1 \gg \Gamma$ is approached, the second term in equation

(16.23) becomes important and the signal becomes double peaked. This can be regarded as the effect of saturation of the magnetic resonance which occurs when the rotating field $\underline{B}_1$ is sufficiently strong that the induced magnetic dipole transitions substantially change the populations of the levels within the excited state lifetime of $1/\Gamma$.

16.3. Discussion of the optical double-resonance method

16.3.1. The effects of hyperfine structure. In their original investigation Brossel and Bitter (1952) also performed double-resonance experiments on the odd mercury isotopes ^{199}Hg and ^{201}Hg which have nuclear spin $I=1/2$ and $3/2$ respectively. The r.f. magnetic field now induces transitions between the hyperfine structure levels $|F,M_F\rangle$ which satisfy the selection rule $\Delta F=0$, $\Delta M_F = \pm 1$. In low magnetic fields the observed resonances allow the g_F factor (equation (15.28)) of a given hyperfine level to be determined and so lead to a determination of the nuclear spin I (Problem 16.5).

In addition to these Zeeman resonances, however, r.f. transitions satisfying the selection rule $\Delta F = \pm 1$, $\Delta M_F = \pm 1$ can also be detected and enable the hyperfine structure intervals in zero field to be obtained. Indeed the investigation of the hyperfine structure of excited energy levels has been one of the main applications of the optical double-resonance technique. However, we reserve a detailed discussion of these experiments until Chapter 18 when the hyperfine structure experiments on both excited and ground state atoms will be considered.

16.3.2. Comparison with the Zeeman effect in optical spectroscopy. In optical double-resonance experiments the Doppler shift due to the motion of an atom through the r.f. magnetic field is negligible in comparison with the natural width of the excited levels. Substituting into equation (16.21) a typical atomic lifetime of 10^{-7} s leads to a line-width for the magnetic resonance signal, at low r.f. power of

3 MHz. If the centre of this resonance can be determined to 1/100th part of the linewidth and the r.f. field employed has a frequency of 150 MHz, we achieve an accuracy 0·02 per cent in the measurement of g_J. Although this precision is poor compared with that attainable in atomic beam experiments on ground or long-lived metastable states, it is nevertheless orders of magnitude better than that obtained when using the techniques of optical spectroscopy (Problem 16.6).

The disadvantage, however, of the optical double-resonance method is that the excited levels can only be investigated one by one, and often only a few levels in any given atom can be excited sufficiently strongly to enable accurate measurements to be made. This should be contrasted with the optical Zeeman effect experiments in which a great deal of information about many different spectral lines can be recorded simultaneously by combining a conventional spectrograph and a Fabry-Perot étalon.

16.3.3. Sensitivity of optical double-resonance experiments. In conventional solid state magnetic resonance experiments the necessary population difference between the states is created by the Boltzmann factor $\exp(\underline{\mu}.\underline{B}/kT)$ and is enhanced by working at low temperatures and high field strengths. The resonance is detected by changes in the bulk magnetization of the sample or in the radio frequency power absorbed. These factors together limit the sensitivity of the method and samples containing 10^{12} spins per cubic centimetre are required in electron resonance experiments.

By contrast optical double-resonance experiments can be performed with vapour densities as low as 10^6 atoms cm^{-3}. This great increase in sensitivity is due to the high atomic polarization achieved by optical excitation combined with the fact that in these experiments the absorption of an r.f. quantum triggers the detection of a visible or ultraviolet quantum whose energy is some 10^6 - 10^7 times greater. Optical double-resonance experiments can therefore be performed on samples containing only a few milligrams of mass-separated isotopes or even on short-lived radioactive isotopes.

16.3.4. Amplitude and frequency of the rotating field. An examination of equation (16.23) reveals that strong resonance signals will be observed only when the condition

$$\omega_1 = \gamma B_1 \approx \Gamma$$

is satisfied. This is the r.f. magnetic field strength that will cause the classical magnetic moment to precess through one radian during a mean lifetime $\tau = 1/\Gamma$. Using typical values of $\tau = 10^{-8}$ s and $\gamma = 10^7\ s^{-1}G^{-1}$ (corresponding to a magnetic moment of approximately one Bohr magneton) we find $B_1 \approx 10$ G. Actually this is an overestimate since in many cases induced magnetic dipole transitions are detectable at much smaller r.f. field strengths. However, it shows clearly that in general r.f. power amplifiers will be necessary to excite the resonance.

The frequencies of the rotating fields must also be chosen so that the Zeeman states are well resolved at resonance. This indicates the use of frequencies in the 10-100 MHz range and even higher frequencies are required for hyperfine structure investigations in the group IIB elements. For levels with very short lifetimes it may be difficult to obtain r.f. magnetic fields of the necessary magnitude. Furthermore the application of high power, high frequency fields to the resonance cell often leads to the excitation of a discharge in the cell. This difficulty can be overcome by improving the vacuum in the resonance cell or by using r.f. loops which are specially designed to produce a node in the electric field distribution.

16.3.5. The linewidth of the magnetic resonance signal. Equation (16.21) predicts that the square of the half-width of the resonance signal should be a linear function of the r.f. power applied. This relationship was verified by Brossel and Bitter (1952) and their results are shown in

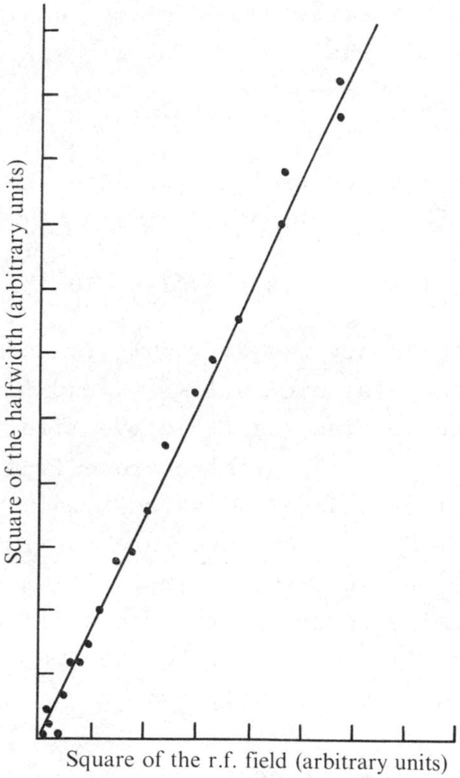

Fig.16.6. Broadening of the Brossel-Bitter optical double-resonance signal as a function of r.f. power. (After Brossel and Bitter (1952).)

Fig.16.6. However, since in their experiments all three Zeeman levels were coupled by the r.f. magnetic field, equation (16.21) must be modified to

$$\Delta\omega_{1/2} \quad = \quad 2\{\Gamma^2 + 5{\cdot}8(\gamma B_1)^2\}^{1/2} \qquad (16.24)$$

which can be derived from equation (16.23) after some straightforward but rather tedious algebra.

Both equations (16.21) and (16.24) show that the natural linewidth Γ may be obtained by extrapolation of the measured linewidths to zero r.f. field strengths. When this extrapolation was performed in the original Brossel-

Bitter experiment, the effective lifetime obtained for the $6\ ^3P_1$ level of mercury was

$$\tau_{eff}(6\ ^3P_1) = (1\cdot 55 \pm 0\cdot 015) \times 10^{-7}\ s.$$

This result is some 30 per cent longer than the currently accepted value of the radiative lifetime:

$$\tau(6\ ^3P_1) = (1\cdot 18 \pm 0\cdot 02) \times 10^{-7}\ s.$$

Initially this discrepancy was difficult to understand since most processes such as atomic collisions and field inhomogeneity tend to shorten the effective lifetime. It was eventually realized that the problem arose from the trapping or re-absorption of resonance radiation within the cell owing to the rather high vapour pressure (2×10^{-4} Torr at 0^oC) used in these initial experiments. It was found later that similar narrowing occurs in the Hanle effect signals described in Chapter 15. We now take the opportunity to discuss this topic in more detail.

16.4. Radiation trapping and coherence narrowing

<u>16.4.1. Experimental measurements using linearly polarized light.</u> The first detailed investigation of the width of optical double-resonance signals over a wide range of pressure were made by Guiochon *et al.*(1957). Both the odd and even isotopes of mercury were studied and similar measurements using the Hanle effect were made by Barrat (1957). These experiments demonstrated a pronounced narrowing of the resonance fluorescence signals as the mercury vapour pressure is increased. This leads to an increase in the value of τ_{eff} deduced from the widths of the measured signals, and results obtained by the two techniques for the even isotope ^{202}Hg are shown in Fig.16.7. There is good agreement between the values of τ_{eff} obtained and further results are included in a review of the experimental data prepared by Barrat (1959a).

This narrowing of the signals is attributed to re-

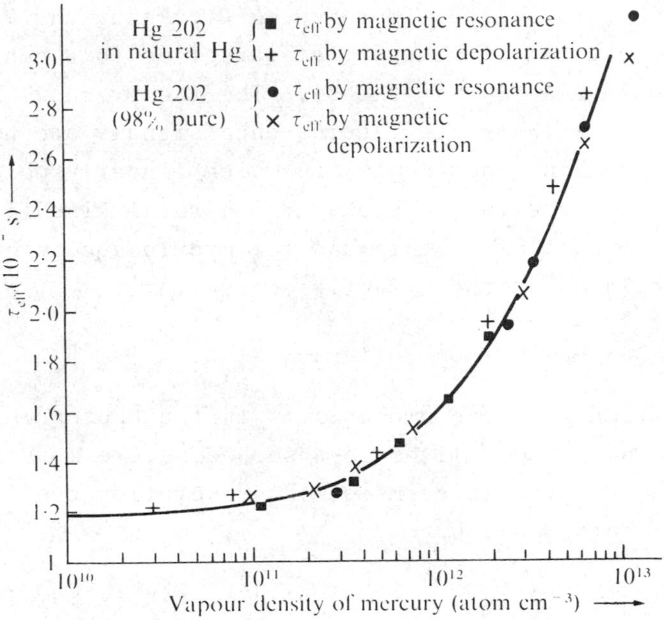

Fig.16.7. Coherence narrowing of Hanle effect and optical double-resonance signals in ^{202}Hg at high vapour densities. (After Barrat (1957).)

absorption or trapping of the emitted resonance fluorescence photons within the cell. If the atoms which are involved in the multiple scattering process are contained in the same magnetic field, then the time development of their excited state wavefunctions will be identical. Consequently information about the polarization of the atom in the excited state can be partially transferred by a photon from one atom to another. The second atom can contribute to the observed signal and the lifetime of the atom has effectively been doubled. The resulting increase in the average lifetime of the ensemble of excited atoms causes the resonance signals to be narrower than expected and the phenomenon is therefore often called *coherence narrowing*.

Barrat (1959b) developed a density matrix theory to

describe the effect of this multiple scattering of resonance radiation, which was later improved by D'yakanov and Perel (1965a). For the moment we restrict ourselves to a consideration of the $J_e=1 \leftrightarrow J_g=0$ case. The theory shows that the shapes of the resonance fluorescence signals are unchanged but that in experiments in which linearly polarized light is used to excite the atoms, the natural linewidth appearing in equations in this and the previous chapter should be replaced by the effective decay rate

$$\Gamma_{eff} = \Gamma(1 - \frac{7}{10} x). \qquad (16.25)$$

In this equation x is the probability that a photon emitted in the resonance line will be re-absorbed before leaving the cell and is given in terms of the absorption coefficient at the line centre, κ_0, by

$$x = 1 - \frac{1}{\sqrt{\pi}} \int_{-\infty}^{\infty} \exp\{-\kappa_0 L \exp(-y^2)\} \exp(-y^2) dy \qquad (16.26)$$

where the length parameter L is of the order of the dimensions of the resonance cell. Theoretical curves obtained by evaluating equation (16.25) are in good agreement with the experimental results over the density range from $10^{10} - 2 \times 10^{13}$ cm^{-3}.

Unfortunately it is difficult to extend these measurements to higher densities because strong multiple scattering is accompanied by a considerable depolarization of the fluorescent light and a consequential reduction in the signal-to-noise ratio. In the limit of very high densities, where the atoms are being isotropically irradiated by scattered photons, the fluorescent light is expected to be completely depolarized. Expressions for the percentage polarization at intermediate densities were obtained by Barrat (1959b) by averaging the coherence transferred over the arbitrary directions of propagation of the intermediate photons. For the case of excitation by linearly polarized light in the presence of a strong static field, the results are

$$\frac{P_\sigma}{100} = \frac{1-x}{1-3x/5} \quad ; \quad \frac{P_\pi}{100} = \frac{1-x}{1-4x/5} \qquad (16.27)$$

for excitation by σ- and π-polarized light respectively. The r.f. magnetic field is assumed to be zero. The polarizations for both σ and π excitation and the amplitude of the Hanle effect signals all tend to zero at high densities, $x \approx 1$, as expected.

16.4.2. Comparison with delayed-coincidence lifetime measurements. It is interesting to note that the effective lifetimes measured by the resonance fluorescence techniques using linearly polarized light reach a theoretically predicted maximum value of $(10/3)\tau$ at high densities. This is in contrast to the lifetimes measured by the time-resolved techniques described in section 6.4 which increase indefinitely as trapping becomes more effective, as shown by Fig.6.13. We must recognize that these two techniques are monitoring different parts of the ensemble density matrix. Those described in sections 6.3 and 6.4 measured only the rate of relaxation of the excited state population, determined by the diagonal elements of the excited state density matrix. On the other hand, the Hanle effect and optical double-resonance experiments using linearly polarized light measure a quantity called the *transverse alignment* which depends on the off-diagonal elements of ρ_e.

16.4.3. Experimental measurements using circularly polarized light. Further investigations of coherence narrowing by Omont (1965a) using circularly polarized light showed that there was yet a third relaxation rate. This characterizes the decay of the *orientation* of the excited state, a quantity which is proportional to the magnetic moment of the ensemble of excited atoms. The terms population, orientation, and alignment are explained in more detail in section 16.8. For the moment they can be identified approximately with the $\Delta m=0$, 1 and 2 terms respectively of the general expression for the Hanle effect signal given by equations (15.35)-(15.37).

The theoretical expression, equation (16.25), for the effective decay rate must now be cast in the more general form

$$\Gamma^{(k)}_{eff} = \Gamma\{1 - \alpha^{(k)}x\}. \qquad (16.28)$$

For the $J_e=1 \rightarrow J_g=0$ case, the parameter $\alpha^{(k)}$ takes the following values:

Population	; k=0	; $\alpha^{(0)}=1$
Orientation	; k=1	; $\alpha^{(1)}=1/2$
Alignment	; k=2	; $\alpha^{(2)}=7/10$

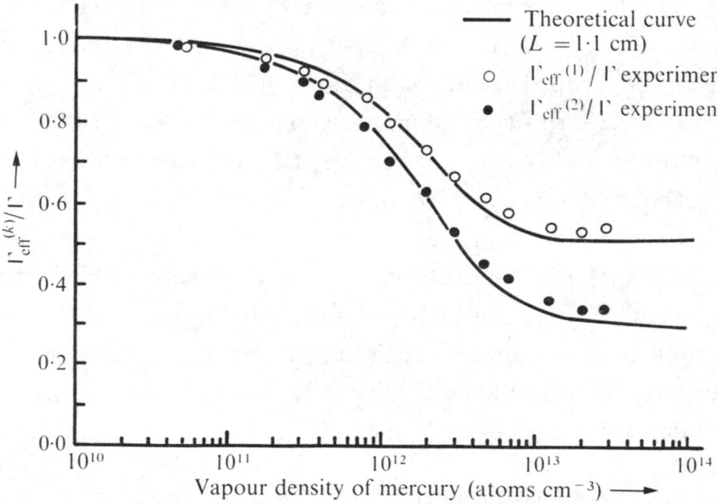

Fig.16.8. Effect of radiation trapping on the relaxation rates of the orientation ($\Gamma^{(1)}$) and alignment ($\Gamma^{(2)}$) of a sample of ^{198}Hg atoms. The alignment lifetime is increased by a factor of 10/3 in the limit of complete trapping while the orientation lifetime is increased by a factor of 2. o , •, experimental points; solid lines, theoretical curves. (After Omont (1965a).)

By using circularly and linearly polarized light alternatively in the same geometry, Omont (1965a) was able to measure both $\Gamma_{eff}^{(1)}$ and $\Gamma_{eff}^{(2)}$ using the Hanle effect in isotopically pure ^{198}Hg. The results obtained are plotted in Fig.16.8 together with theoretical curves derived from equations (16.26) and (16.28) using L=1·1 cm. The agreement between the theory and experiment is obviously very good. The ratio of the orientation and alignment linewidths was checked by plotting $\{\Gamma - \Gamma_{eff}^{(1)}\}/\Gamma$ against $\{\Gamma - \Gamma_{eff}^{(2)}\}/\Gamma$ and an excellent fit to a straight line with a slope of 5/7 was obtained in complete agreement with the theory.

It was then realized that these different decay rates can be measured directly in delayed-coincidence experiments if suitably polarized light is used. Time-resolved experiments by Dodd *et al.*(1970) and also by Deech and Baylis (1971), which are similar to those described in section 15.8, have confirmed the original measurements made by Omont.

16.5. Collision broadening in resonance fluorescence experiments

<u>16.5.1. Introduction.</u> When Hanle effect or optical double-resonance experiments are extended to very high atomic densities, the width of the signal increases very rapidly. This is illustrated in Fig.16.9 for the case of the Hanle signals from the $(6s^2 6p7s)\ ^3P_1^o$ level of lead studied by Happer and Saloman (1967). In the density range 10^{10}-10^{14} cm^{-3} coherence narrowing is important, but above 10^{14} cm^{-3} there is a broadening of the signal due to collisions between the radiating excited atoms and other atoms in the electronic ground state. During a collision we can imagine that a torque is exerted on the classical oscillating dipole moment of the atom and that its direction of oscillation is changed. In terms of a fixed laboratory axis of quantization, this rotation can be described by a mixing of the different Zeeman sub-levels of the excited atom. Thus collisions cause the components of the ensemble-averaged density matrix

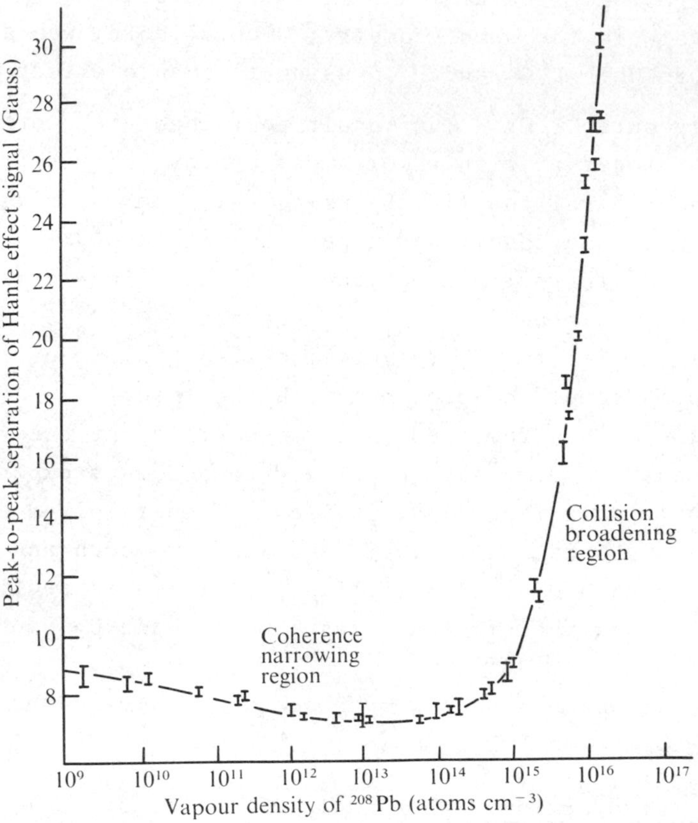

Fig.16.9. Coherence narrowing and collision broadening of the width of the Hanle effect signals on the $^3P_1^o$ state of lead. (After Happer and Saloman (1967).)

to relax more quickly, leading to an increase in the widths of resonance fluorescence signals and to a reduction in the polarization of the fluorescent light observed in zero magnetic field.

As in the case of radiation trapping, collisions also cause the population, orientation, and alignment of the excited state to relax at different rates, the orientation being typically from 5 per cent to 25 per cent more sensitive to collisions than the alignment. In many cases, however, collisional quenching of the excited atoms is negligible and the population relaxation rate is determined solely by the radiative lifetime. Long range electrostatic interactions are responsible for most of the collisional depolarization and Hanle effect studies have become one of the most sensitive probes of this interaction. These experiments have recently stimulated considerable improvements in the theory of collision broadening.

The density range over which most resonance fluorescence experiments can be performed is such that only collisions involving two atoms need be considered. In this case the additional collisional relaxation rate, $\Gamma^{(k)}_{coll}$, may be expressed in terms of a binary collision cross-section $\sigma^{(k)}(v)$, averaged over the distribution of relative velocities v of the perturber and excited atoms:

$$\Gamma^{(k)}_{coll} = N\langle v\,\sigma^{(k)}(v)\rangle_{Ave}$$

where N is the number density of perturbers. It is usual to write this equation in terms of a velocity averaged cross-section, $\sigma^{(k)}(\bar{v})$, by assuming a Maxwellian distribution for the relative velocities, giving

$$\Gamma^{(k)}_{coll} = N\,\bar{v}\,\sigma^{(k)}(\bar{v}) \qquad (16.29)$$

where

$$\bar{v}\,\sigma^{(k)}(\bar{v}) = \langle v\,\sigma^{(k)}(v)\rangle_{Ave}$$

and

$$\bar{v} = \left\{\frac{8kT}{\pi}\,\frac{(M_1+M_2)}{M_1M_2}\right\}^{1/2} .$$

It is found that the effects of collision broadening and resonance trapping are additive, and the effective decay rate for the simple case of an excited atom which has only a single allowed emission line becomes:

$$\Gamma_{\text{eff}}^{(k)} = \Gamma\{1 - \alpha^{(k)}x\} + N\bar{v}\,\sigma^{(k)}(\bar{v}). \qquad (16.30)$$

Thus in the region where binary collisions are dominant, the half-widths of the Hanle effect and optical double-resonance signals are proportional to the density of the perturbing atoms and these experiments provide an accurate means of measuring the velocity-averaged collision cross-sections. Although Mitchell and Zemansky (1966) include accounts of many early investigations of collision broadening, we shall not refer to these experiments again. The effects of resonance trapping, collisional quenching and of hyperfine structure were not sufficiently investigated by these early workers and an accurate interpretation of their measurements is now largely impossible.

16.5.2. Experimental investigations of resonance broadening. The most important term in an expansion of the long-range interatomic potential is the electrostatic dipole-dipole interaction:

$$V(R) = -\frac{1}{4\pi\varepsilon_0 R^3}\{3(\underline{D}_1\cdot\hat{\underline{u}})(\underline{D}_2\cdot\hat{\underline{u}}) - (\underline{D}_1\cdot\underline{D}_2)\} \quad (16.31)$$

where $\underline{R}=R\hat{\underline{u}}$ is the position vector of the perturber (2) with respect to the excited atom (1) and $\underline{D}_a = -\sum_i e\underline{r}_i^a$ is the electric dipole operator for atom a. That part of equation (16.31) which is angular dependent is responsible for the collisional depolarization observed in resonance fluorescence experiments, while the scalar part is responsible for the broadening of the spectral distribution of the resonance line as discussed in section 8.2. When an excited atom in a resonance level collides with an identical atom in the ground state the odd parity operator of equation (16.31) possesses finite matrix elements in first-order perturbation theory (Problem 16.7). During the collision there is a resonant transfer of excitation from atom 1 to atom 2 and the depolarizing cross-sections can be as large as 10^{-12}

for strongly radiating states. These depolarization cross-sections are many hundreds of times larger than the gas kinetic cross-sections and are typical of *resonance* or *self broadening* collisions. As a first example we consider the resonance broadening of Hanle effect signals in lead.

(a) Resonance broadening of the $(6s^2 6p7s)\ ^3P_1^o$ level of lead. The $3P_1^o$ level of lead may be excited from the ground state $(6s^2 6p^2)\ ^3P_0$ by absorption of resonance radiation at 2833 Å, as shown in Fig.16.10. Unlike mercury, the $6\ ^3P_1$ level in lead can also decay by line fluorescence at 3639 Å and 4058 Å

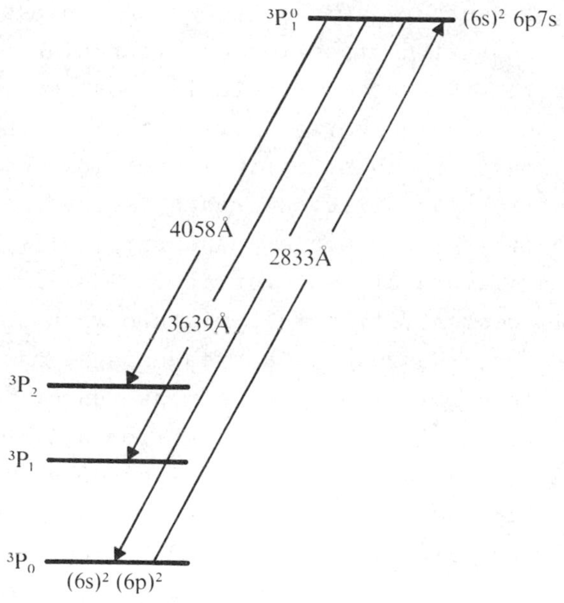

Fig.16.10. Energy-level diagram showing the $^3P_0$ ground state of lead, the two metastable levels $^3P_1$ and $^3P_2$, and the resonance level $^3P_1^o$. The thermal population of the metastable levels is normally very small so that little trapping of the cross fluorescent light at 4058 Å and 3639 Å occurs.

to the $^3P_1$ and $^3P_2$ levels. These metastable levels lie several tenths of an eV above the ground state and are not normally sufficiently populated to cause trapping of the 4058 Å and 3639 Å fluorescent light. Consequently the coherence narrowing and accompanying reduction in signal-to-noise due to depolarization are not nearly as great as they would have been in the absence of these branching transitions. This enabled Happer and Saloman (1967) to measure the depolarizing collision cross-sections using a fused quartz resonance cell containing a few milligrams of ^{208}Pb of 99·75 per cent purity. The results of their experiments are given in Table 16.2 and are discussed in more detail after the following section.

(b) Resonance broadening of the 6 $^3P_1$ level of mercury. By contrast, in mercury at the high vapour pressures required for the observation of collision broadening, the polarization of the fluorescent radiation has become so weak owing to the depolarization caused by resonance trapping that the signal-to-noise ratio is too low for accurate measurements. The direct approach used by Happer and Saloman therefore fails. Fortunately the difficulty may be overcome by using a mixture of two isotopes. Thus if ^{202}Hg is present as a small concentration in a mixture with another isotope, such as ^{204}Hg, the resonance broadening cross-sections for the 6 $^3P_1$ level of ^{202}Hg may be measured by illuminating the cell with resonance radiation from a lamp filled with ^{202}Hg. As Table 16.1 shows, the isotope shift is sufficiently large compared with the Doppler width of the 2537 Å line that this radiation is not trapped by the ^{204}Hg which forms the main vapour in the cell. Nevertheless, since the isotopes have the same electronic structure, ^{204}Hg will depolarize ^{202}Hg on collision in exactly the same way that ^{202}Hg atoms would do, provided that the isotope shift $\Delta E \ll \hbar/T_c$ where T_c is the duration of one collision. This condition is satisfied by a number of different pairs of mercury isotopes.

With this technique Omont and Meunier (1968) were able

TABLE 16.1

Hyperfine structure of Hg 2537 Å *line*

The positions of the components are measured with respect to that of the isotope ^{198}Hg

Isotope	Excited level quantum number F	Position of component (10^{-3} cm^{-1})
201	1/2	+ 229·4
199	3/2	+ 224·6
198	1	0·0
201	3/2	- 22·6
200	1	- 160·4
202	1	- 337·2
201	5/2	- 489·3
204	1	- 511·1
199	1/2	- 514·3

to measure the cross-sections for the destruction of alignment and orientation of the 6 $^3P_1$ level of mercury using the Hanle effect. Their results for ^{202}Hg broadened by ^{204}Hg are shown in Fig.16.11. The non-linear variation of $\Gamma_{eff}^{(k)}$ as a function of density indicates that trapping of the fluorescence from ^{202}Hg has not been entirely eliminated. The measured values were therefore corrected by using equation (16.30) and choosing a concentration for the ^{202}Hg isotope such that the revised data were linear as a function of density. The cross-sections obtained in these experiments with different isotopic mixtures are in good agreement and only the average value is included in Table 16.2.

16.5.3. Comparison with theoretical calculations in resonance broadening. The impact theory of collision broadening in re-

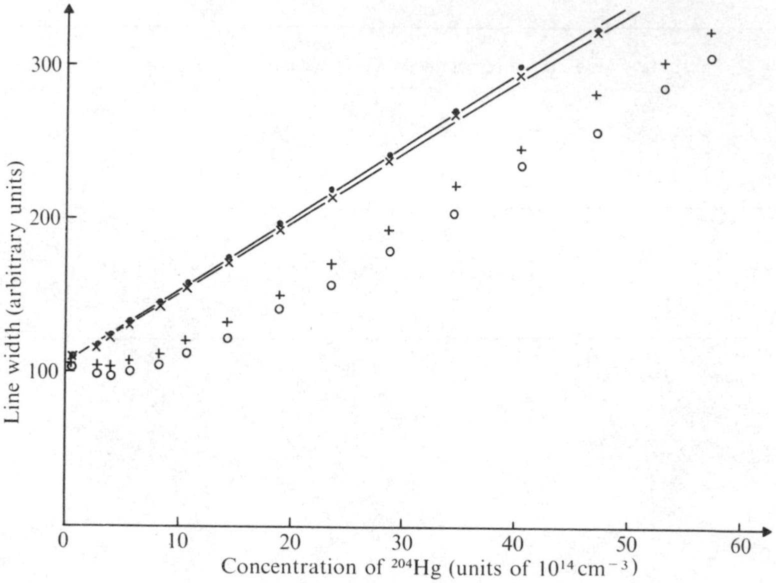

Fig.16.11. Collision broadening of ^{202}Hg Hanle effect signals by ^{204}Hg atoms. Uncorrected measurements : o alignment; + orientation. Measurements corrected for radiation trapping assuming ^{202}Hg concentration as 2×10^{-3} : • alignment ; × orientation. (After Omont and Meunier (1968).)

sonance fluorescence experiments has been developed by Omont (1965b) and D'yakanov and Perel (1965b) using the density matrix approach. In the case of resonance broadening the theories predict that the collisional relaxation rate, $\Gamma^{(k)}_{coll}$, is independent of the relative velocities of the atoms and is proportional to the oscillator strength of the resonance line, $f_{J'J}$:

$$\Gamma^{(k)}_{coll} = \left\{\frac{2\sqrt{3}}{3\pi} K^{(k)}(i)\right\} \frac{\pi e^2 f_{J'J} N}{8\varepsilon_0 m\omega_{J'J}} . \qquad (16.32)$$

The constant $K^{(k)}(i)$ depends on the approximations made in the calculations and on the particular experimental method chosen to monitor a given multipole moment. The values of

TABLE 16.2

Resonance broadening collision cross-sections in lead and mercury

Level studied		Orientation-destroying cross-section, $\sigma^{(1)}$ (10^{-13} cm^2)	Alignment-destroying cross-section, $\sigma^{(2)}$ (10^{-13} cm^2)	$\sigma^{(1)}/\sigma^{(2)}$
Pb(7s $^3P_1^o$) T=1240K	Experiment	—	7·6 ± 1·7	1·21 ± 0·05
	Theory	7·50	6·25	1·20
Hg(6p $^3P_1$) T=350K	Experiment	1·38 ± 0·15	1·46 ± 0·15	0·95 ± 0·03
	Theory	1·45	1·51	0·96

$K^{(k)}(i)$ shown in Table 16.3 were obtained in recent numerical solutions of the coupled differential equations appearing in the broadening theory and are expected to be accurate to between 1 per cent and 2 per cent. Using the f-values of the mercury and lead resonance lines obtained by experiments at low pressures, the theoretical broadening cross-sections determined from equations (16.29) and (16.32) have been evaluated and are given in Table 16.2. There is good agreement between the experimental and theoretical values of the cross-sections, and the agreement between the $\sigma^{(1)}/\sigma^{(2)}$ ratios, in which the experimental uncertainties in N and $f_{J'J}$ are eliminated, is even more impressive. We can therefore be certain that the problem of depolarization in resonant collisions is now well understood.

TABLE 16.3

Constants for collisional relaxation obtained by numerical integration in the resonance broadening case

Relaxation constant	Omont and Meunier (1968)	Carrington, Stacey and Cooper (1973)
$_{ee}K^{(1)}(1)$	2·57	2·551
$_{ee}K^{(2)}(1)$	2·72	2·652
$_{ee}K^{(2)}(1+2)$	2·63	—

16.5.4. Experimental investigations of non-resonance broadening. When the perturbing atoms or molecules are of a different chemical species from that of the emitter, the broadening is known as *foreign or non-resonance broadening*. In this case the dipole-dipole interaction of equation (16.31) has finite matrix elements only in second-order perturbation theory. The energy of interaction is proportional to $1/R^6$ and is thus the angular equivalent of the scalar van der Waals potential discussed in section 8.2. This second-order calculation must also be used in the case of any level which does not possess an allowed electric dipole transition to the ground state. It also applies in cases such as the 4 $^3P_1$ and 5 $^3P_1$ levels of Zn and Cd respectively where the oscillator strength of the resonance line is very small. The resulting depolarization collision cross-sections are typically 10^{-15} - 10^{-14} cm^2, i.e. only slightly larger than the gas kinetic cross-sections. Indeed the cross-sections observed in foreign gas broadening are so small that the effects of the overlap of the wavefunctions of the emitter and perturber which give rise to short range repulsive forces often need to be considered.

(a) Foreign gas broadening of the 6 $^3P_1$ level of mercury.

The depolarization of resonance fluorescence signals by foreign gases has been thoroughly investigated by Barrat *et al.* (1966) for the case of the 6 $^3P_1$ level of ^{202}Hg. In the presence of collisions it can be shown that the polarization of resonance fluorescence excited by π-polarized light, equation (16.27), is modified to

$$\frac{P_\pi}{100} = \frac{\Gamma(1-x) + \Gamma^{(0)}_{coll}}{\Gamma(1-4x/5) + (\Gamma^{(0)}_{coll} + \Gamma^{(2)}_{coll})} . \qquad (16.33)$$

For constant mercury vapour pressure the probability of photon re-absorption, x, is constant. Thus if the collisional quenching is negligible, $\Gamma^{(0)}_{coll} \approx 0$, a plot of $1/P_\pi$ is a linear function of the pressure of the perturbing gas, as shown in Fig.16.12 for the noble gases and N_2. From these graphs the alignment destruction cross-section, $\sigma^{(2)}$, can be obtained directly. Similar plots for the cases of depolarization by O_2, H_2, CO, and CO_2 showed definite curvature and values of $\sigma^{(0)}$ and $\sigma^{(2)}$ were obtained by manipulating equation (16.33) into a more suitable form. The values of the sums of the cross-sections, $(\sigma^{(0)} + \sigma^{(2)})$, were checked by independent measurements of the widths of the Hanle effect signals excited by linearly polarized light, while observations with circularly polarized radiation enabled the sum of the quenching and orientation cross-sections, $(\sigma^{(0)} + \sigma^{(1)})$, to be obtained. The results of these measurements are given in Table 16.4. There is good agreement between the results obtained by the different techniques. It was found that quenching was absent in collisions with the noble gases and very small in the case of N_2. This corresponds to the fact that the electronic states of these perturbers are separated by several eV from the 6 $^3P_1$ level of mercury. For quenching to take place all the electronic energy of the mercury atom would have to be converted into kinetic energy, which is a very unlikely process in two-body collisions.

TABLE 16.4

Collision cross-sections for foreign gas broadening of the 6 $^3P_1$ level of mercury

	Polarization measurements		Hanle effect measurements	
Perturber	$\sigma^{(0)}$ (10^{-16}cm^2)	$\sigma^{(2)}$ (10^{-16}cm^2)	$\sigma^{(0)} + \sigma^{(2)}$ (10^{-16}cm^2)	$\sigma^{(0)} + \sigma^{(1)}$ (10^{-16}cm^2)
He	0	40·2	39·6	45·5
Ne	0	46·5	47·1	59·7
Ar	0	83·3	83·3	101
Kr	0	123	124	145
Xe	0	170	173	179
H_2	25·1	33·0	56·6	61·3
N_2	3<	135	145	145
O_2	62·8	53·4	112	112
CO	20·4	140	160	160
CO_2	15·7	248	270	270

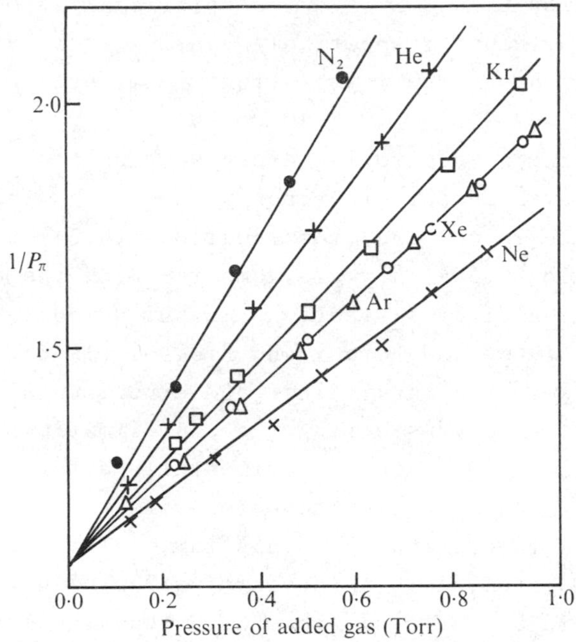

Fig.16.12. Variation of (mercury resonance fluorescence polarization, P)$^{-1}$ as a function of the pressure of perturbing gas. (After Barrat *et al.*(1966).)

<u>16.5.5. Comparison with theoretical calculations of foreign gas broadening.</u> The theoretical evaluation of foreign gas broadening cross-sections contains difficulties which do not occur in the case of resonance broadening. These are associated with the summations over intermediate states which appear in second-order perturbation theory. In reducing these summations to expressions which can be evaluated using available atomic wavefunctions, several approximations must be made. Consequently there is a corresponding uncertainty in the final result of the calculation. However, for the $J_e=1 \rightarrow J_g=0$ case the collision cross-sections may be expressed approximately in the form

$$\sigma^{(1)}(\bar{v}) = 1{\cdot}11\,\sigma^{(2)}(\bar{v}) = \frac{5\pi}{4}\left(\frac{C_{12}\alpha_1\alpha_2}{\bar{v}}\right)^{2/5} . \tag{16.34}$$

In this equation α_1 and α_2 are the polarizabilities of the emitter and perturber respectively and C_{12} is a constant determined by the energy level structure of both of the atoms involved.

The alignment-destroying cross-sections for the 6 $^3P_1$ level of the even mercury isotopes perturbed by the noble gases have been calculated from equation (16.34) and are listed in Table 16.5. They are compared with the mean values of the experimental cross-sections measured independently by Barrat *et al.*(1966) and Faroux and Brossel (1966). The agreement between the theoretical and experimental values is quite reasonable considering the approximations involved. The table also includes the experimentally determined ratio of $\sigma^{(1)}/\sigma^{(2)}$ which should be compared with the value of 1·11 predicted theoretically. This ratio eliminates the uncertainties involved in the determination of C_{12}, α_1, and α_2 and it would appear from Table 16.5 that the theory of collisional depolarization, at least for atoms, is in a satisfactory state.

However, discrepancies appear when the temperature dependence of the cross-sections for the broadening by helium and neon is examined in detail. Faroux and Brossel (1967) observed that the cross-sections for the depolarization of the 6 $^3P_1$ level of mercury are almost independent of temperature in the range 0 to 600°C instead of varying as $(T)^{-1/5}$ as predicted by equation (16.34). Similar results were obtained by Carrington and Corney (1971) for the broadening of the Hanle signals observed on the 2p levels of neon using the technique of alignment in a d.c. discharge referred to in section 15.3. These experiments indicate the necessity of including the effects of short range repulsive interactions when broadening by atoms of low polarizability (weak van der Waals forces) is being considered. So far the theory of these interactions has been worked out only for the very simplest atoms and the experiments therefore suggest a need for the development of new methods of calculation.

TABLE 16.5.

Experimental and theoretical alignment-destroying cross-sections for the 6 $^3P_1$ level of mercury

Perturber	$\sigma^{(2)}$ theory (10^{-16} cm^2)	$\sigma^{(2)}$ experiment (10^{-16} cm^2)	$\sigma^{(1)}/\sigma^{(2)}$ experiment
He	27·5	39	1·17
Ne	37	47	1·25
Ar	72	83	1·20
Kr	99	121	1·14
Xe	121	168	1·06

16.5.6. Comment on collisional depolarization studies. The resonance fluorescence experiments described above have served to stimulate and test the development of recent sophisticated theories of collision broadening. However, the hope that new information on interatomic forces would be obtained by these experiments has not been realized. This is largely because the observed broadening is the mean effect of collisions averaged over all possible relative orientations and over the thermal velocity distribution of the ensemble. It may be that in the future more detailed information on the interatomic potentials could be obtained by combining tunable dye laser excitation and atomic beam scattering techniques.

Unfortunately limitations of space do not permit us to describe further studies of the quenching of resonance fluorescence or experiments involving sensitized fluorescence. However, a very complete discussion of this work is contained in Massey (1971).

16.6. Light modulation in double-resonance experiments

16.6.1. Introduction. So far we have studied only the time-independent signals observed in optical double-resonance experiments. It is now interesting to consider whether there are signals linked with the off-diagonal elements of the excited state density matrix which are time dependent. The classical expression of equation (16.22) clearly includes terms modulated at the angular frequency ω_e but, since the phase of this nutational motion is fixed by the instant of excitation, these terms do not give rise to time-dependent signals in steady state experiments. However, the phase of the forced precession at the angular frequency ω_0 is fixed by the phase of the r.f. magnetic field and is the same for all atoms of the ensemble. When the static field $\underline{B}$ is exactly on resonance, $B=B_0=\omega_0/\gamma$, the classical dipoles may be imagined to be uniformly distributed in a circular disk which spins about the direction of 0z at the angular frequency ω_0, as shown in Fig.16.13(a). If the σ-polarized light emitted in the direction of $\underline{B}$ is observed through a linear polarizer, we would expect the fluorescent intensity to be modulated at twice the frequency of the r.f. field.

Quantum mechanically the modulation of the fluorescent light is associated with the radio frequency coherence of the excited state density matrix. In a standard Brossel-Bitter double-resonance experiment π-polarized light excites the atoms initially to the m=0 state of the excited level, Fig. 16.13(b), and then interaction with the r.f. magnetic field transforms each atom into a coherent superposition of the m=0,±1 states. The relative phases of the probability amplitudes of these states are fixed by the phase of the r.f. field and are the same for every atom of the sample. Thus the r.f. field is able to generate substantial hertzian coherence in the excited state density matrix. The fluorescent light emitted in the direction of $\underline{B}$ is then a coherent

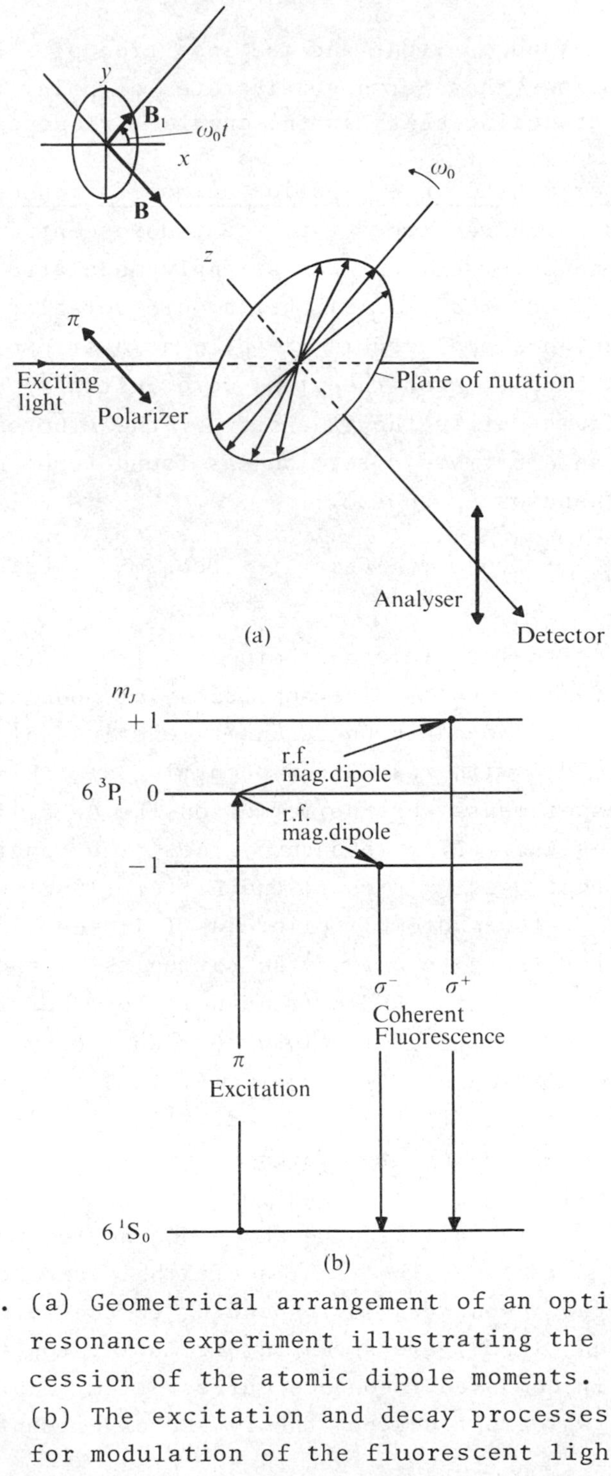

Fig.16.13. (a) Geometrical arrangement of an optical double-resonance experiment illustrating the forced precession of the atomic dipole moments.
(b) The excitation and decay processes responsible for modulation of the fluorescent light.

mixture of σ^+ and σ^- light and the insertion of a linear polarizer allows these orthogonal states of polarization to interfere, producing beats at the angular frequency $2\omega_0$.

16.6.2. Experimental investigation of modulation phenomena. Experiments which confirmed that the fluorescent light in a double-resonance experiment was strongly modulated were first reported by Dodd *et al.*(1959), but a more detailed account of the techniques employed is given in a later publication (Dodd *et al.*1963). When the atoms were initially excited to a pure Zeeman state the intensity of the fluorescent light emitted in an arbitrary direction was found to be modulated at the frequencies ω_0 and $2\omega_0$:

$$I(t) = I_0(A + B\cos\omega_0 t + C\sin\omega_0 t + D\cos 2\omega_0 t + E\sin 2\omega_0 t), \tag{16.35}$$

corresponding to interferences between states differing by $\Delta m=1$ and 2 respectively. The amplitudes of modulation B, C, D, and E were found to be resonant functions of the static magnetic field $\underline{B}$ with resonances occurring at $B=B_0=\hbar\omega_0/g_J\mu_B$.

The experiments were performed on the 6 $^3P_1$ level of mercury using the 2537 Å resonance line. The apparatus was similar to that used by Brossel and Bitter (1952) with the exception that the photomultiplier was followed by a narrow band amplifier tuned to one of the harmonics of the oscillator frequency $\omega_0/2\pi$. Radio frequencies of the order of 10 MHz were used and the signals were recorded by audio frequency modulation of the magnetic field $\underline{B}$ and phase-sensitive detection. The observed signals therefore corresponded to the differentiated forms of the resonance functions B, C, D, and E. Later investigations using low frequency (1 kHz) r.f. fields allowed the shape of the resonances to be observed directly. The results of Kibble and Series (1963) for the resonances corresponding to the terms D and E in equation (16.35) are shown in Fig.16.14, together with theoretically computed resonance curves. There is excellent qualitative agreement between theory and experiment.

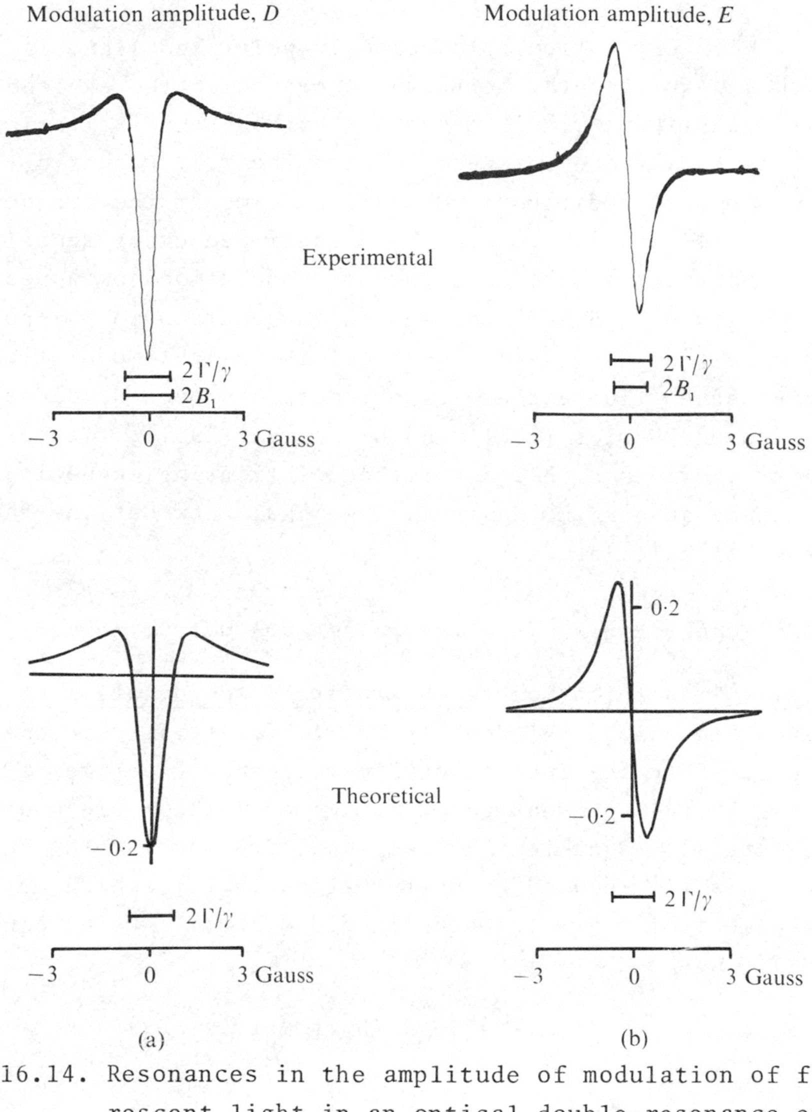

Fig.16.14. Resonances in the amplitude of modulation of fluorescent light in an optical double-resonance experiment for the case $\omega \ll \Gamma$.
(a) The in-phase component D at the angular frequency $2\omega_0$.
(b) The out-of-phase component E at the angular frequency $2\omega_0$.
The upper curves are experimental, the lower ones are computed from theoretical formulae describing the phenomenon. (After Kibble and Series (1963).)

When excitation by coherently polarized light is used in these experiments, hertzian coherence in the excited-state density matrix is generated both by the excitation process and by the interaction with the r.f. magnetic field. This produces additional modulation terms at the frequencies $3\omega_0$ and $4\omega_0$ which have also been confirmed experimentally. The resonances in the amplitude of modulation now appear at fields given by $B \approx B_0/2$ and $\approx 3B_0/2$ in addition to the normal resonances at $B=B_0$. The theory of this light modulation in resonance fluorescence experiments has been developed by Dodd and Series (1961) and by Barrat (1961). Since this theory provides an elegant method of treating magnetic resonance quantum mechanically, we shall now outline the method of calculation.

16.7. Magnetic resonance in the density matrix formalism

<u>16.7.1. Transformation to the rotating coordinate system.</u>

We wish to extend the density matrix formalism developed in Chapter 15 to the optical double-resonance experiments discussed in this present chapter. We must therefore modify the Liouville equations of sections 15.5 and 15.6 by including a time-dependent perturbation, $\mathcal{H}_3(t)$, which describes the effect of the r.f. magnetic field $\underline{B}_1(t)$. Using equation (16.7), we have

$$\mathcal{H}_3(t) = -\underline{\mu}.\underline{B}_1(t) = \gamma\hbar\underline{J}.(B_1\cos\omega_0 t\ \hat{\underline{i}} + B_1\sin\omega_0 t\ \hat{\underline{j}}) \tag{16.36}$$

where $\gamma = g_J\mu_B/\hbar$. As in the case of the classical problem discussed in section 16.2.1, it is easier to solve the quantum-mechanical equation of motion in a coordinate frame $0x'y'z$ which rotates about the z-axis at the angular frequency ω_0. A wavefunction $\Psi(t)$ in the laboratory frame is transformed into the corresponding wavefunction $\Psi'(t)$ in the rotating frame by the application of the quantum-mechanical rotation operator $\exp(iJ_z\omega_0 t)$, giving

$$\Psi'(t) = \exp(iJ_z\omega_0 t)\Psi(t) \qquad (16.37)$$

as shown in Rose (1957). Consequently the density operator in the rotating frame, σ, is given by

$$\sigma = \exp(iJ_z\omega_0 t)\rho\, \exp(-iJ_z\omega_0 t). \qquad (16.38)$$

In the rotating frame the equation of motion for the density operator becomes

$$\frac{d\sigma}{dt} = \frac{1}{i\hbar}[\tilde{\mathcal{H}},\sigma] + \mathcal{L}(\sigma) \qquad (16.39)$$

where the Hamiltonian for the system has been transformed to

$$\tilde{\mathcal{H}} = \mathcal{H}_0 + \hbar(\omega_L-\omega_0)J_z + \hbar\omega_1 J_{x'}. \qquad (16.40)$$

As in the classical case the transformation to the rotating frame has eliminated the time dependence of the r.f. perturbation $\mathcal{H}_3(t)$ and has introduced the effective static magnetic field

$$\underline{B}_e = \hat{\underline{k}}(B - \omega_0/\gamma) + \hat{\underline{i}}'B_1$$

which is shown in Fig.16.4. In equation (16.39) the Liouville operator $\mathcal{L}(\sigma)$ represents the effect of the excitation and relaxation terms, $d^{(1)}/dt$ and $d^{(2)}/dt$ respectively in equation (15.16), when transformed into the rotating frame.

16.7.2. Expansion in the eigenstates of the effective field.

In order to solve the new equation of motion, equation (16.39), it is convenient to perform a further transformation of coordinates so that the axis of quantization Oz' lies parallel to the direction of the effective field $\underline{B}_e$. This is accomplished by a rotation through the angle

$$\beta = \tan^{-1}\left(\frac{B_1}{B_1-\omega_0/\gamma}\right)$$

about the Oy' axis and is represented quantum mechanically by the rotation operator R. The density operator in this frame becomes σ' where

$$\sigma' = R\,\sigma\,R^{-1}. \tag{16.41}$$

After this second transformation the equation of motion takes a very simple form:

$$\frac{d\sigma'}{dt} = \frac{1}{i\hbar}\left[(\mathcal{H}_0 + \hbar\omega_e J_{z'}),\sigma'\right] + \mathcal{L}(\sigma') \tag{16.42}$$

where $J_{z'}$ is the component of the angular momentum operator $\underline{J}$ along the direction of the effective field $\underline{B}_e$. In this frame the equation of motion is identical to that given by equation (15.22). Thus we recognize the fact that the solutions of equation (16.42) represent Larmor precession about the effective field $\underline{B}_e$ at the nutational angular frequency $\omega_e = \{(\omega_L-\omega_0)^2+\omega_1^2\}^{1/2}$. If $|\nu\rangle$ and $|\nu'\rangle$ are eigenstates with respect to the new axis of quantization Oz', having magnetic quantum numbers ν and ν' respectively, then the components of the density matrix in this frame satisfy the equation

$$\frac{d\sigma'_{\nu\nu'}}{dt} + \{i\omega_e(\nu-\nu') + \Gamma\}\sigma'_{\nu\nu'} = \frac{d^{(1)}}{dt}\,\sigma'_{\nu\nu'}\,. \tag{16.43}$$

This follows from equation (16.42), since the radiative decay process represented by $d^{(2)}\sigma'_{\nu\nu'}/dt$ is isotropic and independent of the axis of quantization.

<u>16.7.3. The excited state density matrix.</u> In order to solve equation (16.43) it is necessary to express the excitation term in the laboratory frame of reference. Using the two rotation operators we have

$$\frac{d^{(1)}}{dt}\,\sigma'_{\nu\nu'} = \left\{R\exp(iJ_z\omega_0 t)\,\frac{d^{(1)}\rho}{dt}\,\exp(-iJ_z\omega_0 t)R^{-1}\right\}_{\nu\nu'} \tag{16.44}$$

Multiplying equation (16.43) by the integrating factor $\exp[\{i\omega_e(\nu-\nu')+\Gamma\}t]$ and substituting from equation (16.44), we have

$$\frac{d}{dt}(\sigma'_{\nu\nu'}\exp[\{i\omega_e(\nu-\nu')+\Gamma\}t]) =$$

$$\exp[\{i\omega_e(\nu-\nu')+\Gamma\}t]\sum_{n,n'} R_{\nu n}R^{-1}_{n'\nu'}\langle n|\exp(iJ_z\omega_0 t)\frac{d^{(1)}\rho}{dt}\exp(-iJ_z\omega_0 t)|n'\rangle \quad (16.45)$$

where $|n\rangle$, $|n'\rangle$ are eigenstates referred to the laboratory axis of quantization 0z having magnetic quantum numbers n and n' respectively. The states $|n\rangle$ and $|\nu\rangle$ are related by the linear transformation

$$|\nu\rangle = \sum_n R_{\nu n}|n\rangle = \sum_n \mathscr{D}^{J_e}(0,\beta,0)_{n\nu}|n\rangle \quad (16.46)$$

where $\mathscr{D}^{J_e}(0,\beta,0)_{n\nu}$ are the elements of the rotation matrix (Rose 1957). For the case $J_e=1$ we have

$$R_{\nu n} = \begin{pmatrix} \cos^2\frac{1}{2}\beta & -\sqrt{2}\sin\frac{1}{2}\beta\cos\frac{1}{2}\beta & \sin^2\frac{1}{2}\beta \\ \sqrt{2}\sin\frac{1}{2}\beta\cos\frac{1}{2}\beta & \cos^2\frac{1}{2}\beta-\sin^2\frac{1}{2}\beta & -\sqrt{2}\sin\frac{1}{2}\beta\cos\frac{1}{2}\beta \\ \sin^2\frac{1}{2}\beta & \sqrt{2}\sin\frac{1}{2}\beta\cos\frac{1}{2}\beta & \cos^2\frac{1}{2}\beta \end{pmatrix}. \quad (16.47)$$

We see that the elements of the rotation matrix are functions of B_1 and B which show resonance behaviour in the region $B=B_0=\omega_0/\gamma$. Essentially they describe the effectiveness of the perturbing field $\underline{B}_1$ in mixing the different states of the system.

Making use of the fact that $|n\rangle$ and $|n'\rangle$ are eigenstates of J_z enables equation (16.45) to be reduced to

$$\frac{d}{dt}(\sigma'_{\nu\nu'}\exp[\{i\omega_e(\nu-\nu')+\Gamma\}t]) =$$

$$\exp\{(\Gamma+ix)t\}\sum_{n,n'} R_{\nu n}R^{-1}_{n'\nu'}\frac{d^{(1)}}{dt}\rho_{nn'} \quad (16.48)$$

where

$$x = \omega_e(\nu-\nu') + \omega_0(n-n'). \quad (16.49)$$

The rate at which the element $\rho_{nn'}$ of the density matrix is created by optical excitation is given by equation (15.19) in the form

$$\frac{d^{(1)}}{dt}\rho_{nn'} = \frac{\pi U(\omega)}{\varepsilon_0 \hbar^2}\sum_{\mu} F_{nn'}\,\rho_{\mu\mu} \qquad (16.50)$$

where $F_{nn'}$ is the excitation matrix defined by equation (15.23) and $\rho_{\mu\mu}$ is an element of the ground-state density matrix in the laboratory frame. Using the fact that $\sigma'_{\nu\nu'}(-\infty) = 0$ and integrating equation (16.48) from $-\infty$ to t, assuming that the system of atoms is illuminated by light whose intensity is independent of time, produces the result:

$$\sigma'_{\nu\nu'}(t) = \frac{\pi U(\omega)}{\varepsilon_0 \hbar^2}\sum_{\substack{n,n'\\ \mu}} \frac{R_{\nu n}R^{-1}_{n'\nu'}\,F_{nn'}\,\rho_{\mu\mu}\exp\{i\omega_0(n-n')t\}}{\Gamma + ix}\,. \qquad (16.51)$$

The density matrix in the laboratory frame is obtained by inverting the sequence of the rotations, giving

$$\rho = \exp(-iJ_z\omega_0 t)R^{-1}\,\sigma'\,R\exp(iJ_z\omega_0 t). \qquad (16.52)$$

Finally the elements of the excited state density matrix in the laboratory frame are given by

$$\rho_{mm'} = \sum_{\nu,\nu'}\langle m|\exp(-iJ_z\omega_0 t)R^{-1}|\nu\rangle\sigma'_{\nu\nu'}\langle\nu'|R\exp(iJ_z\omega_0 t)|m'\rangle$$

$$= \frac{\pi U(\omega)}{\varepsilon_0\hbar^2}\sum_{\substack{n,n'\\ \nu,\mu\nu'}}\frac{R^{-1}_{m\nu}\,R_{\nu n}\,R^{-1}_{n'\nu'}\,R_{\nu' m'}\,F_{nn'}\,\rho_{\mu\mu}}{\Gamma + ix}\times$$

$$\times\exp[-i\omega_0\{(m-m')-(n-n')\}t]\,. \qquad (16.53)$$

16.7.4. The intensity of fluorescent light. Introducing the monitoring operator L_F defined by equation (15.26), we obtain an expression for the intensity of fluorescent light observed in optical double-resonance experiments in the form

$$\frac{dI}{d\Omega} = \mathrm{Tr}(\rho, L_F)$$

$$= \frac{U(\;)}{8\pi\varepsilon_0^2}\left(\frac{\omega}{\hbar c}\right)^3 \times$$

$$\times \sum_{\substack{m,m'\\ n,n'\\ \nu,\nu'\\ \mu,\mu'}} \frac{R^{-1}_{m\nu} R_{\nu n} R^{-1}_{n'\nu'} R_{\nu' m'} F_{nn'} G_{m'm} \rho_{\mu\mu} \exp[-i\omega_0(m-m'-n+n')\omega_0 t]}{\Gamma + i\{\omega_e(\nu-\nu') + \omega_0(n-n')\}} \qquad (16.54)$$

where $G_{m'm}$ is the emission matrix defined by equation (15.34). The fluorescent light is clearly modulated at harmonics of the r.f. frequency $\omega_0/2\pi$, the amplitudes of modulation being functions of the magnitudes of the magnetic fields $\underline{B}_1$ and $\underline{B}$. These functions have resonant behaviour which is determined by the detailed form of the rotation matrix elements $R_{m\nu}$ and of the demoninator $(\Gamma+ix)$. In general, modulation frequencies up to the fourth harmonic are predicted but the amplitudes of some of these terms may be small or identically zero in certain experimental situations. We may regard the factor $1/(\Gamma+ix)$ in equation (16.54) as a measure of the hertzian coherence existing in the excited state density matrix. We see that this coherence will only attain appreciable values when $x=\omega_e(\nu-\nu')+\omega_0(n-n') \le \Gamma$. The experimental phenomena predicted by equation (16.54) are exceedingly rich and varied. It is impossible here to give more than an outline of the main features and the reader is urged to consult the original papers by Series and his co-workers. It is convenient to distinguish three main experimental situations.

(a) Excitation to a pure state. When the direction and polarization of the incident light are chosen so that the atoms are excited to a pure eigenstate with respect to the axis of quantization along the field $\underline{B}$, the excitation matrix has only a single diagonal element and in equation (16.54) we have $n=n'$. Modulation of the intensity of the fluorescent light can now occur only at the frequencies ω_0 and $2\omega_0$. For those terms in the general summation satisfying the condition $\nu=\nu'$, the denominator $(\Gamma+ix)$ is independent of the field B and the resonance behaviour in the amplitude of modulation arises solely from the resonant denominators of the elements of the rotation matrix. These resonances occur at fields satisfying the condition $B=B_0$, i.e. when the Larmor precessional frequency matches the frequency of the r.f. magnetic field. The unmodulated component of the fluorescent light, arising from the terms with $m=m'$, has exactly the same shape as that given by the classical calculation, equation (16.23), and describes the familiar Brossel-Bitter resonance curves.

(b) Excitation to a mixture of states. If the direction and polarization of the incident light are chosen so that the atoms are excited to a coherent superposition state we must consider terms for which $n \neq n'$. Additional modulation at the frequency $3\omega_0$ is possible corresponding to terms for which $n-n'=\pm 1$. The condition $x \leq \Gamma$ can now be satisfied only if $\nu \neq \nu'$ and the resonances in the amplitude of modulation occur at fields given by $B \approx \frac{1}{2}B_0$ and $B \approx \frac{3}{2}B_0$.

(c) Excitation to degenerate states. If the frequency of the r.f. magnetic field is such that $\omega_0 \ll \Gamma$ then those terms which satisfy the condition $\nu=\nu'$ in equation (16.54) also automatically fulfil the condition $x \ll \Gamma$ for all values of n and n' Under these conditions additional resonances at the frequency $4\omega_0$ are observed which are not detectable when $\omega_0 \gg \Gamma$.

16.7.5. Modulation of light from incoherent sources. The light beats observed in resonance fluorescence experiments should be clearly distinguished from those observed by Forrester *et al.*(1955). In that experiment an exceedingly weak modulation was detected in the light emitted from a conventional mercury lamp. In this source the atoms are excited by random collisions and the ensemble density matrix possesses zero hertzian coherence. The beat frequencies observed were due to the mixing at the detector of the radiation frequencies emitted by different atoms and consequently the spectrum of modulation frequencies ranged over the entire Doppler width of the observed spectral lines. By contrast the modulation in resonance fluorescence experiments is present in the radiation emitted from each atom, it is confined to harmonics of the frequency $\omega_0/2\pi$, and the depth of modulation can attain 100 per cent in favourable situations.

16.7.6. Optical double-resonance excited by modulated light.
We have seen that in an optical double-resonance experiment the intensity of the fluorescent light is modulated at harmonics of the driving angular frequency ω_0. The nutation which occurs at the angular frequency $\omega_e = \gamma B_e = \gamma\{(B-B_0)^2+B_1^2\}^{1/2}$ is not observed in these steady state experiments because the phase of this motion is random. However, when the atoms are excited by light which is intensity modulated at the frequency $f/2\pi$, it is found that changes in the amplitude of modulation of the fluorescent light indicate resonance at the nutational frequency. These effects were first observed by Corney and Series (1964) and a more detailed account of the work is given by Corney (1968). When the fluorescent light modulated at the frequency $f/2\pi$ is detected, resonances occur whenever the modulation frequency matches an energy separation in the rotating coordinate system:

$$f = (\nu-\nu')\omega_e = (\nu-\nu')\gamma\{(B-B_0)^2 + B_1^2\}^{1/2}. \quad (16.55)$$

One example of these nutational resonances is shown in Fig.16.15 together with the appropriate frequency diagram. Theoretical expressions to cover this experimental situation may be derived by an extension of the light beat calculation detailed above, allowing for the modulated excitation in the manner described in section 15.9.3. When these are evaluated the agreement between theory and experiment is found to be quite satisfactory, as Fig.16.15 shows.

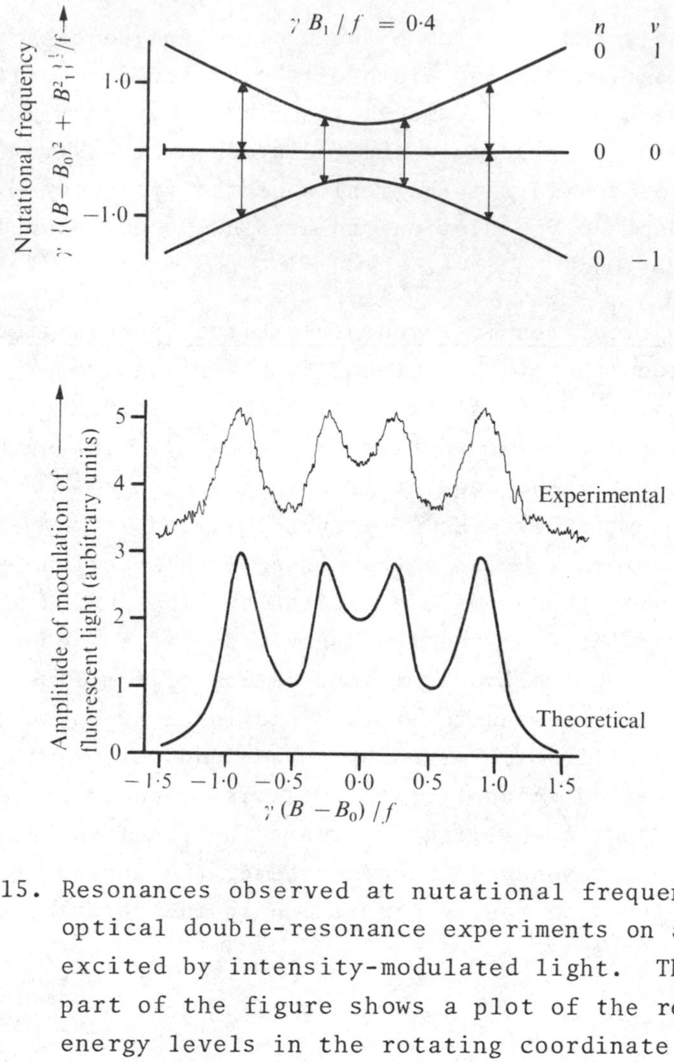

Fig.16.15. Resonances observed at nutational frequencies in optical double-resonance experiments on atoms excited by intensity-modulated light. The upper part of the figure shows a plot of the relevant energy levels in the rotating coordinate system.

16.8. Expansion of the density matrix in terms of irreducible tensor operators

The ensemble density matrix for atoms in an excited level with total angular momentum quantum number J_e has $(2J_e+1)^2$ elements. In general, therefore, $(2J_e+1)^2$ different parameters would be necessary to describe the relaxation of the system in the presence of collisions, radiative trapping etc. Fortunately, however, in resonance fluorescence experiments the atom is effectively in an isotropic environment since the external magnetic fields applied to the atom are generally small compared to other interactions. In this situation it is convenient to expand the density matrix in terms of quantum-mechanical operators which are themselves invariant under rotations of the coordinate system. These operators are called the irreducible tensor operators, $T_q^{(k)}$. They are tensors of rank k and have (2k+1) different components labelled by the projection quantum number q, where $-k \leq q \leq k$. In terms of these operators we may write the density matrix in the form

$$\rho = \sum_{k,q} (-)^q \rho_q^k T_{-q}^{(k)} . \tag{16.56}$$

The parameters ρ_q^k are known as the components of the 2^k-multipole moment of the ensemble density matrix.

Now in order that the relaxation rates should be consistent with the requirements of spherical symmetry, it is necessary that all the (2k+1) components of the kth multipole moment decay at the same rate, that is:

$$\frac{\partial \rho_q^k}{\partial t} = - \Gamma^{(k)} \rho_q^k . \tag{16.57}$$

Thus in an isotropic environment the number of different relaxation rates is reduced from $(2J_e+1)^2$ to $(2J_e+1)$, giving one relaxation rate for each multipole moment of the excited state. Experimentally the number of effective relaxation rates might still be embarrassingly large were it not for the fact that in resonance fluorescence we prepare and monitor the excited atoms through the absorption and emission of electric dipole radiation. The electric-dipole matrix elements have the properties associated with rank one tensors and consequently the observable multipole moments in these experiments are limited to those corresponding to tensors of rank 0, 1, and 2 respectively.

The irreducible tensor operators are defined in terms of the Dirac bra and ket vectors, $|J_e M_e\rangle$ and $\langle J_e, M_e - q|$ respectively, by the expression

$$T_q^{(k)} = \sum_{M_e=-J_e}^{J_e} (-)^{J_e-M_e}(2k+1)^{1/2} \begin{pmatrix} J_e & J_e & k \\ M_e & q-M_e & -q \end{pmatrix} |J_eM_e\rangle\langle J_eM_e-q| \tag{16.58}$$

where the third term in equation (16.58) is a numerical factor called the 3-j symbol whose values may be found in tables. For certain purposes it is convenient to represent the irreducible tensor operators by combinations of the spherical components of the angular momentum operator $\underline{J}$. For the state with $J_e=1$ we have

$$T_0^{(0)} = \underline{I}/\surd 3 \; ; \quad T_0^{(1)} = J_z/\surd 2 \; ; \quad T_{\pm 1}^{(1)} = \mp J_\pm/2$$

$$T_0^{(2)} = (3J_z^2 - \underline{J}^2)/\surd 6 \; ; \; T_{\pm 1}^{(2)} = \mp(J_zJ_\pm + J_\pm J_z)/2 \tag{16.59}$$

$$T_{\pm 2}^{(2)} = (J_\pm)^2/2.$$

These expressions enable the physical significance of the different multipole moments of the ensemble of atoms to be specified more precisely. Thus ρ_0^0 is proportional to the average population of all the sub-levels belonging to the given excited state; ρ_q^1 corresponds to one of the three components of the *orientation* of the sample and is proportional to the spherical components of the magnetic dipole moment $\langle \underline{M}\rangle = \langle -g_J\mu_B\underline{J}\rangle$ of the ensemble; while ρ_q^2 corresponds to one of the five components of the *alignment* or quadrupole moment of the excited atoms. In the limit of vanishingly low atomic densities, these multipole moments all relax at the natural radiative decay rate Γ. At finite densities, however, the effects of radiation trapping and interatomic collisions combine to produce different relaxation rates $\Gamma^{(0)}$, $\Gamma^{(1)}$, and $\Gamma^{(2)}$ for the ensemble population, orientation, and alignment respectively.

Problems

16.1. In intermediate coupling the wavefunction describing

the 6 $^3P_1$ level of mercury may be written as

$$\psi(^3P_1) = \alpha\psi_{LS}(^3P_1) + \beta\psi_{LS}(^1P_1)$$

where α and β are numerical coefficients and $\psi_{LS}(^3P_1)$ and $\psi_{LS}(^1P_1)$ are pure L-S coupled wavefunctions. Show that the Landé g-factor is given by

$$g_J = \alpha^2 g_{LS}(^3P_1) + \beta^2 g_{LS}(^1P_1)$$

and hence calculate the g-factor for the 6 $^3P_1$ level of the even isotopes of mercury in

(a) the L-S coupling approximation (α=1, β=0), and

(b) in intermediate coupling using the values of α and β given in Table 5.3.

Compare your results with the value of $g_J = 1{\cdot}4838 \pm 0{\cdot}0004$ obtained experimentally by Brossel and Bitter (1952).

(Ans: (a) $g_J = 1{\cdot}50116$; (b) $g_J = 1{\cdot}4863$)

16.2. A magnetic dipole moment $\underline{\mu}$, associated with a spin angular momentum $\hbar\underline{J} = -\underline{\mu}/\gamma$, is subjected to an external magnetic field $\underline{B}$. Starting from the equation of motion, equation (16.2), show that $\underline{\mu}^2$ is independent of time and that $\underline{\mu}$ satisfies the equation

$$\ddot{\underline{\mu}} + \omega_L^2\,\underline{\mu} = \text{constant}.$$

Hence prove that $\underline{\mu}$ precesses at the Larmor angular frequency $\omega_L = \gamma B$ about the direction of the applied field.

16.3. An atom having a permanent magnetic dipole moment $\underline{\mu} = -g_J\mu_B\underline{J}$ is acted on by a time-dependent magnetic field $\underline{B}_1(t)$ which rotates counterclockwise in the x-y plane at the angular frequency ω_0. Show that the atom experiences a perturbation given by

$$\mathcal{H}_1' = -\frac{B_1}{2}\{\mu_+\exp(-i\omega_0 t) + \mu_-\exp(i\omega_0 t)\}$$

where $\mu_\pm = \mu_x \pm i\mu_y$. Hence, using the time-dependent perturbation theory developed in section 9.3, prove that the induced magnetic dipole transition probability between two isolated Zeeman sub-levels is givon by equation (16.18) with ω_1^2 everywhere replaced by $|\langle m|B_1\mu_-|m'\rangle|^2/\hbar^2$.

16.4. Derive the Brossel-Bitter formula, equation (16.23), and show that at low r.f. field strengths the full-width of the signal at half-maximum intensity is given by

$$\Delta\omega_{1/2} \cong 2\{\Gamma^2 + 5\cdot 8(\gamma H_1)^2\}^{1/2}.$$

16.5. Low field Zeeman magnetic resonances in the 5s5p $^3P_1$ state of radioactive ^{109}Cd have been studied using the optical double-resonance technique. Three resonance signals were found corresponding to g_F values of 0·600, 0·420 and 0·165. Hence show that the nuclear spin of ^{109}Cd is I=5/2.

16.6. Describe the Zeeman splitting of the D_2 resonance line of sodium (3 $^2S_{1/2}$ - 3 $^2P_{3/2}$) at 5890 Å in a field of 0·2 T. Sketch the energy level diagram and the allowed electric dipole transitions together with the polarization of the radiation observed in directions parallel and perpendicular to the applied field. Calculate the minimum separation between the Zeeman components in terms of the Doppler width of the D_2 line assuming that the effective temperature of the source is 400 K. Optical double-resonance signals are observed on the 3 $^2P_{3/2}$ level in the same magnetic field. Calculate the ratio of the centre frequency of the resonance to the half-width of the signal assuming that the radiative lifetime of the 3 $^2P_{3/2}$ level is 16 ns. The effects of hyperfine structure may be ignored.
(Ans: 0·19 ; $1\cdot 9 \times 10^{-2}$.)

16.7. An excited mercury atom in the 6 $^1P_1$ level collides with a second mercury atom in the ground state, 6 $^1S_0$.

By expressing the total wavefunction of the system as a product of the wavefunctions of atoms 1 and 2, prove that the matrix elements

$$\langle \psi_1(^1P_1)\psi_2(^1S_0)|V(R)|\psi_1(^1S_0)\psi_2(^1P_1)\rangle$$

of the electrostatic dipole-dipole interaction, equation (16.31), are finite. Hence show that in first-order perturbation theory the energy of interaction is proportional to $(e^2 f_{ik}/\varepsilon_0 m\omega_{ki})(1/R^3)$ where f_{ik} and ω_{ki} are respectively the absorption oscillator strength and angular frequency of the 1850 Å resonance line, $6\ ^1S_0 - 6\ ^1P_1$.

References

Barrat, J.P. (1957). *C.r.hebd.Séanc.Acad.Sci., Paris.* 244, 2785.

________ (1959a). *J.Phys.Radium,Paris.* 20, 657.

________ (1959b). *J.Phys.Radium,Paris.* 20, 541, 633.

________ (1961). *Proc.R.Soc.* A263, 371.

________, Casalta, D., Cojan, J.L., and Hamel, J. (1966). *J.Phys.Radium,Paris.* 27, 608.

Brossel, J. and Bitter, F. (1952). *Phys.Rev.* 86, 308.

________ and Kastler, A. (1949). *C.r.hebd.Séanc.Acad.Sci., Paris.* 229, 1213.

Carrington, C.G. and Corney, A. (1971). *J.Phys.B.* 4, 849.

________, Stacey, D.N. and Cooper, J. (1973). *J.Phys.B.* 6, 417.

Corney, A. (1968). *J.Phys.B.* 1, 458.

________ and Series, G.W. (1964). *Proc.phys.Soc.* 83, 331.

Deech. J.S. and Baylis, W. (1971). *Can.J.Phys.* 49, 90.

Dodd, J.N., Fox, W.N., Series, G.W., and Taylor, M.J. (1959). *Proc.phys.Soc.* 74, 789.

Dodd, J.N., Sandle, W.J., and Williams, O.M. (1970). *J.Phys.B.* 3, 256.

Dodd, J.N. and Series, G.W. (1961). *Proc.R.Soc.* A263, 353.

__________ __________, and Taylor, M.J. (1963). *Proc.R.Soc.* A273, 41.

D'yakanov, M.I. and Perel, V.I. (1965a). *Soviet Phys.JETP* 20, 997.

__________ and __________ (1965b). *Soviet Phys.JETP* 21, 227.

Faroux, J.P. and Brossel, J. (1966). *C.r.hebd.Séanc.Acad.Sci., Paris* 263, B612.

__________ and __________ (1967). *C.r.hebd.Séanc.Acad.Sci., Paris* 265, B1412.

Forrester, A.J., Gudmundsen, R.A., and Johnson, P.O. (1955). *Phys.Rev.* 99, 1691.

Guiochon, M.A., Blamont, J.E., and Brossel, J. (1957). *J.Phys.Radium, Paris.* 18, 99.

Happer, W. and Saloman, E.B. (1967). *Phys.Rev.* 160, 23.

Kibble, B.P. and Series, G.W. (1963). *Proc.R.Soc.* A274, 213.

Majorana, E. (1932). *Nuovo Cim.* 9, 43.

Omont, A. (1965a). *C.r.hebd.Séanc.Acad.Sci., Paris.* 260, 3331.

__________ (1965b). *J.Phys.,Paris.* 26, 26, 576.

__________ and Meunier, J. (1968). *Phys.Rev.* 169, 92.

Rabi, I.I. (1932). *Phys.Rev.* 51, 652.

Rose, M.E. (1957). *Elementary theory of angular momentum.* Wiley, New York.

General references and further reading

The development of the optical double-resonance technique is covered in the following four review articles:

Budick, B. (1967). *Adv.at.& Mol.Phys.* 3, 73.

zu Putlitz, G. (1964). *Ergebn.exact.Naturw.* 37, 105.

Series, G.W. (1959). *Rep.Prog.Phys.* 22, 280.

__________ (1970). *Quantum optics.* (eds. S.M. Kay and A. Maitland), p.395. Academic Press, New York.

Collision processes involving excited atoms and molecules are discussed in detail by

Massey, H.S.W. (1971). *Electronic and ionic impact phenomena.* Vol.III, Clarendon Press, Oxford.

Early experiments in resonance fluorescence are described by

Mitchell, A.C.G. and Zemansky, M.W. (1966). *Resonance radiation and excited atoms*, Cambridge University Press, London.

17

Optical pumping experiments

17.1. Introduction

The success of the Hanle effect and optical double resonance experiments as a means of investigating the lifetimes, energy levels, and collisional processes involving excited atoms raises the interesting question: Can similar techniques be applied to a study of free atoms in the ground state? A way of achieving this, now known as *optical pumping*, was proposed by Kastler (1950), although the first experimental demonstrations of the optical pumping and optical double-resonance methods were not published until 1952. Professor Kastler was awarded the Nobel Prize in 1966 for his seminal contributions to both of these fields.

Optical pumping, then, is a method for producing significant changes in the population distributions of atoms by means of irradiation with light. The term, as used in this chapter, refers to changes in the relative populations of either Zeeman or hyperfine levels belonging to the ground or metastable states of an atom. These changes can be monitored by the modification of the intensity of polarization of the resonantly scattered light, as in optical double-resonance experiments, or more directly by means of the change in intensity of the light transmitted through the optically-pumped sample.

The optical pumping technique permits the investigation of ground-state Zeeman intervals using magnetic resonance methods and many g-factors have been determined to high precision. From the width of the magnetic resonance signals and from the decay rate of transient signals the transverse and longitudinal relaxation times, T_2 and T_1 respectively, can be evaluated and mechanisms for the collisional relaxation elucidated. In low-pressure systems the relaxation occurs principally at the walls of the optical pumping cell and thus a study of the relaxation has increased our understanding of surface physics. Optical pumping magnetic resonance signals can be very narrow and this fact has allowed a systematic study to be made of the interactions of atoms with electromagnetic fields in the optical and radio frequency range. This has led to the discovery of shifts in atomic energy levels produced by light and of multiple quantum transitions.

In this chapter we shall be concerned mainly with the principles of the technique, the effect of relaxation processes, and magnetic resonance transitions between Zeeman sub-levels. We shall therefore initially describe the experiments in terms of the populations of the ground state sub-levels. The discussion of the effects of phase coherence (Hertzian coherence) and experiments involving transverse pumping is reserved until section 17.8. Moreover the application of optical pumping methods to the investigation of hyperfine intervals and the measurement of nuclear moments is postponed until Chapter 18, as are the applications of this technique in devices such as magnetometers, atomic clocks, and masers.

17.2. Principles of optical pumping

17.2.1. Zeeman pumping in pure vapours. The aim of this particular pumping technique is to create a non-thermal distribution of population among the different magnetic sub-levels belonging to the atomic ground state. The basic principle of the technique may be understood by reference to Fig.17.1(a) which gives details of the energy levels and resonance lines of rubidium. This structure is typical of all the alkalis; however, we have neglected the nuclear spin and the diagram would apply strictly only to the case I=0, e.g. $^{40}Ca^{+}$. By absorption of optical resonance radiation in either the D_1 or D_2 lines, the rubidium atoms can be excited to the 5 $^2P_{1/2}$ or 5 $^2P_{3/2}$ levels from which they will decay spontaneously after a mean lifetime of $\approx 10^{-8}$s.

In an optical pumping experiment, Fig.17.2, the light from a rubidium lamp is usually passed through a simple

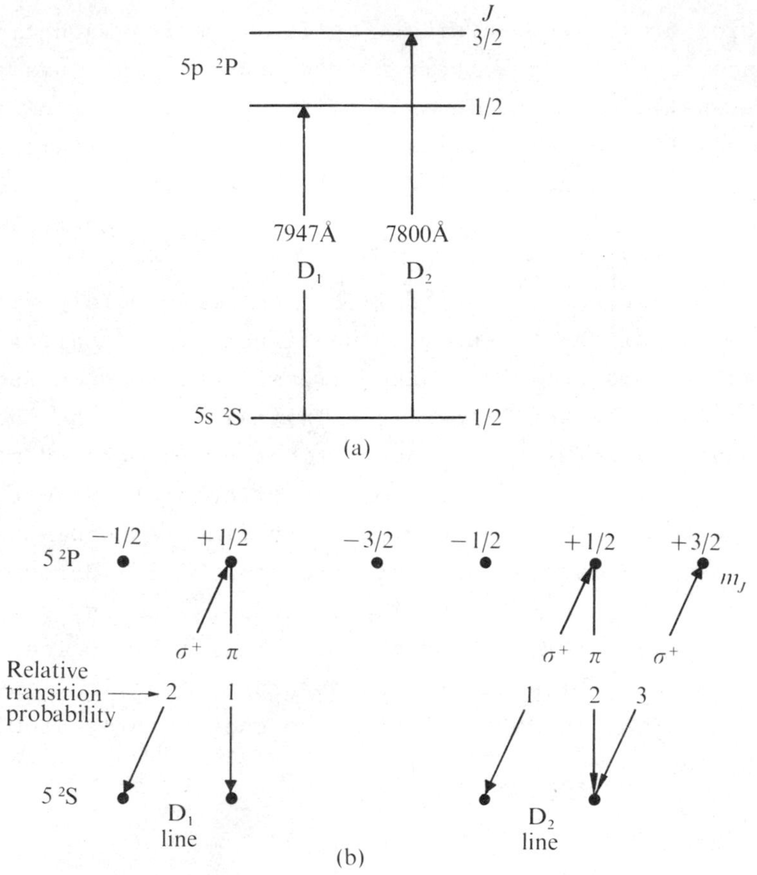

Fig.17.1. (a) Energy levels and resonance lines of rubidium. (b) Zeeman structure of the D_1 and D_2 resonance lines showing the relative transition probabilities of the $\sigma^{\pm}$ and π-polarized components.

interference filter which transmits only the D_1 resonance line. The radiation is then circularly polarized by using a linear polarizer followed by a quarter-wave plate. The light is focussed into a cell which has previously been evacuated and has then had a small quantity of metallic rubidium distilled into it before being sealed off. At room temperature (300 K) the vapour pressure of rubidium ($\approx 10^{-6}$ Torr) is sufficient to ensure that the sample will initially absorb most of the incident D_1 radiation. A static

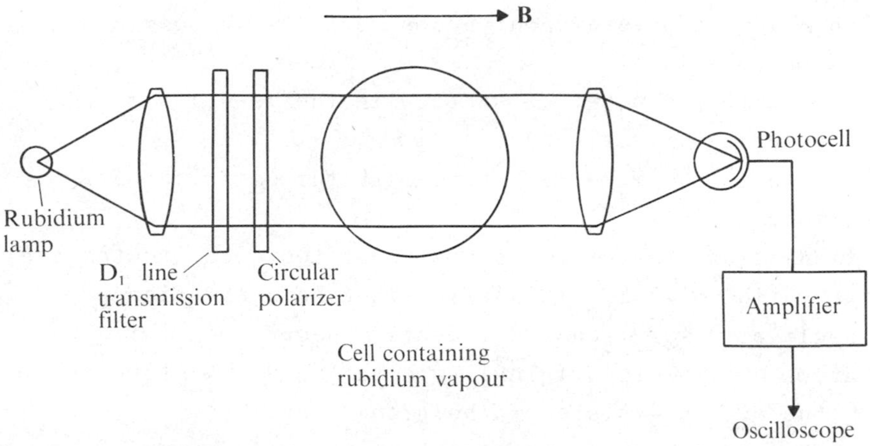

Fig.17.2. Schematic diagram of the apparatus for a rubidium optical pumping experiment.

magnetic field is applied to the cell in a direction parallel to that of the incident light.

We now recall the selection rules for the Zeeman components of an electric dipole transition:

$$\Delta m = m' - m'' = 0, \pm 1$$

where m' and m'' are the magnetic quantum numbers of the upper and lower Zeeman states respectively. The selection rule $\Delta m = +1$ corresponds tothe absorption of σ^+-polarized light. For this polarization the sense of rotation of the electric vector is the same as that of an electric current in the Helmholtz coils which generate the magnetic field. An observer looking towards the lamp from the position of the detector in Fig.17.2 would see the electric vector of σ^+ circularly-polarized light rotating in the anti-clockwise sense and would call it left circularly-polarized radiation.

From Fig.17.1(b) we see that when the atoms in the cell are illuminated with σ^+-polarized D_1 radiation, only those atoms in the $m_J = -\frac{1}{2}$ state of the ground level can absorb a photon. These atoms are excited to the $m_J = +\frac{1}{2}$ state of the 5 $^2P_{1/2}$ level where they spend an average lifetime of $2{\cdot}8 \times 10^{-8}$ s before spontaneous decay. Two possible decay routes exist:

(a) back to the $m_J = -\frac{1}{2}$ state with the re-emission of σ^+ polarized light, and

(b) down to the $m_J = +\frac{1}{2}$ state with the emission of π-polarized light.

The transition probabilities for these two decay routes are in the ratio of 2:1. Thus after they have experienced one cycle of absorption and radiative decay, one third of the atoms which were originally in the $m_J = -\frac{1}{2}$ state end up in the $m_J = +\frac{1}{2}$ state of the ground level.

The atoms have therefore been optically pumped from a lower to a higher m_J state. Simultaneously the z-component of the angular momentum of the atomic system $\langle J_z \rangle$, has been increased. This additional angular momentum has been transferred from the pumping beam to the atomic sample by the absorption of σ^+-polarized photons, each of which possess an angular momentum component of $+1\hbar$ along the direction of propagation. The random initial orientations of the atomic magnetic moments are therefore altered by the pumping process and the sample acquires a net magnetization. The sample is said to be oriented or polarized by the pumping process. Clearly changing the pumping light to σ^--polarization will create an atomic polarization in the reverse direction.

This type of optical pumping which results in a sample possessing a non-thermal distribution of population among the sub-levels of the atomic ground state should be distinguished from the flash lamp pumping used in solid state and liquid lasers. In those systems the aim is to transfer a large fraction of the ions or molecules into a different electronic level.

As Fig.17.1(b) shows, σ^+-polarized D_2 radiation may also be used for optical pumping, the Zeeman component $m_J = -\frac{1}{2} \rightarrow m_J = +\frac{1}{2}$ being the effective transition. Although absorption also occurs in the $m_J=+\frac{1}{2} \rightarrow m_J=+\frac{3}{2}$ component this has no effect on the orientation process. Starting from equal populations in the $m_J=\pm\frac{1}{2}$ levels, the D_2 line will transfer two thirds of the atoms from the lower to the upper state after one pumping cycle. When both D_1 and D_2 lines are used together the optical pumping effect is additive provided that the atoms are not subjected to collisional perturbations whilst in the excited level. We consider the effect of such collisions in the following paragraph.

17.2.2. Zeeman pumping in the presence of a buffer gas. In an evacuated cell atoms of the vapour travel to the cell walls in a time of the order of 10^{-4} s and there the orientation is destroyed. This short effective lifetime of ground state atoms makes it difficult to attain high atomic polarizations and a foreign gas is often added to the optical pumping cell to slow down this relaxation. Most of the atoms on which optical pumping experiments have been performed have ground electronic states with zero orbital angular momentum (e.g. the $^2S_{1/2}$ states of alkalis and the $^1S_0$ states of group IIB metals). The wavefunctions of these states are spherically symmetric and the atoms are relatively insensitive to collisions. For these atoms diamagnetic gases such as H_2, N_2, and the noble gases act as 'buffer gases'. The interatomic collisions not only preserve the orientation of the ground state but also prevent the atoms from colliding with the walls of the cell.

Unfortunately collisions also occur between the excited atoms and those of the added gas and at sufficiently high pressure the mean free time between collisions becomes smaller than the radiative lifetime of the excited atoms. Most atoms then undergo collisions before they return to the ground state and the effect of these collisions is to induce transitions between the Zeeman levels of the excited state. As a result of this mixing the emitted resonance fluorescence is partially depolarized, a phenomenon that we discussed in detail in section 16.5.4.

If the foreign gas pressure is very high the excited

atoms will be uniformly distributed among the Zeeman states of the excited level and atoms will then return at the same rate to all states of the ground level. In this extreme the efficiency of the optical pumping process depends only on the rate at which the incident radiation is absorbed by the different ground state Zeeman levels. We consider again the case of an alkali atom illuminated with σ^+-polarized light. If self-absorption in the lamp causes the intensities of the D_1 and D_2 lines to be equal, then both the $m_J = \pm \frac{1}{2}$ states of the ground level will be emptied at the same rate. This can be verified from Fig.17.1(b) by adding the transition probabilities for the D_1 and D_2 lines. Under these conditions optical pumping can only become efficient if the D_2 radiation is suppressed with a suitable filter. Then only atoms in the $m_J = -\frac{1}{2}$ state will be excited and an orientation of the sample will be achieved as before. We see therefore that optical orientation can be achieved both with and without a buffer gas but that the illumination conditions must be carefully chosen in each case to ensure the optimum efficiency for the pumping process.

17.2.3. Detection of atomic orientation - transmission monitoring. When either the D_1 or the D_2 line is used on its own, the discussion of section 17.2.1 shows that a high atomic polarization can be attained after all the atoms in the $m_J = -\frac{1}{2}$ state have undergone a single optical pumping cycle. If $N_\pm$ are the number densities of atoms in the $m_J = \pm\frac{1}{2}$ states, the polarization is defined by

$$\text{Polarization} = P = \frac{N_+ - N_-}{N_+ + N_-} . \qquad (17.1)$$

From Fig.17.1(b) we deduce that after one cycle the polarization would equal either $\frac{1}{3}$ or $\frac{2}{3}$, depending on whether the D_1 or the D_2 line was used for pumping. As the pumping process continues the cycle is repeated for those atoms left in the $m_J = -\frac{1}{2}$ state and in the absence of relaxation processes all the atoms would eventually be pumped into the $m_J = +\frac{1}{2}$ state. When this situation is reached the sample is no longer able to absorb σ^+-polarized radiation. Thus during the process of orientation, the transparency of the sample to the σ^+-polarized D_1 line increases and the intensity of light transmitted through the cell is directly related to the atomic polarization. A simple photocell followed by an amplifier can be used to monitor the progress of the optical pumping, as shown in Fig.17.2. By the use of phase-sensitive detection the signal-to-noise ratio can be improved and changes of the order of 1 part in 10^4 of the transmitted light intensity can be detected.

<u>17.2.4. Detection of atomic orientation - fluorescence monitoring.</u> An alternative method of monitoring the optical pumping process involves the detection of the fluorescent light emitted by the optically excited atoms in the cell. When the orientation of the sample is complete σ^+-polarized D_1 radiation cannot be absorbed and consequently the atoms no longer re-radiate fluorescent light. Thus the intensity of the fluorescent light directly monitors the population of atoms in the $m_J = -1/2$ level. The polarization of the fluorescent light in this case is constant although generally it too can be used to monitor the orientation. However, when a high pressure of buffer gas is used in the cell, collisions will equalize the populations of the Zeeman levels of the excited state and the polarization technique will no longer be applicable.

Finally it should be noted that when the sample is pumped by the D_1 and D_2 lines simultaneously, the absorption of the D_2 line becomes stronger as the sample orientation increases. Thus in both the transmission and the fluorescence monitoring schemes the sign and magnitude of the signal obtained when radiation in the D_1 and D_2 lines falls on the detector cannot be simply related to the polarization of the sample. It is therefore again desirable to filter out the D_2 component.

<u>17.2.5. Optical pumping in mercury.</u> In order to widen the scope of our discussion of optical pumping experiments, we

recall that mercury has a high vapour pressure at room temperature. It is therefore a very convenient element for optical pumping experiments, especially when the $6\,^1S_0 - 6\,^3P_1$ intercombination line at 2537 Å is used. We are now of course only concerned with the odd mercury isotopes ^{199}Hg and ^{201}Hg which have nuclear spins of $I = \frac{1}{2}$ and $I = \frac{3}{2}$ respectively. To improve the efficiency of the optical pumping cycle it is again desirable to be able to isolate different components of the resonance line. Although the hyperfine structure cannot be resolved by anything as simple as an interference filter, nevertheless the use of separated isotopes in the resonance lamps enables the necessary selection to be obtained. Thus by using ^{204}Hg in the lamp and ^{199}Hg in the cell it is possible to pump on the component connecting the $6\,^1S_0(I=\frac{1}{2})$ ground state with the $6\,^3P_1(F=\frac{1}{2})$ level only. The pumping scheme is then identical to the pumping of alkali atoms using filtered D_1 radiation and much of our preceding discussion applies equally to the experiments on mercury.

There are, however, two very important differences: first the spin angular momentum of the ground state atoms is now associated with a nuclear magnetic moment rather than with an electron moment. Consequently the relaxation time for the nuclear spins is very much longer than that of the electron spins and optical pumping experiments in mercury ought to be easier than those in alkali vapours. Secondly, however, the transition probability for the 2537 Å line, is a factor of ten smaller than those of the resonance lines of the alkalis. This means that the time constant for nuclear orientation by optical pumping in mercury is rather long and indeed many of the initial experiments failed because of the low intensities of the available sources of 2537 Å radiation. The first successful experiments were reported by Cagnac (1958) using intense microwave lamps. This pioneering work is described in more detail in his thesis (Cagnac 1961).

The isotope ^{199}Hg has only two ground state magnetic sub-levels and so has served as an important test case for the basic theory of optical pumping, developed in more detail in sections 17.8 and 17.9. Many detailed comparisons between the theory and experiments using this isotope can be found in another outstanding thesis from the Paris group (Cohen-Tannoudji 1962). Although in ^{201}Hg the ground state has four magnetic sub-levels, it is still true that by illuminating the cell with σ^+-polarized radiation from a ^{204}Hg lamp, the atoms will be concentrated in the states of higher M_I quantum numbers and the sample acquires a net magnetic moment. In this case the ground state nuclear spins are said to be oriented. However, if the cell is illuminated with unpolarized light travelling along the direction of the field $\underline{B}$ both σ^+ and σ^- components are absorbed with equal probability. The ground state sub-levels with large values of $|M_I|$ are then filled at the expense of those with small values of $|M_I|$. Although the sample now has no net magnetic moment, it does have a non-thermal distribution of population and is said to be *aligned*. When we do not wish to distinguish between orientation and alignment we shall say that the sample is polarized.

17.3. *Effect of relaxation processes*

17.3.1. Conditions to be fulfilled for efficient pumping.

The state of complete polarization is, of course, never attained experimentally since relaxation processes tend to restore the spin distribution to thermal equilibrium. In a given experimental situation one relaxation process is generally dominant and can be characterized by a definite longitudinal relaxation time T_1. When the pumping light is suddenly interrupted, for instance by a shutter, the rate of change of the population difference, $n=N_+-N_-$, obeys the equation

$$\left(\frac{dn}{dt}\right)_{relax} = -n/T_1 \ . \tag{17.2}$$

Since the net magnetic moment per unit volume of the system is given by

$$\langle M_z \rangle = |\underline{\mu}|(N_+ - N_-) \tag{17.3}$$

where $|\underline{\mu}| = g_J \mu_B J$ is the magnetic moment of an atom with total angular momentum J, it will decay exponentially with the time constant T_1:

$$\langle M_z(t) \rangle = \langle M_z(0) \rangle \exp(-t/T_1). \tag{17.4}$$

For completeness we note that the magnetization $\langle M_z \rangle$ and the degree of polarization defined by equation (17.1) are proportional to one another, being related by

$$P = \frac{N_+ - N_-}{N} = \frac{\langle M_z \rangle}{|\underline{\mu}|N} \tag{17.5}$$

where N is the total number density of atoms in the sample.

The longitudinal relaxation which takes place in the dark is obviously also present when the cell is illuminated with pumping radiation. However, we must now take into account the transitions stimulated by the absorption of light. The net rate at which atoms are pumped from the $m_J = -\frac{1}{2}$ state to the $m_J = +\frac{1}{2}$ state is given by

$$dN_+ = -dN_- = \alpha P_{-1/2,1/2} N_- dt \tag{17.6}$$

where $P_{\mu m}$ is the absorption probability for electric dipole transitions given by equation (9.40) and α is a numerical factor less than unity which accounts for the fact that only a certain fraction of the excited atoms decay to the $m_J = +\frac{1}{2}$ state. Noting that $N_- = (N-n)/2$ and that the absorption probability is proportional to the intensity of the incident light, I, we find that

$$\left(\frac{dn}{dt}\right)_{pump} = K'I(N-n) \tag{17.7}$$

where K' is a constant. Introducing the pumping time constant $T'_p = 1/K'I$, the total rate of change of the difference in population densities under the simultaneous action of both the pumping and the relaxation processes becomes

$$\frac{dn}{dt} = \frac{N-n}{T'_p} - \frac{n}{T_1} . \tag{17.8}$$

Thus the approach to the equilibrium polarization is described by the equation (Problem 17.1):

$$\frac{n(t)}{N} = P(t) = P(\infty)\{1 - \exp(-t/\tau_1)\} \tag{17.9}$$

where

$$1/\tau_1 = 1/T'_p + 1/T_1, \tag{17.10}$$

and $P(\infty)$ is the steady state polarization given by

$$P(\infty) = (1 + T'_p/T_1)^{-1}. \tag{17.11}$$

We see that to obtain a large steady state polarization we must satisfy the condition $T'_p \leq T_1$, i.e. the pumping rate must be faster than the relaxation rate. Although this condition could have been derived intuitively, the equations developed above will be useful in the subsequent discussion of the experimental determinations of the longitudinal relaxation time T_1.

17.3.2. Light sources for optical pumping. Obviously for a high pumping rate an intense source of resonance radiation is required. With ordinary commercial lamps pumping times of the order of 10^{-3} s can be obtained, but substantial improvements can be made by using small electrodeless lamps which are excited by microwave or high radio frequency power (Problem 17.2). To avoid self-reversal of the resonance lines the lamps are made in the form of a thin disk or a re-entrant sphere as shown in Fig.17.3. The lamps are thoroughly baked under vacuum before being filled with a small quantity of metal and approximately 1 Torr of noble gas, usually argon, to carry the discharge while the lamp is warming up. Because the quantity of metal that is required for these lamps is very small (≈ 5 mg), they are very convenient when using pure isotopes.

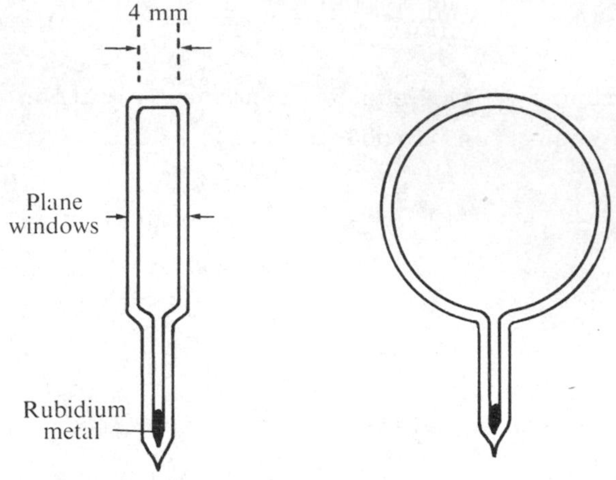

Fig.17.3. Electrodeless lamp for optical pumping experiments in alkali metal vapours.

17.3.3. Optical pumping cells. In order to attain strong steady state polarization it is also necessary to make the relaxation as slow as possible. The first optical pumping experiments of Brossel *et al.* (1952) and Hawkins and Dicke (1953) were made on atomic beams of sodium where relaxation is absent. However, the rapid flight of the atoms through the pumping region severely limited the time that the atoms were illuminated. It was later realized that the experiments could be performed more simply by using low density atomic vapours in sealed-off cells. The relaxation time is then the mean time between two wall collisions and is of the order of 10^{-4} s for a spherical cell having a diameter of 5 cm (Problem 17.3). This relaxation time can be increased considerably either by filling the cell with a suitable buffer gas or by coating the walls of the cell with certain hydrocarbon films. These techniques will be discussed in more detail in the following section.

17.4. Investigation of longitudinal relaxation times

17.4.1. Experimental technique. A convenient method for the

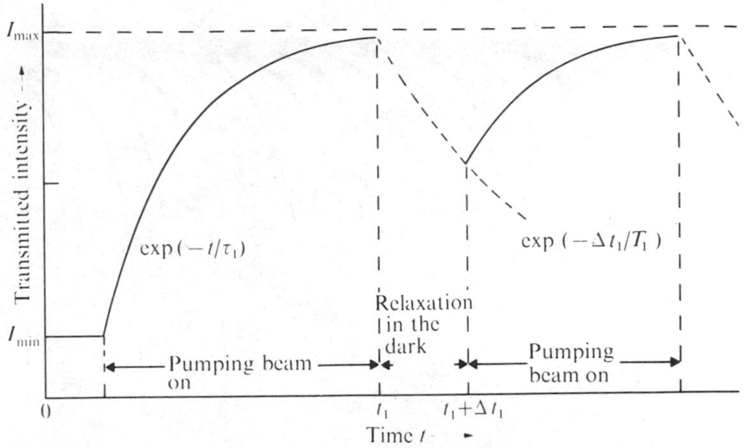

Fig.17.4. The time dependence of the intensity of light transmitted through an optically-pumped atomic vapour.

measurement of longitudinal relaxation times was introduced by Franzen (1959) in a study of the effect of different buffer gases on the efficiency of the optical pumping cycle. The principle of this technique may be understood by reference to Fig.17.4, which shows the signal observed in a transmission monitoring experiment. At time t=0 the optical pumping lamp is switched on but, because there are initially equal numbers of atoms in the $m_J = \pm \frac{1}{2}$ states, the vapour is absorbing and the transmitted intensity is low. The transparency of the vapour increases with time as the pumping proceeds and the signal approaches its maximum value with the exponential time constant τ_1 given by equation (17.10). At some later time t_1, a shutter is inserted to cut off the pumping beam and the atoms of the vapour start to relax in the dark with a time constant T_1, as described by equation (17.4). Thus when the shutter is re-opened at time $t_1 + \Delta t_1$, the transparency of the vapour has been reduced below its maximum value and the transient optical pumping signal starts again from a point lying between I_{max} and I_{min}.

The exponential curve governing the relaxation in the dark can be mapped out by superimposing the signals corresponding to a series of different time intervals Δt_1 on a single oscilloscope photograph. This is illustrated

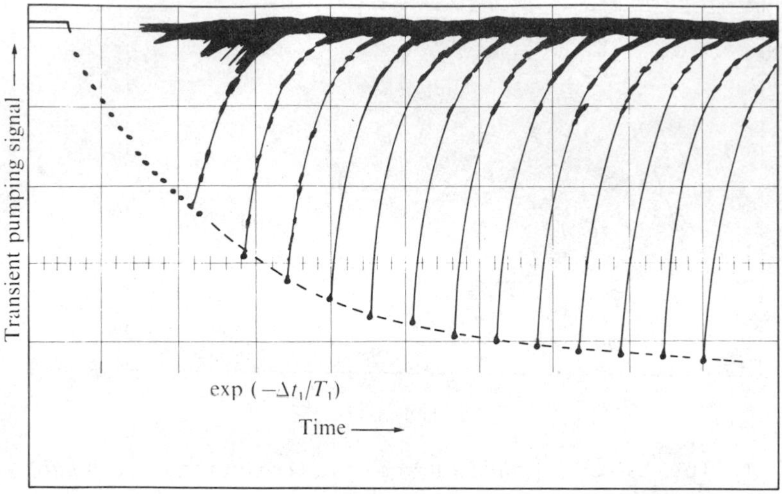

Fig.17.5. Oscillogram showing the longitudinal relaxation of ^{87}Rb vapour observed by Gibbs and Hull (1966). The transients result from reopening the shutter after different time intervals spent relaxing in the dark and show the approach to optical pumping equilibrium. The bright spots from which each trace begins plot out the decay curve, $\exp(-\Delta t_1/T_1)$.

in Fig.17.5 for the relaxation in a ^{87}Rb cell observed by Gibbs and Hull (1966). The starting points of the transient signals lie on the curve

$$I(\Delta t_1) = I_{min} + (I_{max} - I_{min})\exp(-\Delta t_1/T_1)$$

and a detailed analysis enables the longitudinal relaxation time T_1 to be determined. Once the value of T_1 has been measured, the pumping time T_p' for a given light intensity can be obtained by analysis of the transient signals obtained when the cell is illuminated.

17.4.2. Relaxation on the cell walls.

(a) Uncoated cells. The relaxation times measured in pure alkali vapours at low pressures are of the order of 10^{-4} s and are found to be proportional to the linear dimensions of the optical pumping cell. This is to be expected since at pressures below 10^{-3} Torr the atomic mean free path is large compared with the dimensions of the cell. The relaxation time T_1 can then be identified with the mean time of flight T_w of the atoms between two collisions with the walls. In a cell of volume V and surface area S, this time of flight is given by

$$T_w = \frac{N}{\nu} \cdot \frac{V}{S} \qquad (17.12)$$

where ν is the mean number of atoms colliding with unit area of the surface per second. From simple kinetic theory ν is given in terms of the atomic number density N by

$$\nu = \frac{1}{4} N \bar{v} \qquad (17.13)$$

where $\bar{v} = (8kT/\pi m)^{1/2}$ is the mean thermal velocity. For a spherical cell of diameter d or a cubical cell of the same length, the volume to surface ratio is d/6 and thus

$$T_w = \frac{4V}{\bar{v}S} = \frac{2d}{3\bar{v}} \qquad (17.14)$$

For ^{133}Cs at 300 K we have a mean velocity $\bar{v} = 2{\cdot}18 \times 10^4$ cm s^{-1} and we find that $T_w = 0{\cdot}92 \times 10^{-4}$ s for a cell of 3 cm diameter.

Since the observed relaxation times in uncoated cells are of this order of magnitude, we can conclude that at every collision with the walls of the cell the spin of the alkali atom is completely disoriented. In fact the atom is usually absorbed on to the surface and a finite time elapses before it is re-evaporated into the cell. During this time the atom is subjected to the large local magnetic and electric fields produced by the ions and molecules forming the cell walls, and consequently the orientation of its magnetic moment with respect to the external magnetic field is completely destroyed.

These short relaxation times reduce the maximum polarization which can be attained with a given source, equation (17.11), and also lead to rather broad signals in magnetic resonance experiments. It was therefore fortunate when Dehmelt and his co-workers (Robinson *et al.* 1958) discovered that much longer relaxation times could be obtained by covering the glass walls of the pumping cell with coatings of special organic compounds. We now briefly discuss this technique.

(b) Coated cells. Optical pumping cells can be coated with straight-chain saturated hydrocarbons such as eicosane ($C_{20}H_{42}$) or tetracontane ($C_{40}H_{82}$) by melting these paraffins in an atmosphere of noble gas and then randomly shaking the cell as it cools. These compounds have long molecules which may be imagined to project into the cell and thus prevent the oriented atoms from coming within the range of the fields produced at the cell walls. Certain organosilanes such as Drifilm (dimethyl-dichloro-silane) may be used for the same purpose.

A detailed study of wall relaxation in pure rubidium isotopes has been made by Mme. Bouchiat and her colleagues (Bouchiat and Brossel 1962, Bouchiat 1963, Bouchiat and Grossetête 1966). They find that the longitudinal relaxation time T_1 for ^{85}Rb (I=5/2) is approximately twice as long as that for ^{87}Rb (I=3/2). The relaxation times are of the order 0·2 s for ^{87}Rb in spherical cells of 6 cm diameter coated with paraffin or silicone films. By using deuterated paraffins, the relaxation times can be increased by a factor of five giving $T_1(^{87}\text{Rb}) \approx 1$ s and $T_1(^{85}\text{Rb}) \approx 2$ s. Since the nuclear magnetic moment of hydrogen is three times larger than that of deuterium, these results indicate that magnetic dipole-dipole interactions between the electron spin of the alkali atom and the nuclear spins of the wall coating is a significant relaxation mechanism. It also has been shown that collisions of the alkali atoms with the coatings do not perturb the phase of the atomic wavefunction. Thus very narrow Zeeman magnetic resonance signals can be observed having linewidths of only a few Hz. Although similarly narrow resonances can be obtained by using buffer gases in the cell, the low pressure coated cells have the advantage that the moving atoms effectively 'see' only the average magnetic field in the cell. Thus inhomogeneities in the applied magnetic field do not result in a broadening or distortion of the resonance signals and coated cells are preferable for the precision measurements discussed in section 17.7.5.

17.4.3. Spin relaxation in buffer gases.

(a) Diffusion to the cell walls. When the optical pumping cell is filled with a diamagnetic gas, collisions of the gas molecules with the oriented atoms hinder their free flight towards the cell walls. At sufficiently high pressures, $p > 1$ Torr, the atomic mean free path becomes smaller than the cell dimensions and the motion of an atom in the cell is determined by a diffusion process. In the absence of any other relaxation mechanism the decay of the orientation of the sample is determined by the diffusion equation

$$\frac{\partial n}{\partial t} = D\,\nabla^2 n$$

where D is the diffusion coefficient for the optically-pumped atoms in a given buffer gas at the pressure p. The diffusion coefficient may be obtained in terms of its value D_0 at some standard pressure p_0 by the relation $D=D_0p_0/p$. If we assume that the orientation is completely destroyed at the cell walls, then the longitudinal relaxation rate may be identified with the decay rate $1/T_D$ of the lowest-order diffusion mode for the given cell geometry. For a spherical cell of diameter d we have (Problem 17.4):

$$\frac{1}{T_D} = \frac{D_0p_0}{p}\left(\frac{2\pi}{d}\right)^2. \qquad (17.15)$$

The relaxation times for rubidium atoms measured as a function of neon and argon pressures are shown in Fig.17.6. At low pressures we see that T_1 is linearly proportional to the buffer gas pressure p as predicted by equation (17.15)

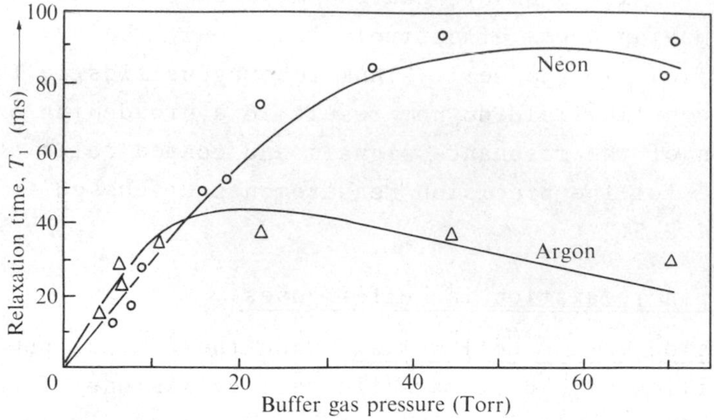

Fig.17.6. Longitudinal relaxation times, T_1, for optically pumped rubidium atoms as a function of buffer gas pressure. (After Franzen (1959).)

but that at higher pressures T_1 passes through a maximum value and then starts to decrease. This is interpreted as the effect of an additional spin relaxation process involving collisions with the buffer gas atoms.

(b) Collisional relaxation by buffer gases. If we assume that only binary collisions are important then the additional relaxation rate can be represented by $N_p\, \sigma(\bar{v}_r)\, \bar{v}_r$, where N_p is the atomic density of the buffer gas at the pressure p, $\bar{v}_r$ is the mean relative velocity, and $\sigma(\bar{v}_r)$ is the velocity averaged spin-disorienting collision cross-section. The total relaxation rate in a cell containing a buffer gas is therefore the sum of the diffusion decay rate and the collisional rate and may be written in the form

$$\frac{1}{T_1} = \frac{D_0 p_0}{p}\left(\frac{2\pi}{d}\right)^2 + N_0 \sigma(\bar{v}_r)\, \bar{v}_r \left(\frac{p}{p_0}\right) \qquad (17.16)$$

for a spherical cell. Detailed analysis of experimental data, such as those shown in Fig.17.6, allows the best values of D_0 and $\sigma(\bar{v}_r)$ to be determined. Results obtained by several different experimentalists are given in Table 17.1.

TABLE 17.1

Diffusion coefficients and spin disorientation cross-sections for rubidium and cesium in foreign gases

Foreign gas	Rb $D_0^{\dagger}$ ($cm^2\ s^{-1}$)	Rb $\sigma(\bar{v}_r)$ (cm^2)	Cs $D_0^{\dagger}$ ($cm^2\ s^{-1}$)	Cs $\sigma(\bar{v}_r)$ (cm^2)
He	0·7	$3{\cdot}3 \times 10^{-25}$	0·20	$2{\cdot}8 \times 10^{-23}$
Ne	0·5	$3{\cdot}3 \times 10^{-24}$	0·15	$9{\cdot}3 \times 10^{-23}$
Ar	0·4	$1{\cdot}1 \times 10^{-22}$	0·13	$1{\cdot}0 \times 10^{-21}$
Kr	0·16	$2{\cdot}3 \times 10^{-20}$	—	$2{\cdot}1 \times 10^{-21}$
Xe	0·13	$1{\cdot}3 \times 10^{-19}$	—	$4{\cdot}6 \times 10^{-20}$
N_2	0·33	$5{\cdot}7 \times 10^{-23}$	0·07	$5{\cdot}5 \times 10^{-22}$
H_2	1·34	$2{\cdot}2 \times 10^{-24}$	—	—

† Values for D_0 refer to gas at 760 Torr pressure.

The values of the diffusion coefficients listed in Table 17.1 are similar to the gas kinetic diffusion coefficients measured by other techniques. However, the disorientation cross-sections are quite remarkable, being many orders of magnitude smaller than gas kinetic cross-sections. In the case of Rb-He collisions, for instance, the rubidium atoms can make over 10^8 collisions with the buffer gas atoms without losing their orientation. The very small size of these collision cross-sections is due to the fact that both the alkali atoms and the atoms of the buffer gas possess spherically symmetric ground state wavefunctions. The relaxation can therefore only proceed through weak magnetic dipole-dipole interactions or through a spin-orbit interaction acting in second order. The detailed nature of the relaxation mechanism is in fact rather complicated and the reader is referred to the excellent review article by Happer (1972) for a detailed discussion.

(c) Effect of buffer gases on magnetic resonance signals. For a cell of given diameter we see from equation (17.16) and Fig.17.6 that there is an optimum buffer gas pressure at which the longitudinal relaxation time T_1 is a maximum. This relaxation time determines the decay of the diagonal elements of the ground state density matrix and hence the magnetization of the sample along the axis of quantization equation (17.5). More generally the density matrix also possesses off-diagonal elements whose relaxation is described by the transverse relaxation time T_2. These elements of the density matrix determine the sample magnetization in the x-y plane and hence the width of optical pumping magnetic resonance signals. In addition to the relaxation caused by collisions which induce changes in the spatial quantum numbers of the atoms concerned, the off-diagonal elements of the density matrix are also destroyed by collisions which merely perturb the relative phases of the atomic sub-states. In spite of this the transverse relaxation time, T_2, remains of the same order of magnitude as T_1, 0·01 - 1·0 s, and consequently very narrow resonance signals may be obtained using cells filled with a buffer gas. In this case, however, the atoms are no longer able to average the applied magnetic field by rapid motion through the whole cell and a very uniform magnetic field is required. For low-frequency Zeeman transitions the buffer gas technique is therefore not as convenient as the use of a suitably coated empty cell.

However, when optical pumping techniques are used to study microwave transitions between different hyperfine levels of the atomic ground state, the Doppler broadening of the magnetic resonance signal in an empty cell becomes quite appreciable. In this case the reduction of the mean free path that occurs in cells filled with buffer gases has the effect of changing the resonance lineshape from the Gaussian to the Lorentzian form and of reducing the linewidth to the value (Problem 17.5)

$$\Delta\omega_{1/2} = 2{\cdot}8\,\frac{L}{\lambda}\,\Delta\omega_{\text{Doppler}} \qquad (17.17)$$

where L is the mean free path of the oriented atom and λ is the wavelength of the microwave radiation. Since the linewidths can be a factor of 300 times narrower than the microwave Doppler width, the buffer gas technique offers a significant advantage in these experiments.

17.5. Spin-exchange collisions

<u>17.5.1. Spin-exchange between like atoms.</u> When measurements of the longitudinal times T_1 are made in coated cells as a function of the alkali vapour pressure, it is found that the relaxation rate is a linear function of the vapour density. Experimental results for cells containing ^{85}Rb and ^{87}Rb respectively are shown in Fig.17.7. In these experiments the additional relaxation at finite pressures is produced by rubidium-rubidium collisions in which the electrons of the two atoms involved are simply exchanged. These collisions are therefore termed *spin-exchange* collisions. Although the electron spin polarization of the sample is preserved in this process, there is a decoupling of the nuclear and electron spins at the instant of exchange which leads to a rapid thermalization of the populations of the two hyperfine levels $F = I\pm\frac{1}{2}$.

Since the forces responsible for spin-exchange are of an electrostatic nature, in contrast to the interactions responsible for relaxation in buffer gases, the spin-exchange collision cross-sections are very large, $\sigma_{ex} \approx 10^{-14}$ cm^2. Experimental results for the alkalis are listed in Table 17.2.

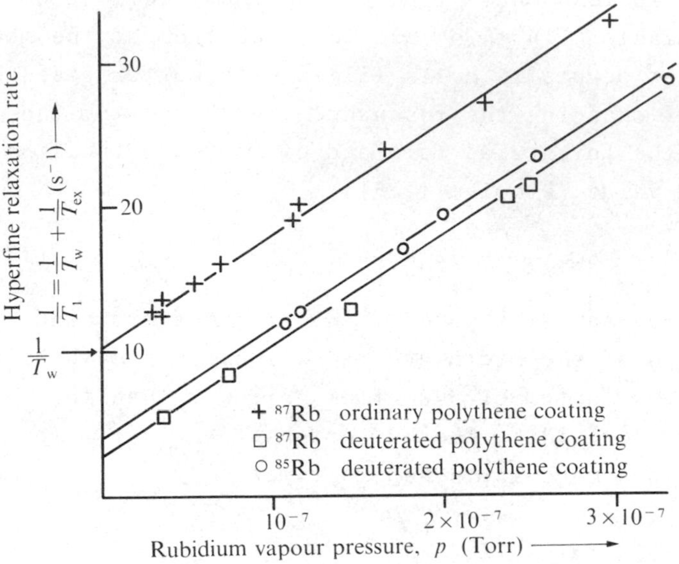

Fig.17.7. Hyperfine relaxation rate of ^{87}Rb and ^{85}Rb as a function of the rubidium vapour pressure. $1/T_{ex}$ is the relaxation rate due to exchange collisions; $1/T_W$ is the relaxation rate due to collisions with the walls of the cell which are coated with polythene. (After Bouchiat and Brossel (1963).)

17.5.2. Spin-exchange between unlike atoms. Spin-exchange collisions have played an important role in optical pumping experiments for they occur not only between like atoms, but also between atoms A and B of two quite different elements. A simple spin-exchange collision can be represented by the equation

$$A(\uparrow) + B(\downarrow) \rightarrow A(\downarrow) + B(\uparrow). \qquad (17.18)$$

Before the collision the atom A, which we assume to have been oriented by direct optical pumping, has its electron spin in the $m_S = +\frac{1}{2}$ state. After the collision the atom B is in the spin-up state and atom A is in the spin-down state. The orientation which originally existed in atom A has now

TABLE 17.2

Experimental values of the spin-exchange cross-sections of alkali atoms (10^{-14} cm^2)

$\sigma_{Na\text{-}Na}$	$\sigma_{K\text{-}K}$	$\sigma_{Rb\text{-}Rb}$	$\sigma_{Cs\text{-}Cs}$	Reference
	2·7 ± 0·7	2·1 ± 0·3	2·20 ± 0·35	Grossetête (1967)
			2·20 ± 0·15	Ernst and Strumia (1968)
1·109 ± 0·005				Moretti and Strumia (1971)
1·03 ± 0·21	1·45 ± 0·21	1·9 ± 0·2	2·06 ± 0·2	Ressler *et al.* (1969)

been transferred to atom B by the spin-exchange collision.

This phenomenon has enabled optical pumping techniques to be applied to systems which, for one reason or another, cannot be oriented directly. For instance, in atomic hydrogen the Lyman α resonance line, $1\ ^2S_{1/2} - 2\ ^2P_{1/2,3/2}$, lies in the vacuum ultraviolet at 1216 Å and the construction of lamps, cells, filters, and polarizers for this region is so difficult that a successful experiment involving the direct optical pumping of atomic hydrogen has not yet been reported. Fortunately atomic hydrogen and many other species can be oriented indirectly by spin-exchange collisions.

The basic apparatus required for spin-exchange optical pumping is identical to that shown in Fig.17.2. The only major difference is that the cell now contains hydrogen at ≈1 Torr pressure in addition to the alkali vapour, say ^{87}Rb. Free hydrogen atoms are created by running a mild discharge in the cell and the orientation of the rubidium atoms, created by pumping with circularly polarized D_1 resonance radiation, is transferred by spin-exchange collisions to the hydrogen atoms. The use of this technique in the measurement of the gyromagnetic ratios of hydrogen, tritium, free electrons, and ^{85}Rb is discussed in section 17.7.4 below. Further applications of spin-exchange optical pumping to the measurement of the ground state hyperfine structure of one-electron atoms will be described in Chapter 18.

17.6. Optical pumping of metastable atoms

All the noble gases have two metastable levels lying 10-20 eV above the ground state. These levels can be populated by electron impact in a weak discharge and by the use of suitable light sources the metastable levels having $J \neq 0$ can be oriented by optical pumping. The arrangement of the apparatus and the energy levels involved in the optical pumping of the $2\ ^3S_1$ metastable level of helium are shown in Figs.17.8(a) and (b) respectively. Triplet metastable atoms at concentrations of 10^{10} - 10^{11} cm^{-3} are produced by running a mild r.f. discharge in the helium gas which fills

(a)

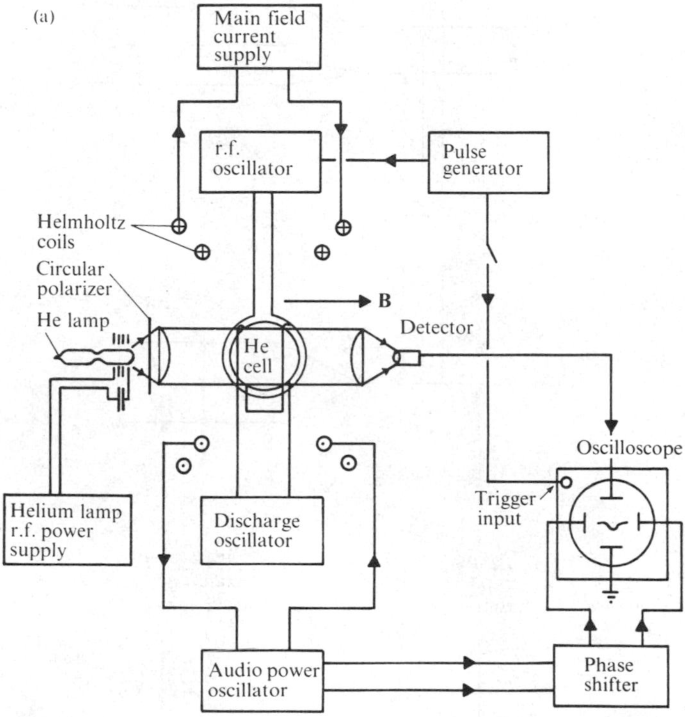

Fig.17.8 (a) Schematic diagram of apparatus used for optical pumping experiments in helium. (After Greenhow (1963).)

the optical pumping cell to a pressure of a few Torr. The ground state helium atoms serve as a buffer gas for the metastables which are then optically pumped by the strong infrared lines of the fine structure multiplet $2\ ^3S_1 - 2\ ^3P_{0,1,2}$ at 10 830 Å. Since photomultiplier tubes have high dark currents and low quantum efficiencies at this wavelength, silicon or lead sulphide photocells are often used for the transmission monitoring. The pumping of helium has an importance comparable to that of the alkali atoms since it is used in the construction of sensitive magnetometers, as a source of polarized electrons, and may be used to polarize ions via Penning reactions with atoms and molecules. Optically-pumped helium atoms have also been used to polarize ^{3}He nuclei as we now explain.

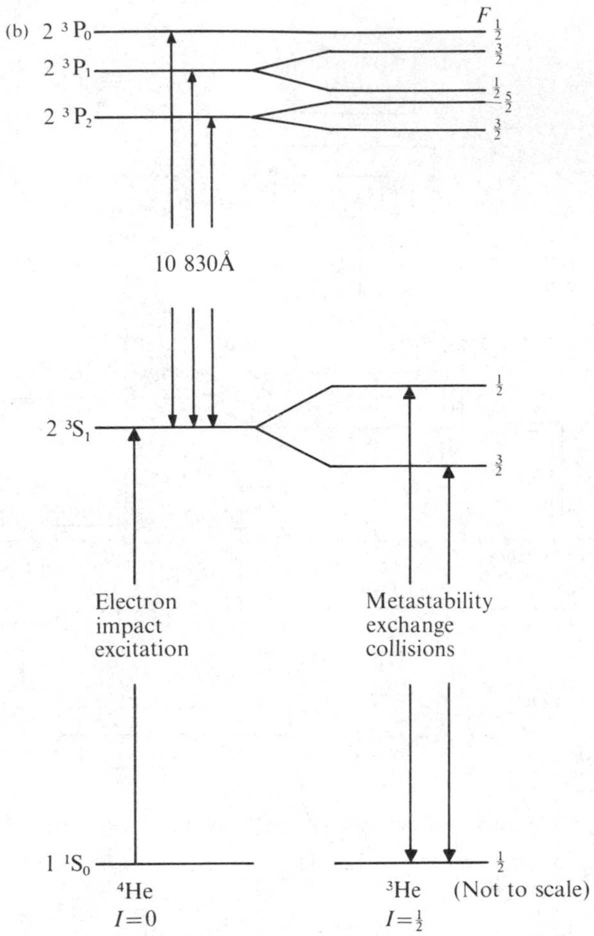

Fig.17.8 (b) Partial energy-level diagram of helium showing the levels involved in the optical pumping cycle. (After Greenhow (1963).)

17.6.1. Optical pumping of ^{3}He. Optical orientation of the metastable atoms can be achieved using either ^{4}He or ^{3}He in the optical pumping cell. In the latter case the nuclear spin of $I = \frac{1}{2}$ gives rise to two hyperfine levels belonging to the $2\ ^3S_1$ state with total angular momentum quantum numbers of $F = \frac{1}{2}$ and $\frac{3}{2}$. Provided that the coupling of the electronic and nuclear spins is not broken down by the application of a large external magnetic field, the ^{3}He nuclei share in the orientation of the metastable state and a large nuclear polarization is created.

Then in helium a special kind of exchange collision occurs in which the energy of a metastable atom is transferred to a colliding ground state atom. In the case ^{4}He it is impossible to detect such a transfer because the atoms are indistinguishable, having zero nuclear spin. In ^{3}He, however, the nuclear polarization produced in the $2\ ^3S_1$ state is preserved and metastability exchange collisions of the type

$$^3\mathrm{He}(^3S_1)\uparrow + {}^3\mathrm{He}(^1S_0)\downarrow \rightarrow {}^3\mathrm{He}(^1S_0)\uparrow + {}^3\mathrm{He}(^3S_1)\downarrow \quad (17.19)$$

transfer the nuclear polarization to the $1\ ^1S_0$ ground state. The metastability exchange collision cross-sections are quite large, $\approx 4 \times 10^{-16}$ cm^2 at 300 K, and this process provides a rapid and efficient method of producing polarized nuclear spins in the ground state. By this method nuclear polarizations of up to 40 per cent have been obtained in ^{3}He gas at 1 Torr pressure.

Because of their large number density it may take several minutes to polarize the ground state ^{3}He atoms at typical optical pumping rates. Consequently the optical pumping transients in ^{3}He are an order of magnitude longer than in any other optically pumped system. Also the orientation of the nuclear magnetic moment in the $^1S_0$ state is extremely resistant to depolarization by either wall or atomic collisions and relaxation times as long as ten hours may be achieved if the discharge in the cell is turned off. With the discharge running, collisions with free electrons destroy the polarization within a few minutes.

17.7 Optical pumping and magnetic resonance

17.7.1. Principles of the magnetic resonance technique. By means of optical pumping techniques large population differences can be generated between the Zeeman levels $|JM_J\rangle$ or $|FM_F\rangle$ of the ground or metastable states of a sample of free atoms. If these atoms are then subjected to a radio frequency magnetic field $\underline{B}_1$ which rotates at the angular frequency ω_0 in a plane perpendicular to the static field $\underline{B}$, magnetic resonance transitions can occur as explained in detail in section 16.2. When the frequency of the applied field equals the Larmor angular frequency,

$$\omega_0 = \omega_L = g_F \mu_B B/\hbar, \tag{17.20}$$

magnetic dipole transitions satisfying the selection rules $\Delta M_J = \pm 1$ or $\Delta M_F = \pm 1$ are stimulated between the Zeeman levels with the emission or absorption of quanta of the radio-frequency field. This is in fact one of the most important applications of the optical pumping technique. We proceed to consider the detection methods used in these experiments before discussing briefly a few of the many results obtained by this technique.

17.7.2. Detection of magnetic resonance transitions.

(a) Transmission monitoring. We return to a consideration of the optical pumping of alkali atoms using σ^+ polarized radiation in the D_1 line. In the steady state a certain population distribution between the $m_J = \pm \frac{1}{2}$ Zeeman levels is established by the combined effects of the optical pumping and relaxation processes. When a radio frequency magnetic field $\underline{B}_1$ is applied to the cell, magnetic dipole transitions are induced and atoms are returned more rapidly to the $m_J = -\frac{1}{2}$ level. A new dynamic equilibrium is reached in which the transparency of the vapour is reduced below the value it had in the absence of $\underline{B}_1$. The reduction in transparency depends on the angular frequency difference $(\omega_L - \omega_0)$ and on the magnitude of the radio frequency field. When the frequency of the applied field is slowly varied with $|\underline{B}|$ and $|\underline{B}_1|$ fixed, the changes in the intensity of the transmitted light trace out a magnetic resonance curve. This can be displayed directly on an oscilloscope connected to the photocell, as shown in Fig.17.9, or may be recorded point by point by using a micro-ammeter.

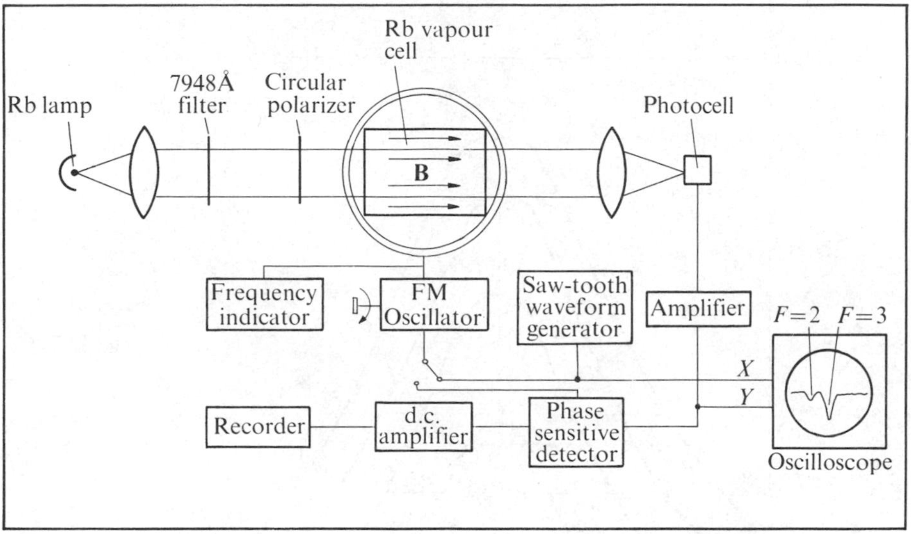

Fig.17.9. Schematic diagram of the apparatus required for optical pumping magnetic resonance experiments.

(b) Fluorescence monitoring. The magnetic resonance transitions may also be monitored using the changes which occur in either the intensity or the polarization of the fluorescent light emitted from the optical pumping cell. Magnetic resonance curves obtained by Cagnac (1961) with this technique on the nuclear spin transition $M_I = +\frac{1}{2} \rightarrow M_I = -\frac{1}{2}$ in the $6\ {}^1S_0$ ground state of ^{199}Hg are shown in Fig.17.10. As in the case of magnetic resonance in excited states discussed in section 16.2.3, the signals here also show a power broadening as the amplitude of the radio frequency field is increased.

These experiments have enabled the nuclear moments of ^{199}Hg and ^{201}Hg to be measured to very high precision:

$$\mu_{199} = +\ 0{\cdot}497\ 865\ (6)\ \mu_n$$

$$\mu_{201} = -\ 0{\cdot}551\ 344\ (9)\ \mu_n$$

where μ_n is the nuclear magneton. The nuclear moments of radioactive mercury isotopes and of ^{111}Cd and ^{113}Cd have also

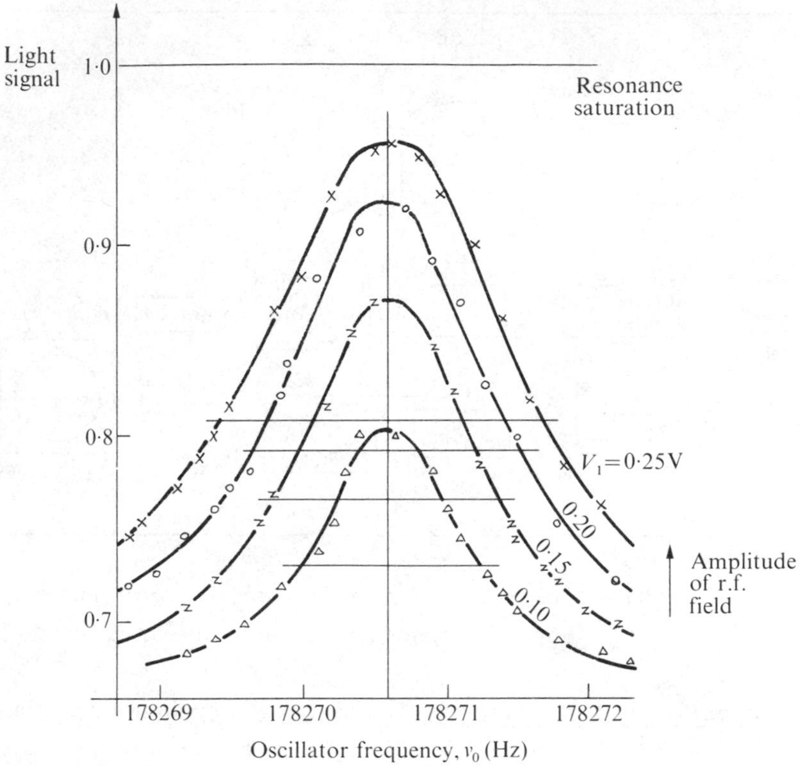

Fig.17.10. Nuclear magnetic resonance in the ground state of ^{199}Hg observed by Cagnac (1961) using optical pumping techniques.

been measured in this way.

17.7.3. Discussion of optical pumping magnetic resonance experiments.

(a) Population differences created by optical pumping. In the absence of optical pumping the ratio of the populations of adjacent Zeeman sub-levels is given by the Boltzmann formula

$$\frac{N_2}{N_1} = \exp(-\hbar\omega_L/kT).$$

At room temperature (300 K) the thermal energy kT is of the order of 0·025 eV ≡ 200 cm^{-1}. Now hyperfine splittings and

Zeeman intervals in easily produced magnetic fields rarely exceed 1 cm^{-1} = 30 000 MHz, thus $\hbar\omega_L/kT \leq \frac{1}{200}$ and

$$\frac{N_1-N_2}{N_1+N_2} \cong \frac{1}{2}\left(\frac{\hbar\omega_L}{kT}\right) \leq \frac{1}{400}$$

These thermal population differences are generally negligible and certainly are not sufficiently large to permit the direct observation of magnetic resonance. Fortunately the distribution of the population among the ground state sub-levels can be changed considerably by optical pumping. Relative population differences $(N_1-N_2)/(N_1+N_2)$ of the order of 0·3 to 0·8 are easily obtained and strong magnetic resonance signals may be observed in the transmitted or the resonantly scattered light.

(b) Magnetic resonance linewidth. The width of the magnetic resonance signal at half-intensity gives information on the rate at which the precessing magnetic dipoles move out of phase with the r.f. field which forces their precessional motion. This process can be described by a relaxation time T_2 called the transverse relaxation time. The measurement of T_2 will be discussed in more detail in section 17.7.5 below; in general, however, T_2 is different from the longitudinal relaxation time T_1 associated with the decay of the z-component of the sample magnetization.

Since the width of the magnetic resonance line is determined by the inverse of the relaxation time for the spins in the system, it is in turn dependent on the method by which the optical pumping cell has been prepared. In evacuated and uncoated cells the relaxation time is of the order of 10^{-4} s and linewidths of approximately 2 kHz are obtained. In cells which have been coated or which are filled with a buffer gas at the optimum pressure, the relaxation time is increased by a factor of between 10^3 and 10^4 and correspondingly narrow resonance signals may be observed.

These narrow resonances enable the centres of the lines

to be measured very accurately and in favourable cases a relative precision $\Delta\nu/\nu = 10^{-6}$ has been obtained for Zeeman resonances. If the magnetic field has been calibrated by means of proton resonance then this precision can be transferred to the measured g-factors. The accuracy of many of the results is comparable with or better than measurements made by the atomic beam technique.

Because of the very long relaxation times associated with atoms in ground or metastable states, the amplitude of the r.f. fields required in these magnetic resonance experiments is correspondingly small. We need only satisfy the condition $\gamma B_1 T_2 \approx 1$ and fields of the order of a milligauss or less are all that is necessary. These can be supplied directly by r.f. oscillators or signal generators, thus eliminating the power amplification which is necessary when short-lived excited levels are being studied.

(c) Sensitivity of optical pumping experiments. Optical pumping experiments share the same high sensitivity that was noted in section 16.3.3 in the case of the optical double-resonance experiments. The detection scheme can be imagined to be a 'trigger' mechanism in which the absorption of a radio frequency photon, $\nu \approx 10^3$ Hz, is followed by the absorption of a photon in the visible or near ultraviolet region where $\nu \approx 10^{15}$. There is therefore a quantum gain of 10^{12} and good resonance signals can be obtained on samples containing as few as 10^{11} atoms.

17.7.4. Spin-exchange magnetic resonance experiments. Atoms whose resonance lines lie in the vacuum ultraviolet can often be polarized by spin-exchange collisions with optically-pumped alkali atoms as discussed in section 17.5. In the case of hydrogen atoms polarized by spin-exchange with rubidium, the application of an r.f. field at the hydrogen Larmor frequency will induce magnetic resonance and so reduce the polarization of the hydrogen atoms in the sample. This magnetic resonance can in fact be monitored by its effect on the transmission of the rubidium resonance radiation through the cell, for spin-exchange collisions with the disoriented hydrogen atoms serve to reduce the polarization of the rubidium atoms. Balling and Pipkin (1965) have used this spin-exchange technique to measure the ratios of the g-factors of free electrons, hydrogen, tritium, and ^{85}Rb to a precision of 1 part in 10^6 or better. In a similar way the metastability exchange collisions in helium allow the nuclear resonance in the ^{3}He($1\ ^1S_0$) ground state to be observed via the absorption of the 10 830 Å line by atoms in the $2\ ^3S_1$ metastable level.

17.7.5. Results of precision measurements of g-factor ratios. The results of a number of precision measurements of different g-factor ratios are given in Table 17.3. In these experiments particular attention has to be paid to the homogeneity and stability of the applied magnetic field in order to obtain symmetric and narrow magnetic resonance signals.

TABLE 17.3

Results of precision measurements of g-factor ratios

Gyromagnetic ratio	Experimental result	Reference
$g_J(^{87}\mathrm{Rb})/g_J(^{85}\mathrm{Rb})$	$1 + (4\cdot1 \pm 6\cdot0) \times 10^{-9}$	White *et al.* (1968)
$g_J(^{85}\mathrm{Rb})/g(e)$	$1 + (6\cdot3 \pm 1\cdot0) \times 10^{-6}$	Balling and Pipkin (1965)
$g_J(^1\mathrm{H})/g(e)$	$1 - (17\cdot4 \pm 1\cdot0) \times 10^{-6}$	Balling and Pipkin (1965)
$g_J(^{85}\mathrm{Rb})/g_J(^1\mathrm{H})$	$1 + (23\cdot74 \pm 0\cdot1) \times 10^{-6}$	Balling and Pipkin (1965)
$g_J(^3\mathrm{H})/g_J(^1\mathrm{H})$	$1 - (0\cdot11 \pm 0\cdot3) \times 10^{-6}$	Balling and Pipkin (1965)
$g_J(^{87}\mathrm{Rb})/g_I'(p)$	$658\cdot234 \pm 0\cdot004$	Bender (1962)
$g_J(\mathrm{He}\ 2\ ^3S_1)/g_I'(p)$	$658\cdot199 \pm 0\cdot012$	Schearer and Sinclair (1968)

It is also found that the position of the centre of the resonance depends on the intensity of the pumping radiation for reasons which are discussed in section 17.9.7. Consequently it is necessary to make a series of measurements at different light intensities and to extrapolate the results to zero intensity in order to find the position of the unperturbed resonance. Although all of the g-factor ratios included in Table 17.3 are of interest since they provide tests of the theoretical calculation of atomic g-factors, the ratio for the free electron is of particular importance in the determination of fundamental atomic constants (Problem 17.6). The application of the optical pumping magnetic resonance technique to the determination of atomic hyperfine structures is equally important and is described in Chapter 18.

17.7.6. Measurement of the transverse relaxation time T_2.

(a) Steady state experiments. When a magnetic resonance signal is observed in a simple optical pumping apparatus, the width of the resonance is often determined by the inhomogeneity of the applied magnetic field. However, by careful design of the field coils this problem can be overcome and the measured linewidth $\Delta\omega_{1/2}$ then depends on the effective transverse relaxation time τ_2 and the radio frequency power used to drive the resonance. The explicit expression for $\Delta\omega_{1/2}$ is

$$\Delta\omega_{1/2} = 2(1/\tau_2^2 + \gamma^2 B_1^2 \tau_1/\tau_2)^{1/2} \qquad (17.21)$$

which reduces to equation (16.21) when the effective longitudinal and transverse relaxation times are equal. Since the phase-memory of the precessing atoms is destroyed by optical excitation, the effective transverse relaxation time is given by

$$1/\tau_2 = 1/T_2 + 1/T_p \qquad (17.22)$$

where $1/T_p = KI$ is the relaxation rate induced by the pumping light and is proportional to the incident intensity. If the resonance half-widths are measured as a function of B_1 and I then two extrapolations to zero r.f. power and zero light intensity can be made and the intrinsic transverse relaxation time T_2 may be obtained. A large number of measurements are required in this technique and the apparatus must operate in a very stable and reproducible manner. Since these are stringent requirements, the simpler transient method described below is often preferred.

(b) Transient experiments. When a radio-frequency magnetic field $\underline{B}_1$, close to the resonance frequency, is suddenly applied to an optically pumped sample at time $t=t_0$, the spins of the atoms begin to nutate in phase. As explained in detail in section 16.2.1 the motion in the rotating frame can be described classically as a precession of the macroscopic magnetization vector $\langle \underline{M} \rangle$ about the effective field

$$\underline{B}_e = \hat{\underline{i}}' B_1 + \hat{\underline{k}}(B - \omega_0/\gamma), \tag{17.23}$$

with a precessional angular frequency $\omega_e=\gamma B_e$. Quantum mechanically this forced precession can be interpreted in terms of induced magnetic dipole transitions between the ground state magnetic sub-levels, and an application of the results of section 16.2.2 leads to an expression for the population excess, $n=N_+-N_-$, in the form

$$n(t) = n(\infty)\left[1 - \frac{2\omega_1^2}{\omega_1^2+(\omega_L-\omega_0)^2}\sin^2\{\omega_e(t-t_0)/2\}\right]. \tag{17.24}$$

On exact resonance $\omega_0 \to \omega_L$ and $B_e \to B_1$ so that this formula reduces to

$$n(t) = n(\infty)\,[1 - 2\sin^2\{\gamma B_1(t-t_0)/2\}] \tag{17.25}$$

and the atoms oscillate between the two spin states at a frequency determined by the magnitude of the radio frequency magnetic field. This gives rise to oscillations in the intensity of the light transmitted or scattered by the cell which can be detected and displayed directly on an oscilloscope. The transient nutation at resonance of the ground-state nuclear spins on an optically-pumped ^{3}He sample is shown in Fig.17.11. The oscillations are obviously damped.

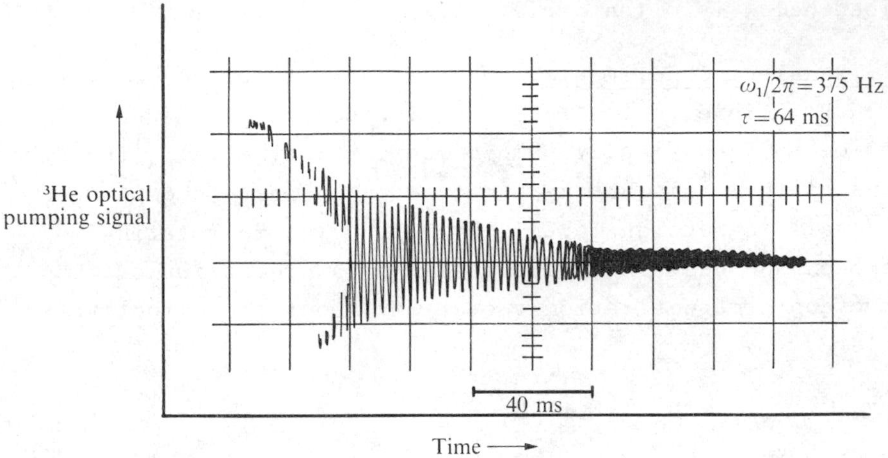

Fig.17.11. Transient nutation at resonance of ^{3}He ($I = \frac{1}{2}$) atoms in the ground state, $1\ ^1S_0$. The time constant for relaxation can be obtained from the damping of the signal. (After Greenhow (1963).)

This damping occurs because the nutating atoms relax and because atoms that are optically pumped after the time t_0 begin to nutate with different phases. It has been shown by Cagnac (1961) that the damping constant for the sinusoidal oscillations is given by

$$1/\tau = (1/\tau_1 + 1/\tau_2)/2. \qquad (17.26)$$

It depends on the longitudinal relaxation time τ_1 (equation (17.10)) and on the transverse relaxation time τ_2 (equation (17.22)), both measured in the presence of the pumping light. By combining this type of transient measurement with that described in section 17.4 the intrinsic transverse relaxation time T_2 can be obtained. At resonance the nutational frequency also permits an absolute measurement of the magnitude of the r.f. magnetic field $\underline{B}_1$ as shown by equation (17.25).

17.8. Transverse magnetization and Hertzian coherence in optical pumping experiments

17.8.1. Distinction between longitudinal and transverse quantities. In this chapter so far we have described the ensemble of atoms in the optical pumping cell in terms of the populations of the different Zeeman sub-levels $|\mu\rangle$ of the atomic ground state. The optical pumping technique allows the experimenter to create large differences between the Zeeman state populations and to study the behaviour of physical quantities, such as the macroscopic longitudinal magnetization of the sample $\langle M_z\rangle$, which depend directly on these population differences. These experiments can be described quantum mechanically by means of a ground-state density matrix ρ_g which possesses only diagonal matrix elements, i.e. $\rho_{\mu\mu'}=0$ for $\mu\neq\mu'$.

However, recalling the extensive discussion of the density matrix in section 15.5.5, we recognize that it may be possible in certain optical pumping experiments to prepare atoms in such a way that the relative phases of the probability amplitudes $a_{\mu}^{(i)}(t)$ and $a_{\mu'}^{(i)}(t)$ of the states μ and μ' do not vary randomly from one atom to another. The ground-state density matrix then has finite off-diagonal elements. We recall that these off-diagonal elements evolve in time according to

$$\rho_{\mu\mu'}(t) = \rho_{\mu\mu'}(t_0)\exp\{-i\omega_L(\mu-\mu')(t-t_0)\} \tag{17.27}$$

where ω_L is the Larmor angular frequency of the ground-state atoms in the applied external magnetic field and all other external perturbations have been neglected. Since the characteristic frequencies $\omega_L(\mu-\mu')/2\pi$ lie in the radio frequency range, the density matrix in this case is said to possess finite Hertzian coherence. When the off-diagonal elements of the ground-state density matrix are finite we may expect to observe new phenomena in optical pumping experiments, such as zero field level-crossing signals and radio frequency modulation of the transmitted light intensity. As we shall now demonstrate, these new phenomena are closely connected with the existence, in a plane perpendicular to $\underline{B}$, of a finite component of the ensemble magnetization called the transverse magnetization, $\langle M_\perp \rangle$.

17.8.2. Transverse pumping in a weak magnetic field. At thermal equilibrium the transverse magnetization $\langle M_\perp \rangle$ of the sample is zero due to the fact that there is no privileged direction in the transverse plane. The projections of the atomic magnetic moments in this plane vary randomly in direction from one atom to another, so that the resultant $\langle M_\perp \rangle$ is zero. This situation remains unchanged if the incident radiation is in a pure state of polarization (σ^+, σ^-, or π) with respect to the field $\underline{B}$ for there is still no privileged direction in the plane perpendicular to $\underline{B}$. However, a finite transverse magnetization can be created by arranging the pumping beam so that it propagates along the 0x direction, perpendicular to the applied magnetic field $\underline{B}$, as shown in Fig.17.12. If the light is circularly polarized with respect to the 0x direction then, with respect to the axis of quantization 0z, it appears to be a coherent mixture of σ^+, σ^-, and π polarizations and a privileged transverse direction has been determined.

In this experiment the magnetic dipole moment of an atom which has just completed the optical pumping cycle at time t_0 finds itself pointing along the 0x axis. Immediately afterwards it starts to precess about the magnetic field $\underline{B}$ at the Larmor angular frequency ω_L of the ground-state atoms. This precession does not last indefinitely for after a mean time τ the orientation of the atom is destroyed by a collision or by absorption of another photon. In order to

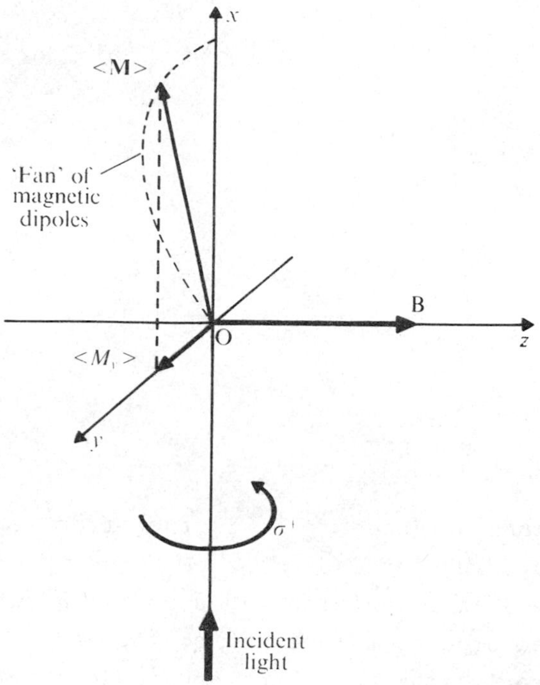

Fig.17.12. Geometrical arrangement for transverse optical pumping experiments.

calculate the transverse magnetization of the vapour at some time t, it is necessary to find the resultant of the dipoles which were created at all previous times t_0 and which have not yet experienced a disorienting process. These dipoles are spread out in the form of a fan starting from the 0x axis and having an angular spread of the order of $\omega_L\tau$ in the plane perpendicular to 0z. If the applied magnetic field is very small, so that $\omega_L\tau \ll 1$, then the angular spread is also small and the transverse magnetization has a finite value. When the magnetic field is increased, so that $\omega_L\tau \gg 1$, the angular distribution in the x0y plane becomes isotropic and the transverse magnetization is reduced to zero.

The components of the transverse magnetization can be obtained as functions of the applied field by using the techniques employed in the classical Hanle effect calculation detailed in section 15.2.3. We obtain

$$\langle M_x \rangle \propto \Gamma_g \int_{-\infty}^{t} \cos\{\omega_L(t-t_0)\} \exp\{-\Gamma_g(t-t_0)\}dt_0$$

$$= C \frac{\Gamma_g^2}{\omega_L^2 + \Gamma_g^2} \tag{17.28}$$

and

$$\langle M_y \rangle \propto \Gamma_g \int_{-\infty}^{t} \sin\{\omega_L(t-t_0)\} \exp\{-\Gamma_g(t-t_0)\}dt_0$$

$$= -C \frac{\omega_L \Gamma_g}{\omega_L^2 + \Gamma_g^2} \tag{17.29}$$

where C is a constant and $\Gamma_g = 1/\tau$ is the effective width of the ground-state sub-levels. Thus $\langle M_x \rangle$ and $\langle M_y \rangle$ are predicted to vary as a Lorentzian and dispersion curve respectively when measured as a function of the applied field $\underline{B}$.

These predictions have been confirmed experimentally by Lehmann and Cohen-Tannoudji (1964) using ^{111}Cd (I=1/2) pumped by the $5\,^1S_0 - 5\,^1P_1$ transition at 2288 Å. The experimental signals were obtained by monitoring the polarization of the fluorescent light and are very narrow owing to the very long relaxation times ($\tau = 1/\Gamma_g$) of the ground-state atoms. This technique therefore provides an extremely sensitive method of measuring very weak magnetic fields.

These experiments are in fact entirely analagous to the Hanle effect or zero-field level-crossing experiments involving excited atoms discussed in Chapter 15. The coherent polarization of the pumping light referred to the quantization axis 0z in Fig.17.12 prepares the atoms in a coherent superposition of ground-state Zeeman sub-levels. The ensemble density matrix now has finite off-diagonal elements $\rho_{\frac{1}{2}-\frac{1}{2}} = \rho^*_{-\frac{1}{2}\frac{1}{2}}$, the modulus and argument of which represent respectively the magnitude and orientation of $\langle M_\perp \rangle$ in the plane perpendicular to $\underline{B}$.

17.8.3. Transverse pumping with modulated light. In our discussion of resonance fluorescence experiments involving modulated excitation, section 15.9, we showed how large oscillating coherences could be created in the density matrix of the excited state at fields which are considerably different from zero. This technique had, in fact, been developed three years earlier by Bell and Bloom (1961) in optical pumping experiments on alkali vapours. The pumping is again effected by a circularly-polarized beam propagating in the 0x direction, as shown in Fig.17.13, but the pumping light

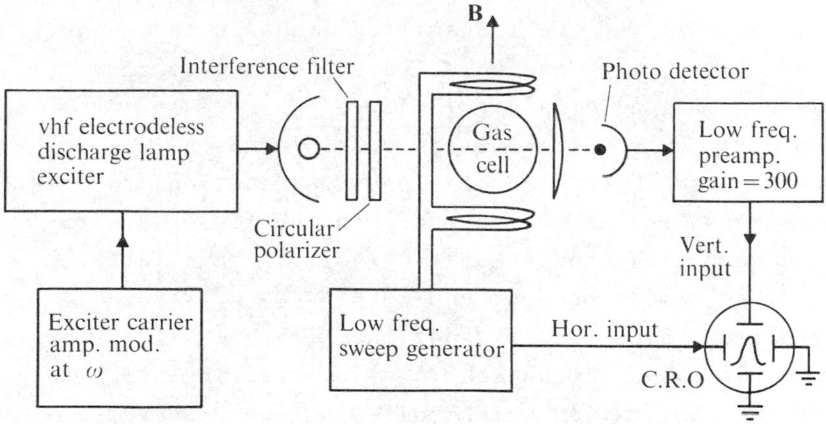

Fig.17.13. Production of oscillating coherence by transverse optical pumping with modulated light. When the pumping light is modulated at the Larmor frequency coherence may be generated between the ground state Zeeman sub-levels. The coherence causes a change in the mean intensity of the transmitted light. (After Bell and Bloom (1961).)

is now modulated at the angular frequency f by amplitude modulation of the r.f. power supplied to the lamp.

The effect of the modulated excitation becomes clear when we consider the behaviour of the dipoles in a frame of reference 0XYZ which rotates about the 0z axis at the frequency $f/2\pi$. In this frame, the OZ axis of which coincides with the z-axis of the laboratory frame, the majority of the dipoles are created with their axes along the OX direction. They then start to precess at the angular frequency $(\omega_L - f)$ characteristic of the rotating frame. Thus in this frame we have the same situation as that discussed in the previous section with the exception that the angular frequency ω_L is now replaced by $(\omega_L - f)$. If the frequency of modulation of the pumping light is close to the Larmor precessional frequency, then $|\omega_L - f|\tau \leq 1$ and a large precessing transverse magnetization is built up in the ground state. This precessing magnetization causes a resonant change in the mean intensity of the transmitted light as the magnetic field is swept through the value corresponding to $\omega_L = f$, as shown schematically in Fig.17.13.

We have now seen that it is possible, using the techniques discussed in this and the previous section, to create a global transverse magnetization. In the former experiments the magnetization was static while in this latter example it is time dependent and is associated with an optically-driven spin-precession at the frequency $f/2\pi$. We shall now discuss a third technique for creating a finite transverse magnetization which relies not on transverse pumping but on the ground-state coherence generated in magnetic resonance experiments.

17.8.4. Transverse magnetization created by magnetic resonance experiments. In a conventional optical pumping experiment the circularly-polarized pumping beam propagates in the direction of the static magnetic field $\underline{B}$ and the vapour acquires a longitudinal magnetization $\langle M_z \rangle$. If in addition a magnetic field $\underline{B}_1$ is applied which rotates in the plane perpendicular to $\underline{B}$ at the angular frequency ω_0, then magnetic resonance can be induced in the ground state as ω_0 approaches the Larmor angular frequency ω_L. The classical theory of magnetic resonance discussed in section 16.2 shows that the magnetization of the sample starts to nutate about the direction of the effective field in a co-ordinate system rotating about Oz at the frequency $\omega_0/2\pi$. In the steady state experiments considered here, the competition between nutation and relaxation processes leads to the creation of a finite transverse magnetization $\langle M_\perp \rangle$ in the rotating frame. The amplitude and phase of $\langle M_\perp \rangle$ relative to these of $\underline{B}_1$ depend on the frequency difference $(\omega_0 - \omega_L)$ and on the effective longitudinal and transverse relaxation times τ_1 and τ_2 respectively. In the laboratory frame the transverse magnetization appears to undergo a forced precession at the frequency $\omega_0/2\pi$, and consequently those physical observables of the system which depend on $\langle M_x \rangle$ and $\langle M_y \rangle$ are modulated at the same frequency.

<u>17.8.5. Detection of transverse magnetization using the crossed beam technique.</u> We now describe an optical detection technique proposed by Dehmelt (1957) and applied first in the optical pumping of sodium and potassium by Bell and Bloom (1957). This method is extremely useful since it provides optical signals which are directly proportional to the transverse magnetization of the sample. The technique consists of adding a second light beam to a conventional optical pumping magnetic resonance experiment. The first beam, labelled 1 in Fig.17.14, optically pumps the vapour in the resonance cell while the cross beam 2, propagating in a perpendicular direction, monitors the polarization of the sample. The intensity of the lamp in the detection beam is arranged to be sufficiently weak that it does not appreciably perturb the ensemble of atoms, and we also assume that the static magnetic field is sufficiently large that no transverse pumping can occur, i.e. $\omega_L \tau \gg 1$.

We now consider the application of the crossed beam technique to magnetic resonance experiments in which the optical pumping cell is filled with ^{199}Hg $(I=1/2)$. When the resonance lamp for the detection beam 2 contains ^{204}Hg, only the $6\,^1S_0(I=1/2) \rightarrow 6\,^3P_1(F=1/2)$ component of the 2537 Å line of ^{199}Hg can be excited. The optical transitions

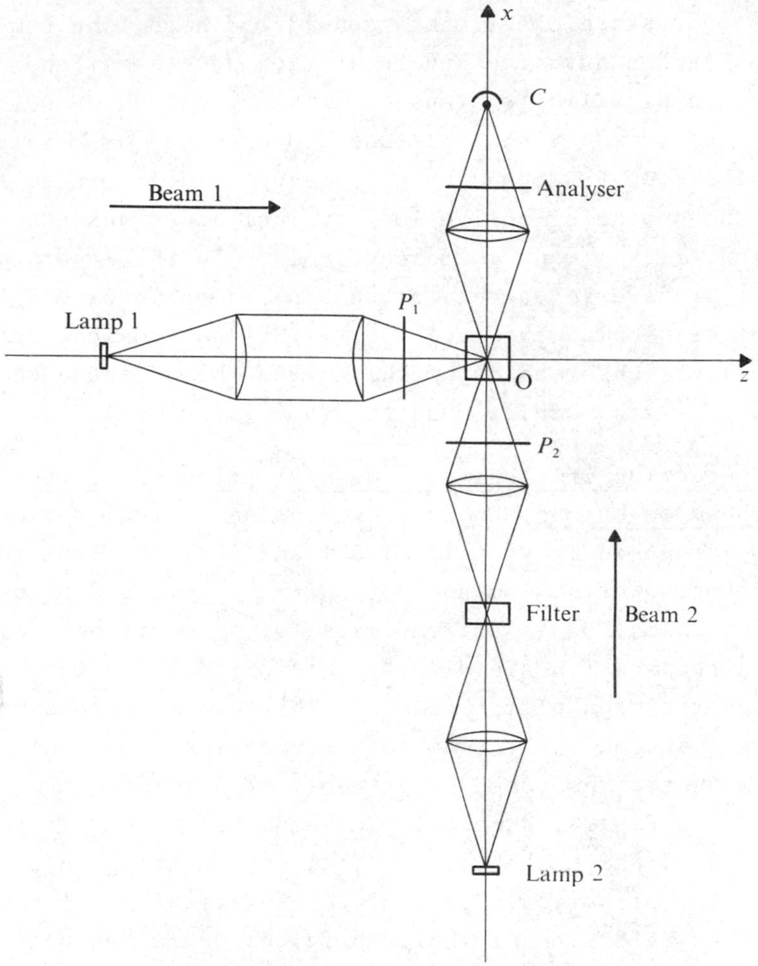

Fig.17.14. Detection of transverse magnetization using the crossed beam technique. (After Cohen-Tannoudji and Kastler (1966).).

possible for the detection beam are therefore identical to those of the D_1 line of an alkali atom having zero nuclear spin. When the precessing transverse magnetization $\langle M_\perp \rangle$ of the sample points in the 0x direction, the atoms are effectively in the $m_I = 1/2$ sub-level with respect to a quantization axis taken parallel to 0x. Referring to Fig.17.1 we see that the absorption on the F=1/2 hyperfine component of the 2537 Å line will be small. One half-period later the precessing magnetization $\langle M_{\perp} \rangle$ will be pointing in the negative 0x direction and, with respect to 0x as the quantization axis, the atoms are in the $m_I = -1/2$ state. The absorption is now very strong. We conclude that on the cross beam 2 the absorption signal will be modulated at the frequency $\omega_0/2\pi$ characteristic of the forced precession of $\langle M_{\perp} \rangle$.

Magnetic resonance curves obtained in ^{199}Hg by Cohen-Tannoudji (1962) using the cross-beam technique are shown in Fig.17.15. Each curve corresponds to a fixed value of the amplitude of the r.f. field $\underline{B}_1$, the signal intensity in-

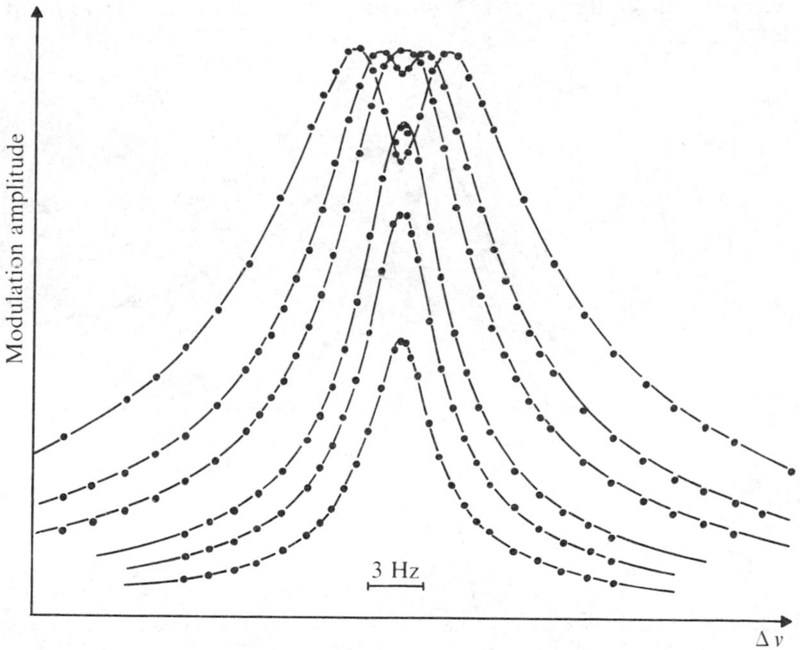

Fig.17.15. Nuclear magnetic resonance curves for ^{199}Hg obtained by Cohen-Tannoudji (1962) using the crossed beam technique. (After Cohen-Tannoudji and Kastler (1966).)

creasing with increasing values of $|\underline{B}_1|$. In this experiment the amplitude of modulation of the photoelectric current from detector C was measured as a function of the applied magnetic field $\underline{B}$. The measured signal is therefore proportional to $|\langle M_\perp \rangle|$.

Using phase-sensitive detection of the radio frequency modulation of the cross-beam absorption signal both the in-phase and quadrature components of $\langle M_\perp \rangle$ may be studied separately. The in-phase component gives a signal proportional to

$$\langle M_x \rangle = + M_\infty \frac{\omega_1(\omega_L - \omega_0)}{\frac{1}{\tau_2^2} + \omega_1^2 \frac{\tau_1}{\tau_2} + (\omega_L - \omega_0)^2} \qquad (17.30)$$

which varies as a dispersion curve, as shown in Fig.17.16.

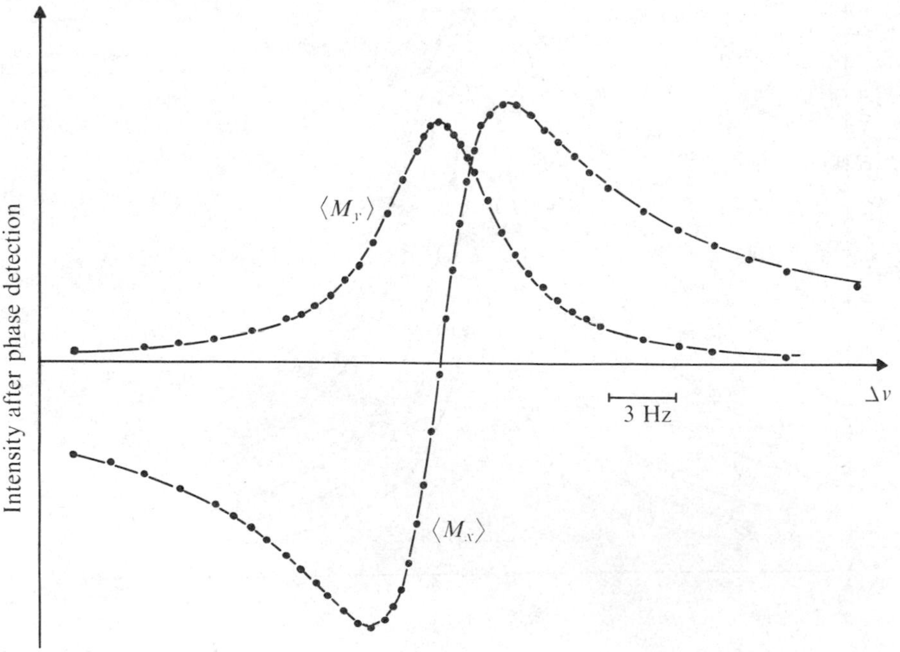

Fig.17.16. In-phase and quadrature components of the transverse magnetization, $\langle M_\perp \rangle$, obtained by phase-sensitive detection using the cross-beam technique in ^{199}Hg. (After Cohen-Tannoudji and Kastler (1966).)

The signal in quadrature is proportional to

$$\langle M_y \rangle = - M_\infty \frac{\omega_1/\tau_2}{\frac{1}{\tau_2^2} + \omega_1^2 \frac{\tau_1}{\tau_2} + (\omega_L - \omega_0)^2} \tag{17.31}$$

and displays an absorption profile when plotted as a function of $(\omega_L - \omega_0)$. The cause of the central dip in some of the curves displayed in Fig.17.15 is now obvious since the signal recorded without phase-sensitive detection is proportional to $(\langle M_x \rangle^2 + \langle M_y \rangle^2)^{1/2}$.

The signal-to-noise ratios obtained in cross-beam experiments are obviously excellent and demonstrate clearly the importance of this method. This technique can also be used in pulsed experiments to measure the transient relaxation rate $1/\tau_2$ of the off-diagonal elements of the ground-state density matrix. Finally it should be noted that the light used in the detection beam 2 of Fig.17.14 is coherently polarized and monitors the atomic coherence created in the ensemble of ground-state atoms by magnetic resonance. These experiments are therefore the modulated absorption analogues of the light beat experiments of Series and co-workers described in section 16.7.

17.9. Quantum theory of the optical pumping cycle

<u>17.9.1 Introduction.</u> That a quantum theory of the optical pumping cycle is necessary is clearly illustrated by the experiments discussed in the previous section. These can only be interpreted quantitatively if the time-development of the non-diagonal elements of the density matrix is known. We also recall that the ground-state magnetic resonance curves are observed to be broader at higher intensities of the optical pumping beam (section 17.7.6), demonstrating that the atoms in the cell are perturbed by the very light which enables the atomic orientation to be created. This perturbation must be evaluated quantitatively to discover whether the broadening of atomic energy levels is the only

result of the interaction of atoms with the pumping beam.

The density matrix theory of the optical pumping cycle was developed by Barrat and Cohen-Tannoudji (1961) and by Cohen-Tannoudji (1962) in his doctoral thesis. The reader should refer to these papers for detailed discussions of the main results for we shall have space enough here to give only an outline of their derivation and physical interpretation. The theory is an extension of the density matrix formalism laid down in Chapters 15 and 16 during our discussion of resonance fluorescence experiments involving excited atoms. In the optical pumping experiments considered here, the light source is again assumed to be a conventional resonance lamp emitting a Doppler-broadened spectral line. In the optical sense the source is incoherent and the total density matrix of the ensemble of atoms can therefore be separated into density matrices ρ_g and ρ_e belonging to the ground and excited states, respectively. In Chapters 15 and 16 we have shown how ρ_e can be calculated in a number of different physical situations. In the present experiments the calculation of ρ_e may need to be modified slightly to take account of finite ground-state coherences, $\rho_{\mu\mu'} \neq 0$, since ρ_g can no longer be assumed to be diagonal.

Following a discussion similar to that given in section 15.5 it can be shown that the Liouville equation for the ground-state density matrix is given by

$$\frac{\partial \rho_g}{\partial t} = \frac{1}{i\hbar}[\mathcal{H}_0', \rho_g] + \frac{d^{(1)}}{dt}\rho_g + \frac{d^{(2)}}{dt}\rho_g + \frac{d^{(3)}}{dt}\rho_g. \tag{17.32}$$

The commutator bracket $[\mathcal{H}_0', \rho_g]$ describes the time-development of the system under the effects of the atomic Hamiltonian $\mathcal{H}_0$ and of any static external magnetic field, while the last three terms give the evolution under the influence of depopulation pumping, repopulation pumping, and relaxation processes respectively. These processes are considered separately in the sections which follow.

Detailed solutions of equation (17.32) result not only in expressions for the populations of the different ground-state sub-levels which are equivalent to those discussed previously in this chapter, but also demonstrate clearly that finite hertzian coherence is generated in certain optical pumping experiments. The theory enables explicit expressions to be obtained for the signals observed in both absorption and fluorescence monitoring experiments. Furthermore the quantum theory of the optical pumping cycle leads to the prediction that the atomic energy levels are not only broadened but are also shifted by interaction with the pumping light. These light shifts are discussed in detail in section 17.9.7.

17.9.2. Depopulation pumping. The term $d^{(1)}\rho_g/dt$ in equation (17.32) represents the effect of optical excitation of ground-state atoms by the pumping light and is therefore termed depopulation pumping. Depopulation pumping occurs when certain ground-state sub-levels absorb light more strongly than others. Since atoms are removed more rapidly from the strongly absorbing sub-levels, an excess population will tend to build up in the weakly absorbing levels, thus creating an orientation of the sample. Depopulation pumping can occur if the pumping light is anisotropic or polarized, or if the frequency spectrum of the light is non-uniform in the neighbourhood of the atomic absorption lines.

For the moment we assume that no coherence exists in the ground state and that the atoms are illuminated with light of broad spectral profile. Then the rate, $P_{\mu m}$, at which atoms in the ground-state sub-level $|\mu\rangle$ absorb light and are raised to the sub-level $|m\rangle$ of the excited state is given by the result of the perturbation theory calculation detailed in section 9.3. From equation (9.40) we have

$$P_{\mu m} = \frac{\pi U(\omega)}{\varepsilon_0 \hbar^2} |\langle m|\underline{D}.\hat{\underline{e}}|\mu\rangle|^2 \qquad (17.33)$$

where $U(\omega)$ is the energy density of radiation at the atomic absorption frequency, $\underline{D} = \sum_i -e\underline{r}_i$ is the atomic electric dipole moment operator and $\hat{\underline{e}}$ is the polarization vector of the incident light. The total rate at which the population of the ground-state sub-level $|\mu\rangle$ is depleted as a result of depopulation pumping is obtained by summing over the accessible excited state sub-levels, with the result

$$\frac{d^{(1)}}{dt}\rho_{\mu\mu} = -\delta\Gamma_{\mu\mu}\rho_{\mu\mu} = -\sum_m P_{\mu m}\rho_{\mu\mu}. \qquad (17.34)$$

The mean pumping rate R is given by

$$R = 1/T_p = \frac{1}{(2J_g+1)} \sum_{\mu,m} P_{\mu m} \qquad (17.35)$$

and plays a role for the ground-state atoms analogous to that of the spontaneous decay rate in the case of excited atoms. Equation (17.34) is applicable only when there is no coherence in the ground state. Barrat and Cohen-Tannoudji (1961) have shown that the correct generalization of equation (17.34) for the case of light whose spectral profile is independent of frequency in the neighbourhood of the absorption lines is given by

$$\frac{d^{(1)}}{dt}\rho_{\mu\mu'} = -\frac{1}{2}[\delta\Gamma,\rho_g]_{\mu\mu'} \qquad (17.36)$$

where

$$\delta\Gamma = \frac{\pi U(\omega)}{\varepsilon_0\hbar^2}\sum_m \hat{\underline{e}}^*.\underline{D}|m\rangle\langle m|\underline{D}.\hat{\underline{e}} \quad . \qquad (17.37)$$

We shall consider depopulation pumping for the case of light of non-uniform spectral distribution in section 17.9.7, where we discuss the problem of energy level shifts produced by optical pumping.

17.9.3. Repopulation pumping. For an excited atom in the optical pumping cell there is a total probability per unit time, Γ_{ki}, that it will return to the ground state by spontaneous emission. From equation (4.18) the transition probability between non-degenerate sub-levels can be written in the form

$$A_{m\mu} = \frac{\omega^3}{3\pi\varepsilon_0\hbar c^3} |\langle m|\underline{D}|\mu\rangle|^2 \qquad (17.38)$$

and the rate at which the sub-level $|\mu\rangle$ of the ground state is repopulated is therefore given by

$$\frac{d^{(2)}}{dt} \rho_{\mu\mu} = \sum_m A_{m\mu} \rho_{mm}. \qquad (17.39)$$

We shall not consider the generalization of equation (17.39) which must be made when there is atomic coherence in the excited state. Details of this are given by Happer (1972) and by Barrat and Cohen-Tannoudji (1961). It should be noted, however, that the polarization of the atomic ground state may be partially conserved during the absorption and re-emission of a photon. This circulation of coherence in the optical pumping cycle combined with the long relaxation times of ground-state atoms produces many interesting new effects which are not observable in experiments on excited atoms.

17.9.4. Effect of relaxation processes. In an optical pumping experiment the vapour in the cell eventually reaches a time-independent equilibrium state provided that the intensity of the pumping light and the magnitude of the external fields are maintained constant. When any one of these experimental parameters is suddenly changed the ensemble of atoms relaxes towards a new equilibrium state at a rate determined by the combined effects of optical pumping, diffusion to the cell walls, and by interatomic collisions. Unfortunately this relaxation is seldom described by a single exponential time constant and is in general a rather complicated process, details of which are given in the review article by Happer (1972). Here we make the simplifying assumption that only two relaxation times T_1 and T_2 are involved, describing respectively the collisional relaxation of the diagonal and off-diagonal elements of the ground-state density matrix. We have

$$\frac{d^{(3)}}{dt} \rho_{\mu\mu} = \frac{1 - \rho_{\mu\mu}}{T_1} \tag{17.40}$$

and

$$\frac{d^{(3)}}{dt} \rho_{\mu\mu'} = - \frac{\rho_{\mu\mu'}}{T_2} \quad \text{for } \mu \neq \mu'. \tag{17.41}$$

This hypothesis is justified for systems such as ^{199}Hg or alkali atoms with zero nuclear spin since these atoms possess only two magnetic sub-levels in the ground state. For most atoms, however, many more sub-levels are involved, necessitating the introducion of further relaxation times.

17.9.5. Density matrix formulation of the monitoring operators. The intensity of light emitted by a sample of excited atoms into a solid angle dΩ about the direction specified by the polar angles (θ,ϕ) has already been derived in detail in section 15.6.2. We have

$$\frac{dI}{d\Omega} = \mathrm{Tr}(\rho_e \, L_F) \tag{17.42}$$

where L_F is the fluorescence monitoring operator defined by

$$L_F = \frac{\omega^3}{8\pi^2 \varepsilon_0 \hbar c^3} \sum_{\mu'} \underline{\hat{e}}'.\underline{D}|\mu'\rangle\langle\mu'|\underline{D}.\underline{\hat{e}}'^* \tag{17.43}$$

and $\underline{\hat{e}}'$ is the polarization vector of the emitted radiation.

In absorption monitoring the change, ΔI, in the intensity of light transmitted through an optically thin sample of length Δl depends on the rate at which atoms are stimulated to leave the ground state:

$$\Delta I = - \hbar\omega \sum_m P_{\mu m} \, \rho_{\mu\mu} \, \Delta l. \tag{17.44}$$

The absorption signal may be expressed in a form similar to that of equation (17.42) by the introduction of a light absorption monitoring operator, L_A, defined by

$$- \frac{dI}{dl} = \mathrm{Tr}(\rho_g \, L_A) \; . \tag{17.45}$$

For light of energy density $U(\omega)$ and the polarization vector $\underline{\hat{e}}$, we have from equations (17.33) and (17.44)

$$L_A = \frac{\pi\omega U(\omega)}{\varepsilon_0\hbar} \sum_m \underline{\hat{e}}.\underline{D}|m\rangle\langle m|\underline{D}.\underline{\hat{e}}^* \quad . \tag{17.46}$$

We see that the fluorescence and absorption monitoring operators have very similar forms. However, because equation (17.45) involves the ground-state part of the ensemble density matrix, the absorption monitoring technique offers a more direct way of measuring the parameters of interest in many optical pumping experiments.

<u>17.9.6. Application to optical pumping of ^{199}Hg.</u> As an example of the previous discussion we consider now the application of the density matrix theory to magnetic resonance experiments in the ground state of ^{199}Hg, $I=1/2$. The atoms in the cell are subjected to an external magnetic field $\underline{B}$ which defines the axis of quantization $0z$ and the static Hamiltonian takes the form

$$\mathcal{H}_0' = g_I\mu_n BI_z = \hbar\omega_L I_z \tag{17.47}$$

where μ_n and g_I are the nuclear magneton and g-factor respectively. In addition there exists a time-dependent field $\underline{B}_1$ rotating in a plane perpendicular to $\underline{B}$ at the angular frequency ω_0. We must therefore add to equation (17.32) the perturbation operator

$$\begin{aligned}\mathcal{H}_1 &= g_I\mu_n B_1(I_x\cos\omega_0 t + I_y\sin\omega_0 t)\\ &= \frac{\hbar\omega_1}{2}\{I_+\exp(-i\omega_0 t)+ I_-\exp(+i\omega_0 t)\}\end{aligned} \tag{17.48}$$

where $I_\pm = I_x \pm iI_y$. It is convenient to remove the time dependence of equation (17.48) by a transformation to the rotating frame. The density operator transforms as

$$\sigma_g = \exp(i\omega_0 I_z t)\rho_g \exp(-i\omega_0 I_z t) \tag{17.49}$$

and the Liouville equation in the rotating frame becomes

$$\frac{\partial\sigma_g}{\partial t} = \frac{1}{i\hbar}[\tilde{\mathcal{H}},\sigma_g] + \frac{d^{(1)}}{dt}\sigma_g + \frac{d^{(2)}}{dt}\sigma_g + \frac{d^{(3)}}{dt}\sigma_g. \tag{17.50}$$

The operator $\tilde{\mathcal{H}}$ is determined by the effective field in the rotating frame and takes the form

$$\tilde{\mathcal{H}} = \hbar(\omega_L-\omega_0)I_z + \frac{\hbar\omega_1}{2}(I_+ + I_-). \tag{17.51}$$

Since the depopulation, repopulation, and relaxation terms in equation (17.32) do not depend explicitly on time and are rotationally invariant, they have the same form in the rotating frame as they had in the laboratory frame of reference.

We now assume that the ^{199}Hg atoms in the cell are optically pumped by circularly-polarized σ^+ radiation propagating in the direction of the field $\underline{B}$. If the light is emitted by a ^{204}Hg lamp, then only the F=1/2 component of the 2537 Å line is absorbed by the atoms in the cell and it can be shown (Problem 17.7) that the depopulation and repopulation terms in equation (17.50) are given by

$$\frac{\partial^{(1)}}{\partial t}\sigma_g + \frac{\partial^{(2)}}{\partial t}\sigma_g = \begin{pmatrix} \frac{2}{3T_p}\sigma_{-\frac{1}{2}-\frac{1}{2}} & -\frac{1}{T_p}\sigma_{\frac{1}{2}-\frac{1}{2}} \\ -\frac{1}{T_p}\sigma_{-\frac{1}{2}\frac{1}{2}} & \left(-\frac{2}{T_p}+\frac{4}{3T_p}\right)\sigma_{-\frac{1}{2}-\frac{1}{2}} \end{pmatrix} \tag{17.52}$$

where T_p is the mean pumping time defined by equation (17.35).

The Liouville equation can then be separated into a set of four linear differential equations for the components of the ground-state density matrix in the rotating frame. Not all of these equations are independent because of the normalization condition which we write here in the form

$$\sigma_{\frac{1}{2}\frac{1}{2}} + \sigma_{-\frac{1}{2}-\frac{1}{2}} = 2. \tag{17.53}$$

The equations can be simplified (Problem 17.8) by defining three new variables

$$\left.\begin{aligned} M_x &= |\underline{\mu}|\,(\sigma_{\frac{1}{2}-\frac{1}{2}} + \sigma_{-\frac{1}{2}\frac{1}{2}}) \\ M_y &= i|\underline{\mu}|\,(\sigma_{\frac{1}{2}-\frac{1}{2}} - \sigma_{-\frac{1}{2}\frac{1}{2}}) \\ M_z &= |\underline{\mu}|\,(\sigma_{\frac{1}{2}\frac{1}{2}} - \sigma_{-\frac{1}{2}-\frac{1}{2}}) \end{aligned}\right\} \qquad (17.54)$$

which are in fact the components of the sample magnetization in the rotating frame. For convenience the angular brackets on M_x, M_y, and M_z have been dropped and $|\underline{\mu}| = g_I \mu_n I$ is just the nuclear magnetic moment of one atom. Finally, after some manipulation, we have

$$\left.\begin{aligned} \dot{M}_x &= -M_x/\tau_2 - (\omega_L - \omega_0)M_y \\ \dot{M}_y &= (\omega_L - \omega_0)M_x - M_y/\tau_2 - \omega_1 M_z \\ \dot{M}_z &= + \omega_1 M_y + (M_\infty - M_z)/\tau_1 \end{aligned}\right\} \qquad (17.55)$$

The effective longitudinal and transverse relaxation times in the presence of the pumping light are given by

$$\left.\begin{aligned} \frac{1}{\tau_1} &= \frac{1}{T_1} + \frac{2}{3T_p} \\ \frac{1}{\tau_2} &= \frac{1}{T_2} + \frac{1}{T_p} \end{aligned}\right\} \qquad (17.56)$$

and $M_\infty = 4|\underline{\mu}|\,\tau_1/3T_p$ is the equilibrium magnetization of the sample. In the limit of very weak pumping light, $T_p \to \infty$ and equation (17.55) becomes identical with the Bloch (1946) equations which are so well known in solid state magnetic resonance experiments.

We complete our illustration of the power of the density matrix treatment by seeking the stationary solutions of equation (17.55) in the rotating frame. By setting all time derivatives equal to zero, we obtain

$$M_z = M_\infty \left\{ 1 - \frac{\omega_1^2 \tau_1/\tau_2}{\frac{1}{\tau_2^2} + \omega_1^2 \frac{\tau_1}{\tau_2} + (\omega_L - \omega_0)^2} \right\} \tag{17.57}$$

and M_x and M_y are given by equations (17.30) and (17.31) respectively. From equation (17.42) it can be shown that the magnetic resonance signals obtained by fluorescence monitoring and displayed in Fig.17.10 are proportional to

$$\frac{dI}{d\Omega} \propto \sigma_{-\frac{1}{2}-\frac{1}{2}} = 1 - \frac{M_z}{2} . \tag{17.58}$$

The derivation of equation (17.21) for the width of the magnetic resonance signals observed in an optical pumping experiment is now quite straight-forward. A study of the time-dependent solutions of equation (17.55) (Problems 17.9 and 17.10) leads to equations which we have discussed earlier in connection with the measurement of the longitudinal and transverse relaxation times in transient experiments.

17.9.7. Light shifts in optical pumping experiments.

(a) Generalization of the theory of depopulation pumping.

When the atoms in the optical pumping cell are illuminated with light whose spectral profile is no longer uniform over the region about the absorption lines, a small shift in the atomic energy levels can be observed in addition to the usual level broadening given by equation (17.56). The theoretical expression for this light shift may be derived using a semi-classical argument due to Pancharatnam (1966). This generalizes the expression for the energy of $-\frac{1}{2}\alpha E^2$ acquired by an atom of polarizability α in an external electric field E. For the Fourier component $\underline{\hat{e}}E(\omega)\exp(-i\omega t)$ of the incident radiation field, the frequency-dependent atomic polarizability tensor has the form

$$\alpha = -\sum_m \frac{\underline{D}|m\rangle\langle m|\underline{D}}{4\pi\varepsilon_0\hbar(\omega-\omega_{m\mu} + \frac{i\Gamma}{2})} \qquad (17.59)$$

where $\omega_{m\mu} = (E_m - E_\mu)/\hbar$ is the angular frequency of the optical absorption line and Γ is the radiative width of the excited states $|m\rangle$. The energy of interaction with electromagnetic radiation of density $U(\omega)$ is then determined by the complex operator

$$\delta\mathcal{H} = -2\pi\int_0^\infty \underline{\hat{e}}^*.\alpha.\underline{\hat{e}}\ U(\omega)\ d\omega$$

$$= \int_0^\infty U(\omega)\ d\omega \sum_m \frac{\underline{\hat{e}}^*.\underline{D}|m\rangle\langle m|\underline{D}.\underline{\hat{e}}}{2\varepsilon_0\hbar(\omega-\omega_{m\mu} + \frac{i\Gamma}{2})} \qquad (17.60)$$

The effective Hamiltonian can be expressed in terms of its real and imaginary parts:

$$\delta\mathcal{H} = \delta\mathcal{E} - i\hbar\frac{\delta\Gamma}{2} \qquad (17.61)$$

giving a Hermitian light shift operator $\delta\mathcal{E}$ and a Hermitian light absorption operator $\delta\Gamma$. From equations (17.60) and (17.61) we have

$$\delta\Gamma = \sum_m \frac{\underline{\hat{e}}^*.\underline{D}|m\rangle\langle m|\underline{D}.\underline{\hat{e}}}{\varepsilon_0\hbar^2} \int_0^\infty U(\omega)\ d\omega \frac{\Gamma/2}{(\omega-\omega_{m\mu})^2 + \frac{\Gamma^2}{4}} \qquad (17.62)$$

and

$$\delta\mathcal{E} = \sum_m \frac{\underline{\hat{e}}^*.\underline{D}|m\rangle\langle m|\underline{D}.\underline{\hat{e}}}{2\varepsilon_0\hbar} \int_0^\infty U(\omega)\ d\omega \frac{(\omega-\omega_{m\mu})}{(\omega-\omega_{m\mu})^2 + \frac{\Gamma^2}{4}} \qquad (17.63)$$

Finally the depopulation pumping rate may be expressed in terms of the effective Hamiltonian by

$$\frac{d^{(1)}}{dt}\rho_g = \frac{1}{i\hbar}[\delta\mathcal{H}\rho_g - \rho_g\,\delta\mathcal{H}^+]. \qquad (17.64)$$

This completes the generalization of equation (17.36) to the case of light of non-uniform spectral profile.

The evaluation of the integrals in equations (17.62) and (17.63) is illustrated in Figs. 17.17(a) and (b) respectively. The spectral distribution of the incident pumping light is assumed to be centred at ω_p and to have a Doppler-broadened profile of width $\Delta\omega_p$. The integrals in equations (17.62) and (17.63) therefore represent the folding of this spectral profile with the atomic absorption and dispersion curves which are centred at the resonance frequency $\omega_{m\mu}/2\pi$. We have assumed that the atoms in the cell are stationary so that the width of the atomic response curves, Γ, is determined by the natural width of the excited level and $\Gamma \ll \omega_p$. The width and shift of the ground-state energy levels are then obviously proportional to the intensity of the incident light. The variation of $\delta\Gamma$ and $\delta\mathcal{E}$ with the angular frequency separation $(\omega_p-\omega_{m\mu})$ may also be simply predicted. Since $\Delta\omega_p$ is large compared with Γ, we see that $\delta\Gamma$ will have the same shape as $U(\omega)$ and will reach a maximum value when $\omega_p=\omega_{m\mu}$. On the other hand, $\delta\mathcal{E}$ is zero by reason of symmetry when $\omega_p=\omega_{m\mu}$ and increases with $(\omega_p-\omega_{m\mu})$, eventually reaching a maximum when $(\omega_p-\omega_{m\mu}) \approx \Delta\omega_p$. As a function of $(\omega_p-\omega_{m\mu})$ the light-shift operator should be asymmetric like the atomic dispersion curve.

This difference in behaviour between $\delta\Gamma$ and $\delta\mathcal{E}$ suggests that these quantities can be associated with two different kinds of radiative transitions. For instance $\delta\Gamma$ reaches a maximum when $\omega_p=\omega_{m\mu}$, at which point photons in the light beam can be absorbed by an atom. The atom makes a real transition between two energy levels and the total energy is conserved in the atom-photon system. By contrast, $\delta\mathcal{E}$ reaches a maximum when $\omega_p-\omega_{m\mu} \approx \Delta\omega_p$ and real absorption can no longer occur since energy would not be conserved. However, because of the uncertainty relation $\Delta E\ \Delta t \approx \hbar$, the atom can make a virtual transition, the effect of which is to slightly modify the energy level of the atomic ground state.

The effect of these real and virtual transitions on the propagation of light through the medium has been known

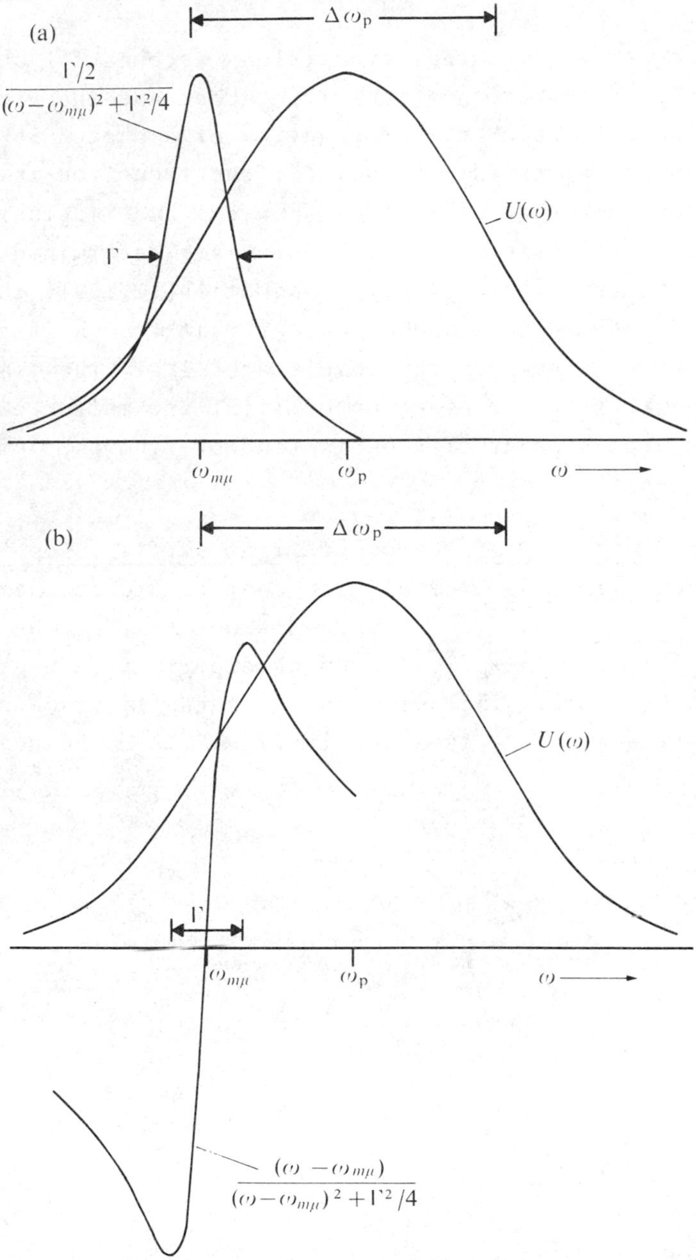

Fig.17.17. Evaluation of the broadening and shift of atomic levels due to optical excitation. (a) The integral for the absorption operator, $\delta\Gamma$.(b) The integral for the light shift operator $\delta\mathcal{E}$. The operators $\delta\Gamma$ and $\delta\mathcal{E}$ are proportional to the folding integrals of the plotted functions. (After Cohen-Tannoudji and Kastler (1966).)

for a long time. The real transitions account for absorption or attenuation of the light beam and the virtual transitions, in which the propagation of a photon is instantaneously arrested, account for the reduction in the velocity of propagation. Since the real and imaginary parts of the refractive index of the medium are determined by the complex atomic polarizability of equation (17.59), this discussion should have made it clear that the shift and broadening of atomic energy levels are simply the complementary aspect of the interaction of light and matter observed from the atom's point of view instead of from that of the photon.

(b) Experimental verification of light shifts in optical pumping experiments. The displacement of atomic energy levels by light was first observed experimentally by Cohen-Tannoudji (1961) in optical pumping experiments on ground state ^{199}Hg atoms. The shift in the magnetic resonance signal $M_I = \frac{1}{2} \rightarrow M_I = -\frac{1}{2}$ is shown in Fig.17.18. The centre

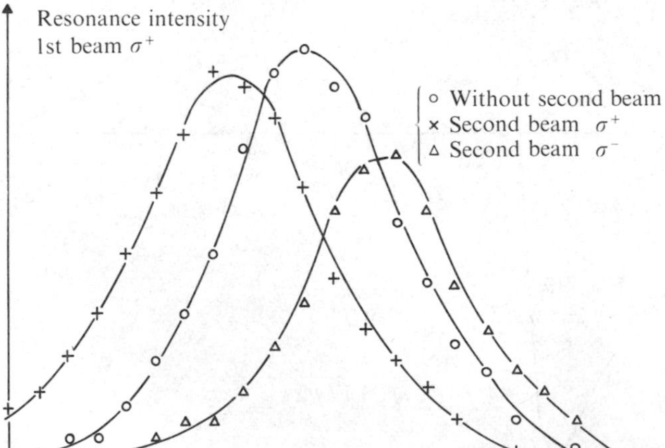

Fig.17.18. Experimental proof of the shift of the line centre of the nuclear magnetic resonance signal in ^{199}Hg induced by virtual optical transitions. (After Cohen-Tannoudji and Kastler (1966).)

curve gives the results obtained under ordinary optical pumping conditions using circularly-polarized light from a ^{199}Hg lamp. The curves on either side show the effect of simultaneously irradiating the pumping cell with either σ^+ or σ^- polarized light from a ^{201}Hg lamp. This isotope was selected since one of the hyperfine components emitted by the ^{201}Hg lamp is about a Doppler width away from the $F=\frac{1}{2}$ component of the ^{199}Hg 2537 Å resonance line. The use of circularly-polarized light ensures that in the first case only the $M_I = -\frac{1}{2}$ sub-level is shifted by the virtual transitions while in the second case only the $M_I = \frac{1}{2}$ sub-level is displaced.

To measure the shift in the resonance frequency as a function of $\omega_p - \omega_{m\mu}$, the ^{201}Hg lamp was removed and the previously symmetric profile of the ^{199}Hg lamp was distorted by passing the light through a magnetically-scanned absorption cell containing ^{198}Hg. The results shown in Fig.17.19

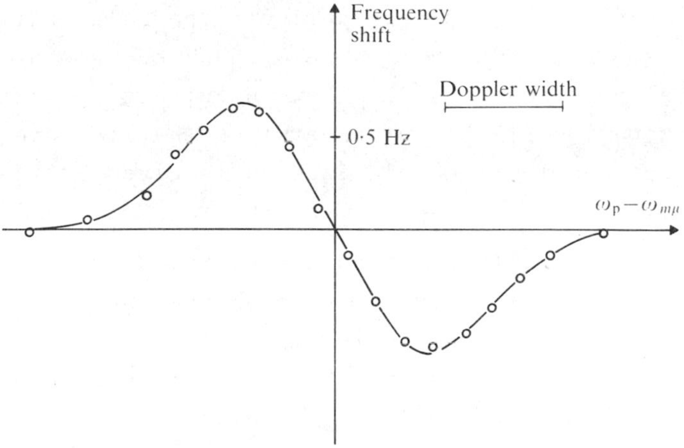

Fig.17.19. Frequency dependence of the light shift due to virtual optical transitions. The shift is proportional to the refractive index of the vapour. (After Cohen-Tannoudji and Kastler (1966).)

clearly demonstrate that $\delta\mathcal{E}$ varies as a dispersion curve in complete agreement with the theory.

Additional contributions to the shift in atomic energy levels by light come from the circulation of coherence in the optical pumping cycle and are described theoretically by generalizations of equation (17.39). Although the shifts involved in all these experiments are usually less than 10 Hz and are small compared with the Zeeman resonance frequencies of 1-100 kHz, nevertheless they are significant in precision experiments. Systematic errors can be avoided by extrapolating the resonance frequency measurements to zero intensity of the optical pumping lamp. Considerably larger energy level shifts have been observed by Aleksandrov *et al.* (1966) and Bonch-Bruevich *et al.* (1966) using high power lasers, but the experimental and theoretical analysis of this phenomenon is still far from complete.

Problems

17.1. In an optical pumping experiment the polarization of the atoms in the cell is zero when the pumping lamp is switched on at $t=0$. Show, by direct substitution in equation (17.8), or otherwise, that the approach to the equilibrium polarization is given by equation (17.9).

17.2. The flux of circularly-polarized radiation in the D_1 resonance line incident on an optical pumping cell from a conventional sodium lamp is measured as $8{\cdot}12 \times 10^{14}$ photons cm^{-2} s^{-1}. The linewidth of the D_1 line is found to be 10^{10} Hz, which is about 20 times the Doppler width of the atomic absorption line in the cell. By using equations (9.50) and (9.51) and assuming that the absorption oscillator strength for σ^+ polarized radiation is 1/6, show that the mean photon absorption rate per atom, $1/T_p'$, is equal to $3{\cdot}59 \times 10^2$ s^{-1}. If the longitudinal relaxation time T_1 is $1{\cdot}5 \times 10^{-4}$ s, show that the maximum polarization attainable is only

5·1 per cent. The use of microwave lamps enables the pumping time to be reduced by a factor of 10 and in this case the equilibrium polarization would approach 35 per cent.

17.3. An optical pumping cell, 5 cm in diameter, is filled with ^{133}Cs vapour. Show that the mean thermal velocity of the atoms at 300 K is $2{\cdot}18 \times 10^4$ cm s^{-1} and hence that the mean time between two collisions with the walls is approximately 2×10^{-4} s.

17.4. Using the techniques developed in Problem 7.10, separate the diffusion equation

$$\frac{\partial n}{\partial t} = D\,\nabla^2 n$$

in spherical polar coordinates, (r,θ,ϕ). Hence show that the spherically-symmetric solutions are given by

$$n(r) = \frac{A}{r}\left\{\begin{matrix}\sin\\ \cos\end{matrix}\; kr\right\}$$

where A and k are arbitrary constants. If the density of oriented atoms is finite at the centre of a spherical optical pumping cell and is zero at the cell walls, show that the decay rate of the lowest-order diffusion mode is correctly given by equation (17.15).

17.5. The microwave transition between the ground-state hyperfine levels of ^{87}Rb occurs at 6835 MHz. Show that the Doppler width of this transition at 300 K is $\Delta\omega_{Doppler} = 5{\cdot}71 \times 10^4\ s^{-1}$. If, instead of moving freely, the radiating atoms are constrained to oscillate about their mean positions with simple harmonic motion of frequency $f/2\pi$ and amplitude L, show that the emitted wave is frequency modulated. By qualitative arguments show that the line width of the radiation is given approximately by equation (17.17). Assuming a gas kinetic collision cross-section of $1{\cdot}08 \times 10^{-16}\ cm^2$, estimate the factor by which the Doppler width of the ^{87}Rb microwave line is reduced in a cell containing a buffer gas at a pressure of 50 Torr.

$$[\exp(ia\ \sin ft) = \sum_{n=-\infty}^{\infty} J_n(a)\exp(inft)]$$

17.6. The gyromagnetic ratio of the proton in a spherical water sample, $\gamma_p = g_I' \mu_B/\hbar$, has been determined to be $(2{\cdot}67513 \pm 0{\cdot}00001) \times 10^8\ (\mathrm{sT})^{-1}$. By combining this value with the gyromagnetic ratios $g_J(^{85}\mathrm{Rb})/g(e)$; $g_J(^{87}\mathrm{Rb})/g_J(^{85}\mathrm{Rb})$, and $g_J(^{87}\mathrm{Rb})/g_I'(p)$ given in Table 17.3, show that these lead to the result

$$g(e)\ \mu_B/\hbar = 1{\cdot}760\ 850 \times 10^{11}\ \mathrm{C\ kg}^{-1}.$$

Hence, using the theoretical quantum-electrodynamical value of $g(e)/2 = 1{\cdot}001\ 159\ 6$, obtain a value for e/m and compare your result with the currently recommended value.

17.7. Using the selection rules for σ^+ polarized radiation and equations (17.34) - (17.37), show that the depopulation pumping term for $^{199}\mathrm{Hg}$ atoms illuminated with the F=1/2 component of the 2537 Å line is given by

$$\frac{d^{(1)}}{dt}\rho_g = -\frac{1}{T_p} \times \begin{pmatrix} 0 & \rho_{\frac{1}{2}-\frac{1}{2}} \\ \rho_{-\frac{1}{2}\frac{1}{2}} & 2\rho_{-\frac{1}{2}-\frac{1}{2}} \end{pmatrix}.$$

By equating the rate at which atoms enter and leave the $F=\frac{1}{2}$, $M_F=\frac{1}{2}$ sub-level of the excited state, and using the relative transition probabilities indicated in Fig.17.1, show that the repopulation pumping term is given by

$$\frac{d^{(2)}}{dt}\rho_g = \frac{2}{3T_p}\begin{pmatrix} \rho_{-\frac{1}{2}-\frac{1}{2}} & 0 \\ 0 & 2\rho_{-\frac{1}{2}-\frac{1}{2}} \end{pmatrix}.$$

17.8. Show that the matrix of the effective Hamiltonian operator, equation (17.51), is given by

$$\tilde{\mathcal{H}} = \frac{\hbar}{2}\begin{pmatrix}(\omega_L-\omega_0) & \omega_1\\ \omega_1 & -(\omega_L-\omega_0)\end{pmatrix} .$$

Hence show that in the absence of pumping and relaxation terms the time evolution in the rotating frame is given by

$$\frac{1}{i\hbar}[\tilde{\mathcal{H}},\sigma_g] =$$

$$\begin{pmatrix}\frac{i\omega_1}{2}(\sigma_{\frac{1}{2}-\frac{1}{2}}-\sigma_{-\frac{1}{2}\frac{1}{2}}) & \frac{i\omega_1}{2}(\sigma_{\frac{1}{2}\frac{1}{2}}-\sigma_{-\frac{1}{2}-\frac{1}{2}})-i(\omega_L-\omega_0)\sigma_{\frac{1}{2}-\frac{1}{2}}\\ -\frac{i\omega_1}{2}(\sigma_{\frac{1}{2}\frac{1}{2}}-\sigma_{\frac{1}{2}-\frac{1}{2}})+i(\omega_L-\omega_0)\sigma_{-\frac{1}{2}\frac{1}{2}} & -\frac{i\omega_1}{2}(\sigma_{\frac{1}{2}-\frac{1}{2}}-\sigma_{-\frac{1}{2}\frac{1}{2}})\end{pmatrix}$$

Finally, introducing the detailed expressions for the pumping terms into equation (17.50), write out the complete differential equation for each component of the density matrix. Using equations (17.53) and (17.54) show that this set of linear differential equations may be reduced to the form given by equation (17.55).

17.9. Optical pumping experiments in which there is no applied r.f. magnetic field are described by setting $\omega_1=0$ in equation (17.55). Show that the transient signal observed under the action of the pumping light alone is determined by

$$M_z(t) = M_\infty + \{M_z(0) - M_\infty\}\exp(-t/\tau_1).$$

This reduces to equation (17.9) when the initial magnetization $M_z(0)$ is zero.

17.10. At resonance, $\omega_L=\omega_0$ in equation (17.55). Show that in this situation the x-component of the sample magnetization in the rotating frame decays exponentially with the time-constant τ_2.

Show also that on resonance, M_z is given by the solutions of the equation

$$\ddot{M}_z + \left(\frac{1}{\tau_1}+\frac{1}{\tau_2}\right)\dot{M}_z + (\omega_1^2 + \frac{1}{\tau_1\tau_2})M_z = \frac{M_\infty}{\tau_1\tau_2} .$$

When $\omega_1 \gg \frac{1}{2}\left(\frac{1}{\tau_1} + \frac{1}{\tau_2}\right)$, prove that $M_z(t)$ oscillates sinusoidally at the angular frequency ω_1 and is exponentially damped with the time constant

$$\frac{1}{\tau} = \frac{1}{2}\left(\frac{1}{\tau_1} + \frac{1}{\tau_2}\right) ,$$

in agreement with equation (17.26).

References.

Aleksandrov, E.B., Bonch-Bruevich, A.M., Kostin, N.N., and Khodovoi, V.A. (1966). *Soviet Phys.Lett.JETP* 3, 53.

Balling, L.C. and Pipkin, F.M. (1965). *Phys.Rev.* 139, A19.

Barrat, J.P. and Cohen-Tannoudji, C. (1961). *J.Phys., Paris* 22, 329, 443.

Bell, W.E. and Bloom, A.L. (1957). *Phys.Rev.* 107, 1559.

——— and ——— (1961). *Phys.Rev.Lett.* 6, 280.

Bender, P.L. (1962). *Phys.Rev.* 128, 2218.

Bloch, F. (1946). *Phys.Rev.* 70, 460.

Bonch-Bruevich, A.M., Kostin, N.N., and Khodovoi, V.A. (1966). *Soviet Phys.Lett. JETP*, 3, 279.

Bouchiat, M.A. (1963). *J.Phys., Paris*, 24, 379, 611.

——— and Brossel, J. (1962). *C.r.hebd.Séanc.Acad. Sci., Paris.* 254, 3650, 3828.

——— and ——— (1963). *C.r.hebd.Séanc.Acad. Sci., Paris.* 257, 2825.

——— and Grossetête, F. (1966). *J.Phys., Paris.* 27, 353.

Brossel, J., Kastler, A., and Winter, J. (1952). *J.Phys., Paris.* 13, 668.

Cagnac, B. (1958). *J.Phys., Paris.* 19, 863.

——— (1961). *Annls.Phys.* 6, 467.

Cohen-Tannoudji, C. (1961). *C.r.hebd.Séanc.Acad.Sci., Paris.* 252, 394.

Cohen-Tannoudji, C. (1962). *Annls.Phys.* 7, 423, 469.

Dehmelt, H.G. (1957). *Phys.Rev.* 105, 1924.

Ernst, K. and Strumia, F. (1968). *Phys.Rev.* 170, 48.

Franzen, W. (1959). *Phys.Rev.* 115, 850.

Gibbs, H.M. and Hull, R.J. (1966). *Phys.Rev.* 153, 132.

Greenhow, R.C. (1963). D.Phil. Thesis, University of Oxford.

Grossetête, F. (1967). Thesis, University of Paris, (unpublished).

Happer, W. (1972). *Rev.mod.Phys.* 44, 169.

Hawkins, W.B. and Dicke, R.H. (1953). *Phys.Rev.* 91, 1008.

Kastler, A. (1950). *J.Phys., Paris.* 11, 255.

Lehmann, J.C. and Cohen-Tannoudji, C. (1964). *C.r.hebd.Séanc. Acad.Sci., Paris.* 258, 4463.

Moretti, A. and Strumia, F. (1971). *Phys.Rev.* 3A, 349.

Pancharatnam, S. (1966). *J.opt.Soc.Am.* 56, 1636.

Ressler, N.W., Sands, R.H., and Stark, T.E. (1969). *Phys.Rev.* 184, 102.

Robinson, H.G. Ensberg, E.S., and Dehmelt, H.G. (1958). *Bull.Am.phys.Soc.* 3, 9.

Schearer, L.D. and Sinclair, F.D. (1968). *Phys.Rev.* 175, 36.

White, C.W., Hughes, W.M., Hayne, G.S., and Robinson, H.G. (1968). *Phys.Rev.* 174, 23.

General references and further reading

Balling, L.C. (1975). *Advances in quantum electronics.* Vol.3. (ed. D.W. Goodwin), Academic Press, New York.

Bernheim, R.A. (1965). *Optical pumping - an introduction.* W.A. Benjamin, Inc., New York.

This book includes a useful collection of reprints of early papers on optical pumping.

However, in our opinion the most readable and satisfying review of the subject remains that by

Cohen-Tannoudji, C. and Kastler, A. (1966). *Progress in optics*. V, p.3. (ed. E. Wolf). North Holland, Amsterdam.

Happer, W. (1972). *Rev.mod.Phys.* 44, 169. This gives an excellent comprehensive review of optical pumping but some readers may well find this article rather too advanced for introductory reading.

Major, F.G. (1968). *Methods of experimental physics*. 7B. Academic Press, New York.

Skrotskii, G.V. and Izyumova, T.G. (1961). *Soviet Phys.Usp.* 4, 177.

Slichter, C.P. (1963). *Principles of magnetic resonance*. Harper and Row, New York.

18

The hyperfine structure of atoms and its investigation by magnetic resonance methods

When an optical spectral line is isolated and examined under high resolution it is usually found to consist of a partially resolved blend of many different components lying within 10^{-3} - 1 cm^{-1} of the line centre. This complex of components is generally the result of two quite different effects: isotope shift and hyperfine structure. As we have already noted several times in Chapters 15-17, the centres of gravity of the spectral lines emitted by different isotopes are displaced from one another by the effect of the different masses and charge distributions of their nuclei. Although isotope shifts contain valuable information about some properties of the atomic nucleus, measurements of these shifts by magnetic resonance techniques are impossible because the energy levels involved belong to two quite distinct atoms. We shall therefore not discuss isotope shifts any further but refer the reader to the review article by Stacey (1966) for additional information. The effect can be eliminated when it is experimentally undesirable by using a source containing a single pure isotope of the element of interest.

On the other hand when the spectra of each of the odd isotopes of an element are examined in detail, it is found that there is a residual structure known as the *hyperfine structure*. Pauli (1924) suggested that this was due to a magnetic moment $\underline{\mu}_I$ associated with an intrinsic nuclear spin angular momentum of $\hbar\underline{I}$. The interaction of the magnetic moment of the nucleus with the magnetic field produced by the valence electrons causes a small splitting of the energy levels of an atom and provides the largest contribution to the observed hyperfine structure.

The small scale of the hyperfine splittings and the large Doppler width of spectral lines makes accurate measurements by optical spectroscopy very difficult. Magnetic dipole transitions, however, can be induced directly between the different hyperfine-structure levels of a given configuration and the precision of the measurements by radio-frequency spectroscopy is then generally limited only by the natural width of the energy levels involved. This precision can be extremely high for the ground states and long-lived metastable states of free atoms, as we saw in Chapter 17.

Magnetic resonance measurements on both ground state and excited levels of atoms yield values of the nuclear spin, the nuclear magnetic moment, and of the nuclear electric quadrupole moment. In this chapter we shall discuss the investigation of the hyperfine structure of atoms by the optical pumping, atomic beam, and optical double-resonance techniques. Although the data obtained are of fundamental interest, the application of these techniques to the construction of atomic clocks and maser oscillators is perhaps of more practical importance. Before moving on to the discussion of these topics, we briefly outline the basic theory of atomic hyperfine structure and the associated Zeeman effect closely following the more detailed account given in Woodgate (1970).

18.1. Theory of hyperfine structure

18.1.1. Magnetic dipole interaction. We consider first the interaction of the nuclear magnetic moment $\underline{\mu}_I$ with the magnetic field $\underline{B}_{el}(0)$ produced at the nucleus by the valence electrons. Consideration of the classical energy of orientation in the field allows the Hamiltonian describing the interaction to be written in the form:

$$\mathcal{H}_1 = -\underline{\mu}_I \cdot \underline{B}_{el}(0). \qquad (18.1)$$

We assume that this term can be treated as a small perturbation when compared with the zeroth-order Hamiltonian, $\mathcal{H}_0$, which describes the central electrostatic field of the atom, the repulsion between electrons, and the spin-orbit interaction. We need therefore only consider the effect of $\mathcal{H}_1$ on the states labelled by the quantum numbers (γ LSJ). It follows that J and I are good quantum numbers and that the nuclear magnetic moment may be written in the form

$$\underline{\mu}_I = g_I \mu_n \underline{I} \tag{18.2}$$

where g_I is the nuclear g-factor and $\mu_n = e\hbar/2M$ is the nuclear magneton which depends on the proton mass, M. The nuclear magnetic moment of the atom, μ_I, is defined as the largest observable component of $\underline{\mu}_I$:

$$\mu_I = \langle I, M_I{=}I|\mu_z|I, M_I{=}I\rangle = g_I \mu_n I.$$

Two points should be noted in connection with equation (18.2): firstly, the sign adopted in this equation differs from that of the corresponding equation (3.85) for electronic magnetic moments because the nucleus is positively charged; and secondly, nuclear magnetic moments are typically two thousand times smaller than atomic magnetic moments since $\mu_n/\mu_B = m/M \approx 1/1836$. The nuclear g-factor, which is of the order of unity, takes account of the way the resultant nuclear moment is built up from the magnetic moments of individual nucleons. It therefore contains information about the detailed internal structure of the nucleus.

Since the magnetic field at the nucleus, $\underline{B}_{el}(0)$, is determined by the motion of the atomic electrons it follows that equation (18.1) may be rewritten as

$$\mathcal{H}_1 = A_J \, \underline{I}.\underline{J} \tag{18.3}$$

where A_J is called the magnetic hyperfine structure constant and is the quantity which is determined directly from experimental measurements. For a hydrogenic atom, or an atom with just one electron outside closed shells, $\underline{B}_{el}(0)$ and $\underline{J}$ are antiparallel and the constant A_J will be positive if the nucleus has a positive value of g_I. The magnetic field at the nucleus will be zero for atoms with closed sub-shells and will be largest for those atoms with electrons in penetrating orbits. We would therefore expect to observe the largest hyperfine splitting in atoms with unpaired s-electrons, e.g. hydrogen and the alkalis.

It can be shown that, of the possible higher-order multipole moments, the nuclear magnetic moments of even order vanish because of symmetry requirements with respect to the nuclear equatorial plane. The next non-vanishing moment, the nuclear magnetic octupole moment, is very small and will be ignored here.

18.1.2. Electric quadrupole interaction. Although the magnetic dipole interaction is responsible for the largest contribution to the observed hyperfine structure, the finite extent of the nuclear electric charge distribution is also significant in many cases. We therefore consider the electrostatic interaction between a proton at the point $\underline{r}_n$ and an electron at the point $\underline{r}_e$, given by

$$\mathcal{H}_2' = -\frac{e^2}{4\pi\varepsilon_0|\underline{r}_e-\underline{r}_n|} \tag{18.4}$$

where the origin is taken as the centre of mass of the nucleus. To account for the finite extent of the nuclear charge distribution, we assume $r_e > r_n$ and expand equation (18.4) in ascending powers of r_n/r_e, giving

$$\begin{aligned}\mathcal{H}_2' &= -\frac{e^2}{4\pi\varepsilon_0}(r_e^2 + r_n^2 - 2r_e r_n \cos\theta_{en})^{-1/2} \\ &= -\frac{e^2}{4\pi\varepsilon_0}\sum_k \frac{r_n^k}{r_e^{k+1}} P_k(\cos\theta_{en})\end{aligned} \tag{18.5}$$

where $P_k(\cos\theta_{en})$ is the Legendre polynomial of order k and θ_{en} is the angle between $\underline{r}_e$ and $\underline{r}_n$ shown in Fig.18.1.

The first term in the summation of equation (18.5) represents a monopole interaction and when summed over all the protons and electrons gives the familiar Coulomb interaction $-\sum_i Ze^2/4\pi\varepsilon_0 r_{ei}$. The second term represents a nuclear electric dipole interaction which, by the application of parity and time-reversal symmetry arguments, can be shown to be identically zero, as are the higher electric multipole moments of odd order. Finally the term with k=2 corresponds to an electric quadrupole interaction as we

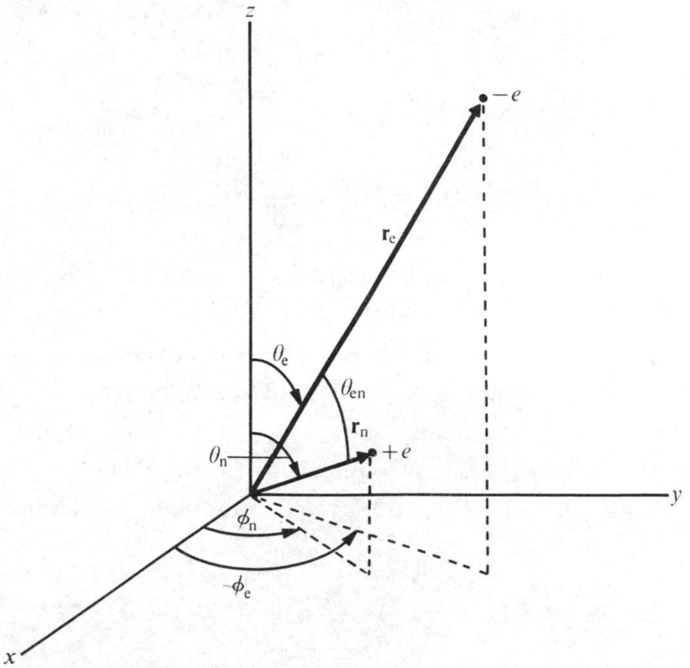

Fig.18.1. Definition of the angles (θ_n, ϕ_n) for a proton charge +e and of (θ_e, ϕ_e) for an electron of charge -e, together with the angle θ_{en} between $\underline{r}_e$ and $\underline{r}_n$.

shall now demonstrate. We complete the separation of electric and nuclear coordinates in equation (18.5) by applying the spherical harmonic addition theorem:

$$P_k(\cos\theta_{en}) = \frac{4\pi}{2k+1} \sum_{q=-k}^{k} (-)^q \, Y_k^{-q}(\theta_n, \phi_n) \, Y_k^q(\theta_e, \phi_e) \tag{18.6}$$

where $Y_k^q(\theta,\phi)$ is the qth component of the spherical harmonic of order k and the spherical polar coordinates (θ,ϕ) are defined with respect to an arbitrary z-axis fixed in space (see Fig.18.1). The electric quadrupole interaction then becomes

$$\mathcal{H}_2 = \frac{1}{4\pi\varepsilon_0} \sum_{q=-2}^{2} (-)^q \{(\tfrac{4\pi}{5})^{1/2} e r_n^2 Y_2^{-q}(\theta_n,\phi_n)\} \times$$

$$\times \{(\tfrac{4\pi}{5})^{1/2} (\tfrac{-e}{r_e^3}) Y_2^q(\theta_e,\phi_e)\}$$

$$= \frac{1}{4\pi\varepsilon_0} \sum_q (-)^q Q_2^{-q}(n) F_2^q(e). \tag{18.7}$$

When summed over all the protons and electrons, this equation gives the complete nuclear electric quadrupole interaction. We observe that it has the form of the scalar product of a nuclear electric quadrupole tensor and an electric field gradient tensor, each of rank two.

The expression given in equation (18.7) may be simplified by defining the nuclear electric quadrupole moment Q as

$$Q = \frac{2}{e} \langle I, M_I=I | \sum_j Q_2^0(n_j) | I, M_I=I \rangle$$

$$= \langle II | \sum_j r_{n_j}^2 (3\cos^2\theta_{n_j} - 1) | II \rangle \tag{18.8}$$

where the summation extends over all the protons in the nucleus (Problem 18.1). In the same way the electronic part of equation (18.7) may be simplified by introducing the average field gradient at the nucleus defined by

$$\langle \frac{\partial^2 V_e}{\partial z^2} \rangle = \frac{2}{4\pi\varepsilon_0} \langle JJ | \sum_i F_2^0(e_i) | JJ \rangle$$

$$= - \langle JJ | \sum_i \frac{e(3\cos^2\theta_{e_i} - 1)}{4\pi\varepsilon_0 r_{e_i}^3} | JJ \rangle \tag{18.9}$$

where the summation is now over the valence electrons of the atom. Using equation (18.8) and (18.9), it can be shown that the electric quadrupole interaction can be written in the form

$$\mathcal{H}_2 = \frac{B_J}{2I(2I-1)J(2J-1)} \{3(\underline{I}.\underline{J})^2 + \tfrac{3}{2}(\underline{I}.\underline{J}) - I(I+1)J(J+1)\} \tag{18.10}$$

where B_J is the electric quadrupole interaction constant defined by

$$B_J = eQ \langle \frac{\partial^2 V_e}{\partial z^2} \rangle \ . \tag{18.11}$$

Hyperfine structure measurements on free atoms enable the values of B_J to be determined experimentally. However, the nuclear electric quadrupole moment can then only be calculated if theoretical estimates of the field gradient exist. Since the values of $\langle \partial^2 V_e/\partial z^2 \rangle$ are seldom known to an accuracy of better then 25 per cent, precise values of the electric quadrupole moments have only been determined in a few special cases. Finally we note that the quadrupole moment Q is positive if the nuclear charge distribution is elongated along the direction of the nuclear spin $\underline{I}$ (prolate) and is negative if the distribution is flattened in this direction (oblate).

18.1.3. Hyperfine structure Hamiltonian. The energy of an atom with finite nuclear spin in a uniform external magnetic field $\underline{B}$ is then determined by the sum of the zeroth-order Hamiltonian $\mathcal{H}_0$ and the hyperfine structure Hamiltonian:

$$\mathcal{H}_{HFS} = A_J \, \underline{I}.\underline{J} + \frac{B_J}{2I(2I-1)J(2J-1)}\{3(\underline{I}.\underline{J})^2 + \tfrac{3}{2}(\underline{I}.\underline{J}) - I(I+1)J(J+1)\} +$$

$$+ g_J\mu_B \, \underline{J}.\underline{B} - g_I\mu_n \, \underline{I}.\underline{B} \ . \tag{18.12}$$

The third and fourth terms in this equation represent the interaction of the electronic and nuclear magnetic moments with the external magnetic field, the difference in sign arising from the difference in the conventions used in defining g_J and g_I as noted above. The interaction constant B_J is identically zero if I=0 or $\frac{1}{2}$, for the nuclear charge distribution is spherically symmetric in these cases: similarly B_J is also zero if J=0, $\frac{1}{2}$ for then the electron charge distribution is spherically symmetric and $\langle \partial^2 V_e/\partial z^2 \rangle$ vanishes. It follows that, even when the nucleus possesses a finite quadrupole moment, this cannot be detected by measurements on the ground states of elements of groups I, II, or III of the periodic table. The information obtained by optical pumping and atomic beam methods is therefore incomplete and this provided one of the major incentives for the development of the optical double-resonance technique, for states with $J > \frac{1}{2}$ can then be investigated.

18.1.4. Hyperfine structure in zero magnetic field. In the absence of nuclear magnetic dipole or electric quadrupole interactions the zeroth-order wavefunctions $|\gamma JIM_JM_I\rangle$ are $(2I+1)(2J+1)$-fold degenerate in the nuclear and electronic magnetic quantum numbers. To evaluate the energy shift arising from the hyperfine interaction in zero magnetic field, we take linear combinations of the functions $|\gamma JIM_JM_I\rangle$ to form new zeroth-order wavefunctions $|\gamma JIFM_F\rangle$ for which the total angular momentum F and the projection $M_F=M_J+M_I$ are good quantum numbers. On the vector model the magnetic interaction $A_J\,\underline{I}.\underline{J}$ may be imagined to cause $\underline{I}$ and $\underline{J}$ to precess rapidly about the resultant total angular momentum

$$\underline{F} = \underline{I} + \underline{J}. \tag{18.13}$$

The energy of a particular hyperfine level is then given by

$$\begin{aligned} E_F &= \langle \gamma JIFM_F|\mathcal{H}_0+\mathcal{H}_{HFS}|\gamma JIFM_F\rangle \\ &= E_J + \frac{1}{2}A_JK + \frac{B_J}{8I(2I-1)J(2J-1)}\{3K(K+1)-4I(I+1)J(J+1)\} \end{aligned} \tag{18.14}$$

where E_J is the energy of the fine-structure multiplet level with total electronic angular momentum quantum number J, and K is given by

$$K = F(F+1) - I(I+1) - J(J+1). \tag{18.15}$$

From equation (18.14) we see that the magnetic and electric interaction between the nucleus and the orbital electrons splits a given fine-structure level into a hyperfine-structure multiplet. The number of levels produced is equal to the number of possible orientations of the angular momentum vectors $\underline{I}$ and $\underline{J}$, i.e. $2I+1$ if $J \geq I$ and $2J+1$ if $I \geq J$. Although the quadrupole term vanishes identically in only a few cases, it has generally been found to be small compared with the magnetic dipole term. Thus, in the limit $B_J \ll A_J$, we find that there is an exact analogy between the magnetic interactions giving rise to fine structure on the one hand and hyperfine structure on the other, the quantum numbers (L,S,J) being replaced by the quantum numbers (J,I,F) respectively. From equations (18.14) and (18.15) it follows that in the limit $B_J \ll A_J$ there is a hyperfine-structure interval rule (Problem 18.2):

$$\Delta E = E_F - E_{F-1} = A_J F$$

corresponding to the interval rule for fine structure derived in section 3.9.5.

A consideration of the additional energy shift produced when B_J is finite shows that each hyperfine level is affected in a different way and this upsets the interval rule. For the case of nuclear spin I=3/2, which applies to Na^{23}, K^{39}, Rb^{87}, Hg^{201} amongst others, the zero-field hyperfine splittings are predicted to have the following values (Problem 18.3):

$$\left.\begin{array}{ll} J = \frac{1}{2} & E_2 - E_1 = 2A_{1/2} \\ J = \frac{3}{2} & E_3 - E_2 = 3A_{3/2} + B_{3/2} \\ & E_2 - E_1 = 2A_{3/2} - B_{3/2} \\ & E_1 - E_0 = A_{3/2} - B_{3/2} \end{array}\right\} \quad (18.16)$$

It follows that in the state J=3/2, the order of the hyperfine levels depends critically on the sign and magnitude of the ratio A_J/B_J.

<u>18.1.5. Zeeman effect in weak magnetic field.</u> In magnetic fields of less than 10^{-3} T the condition $g_J\mu_B B \ll A_J$ is usually

satisfied and the nuclear spin $\underline{I}$ and the electronic angular momentum $\underline{J}$ remain strongly coupled. However, the third and fourth terms in equation (18.12) introduce an extra energy, E_{FM_F}, which must be added to that given by equation (18.14). Using the techniques developed in section 3.9.6 we find that

$$E_{FM_F} = g_F \mu_B \underline{B}.\underline{F} = g_F \mu_B BM_F \tag{18.17}$$

where g_F is the effective g-value given by

$$g_F = g_J \left\{ \frac{F(F+1)+J(J+1)-I(I+1)}{2F(F+1)} \right\} - g_I \frac{m}{M} \left\{ \frac{F(F+1)+I(I+1)-J(J+1)}{2F(F+1)} \right\} \tag{18.18}$$

and the magnetic quantum number M_F takes the 2F+1 values F, F-1, ... , -F. The interaction with the external magnetic field therefore removes the previously existing (2F+1)-fold degeneracy of each hyperfine level $|\gamma JIF\rangle$ and causes the vector $\underline{F}$ to precess about the field direction at the Larmor angular frequency $\omega_L = g_F \mu_B B/\hbar$, as shown in Fig.18.2.

In equation (18.18) the factor $g_I m/M$ is three orders of magnitude smaller than g_J and can often be neglected, especially in experiments involving excited atoms. We are then ignoring the direct interaction of the nuclear magnetic moment with the *laboratory* field while retaining its interaction with the *internal* atomic field, the effects of which are described by the first term in equation (18.18).

18.1.6. Zeeman effect in strong magnetic fields. In the Zeeman splitting of fine-structure levels it is rare to find a case in which the applied field is sufficiently large that $\underline{L}$ and $\underline{S}$ are completely decoupled and the Paschen-Back effect observed. On the other hand, in hyperfine levels where the largest values of A_J/h have typical values of 1000 MHz, then a field of only 0·1 T must be considered quite strong. In the high field limit $g_J \mu_B B \gg A_J$, the complete hyperfine Hamil- tonian is evaluated in the uncoupled representation $|\gamma JIM_J M_I\rangle$

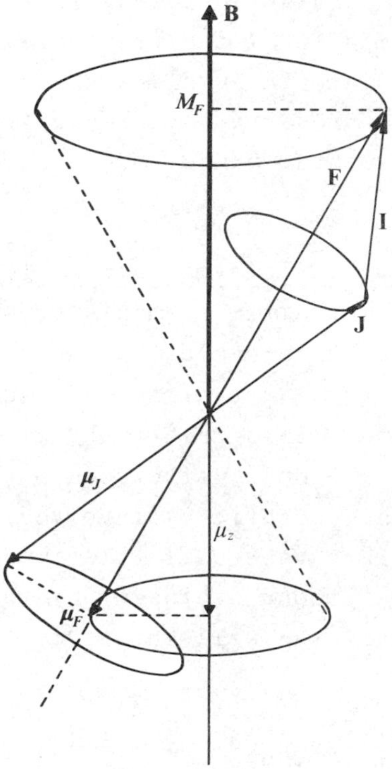

Fig.18.2. Vector model for the atomic magnetic moment $\underline{\mu}$ showing the projections first on the direction of $\underline{F}$ and then on the z-axis. The interaction between $\underline{\mu}_I$ and the external field is neglected.

and the energy obtained using first-order perturbation theory is given by

$$E_{M_J M_I} = g_J \mu_B B M_J - g_I \mu_n B M_I + A_J M_J M_I + \\ + \frac{B_J}{4I(2I-1)J(2J-1)} \{3M_J^2 - J(J+1)\}\{3M_I^2 - I(I+1)\}. \qquad (18.19)$$

The first term in this equation represents the Zeeman splitting of the multiplet level characterized by the quantum number J while the third term causes each Zeeman sub-level M_J to be further sub-divided into (2I+1) hyperfine

Zeeman levels corresponding to the different possible values of M_I. For any Zeeman level the transition from weak to strong field takes place in such a way that the magnetic quantum number M_q is preserved. In weak fields M_q is identical with the quantum number M_F, and in strong fields we have $M_q = M_J + M_I$.

18.1.7. Zeeman effect in intermediate fields. In intermediate fields all the matrix elements of the hyperfine interaction, $\mathcal{H}_{HFS}$, must be evaluated and the energy levels obtained by the solution of the secular equation. For large values of I and J this is a task best left to a computer, but if I or J does not exceed $\frac{1}{2}$ then no equation of order higher than quadratic is involved. The solution for J=1/2 and arbitrary values of I is important because it applies to the $^2S_{1/2}$ ground states of hydrogen and all the alkalis. The result is the well-known Breit-Rabi formula (Problem 18.4):

$$E_{M_J M_I} = -\frac{h\nu_{HFS}}{2(2I+1)} - g_I \mu_n B M_q \pm \frac{h\nu_{HFS}}{2}\left\{1 + \frac{4M_q x}{2I+1} + x^2\right\}^{1/2} \tag{18.20}$$

where $h\nu_{HFS} = A_J(I + \frac{1}{2})$ is the energy separation between the levels $F = I \pm \frac{1}{2}$ in zero magnetic field, $M_q = M_I \pm \frac{1}{2}$, and the dimensionless parameter x is given by

$$x = \frac{(g_J + g_I m/M)\mu_B B}{h\nu_{HFS}} . \tag{18.21}$$

In equation (18.20) the plus sign is to be used for states originating from the zero-field hyperfine level $F=I + \frac{1}{2}$ and the minus sign for states originating from $F=I - \frac{1}{2}$. In many ground-state magnetic resonance experiments the precision is such that intermediate coupling and formulae like equation (18.20) provide the only satisfactory basis in which to analyse the observed resonance frequencies.

The quadrupole interaction vanishes for $J = \frac{1}{2}$ so that the Breit-Rabi formula gives an exact description of the hyperfine structure in this case. If the small term $g_I \mu_n B M_q$ is omitted, a universal plot can be made of $E_{M_J M_I}$ as a function of the parameter x, which is in turn proportional to the applied magnetic field $\underline{B}$. Such a plot is shown in Fig.18.3 for an atom having $J = \frac{1}{2}$ and $I = \frac{3}{2}$ and a further example is shown in Fig.18.14. The construction of additional diagrams is left as an exercise for the student (Problem 18.5). It

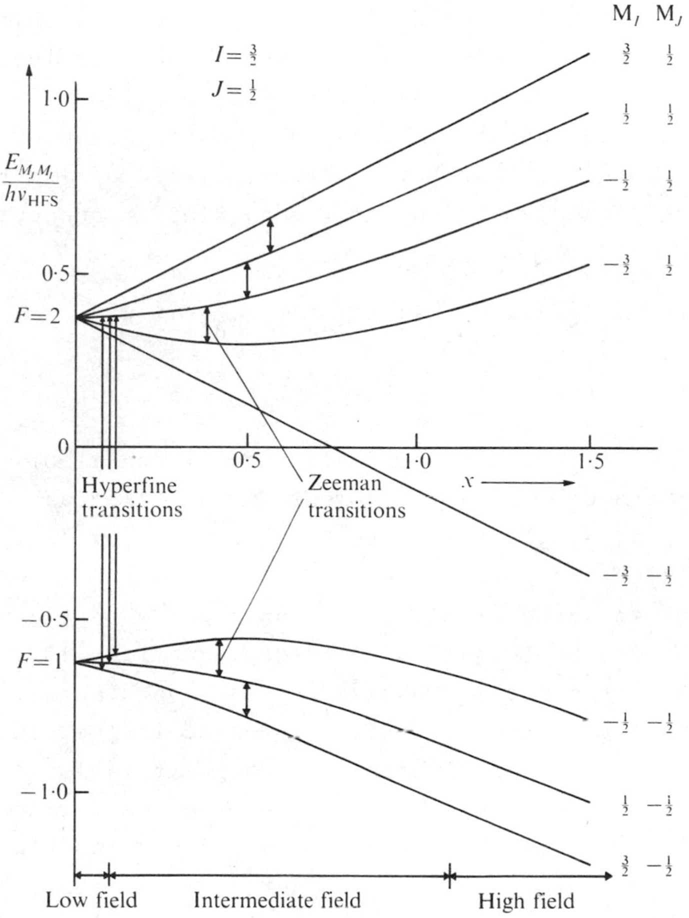

Fig.18.3. A plot of the Breit-Rabi formula for the case $I = \frac{3}{2}$. The abscissa is given by $x = (g_J + g_I m/M)\,\mu_B B/h\nu_{HFS}$ where $h\nu_{HFS}$ is the energy difference between the levels F=2 and F=1 in zero field.

is useful to remember that M_q remains a good quantum number in intermediate fields, states for which $M_q = M_F$ in low fields, going over into states with $M_q = M_I + M_J$ in strong fields.

<u>18.1.8. Selection rules for magnetic dipole transitions.</u> In the following sections we shall be mainly interested in magnetic dipole transitions induced between the different hyperfine sub-levels by a time-dependent magnetic field $\underline{B}_1(t)$. The effect of this field on the atoms may be calculated using the techniques described in section 9.3, provided only that the perturbation operator $\mathcal{H}_1 = -\underline{p}.\underline{E}(t)$ is replaced by the operator $\mathcal{H}_1' = -\underline{\mu}.\underline{B}_1(t)$. For a magnetic field of amplitude B_1 rotating in the x-y plane at the angular frequency ω_0 we have

$$\mathcal{H}_1' = -\underline{\mu}.(\cos\omega_0 t\ \hat{\underline{i}} + \sin\omega_0 t\ \hat{\underline{j}})\ B_1$$

$$= -\frac{B_1}{2}\{\mu_+ \exp(-i\omega_0 t) + \mu_- \exp(i\omega_0 t)\} \quad (18.22)$$

where $\mu_\pm = \mu_x \pm i\mu_y$. If the small nuclear contribution is ignored, the atomic magnetic moment may be taken as $\underline{\mu} \cong \underline{\mu}_J = -g_J\mu_B\ \underline{J}$. Then we see, from equation (9.35), that the transition probability for the induced emission or absorption of magnetic dipole radiation depends on the matrix elements of the perturbation operator, equation (18.22), taken between the states involved. In a weak magnetic field $\underline{B}$ the states are correctly expressed in the coupled representation $|\gamma JIFM_F\rangle$, while in strong external fields the uncoupled representation $|\gamma JIM_JM_I\rangle$ is more appropriate. By applying the general considerations discussed in Chapters 5 and 7, the selection rules for magnetic dipole transitions may be obtained and are listed in the second column of Table 18.1. These transitions are labelled π since the polarization vector of the electric field associated with $\underline{B}_1(t)$ is, in this case, parallel to the axis of quantization. On the other hand, when the time-dependent field $\underline{B}_1(t)$ has a component parallel to 0z, the transitions in Table 18.1

TABLE 18.1

Selection rules for magnetic dipole transitions between hyperfine levels belonging to states with the same electronic angular momentum J

Static magnetic field $\underline{B}$	π-polarization ($\underline{B}_1 \perp \underline{B}$)	σ-polarization ($\underline{B}_1 \parallel \underline{B}$)
Weak	$\Delta M_F = \pm 1$	$\Delta M_F = 0$
Weak	$\Delta F = 0, \pm 1$	$\Delta F = \pm 1$
Strong	$\Delta M_J = \pm 1$	$\Delta M_J = 0$
Strong	$\Delta M_I = 0$	$\Delta M_I = 0$

labelled 'σ-polarization' may be induced. The relative intensities of the different resonance lines may be obtained from the matrix elements given in Table 5.1 provided that the substitutions $(j,m) \rightarrow (F,M_F)$ and $\underline{p} \rightarrow \underline{\mu}$ are made.

Additional transitions satisfying the selection rules

$$\Delta M_J = 0; \quad \Delta M_I = 0, \pm 1$$

are possible if the interaction of $\underline{B}_1(t)$ with the nuclear magnetic moment is included. However, these will generally be too weak to be observable unless the amplitude of the r.f. field is increased by a factor of the order of $\mu_J/\mu_I \approx 10^3$.

18.2. Investigation of hyperfine structure of ground-state atoms by optical pumping

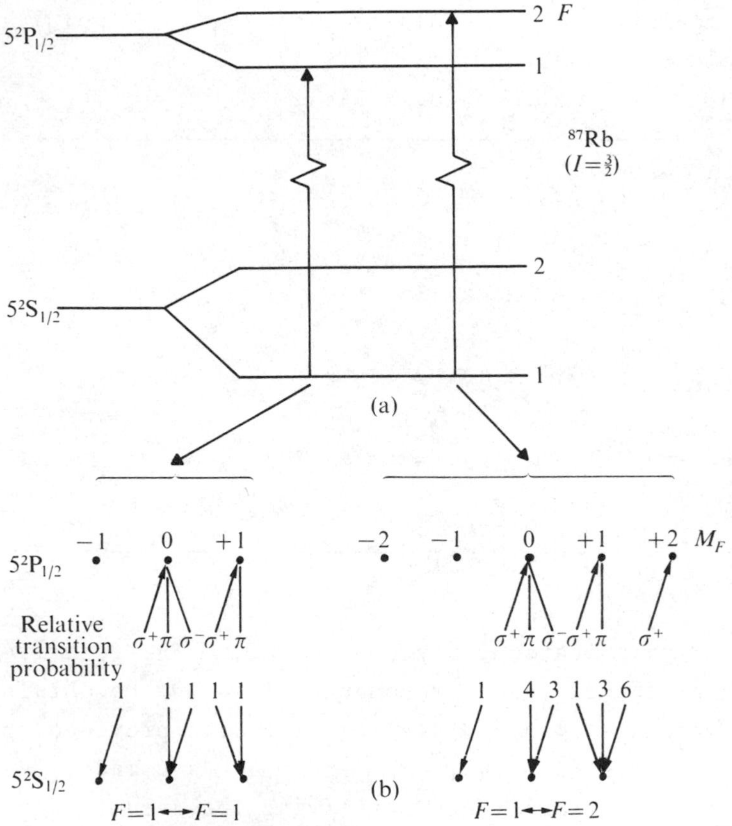

Fig.18.4. Effect of hyperfine structure on the optical pumping cycle.
(a) Hyperfine structure of the $5\ ^2S_{1/2}$ and $5\ ^2P_{1/2}$ levels of ^{87}Rb. (b) Zeeman structure of the transitions excited by σ^+ polarized light from the hyperfine level F=1 of the ground state of ^{87}Rb, $5\ ^2S_{1/2}$.

18.2.1. Effect of hyperfine structure on efficiency of pumping cycle. As an example of the effect of hyperfine structure on the optical pumping process, we consider the case of ^{87}Rb which has a nuclear spin $I = \frac{3}{2}$. The electronic ground state $5\ ^2S_{1/2}$, which we previously assumed to consist of a single level, is in reality split into the two hyperfine levels F=1 and F=2 shown in Fig.18.4(a). The excited state $5\ ^2P_{1/2}$ also consists of two hyperfine levels with the same values of F but separated by a much smaller interval than those of the ground state. In a weak magnetic field, the selection rules for electric dipole transitions between these hyperfine levels of opposite parity are the same as those for the (JM_J) quantum numbers:

$$\Delta F = 0, \pm 1 \quad ; \quad (F=0 \not\leftrightarrow F=0)$$

and

$$\Delta M_F = 0, \pm 1 \qquad (18.23)$$

Consequently when an optical pumping cell filled with ^{87}Rb is illuminated with σ^+ circularly-polarized D_1 radiation from a rubidium resonance lamp, all the transitions shown in Fig.18.4(b) will be excited for those atoms in the F=1 level of the ground state. The transitions for atoms in the F=2 hyperfine ground-state level are even more complex and are not shown for this reason. We see that there are many more states involved than in the case shown in Fig.17.1, where the effect of nuclear spin was ignored, and that the transition probabilities of the different decay routes are also more nearly equal. The net result is that hyperfine structure makes the Zeeman pumping process much less efficient. This point is discussed in more detail in early papers by Hawkins (1955) and Franzen and Emslie (1957).

However, with irradiation by σ^+ polarized light there is still a tendency for atoms to be transferred from states of low values of M_F to states of higher values of M_F, and eventually a steady state is reached in which the populations of the ground-state sub-levels are related by:

$$F=1 \qquad N_1 > N_0 > N_{-1}$$

$$F=2 \qquad N_2 > N_1 > N_0 > N_{-1} > N_{-2} \,.$$

Illumination of the cell with σ^- polarized light reverses the sense of the atomic polarization and pumping with linearly-polarized σ radiation transfers the atomic population symmetrically towards states of higher $|M_F|$ values and results in alignment rather than orientation of the sample.

<u>18.2.2. Zeeman magnetic resonance in ^{85}Rb and ^{87}Rb.</u> These differences of population created between hyperfine Zeeman sub-levels enable magnetic resonance signals to be observed in optical pumping experiments. As an example we now consider the low frequency Zeeman transitions, $\Delta F=0$, $\Delta M_F = \pm 1$, in an optical pumping cell containing a natural mixture of the two rubidium isotopes, ^{85}Rb and ^{87}Rb. The basic properties of these isotopes are listed in Table 18.2 along with those of other alkali atoms required later in this chapter.

TABLE 18.2

Basic atomic parameters of alkali atoms

Atom	Natural abundance (per cent)	I	μ_I (nuclear magnetons)	ν_{HFS} (MHz)
^{6}Li	7·5	1	0·8220	228·2
^{7}Li	92·5	3/2	3·2564	803·5
^{23}Na	100	3/2	2·2176	1772
^{39}K	93·2	3/2	0·3914	461·7
^{41}K	6·8	3/2	0·2148	254·0
^{85}Rb	72·2	5/2	1·3527	3036
^{87}Rb	27·8	3/2	2·7506	6835
^{133}Cs	100	7/2	2·579	9193

(a) Resonances observed in low-resolution experiments. When the optical pumping cell is prepared without a wall coating and contains no buffer gas, the magnetic resonance signals have a line width of 10-20 kHz. The resolution is then relatively low and at fields of the order of 1 G = 10^{-4} T only two resonance signals are observed, as shown schematically in Fig.18.5(a). The reason for this is as follows: in low fields the Zeeman splitting is described by the hyperfine Landé g-factor, equation (18.18), which may be written approximately as

$$g_F \cong g_J \left\{ \frac{F(F+1)+J(J+1)-I(I+1)}{2F(F+1)} \right\}. \qquad (18.24)$$

Since the ground-state hyperfine levels of a particular rubidium isotope are given by $F=I \pm \frac{1}{2}$, we find that equation (18.24) leads to identical g-values, differing in sign:

$$g_F(F=I \pm \tfrac{1}{2}) = \pm g_J/(2I+1). \qquad (18.25)$$

Thus, in this approximation, the magnetic resonance transitions $\Delta M_F = \pm 1$ between the hyperfine levels of a given isotope all occur at the same value of the static field $\underline{B}$ determined by

$$\hbar\omega_0 = g_F \mu_B B \cong \frac{g_J \mu_B B}{2I+1} \qquad (18.26)$$

where $\omega_0/2\pi$ is the frequency of the r.f. field. In this rubidium optical pumping experiment the observation of two well-resolved resonances is just due to the presence in the cell of two isotopes having different values of the nuclear spin I. From equation (18.26) it follows that these low-field Zeeman resonances provide a simple means of determining the nuclear spin I provided that the Landé factor g_J is known, which is usually the case. In a fixed field of 1 G the resonances in rubidium occur at the frequencies $\omega_0/2\pi$ given by

$$\nu_0(^{85}\text{Rb}) = 467{\cdot}081 \text{ kHz}$$

$$\nu_0(^{87}\text{Rb}) = 700{\cdot}621 \text{ kHz}$$

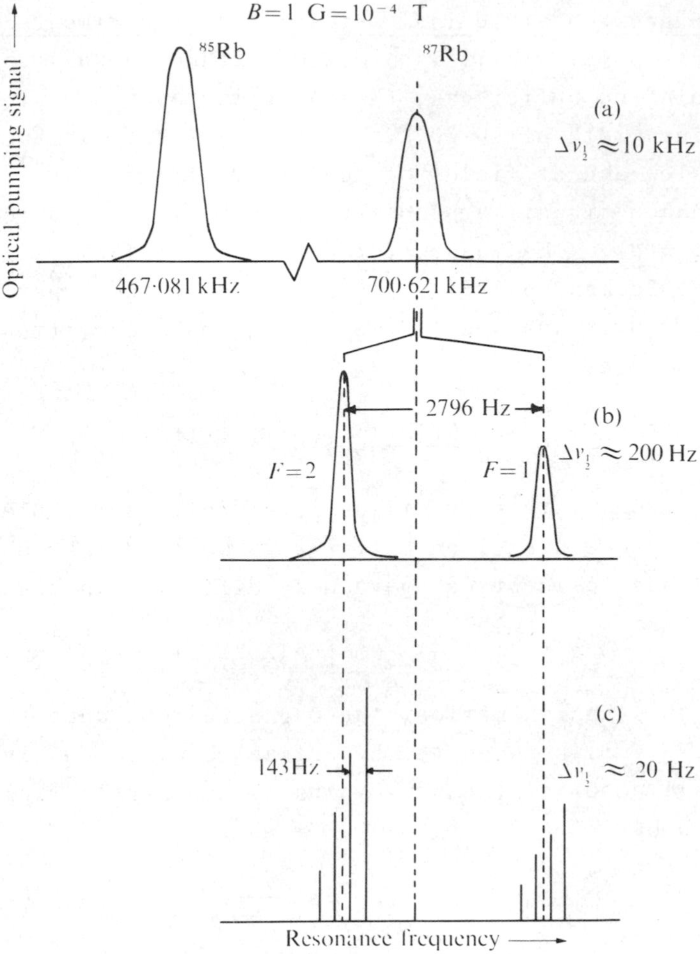

Fig.18.5. Schematic diagram showing the effect of hyperfine structure on optical pumping magnetic resonance experiments in rubidium vapour.
(a) Signal observed with uncoated cell filled with natural isotopic mixture of ^{85}Rb and ^{87}Rb; low-resolution experiment. (b) Signal observed with coated or buffered cells containing ^{87}Rb only; moderate-resolution experiment. (c) Signal observed with coated or buffered cell in a uniform magnetic field; high-resolution experiment on ^{87}Rb.

and it is then a straightforward exercise (Problem 18.6) to show that ^{85}Rb and ^{87}Rb have nuclear spins of I=5/2 and 3/2 respectively.

(b) Resonances observed in medium-resolution experiments. When the width of the magnetic resonance signals is reduced to ≈ 200 Hz by the use of suitably coated or buffered optical pumping cells, it is found that each of the broad resonances discussed in the previous section is resolved into two separate signals, as shown in Fig.18.5(b) for the case of ^{87}Rb. Now we can no longer ignore the term in equation (18.18) which involves the nuclear g-factor, g_I, and we find that the hyperfine levels $F=I \pm \frac{1}{2}$ belonging to a given isotope have slightly different effective g-factors, g_F. In a fixed magnetic field B the resonance frequencies are given by

$$\hbar\omega_0(F=I+\tfrac{1}{2}) = \frac{\mu_B B}{(2I+1)}\left(g_J - g_I 2I \frac{m}{M}\right)$$

and (18.27)

$$\hbar\omega_0(F=I-\tfrac{1}{2}) = \frac{\mu_B B}{(2I+1)}\left\{g_J + g_I 2(I+1)\frac{m}{M}\right\} .$$

In ^{85}Rb and ^{87}Rb these resonances are separated by 824·9 Hz and 2796·1 Hz respectively in a field of 1 G. From these splittings of the low-field Zeeman resonances, approximate values of the nuclear magnetic moments may be determined (Problem 18.7) and the results compared with those given in Table 18.2.

(c) Resonances observed in high-resolution experiments. In optical pumping experiments with rubidium it is possible to reduce the magnetic resonance linewidth still further by careful design of the magnetic field coils and preparation of the optical pumping cell. When $\Delta\nu_{1/2} < 20$ Hz it is found that all the different transitions, $\Delta F=0$, $\Delta M_F = \pm 1$, within a single Zeeman multiplet are well resolved even in fields as low as 1 G, as shown in Fig.18.5(c). We are now observing the very first effects of the decoupling of the nuclear and electronic angular momenta which occurs at high fields. The Zeeman sub-levels for a given total angular momentum quantum number F are no longer equally spaced, as Fig.18.3 clearly demonstrates, and consequently each of the composite resonances observed in the case of moderate linewidth is resolved into 2F separate resonance signals. By expanding the last term in equation (18.20) for the case of small magnetic fields, $g_J\mu_B B \ll A_J$, it can be shown (Problem 18.8) that each resonance is separated from the next by an amount

$$\Delta\nu_0 = \frac{2\nu_{HFS}\, x^2}{(2I+1)^2} = \frac{2\nu_0^2}{\nu_{HFS}}\,. \tag{18.28}$$

In ^{85}Rb and ^{87}Rb $\Delta\nu_0$ has the values of 144 Hz and 143 Hz respectively in a field of 10^{-4} T, and from this information very approximate values of the zero-field hyperfine separation $h\nu_{HFS}$ can be calculated.

The hyperfine separations and the hyperfine structure constants $A_{1/2}$ can be determined much more accurately either by increasing the magnetic field B and following the $\Delta F=0$ resonances into the intermediate field region, or more directly by studying the $\Delta F=\pm 1$ transitions between the different hyperfine levels as we shall now describe.

18.3. Hyperfine pumping and the measurement of ν_{HFS}.

<u>18.3.1. Hyperfine pumping.</u> As we emphasized in Chapter 17, optical pumping was initially developed to produce differences of populations among the Zeeman sub-levels of the ground state. When this state also possesses hyperfine structure, a detailed analysis of the pumping process shows that there is simultaneously a tendency to concentrate atoms in the upper hyperfine level when σ^+ polarized light is used. However, if the atoms are illuminated by a beam of radiation containing all the hyperfine components of the resonance line this hyperfine pumping is very inefficient, and the population differences created between levels with different values of F are always rather small owing to the rapidity of the

competing relaxation processes.

Much larger differences of populations between the different hyperfine levels can be created by suitably altering the relative intensities of the hyperfine components of the pumping lamp. Since these components are very close together conventional interference filters cannot be used, but fortunately in many cases the natural displacement between the resonance lines of different isotopes of an element enables selective light sources to be constructed. This is especially true in mercury, as shown by Table 16.1, and in the case of rubidium which we consider now.

The energy levels and hyperfine structure of D_1 resonance line of rubidium at 7947 Å are shown in Fig.18.6 for the naturally occurring isotopes ^{85}Rb and ^{87}Rb. The hyperfine intervals in the excited state $5\ ^2P_{1/2}$ are small compared with the Doppler width and the resonance line of each isotope consists of only two components labelled A, B for ^{85}Rb and a, b for ^{87}Rb. The relative positions of these components are shown in Fig.18.6(b). Components A and a are nearly coincident and overlap because of Doppler broadening, while components B and b are well resolved. Hyperfine pumping is now possible with a number of different combinations of isotopically separated lamps, filters, and pumping cells.

For instance, if a cell containing ^{87}Rb is illuminated with light from a ^{85}Rb lamp only the component A will be absorbed, pumping atoms from the upper hyperfine level F=2 of the ground state. The population of the hyperfine level F=2 of the ^{87}Rb atoms in the cell will decrease while the population of the lower level F=1 increases.

Alternatively if light from a ^{87}Rb lamp is passed through an absorption cell containing ^{85}Rb, only the b component of the lamp will be transmitted by the filter. When this light is focussed into an optical pumping cell containing ^{87}Rb atoms the b component preferentially pumps the atoms from the F=1 to the F=2 ground-state hyperfine level.

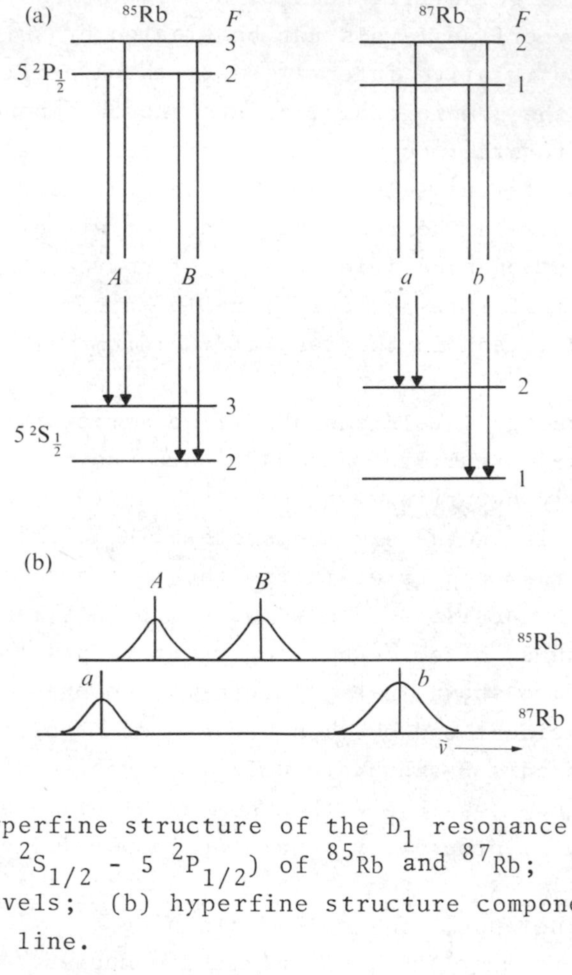

Fig.18.6. Hyperfine structure of the D_1 resonance line ($5\ ^2S_{1/2}$ - $5\ ^2P_{1/2}$) of ^{85}Rb and ^{87}Rb; (a) energy levels; (b) hyperfine structure components of the D_1 line.

In either case the resulting population difference enables the hyperfine magnetic resonance transitions $\Delta F = \pm 1$ to be detected when microwave power at 6835 MHz is fed into the optical pumping cell. Analagous techniques can be used to achieve hyperfine pumping in the mercury isotopes ^{199}Hg and ^{201}Hg as inspection of Table 16.1 reveals.

18.3.2. Measurement of the hyperfine interval $h\nu_{HFS}$. The magnetic resonance transition $\Delta F = \pm 1$, $M_F = 0 \leftrightarrow M_F = 0$ is usually chosen for the measurement of the hyperfine interval

$h\nu_{HFS}$ since this particular transition depends on the magnetic field only through a second-order term which is very small in weak fields. According to the Breit-Rabi formula, equation (18.20), the frequency of this transition is given by (Problem 18.9)

$$\nu_0 = \nu_{HFS}(1 + x^2/2). \qquad (18.29)$$

The frequencies of the nearest other hyperfine transitions $\Delta F=1$, $M_F = 0 \leftrightarrow M_F = \pm 1$ are strongly field dependent:

$$\nu_0 = \nu_{HFS} \pm \frac{g_J \mu_B B}{h(2I+1)} \qquad (18.30)$$

and may be separated from the main (0,0) resonance by applying an external magnetic field of the order of $10^{-2} - 10^{-1}$ G. In fact the frequency of these field-dependent transitions can be used to make an exact evaluation of the quadratic term in equation (18.29), thus leading to a precise measurement of ν_{HFS}. In these experiments the uncertainty in the measured field values leads to a relative error in the determination of the hyperfine interval of $\Delta\nu_0/\nu_{HFS} \approx 10^{-11}$, provided that the effects of the light and pressure shifts discussed below are eliminated.

<u>18.3.3. Pressure shifts of the hyperfine resonance lines.</u> As explained in section 17.4.3 buffer gases are used in optical pumping cells not only to increase the efficiency of the pumping cycle by increasing the relaxation time but also to reduce the Doppler width of the microwave hyperfine transitions. Unfortunately detailed studies of the $\Delta F=0$, $M_F = 0 \leftrightarrow M_F = 0$ transitions have revealed that there are also important shifts of the resonance frequency which are proportional to the buffer gas pressure and which vary with different gases, as shown in Fig.18.7. These shifts are caused by collisions between paramagnetic alkali atoms and the atoms of the diamagnetic buffer gas. Both positive and negative frequency shifts occur because of the competing effects of the repulsive and attractive parts of the interatomic force; the Pauli exchange repulsion tends to

TABLE 18.3

Comparison of measurements of ground-state hyperfine intervals by the optical pumping and atomic beam magnetic resonance techniques

Atom	Nuclear spin I	Hyperfine interval ν_{HFS}: Optical pumping (a) (MHz)	Hyperfine interval ν_{HFS}: Atomic beam (b) (MHz)	Reference
^{1}H	$\frac{1}{2}$	1420·405 749 (6) †	1420·405 73 (5)	(a) Pipkin and Lambert (1962) (b) Kusch (1955)
^{2}H	1	327·384 349 (5)	327·384 302 (30)	(a) Pipkin and Lambert (1962) (b) Kusch (1955)
^{3}H	$\frac{1}{2}$	1516·701 477 (6)	1516·701 70 (7)	(a) Pipkin and Lambert (1962) (b) Prodell and Kusch (1957)
^{6}Li	1	—	228·205 28 (10)	(b) Schlect *et al.* (1962)
^{7}Li	$\frac{3}{2}$	—	803·504 04 (50)	(b) Schlect *et al.* (1962)
^{23}Na	$\frac{3}{2}$	1771·626 2 (1)	1771·631 (2)	(a) Arditi (1960) (b) Logan and Kusch (1951)
^{39}K	$\frac{3}{2}$	461·719 690 (30)	461·719 71 (15)	(a) Bloom and Carr (1960) (b) Kusch and Hughes (1959)

^{40}K	4	—	1285·790 (7)	(b) Eisinger *et al.* (1952)
^{85}Rb	$\frac{5}{2}$	—	3035·732 439 (5)	(b) Penselin *et al.* (1962)
^{87}Rb	$\frac{3}{2}$	6834·682 608 (7)	6834·682 614 (3)	(a) Bender *et al.* (1958) (b) Penselin *et al.* (1962)
^{133}Cs	$\frac{7}{2}$	9192·6320 (5)	9192·631 770 (20)	(a) Arditi and Carver (1958) (b) Markowitz *et al.* (1958)

† The numbers in parentheses indicate the uncertainty in the last digits of each quoted value.

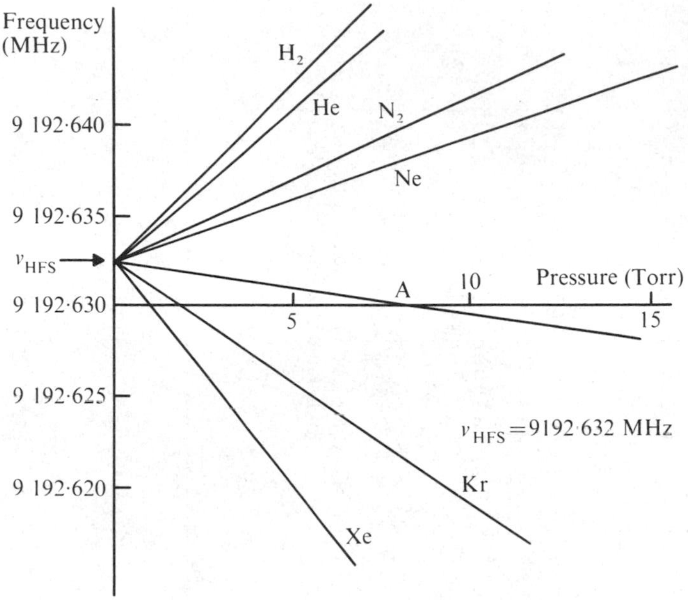

Fig.18.7. Pressure shift of zero-field hyperfine resonance line of ^{133}Cs for different buffer gases. (After Arditi and Carver (1961).)

increase the electron density at the nucleus resulting in an increase in the hyperfine splitting, while the van der Waals attraction tends to reduce the value of $|\psi(0)|^2$ and produces a negative contribution to the observed frequency shift. This latter effect is proportional to the polarizability of the buffer gas molecules and is predominant in Kr and Xe.

Thus, in addition to the extrapolation to zero intensity of the pumping light which must be made to eliminate systematic errors due to the light shifts discussed in section 17.9.7, a further extrapolation of the measured hyperfine resonance frequencies to zero buffer-gas density must be made. Zero-field hyperfine intervals measured by the optical pumping method are listed in Table 18.3, together with a selection of the results obtained by the atomic beam technique. Generally the optical pumping results are more precise than the older atomic beam results. The table also includes values for hydrogen, deuterium, and tritium obtained by the spin-exchange magnetic resonance technique discussed in sections 17.5.2 and 17.7.4. These results are of particular importance because the magnetic interaction constant A_J can be calculated with considerable accuracy in a one-electron system, as we shall discuss later in section 18.8.1. In rubidium the hyperfine frequency is of considerable practical interest since the optical pumping technique permits the design and construction of a simple and compact atomic clock which is described in the following section.

18.4. *Optically pumped rubidium frequency standards*

The rubidium hyperfine resonance fulfills three of the requirements necessary for a frequency standard, namely a linewidth which is narrow compared with the frequency of the line, a transition frequency in the high r.f. range, and a high signal-to-noise ratio. Following early development work by Carpenter *et al.* (1960) and Packard and Swartz (1962), compact optically-pumped frequency standards became available commercially and are now widely used, (Hellwig 1975).

A block diagram of a typical rubidium atomic clock is shown in Fig.18.8. The light from an ^{87}Rb lamp is passed through an ^{85}Rb filter and is used to pump the hyperfine levels of ^{87}Rb atoms in the cell. The cell is enclosed in a microwave cavity which is in turn surrounded by a Mu-metal shield which reduces the earth's magnetic field to ≈ 1 mG. The microwave power required to induce the resonance is obtained by multiplication of the output frequency of a temperature stabilized, voltage tunable 5 MHz quartz oscillator. When the quartz oscillator is adjusted so that the 1368th harmonic of the output frequency is close to 6835 MHz, magnetic dipole transitions are induced between the hyperfine levels of the ^{87}Rb ground state. The optical transmission monitoring technique is used to detect the magnetic resonance and to provide the input to the frequency control servo system. By phase-modulation of the microwave

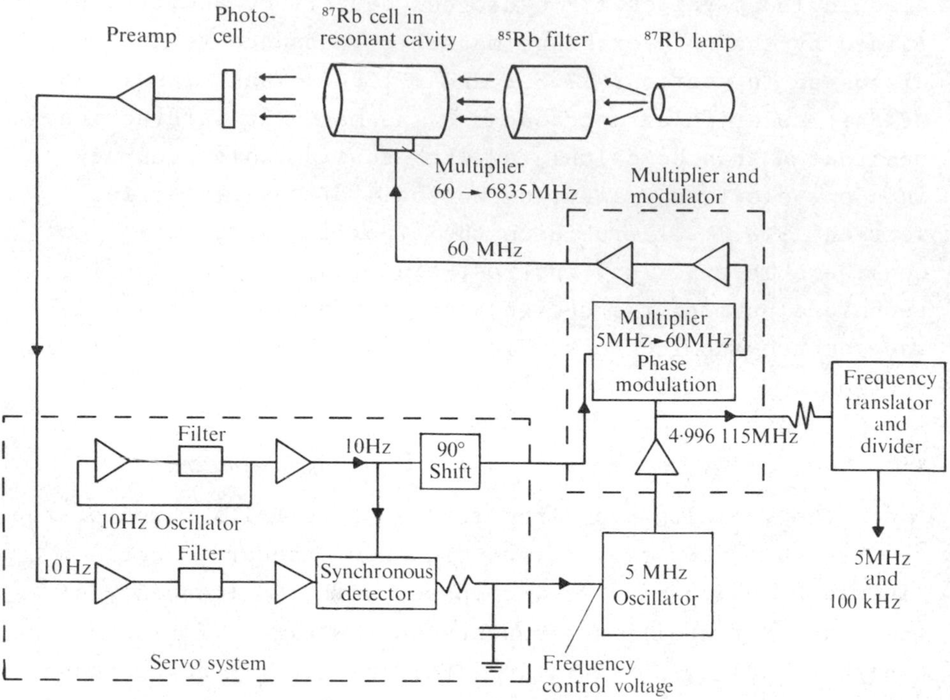

Fig.18.8. Block diagram of optically pumped ^{87}Rb frequency standard. (After Carpenter *et al.* (1960).)

power at a frequency of approximately 20 Hz, an error signal is generated, as shown in Fig.18.9, whose amplitude and phase are used to adjust the frequency of the quartz oscillator. Thus the quartz oscillator is servo-controlled to a sub-harmonic of the atomic rubidium (0,0) hyperfine frequency.

When the whole device is carefully temperature stabilized, a frequency stability, $\Delta\nu/\nu$, of 5×10^{-11} over periods of a year has been demonstrated, and the degree of short-term stability is even higher. However, it should be noted that we quote stability rather than accuracy since the actual operating frequency of each device depends on the composition and pressure of the buffer gas and on the illumination conditions in the optical pumping cell, as explained in sections 17.9.7 and 18.3. Provided that these conditions

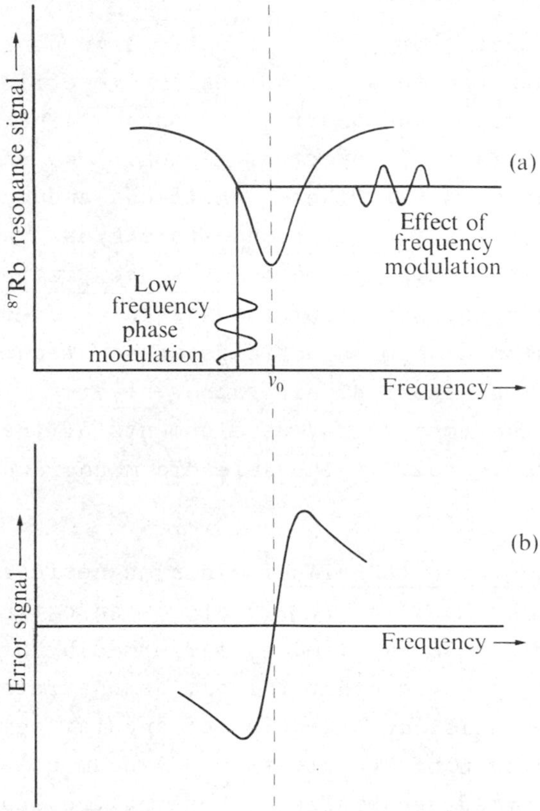

Fig.18.9. (a) Resonance signal showing the effect of low frequency phase-modulation of the microwave power; (b) Amplitude of modulation of ^{87}Rb magnetic resonance signal as a function of microwave frequency. The dispersion-shaped curve provides the necessary frequency-error feed-back signal.

do not change, the frequency stability of the device is assured and it can be used as a compact, light-weight secondary frequency standard or alternatively as an atomic clock. At present the primary frequency standard is based on the atomic beam technique which we study in the following sections.

18.5. The atomic beam magnetic resonance technique

The requirements for efficient optical pumping are that an element must possess a spherically-symmetric electronic ground state and a strong resonance line in the visible or near ultraviolet spectral region. Few easily vaporizable elements satisfy these conditions and consequently the investigation of ground-state hyperfine structures by optical pumping methods has been rather limited. A far greater number of results has been obtained by the powerful atomic beam magnetic resonance technique which we shall now discuss, albeit rather briefly. Fortunately all but the most recent developments of the subject are covered in detail in the splendid monograph by Ramsey (1956).

18.5.1. Principle of the technique. This magnetic resonance method, developed by Rabi *et al.* (1938), is an extension of the early atomic beam experiments performed by Stern and Gerlach which, amongst other things, demonstrated the existence of the spatial quantization of angular momentum and the intrinsic spin of the electron. A beam of atoms or molecules is formed, generally by heating the substance of interest in an oven until the vapour pressure is about 10^{-2} Torr. The atoms effuse through a fine slit and a sequence of collimating apertures into a highly evacuated enclosure, as shown in Fig.18.10. In the main chamber the pressure is of the order of 5×10^{-7} Torr and is sufficiently low that most of the atoms travel the entire length of the apparatus, 0·5 - 2·0 m, without suffering a collision. At the end of the apparatus the beam of atoms is monitored by either a surface ionization detector or a simple mass spectrometer.

The beam is defined by means of a collimating slit S, of the same width as the source and detector slits, which is usually placed in the centre of the apparatus. In the absence of perturbations the full beam intensity will only be detected when the source, collimator, and detector slits

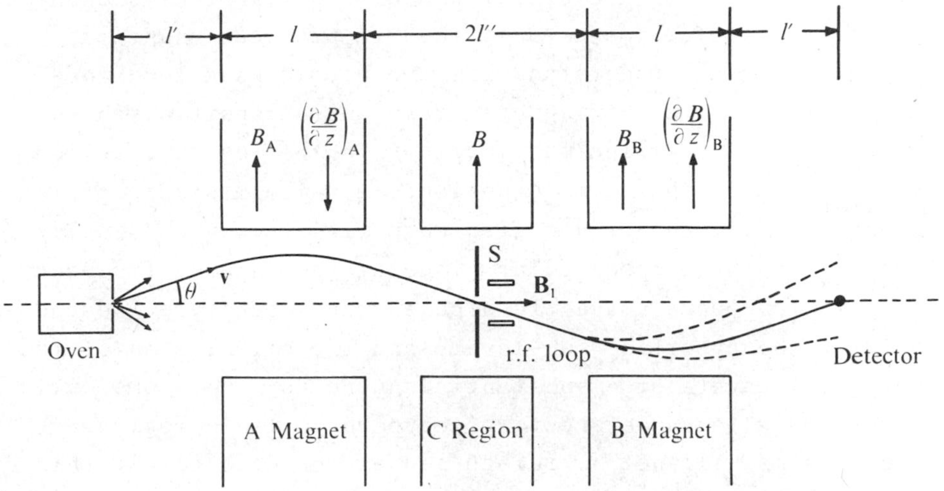

Fig.18.10. Atomic beam magnetic resonance apparatus. The broken curves in the B magnet show the paths of atoms which have undergone a magnetic dipole transition in the field of the C magnet due to the influence of the applied r.f. field.

are in line. However, between the source and the detector the atoms pass through regions of inhomogeneous magnetic fields labelled A and B in Fig.18.10. Atoms which have a permanent magnetic moment will be deflected by the transverse force that they experience in the region A and will only pass through the collimator slit if the trajectory on which they left the oven possesses the correct combination of velocity v and angle of inclination θ to the axis of the apparatus. In the absence of the inhomogeneous field B these atoms would then miss the detector altogether. However, when the field gradient in the B-region is opposite to that in A and is adjusted to have the correct value, the deflected beam will be refocussed at the detector. This refocussing will only occur if the magnitude and orientation of the atomic magnetic moment remain constant throughout the apparatus. The inhomogeneous fields therefore serve the same purpose as the polarizer and analyser in the resonance fluorescence experiments described in previous chapters.

We can now understand how the atomic beam technique may be used to detect magnetic resonance transitions between ground-state hyperfine sub-levels. In the region labelled C in Fig.18.10 the atoms experience a uniform static field $\underline{B}$ and in addition a radio frequency field $\underline{B}_1$ which usually rotates in the perpendicular plane at the angular frequency ω_0. At resonance, magnetic dipole transitions are induced between the hyperfine sub-levels and the orientation of the magnetic moments of a substantial number of the atoms will have been altered before they leave the C-field region. These atoms will not be refocussed at the detector in this so-called 'flop-out' arrangement of the 'polarizer' and 'analyser' fields. Consequently when the detector current is recorded as a function of ω_0 for a fixed value of $\underline{B}$, a series of sharply defined minima will be observed. The width of these magnetic resonance lines is often as narrow as 300 Hz and permits very precise measurements of the spins and magnetic moments of both atoms and nuclei. We shall now discuss some aspects of the atomic beam apparatus in more detail before considering the application of this technique to hyperfine structure measurements in one-electron atoms.

18.5.2. Details of the atomic beam apparatus.

(a) Deflection in inhomogeneous fields. The perturbation experienced by a paramagnetic atom in a magnetic field B, which we discussed in detail in section 18.1, may now be regarded as an effective potential energy, E(B). If the field is non-uniform the atom or molecule will then experience a net force given by

$$\underline{F} = -\nabla E = -\frac{\partial E}{\partial B}\nabla B.$$

The component of this force in the z-direction, transverse to the axis of the apparatus, is

$$F_z = -\frac{\partial E}{\partial B}\cdot\frac{\partial B}{\partial z} = \mu_{eff}\frac{\partial B}{\partial z}\ . \qquad (18.31)$$

This equation has the same form as the force experienced by a constant magnetic dipole moment in a non-uniform field. However, the effective magnetic moment defined by $\mu_{eff} = -\partial E/\partial B$ is generally field dependent. For the case $J=1/2$, using the detailed expression of E(B) given in equation (18.20), we have

$$\mu_{eff} \cong \mp \frac{g_J\mu_B(x + \frac{2M_q}{2I+1})}{2\left\{1 + \frac{4M_q x}{2I+1} + x^2\right\}^{1/2}}\ . \qquad (18.32)$$

The effective magnetic moment is plotted in Fig.18.11

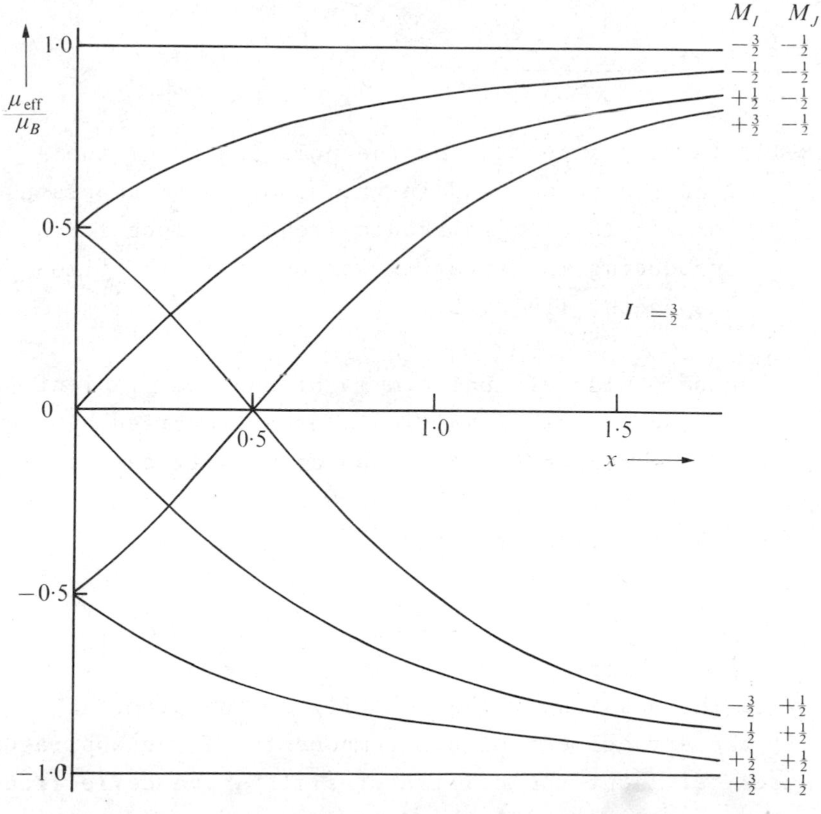

Fig.18.11. Variation of the effective magnetic moment with magnetic field. The nuclear magnetic moment is assumed to be positive.

for the case $I = \frac{3}{2}$ and is clearly the negative slope of the energy levels plotted in Fig.18.3.

The necessary inhomogeneous fields are created by magnets with pole pieces in the shape of arcs of circles of different radii, as shown in Fig.18.12(a). These pole

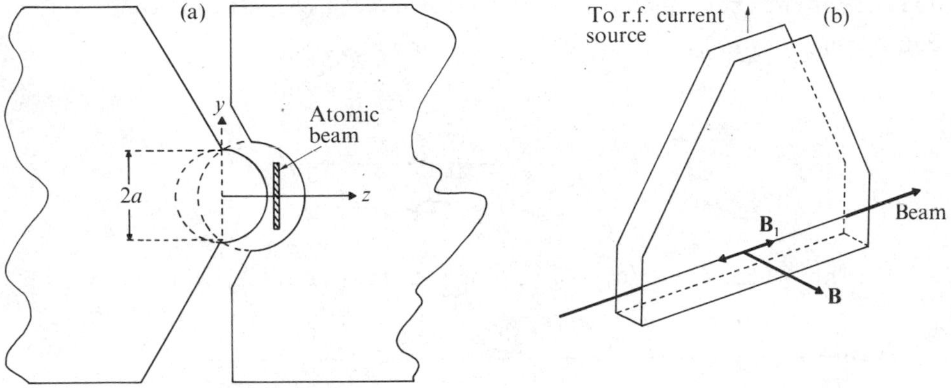

Fig.18.12. (a) Cross section of the pole pieces of the A and B magnets used in the atomic beam apparatus of Fig.18.10. (b) Radio-frequency loop for producing the rotating component of the time-dependent field, $\underline{B}_1$.

pieces produce fields of uniform magnitude and gradient over the rectangular cross-sectional area traversed by the beam. The deflection suffered by an atom as it passes through a non-uniform field of length l is obtained by integration of the equation of motion $M\ddot{z} = F_z$, giving

$$z = \frac{\mu_{eff}\, l^2}{2Mv^2} \cdot \frac{\partial B}{\partial z}$$

where M is the mass and v the velocity of the atom.

If the arrangement of all components of the apparatus is symmetrical about the collimator slit C, the deflection of an atom in the plane of the detector, a distance l' beyond the end of the B-field region, is given by

$$z = \frac{(\mu_{eff})_A \mp (\mu_{eff})_B}{2Mv^2} \cdot \frac{\partial B}{\partial z} l(l + 2l'). \quad (18.33)$$

The negative sign in equation (18.33) refers to the arrangement in which the field gradients in the A and B regions are in opposite directions. In this case all atoms which have passed through the C-field region without making a transition will be refocussed at the detector, independent of their velocities. On the other hand, the gradients in the A and B regions are often arranged to be in the same direction. The deflections in the two magnets now reinforce one another as the positive sign in equation (18.33) demonstrates and, in the absence of r.f. transitions, only a few atoms which travel with very high velocities will reach the detector. This 'flop-in' arrangement of the 'polarizer' and 'analyser' fields has the advantage that the ratio of signal to background is larger than in the 'flop-out' arrangement discussed previously. In a compact apparatus where $l \approx l' \approx 10$ cm, fields of the order of 0·25 T with gradients of 50 T m^{-1} are required to obtain atomic deflections of the order of a millimetre (Problem 18.10).

(b) Detectors. The atoms passing through the final slit are usually detected by an ionization technique which, in the case of the alkalis and other elements of low ionization potential involves surface ionization on a heated tungsten wire about 0·015 mm diameter. An atom such as potassium (ionization potential I = 4·3 eV) striking the wire is adsorbed on to the surface and has a certain probability of releasing its valence electron to the tungsten metal (work function ϕ = 4·5 eV). These atoms are re-evaporated as ions, the ratio of ions to neutral atoms released per unit time being given by

$$N^+/N^0 = \exp\{-e(I - \phi)/kT\}.$$

Provided that the work function of the tungsten wire is not reduced by contamination of the surface, the detector efficiency, $N^+/(N^+ + N^0)$, will be over 80 per cent in the case of potassium. The ions are drawn to a collector electrode held at a negative potential and the detector current is amplified and measured finally with a microammeter or galvanometer.

For atoms with high ionization potentials the atomic beam is usually ionized by electron bombardment and the ions so formed are accelerated into a simple magnetic-deflection mass spectrometer. Selective chopping of the atomic beam with a mechanical shutter enables the ion signal of interest to be separated from the mass spectrum of ions formed from the background gas in the apparatus. The ions are usually counted individually using an electron multiplier, the first dynode of which produces secondary electrons when bombarded with energetic ions. The number of atoms collected per unit time is determined by the solid angle subtended by the detector at the source slit and rarely exceeds $5 \times 10^9\ s^{-1}$.

<u>(c) Radio-frequency fields and the resonance line shape.</u> In the C-region the atoms experience a uniform field $\underline{B}$ in the z-direction and a radio-frequency field of amplitude B_1 which, for the moment, we assume to be rotating in the x-y plane at the angular frequency ω_0. This field is usually created by feeding radio-frequency power into a conductor bent in the shape of a U, as shown in Fig.18.12(b). An oscillating magnetic field $2B_1\cos\omega_0 t$ is produced in the x-y plane which can be decomposed into two counter-rotating fields each of amplitude B_1.

The observed signal is proportional to the magnetic dipole transition probability, equation (16.18), which we write as

$$P_{M_q M'_q}(\omega_0) = \frac{\omega_1^2}{\omega_1^2 + (\omega_L - \omega_0)^2} \sin^2[\{(\omega_L - \omega_0)^2 + \omega_1^2\}^{1/2} \frac{\tau}{2}] \tag{18.34}$$

where $\omega_1 = \gamma B_1$ and $\tau = \Delta l/v$ is the time that the atoms spend in the r.f. field whose effective length is Δl. Here $\omega_L/2\pi$ is to be interpreted as the frequency of any one of the allowed transitions between the hyperfine Zeeman sub-levels.

The resonance is a symmetrical bell-shaped curve which peaks at the resonance frequency $\omega_0=\omega_L$. The amplitude of the peak signal is a maximum when the magnitude of the r.f. field satisfies the condition $\omega_1\tau = \pi$, corresponding to a transition probability of unity for a mono-energetic beam of atoms.

As the oscillator frequency is tuned off-resonance the signal drops and, in the limit of low r.f. field strengths $\omega_1 \ll (\omega_L-\omega_0)$, the intensity is zero when $(\omega_L-\omega_0)\tau = \pm 2\pi$. For a mono-energetic beam the half-width of the resonance is given by $\Delta\omega_{1/2} \approx 2\pi v/\Delta l$, in accord with the uncertainty principle, although when equation (18.34) is correctly averaged over the velocity distribution of atoms in the beam, the full width at half the peak intensity becomes

$$\Delta\nu_{1/2} = \Delta\omega_{1/2}/2\pi = 1{\cdot}072\ v_\alpha/\Delta l \qquad (18.35)$$

where

$$v_\alpha = (2kT/M)^{1/2}$$

is the most probable velocity of atoms in the beam source. In a simple apparatus the effective length of the r.f. field would be about 3 cm, leading to a resonance linewidth of approximately 15 kHz (Problem 18.11).

When the length of the r.f. loop is increased it is usually found that the reduction in the width of the resonance is limited by inhomogeneities in the static magnetic field in the C-region. There are great practical difficulties in producing a very homogeneous field but these are avoided in the technique, introduced by Ramsey (1949), in which two separate oscillating fields are used at either end of an extended C-magnet. The resonance line shape, shown in Fig.18.13, then has the appearance of a double-slit diffraction pattern in which the centre frequency $\omega_0/2\pi$ corresponds to the average precessional frequency of the atoms between the two Ramsey r.f. loops. The width of the narrow central peak of the resonance is determined by the separation L of the loops, while the width of the overall

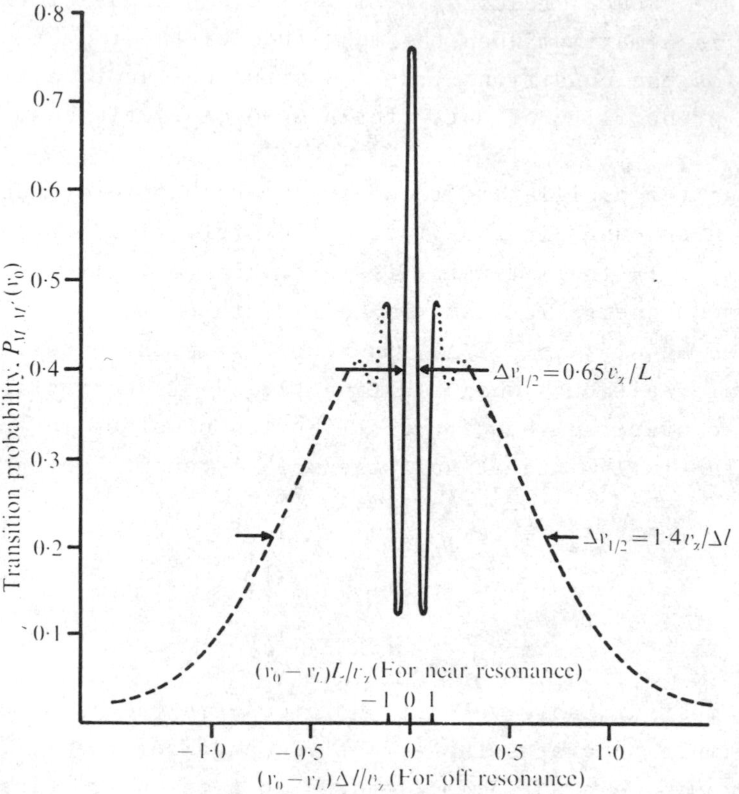

Fig.18.13. Transition probability as a function of frequency for the Ramsey double-loop technique. The full line is calculated near resonance, the dashed line is calculated off resonance and the dotted line is interpolated between these. (After Ramsey (1956).)

signal is determined by the time $v/\Delta l$ spent in just one of the r.f. field regions.

(d) Advantages of the atomic beam method. The atomic beam magnetic resonance technique has several important advantages. The atoms are essentially 'free' and are unperturbed by collisions with other atoms, either in the beam or in the background gas. The method has a high sensitivity and tiny quantities of material are quite sufficient for most experiments. The method is applicable to most systems from highly refractory materials like tungsten to radioactive isotopes with half lives as short as 10 min. The technique is also applicable to molecules and radio or microwave electric dipole transitions may be studied in addition to the magnetic dipole transitions emphasized in this chapter. Finally the precision of the method, $\nu_0/\Delta\nu_{\frac{1}{2}}$, is intrinsically very high, especially when the Ramsey double-loop technique is used to induce the resonance.

18.6. *Hyperfine structure investigations by the atomic beam technique*

The design of the A and B deflecting fields of a given atomic beam apparatus determines which of the magnetic dipole transitions occurring in the C-field region are actually observable. If the field strength in the inhomogeneous regions is so high that $\underline{I}$ and $\underline{J}$ are decoupled, then all states with the same value of M_J effectively have the same value of μ_{eff}. The beam at the collimator slit consists of (2J+1) components corresponding to the allowed orientations of the atomic magnetic moment $\underline{\mu}_J$, each component propagating at a slightly different angle to the beam axis. In this situation only those transitions which involve a change of M_J can be observed. Thus for potassium-39, $I = \frac{3}{2}$, in low C-field only one low-frequency transition

$$\Delta F=0, \quad (F=2,\ M_F= -2 \leftrightarrow F=2,\ M_F= -1)$$

is observable, while of the high-frequency transitions, that corresponding to

$$\Delta F=1, \quad (F=2,\ M_F= -2 \leftrightarrow F=1,\ M_F= -1)$$

will be absent. The observed radio-frequency spectrum is consequently simplified, although there is no overall loss of information.

However, if the beam intensity and detector efficiency are high, the deflecting magnets of the apparatus can be made longer and operated at lower fields corresponding to the intermediate coupling situation. It is then apparent from equation (18.32) and Fig.18.3 that any one of the allowed transitions can be detected. Whatever the situation in the A and B magnets, it should be noted that the magnetic dipole transitions in the C-region can be induced over a wide range of values of $\underline{B}$, and for the sake of conciseness we restrict ourselves to the particular cases of low and high fields.

18.6.1. Measurements at low fields, $g_J\mu_B B \ll A_J$. The measurement of the frequencies of the $\Delta F=0$, $\Delta M_F= \pm 1$ transitions in low magnetic fields leads to an unambiguous determination of the nuclear spin I, as equation (18.24) demonstrates. Then, once the nuclear spin is known, the electronic Landé factor, g_J, can be determined to high precision. In the early atomic beam experiments there was, however, little interest in accurate measurements of g_J since it was presumed that, for the $^2S_{1/2}$ ground states of the atoms used in most experiments, the magnetic moment would have the Dirac value of exactly one Bohr magneton. Later, however, the measurements of Kusch and Foley (1947, 1948) showed that the electron spin g-factor is not exactly 2 but has the experimentally determined value of

$$g_s = 2 \times (1{\cdot}00119 \pm 0{\cdot}00005).$$

Relativistic quantum-electrodynamic calculations then showed that the anomalous electron g-factor is modified by the vacuum fluctuations of the radiation field and should be given by

$$g_s = 2 \times (1 + \alpha/2\pi - 0{\cdot}328\, \alpha^2/\pi^2 + \ldots\ldots). \quad (18.36)$$

Thus atomic beam experiments provide important information about the value of the fine-structure constant α.

When the high-frequency transitions $\Delta F= \pm 1$, $\Delta M_F= \pm 1$ are induced in a weak magnetic field, the Zeeman pattern of the hyperfine multiplet has the form shown in Fig.18.14. In this

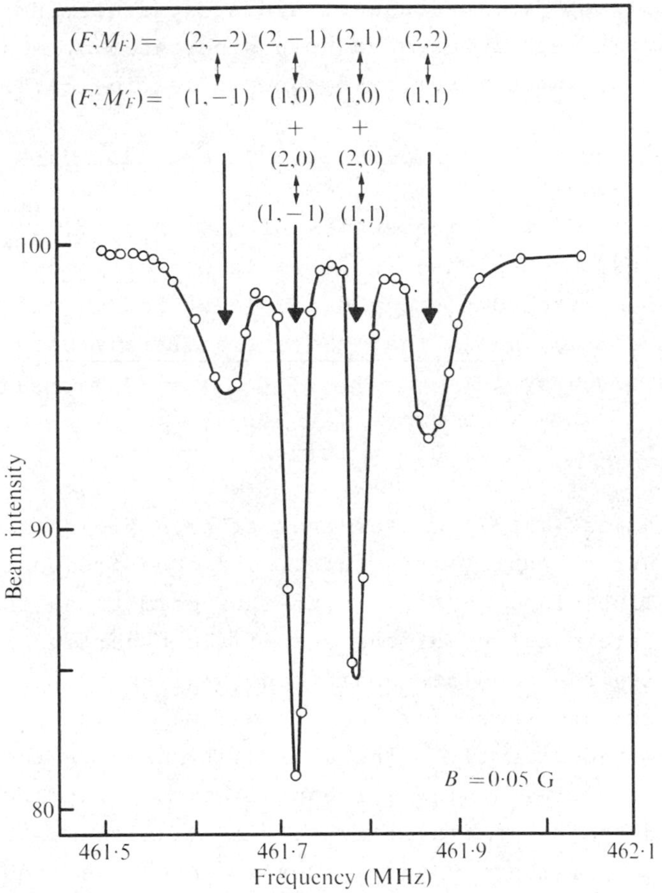

Fig.18.14. The Zeeman pattern of the line $|\Delta F| = 1$ of ^{39}K observed for very low amplitude of the r.f. field. Only π-polarized components of the line are observed. (After Kusch *et al.* (1940).)

experiment all possible π-polarized components of the line are observable although two pairs of transitions are not in fact resolved. From measurements of this type the hyperfine separation, ν_{HFS}, in zero magnetic field can be calculated with the aid of equations (18.14) and (18.17). A selection of the many values of ν_{HFS} obtained by the atomic beam technique is included in Table 18.3. There is excellent agreement with the results obtained by the optical pumping method although in the case of hydrogen, deuterium, and tritium both techniques have been superceded by measurements using the hydrogen maser described in section 18.8.2.

18.6.2. Measurements at high fields, $g_J\mu_B B \gg A_J$. Hyperfine structure separations, ν_{HFS}, can also be obtained from measurements of the frequencies of the $\Delta M_I = \pm 1$, $\Delta M_J = 0$ transitions induced when the C-field is sufficiently strong that $\underline{I}$ and $\underline{J}$ are completely decoupled. For an alkali atom equation (18.19) shows that these transitions occur at frequencies given by

$$\nu_{M_I \pm 1, M_I} = \frac{\nu_{HFS}}{2I+1} \mp \frac{g_I \mu_n B}{h} . \qquad (18.37)$$

From equation (18.37) we see that at high fields the resonance has a doublet structure. The mean frequency of the two lines is $\nu_{HFS}/(2I+1)$ and thus permits an extremely precise determination of the zero-field hyperfine interval. In addition the doublet separation is given by $2g_I\mu_n B/h$ and an accurate value for the nuclear magnetic moment may be determined directly. These transitions have the advantage of occurring at lower frequencies than the $\Delta F = \pm 1$ low-field hyperfine transitions, a fact which was important in early experiments when microwave sources were not easily available. The application of the atomic beam technique to measurements of the hyperfine structure of atoms is illustrated in more detail in Problems 18.12 - 18.15. The method is not restricted to ground-state atoms: a considerable number of experiments have been performed on the metastable states of elements of Group IIB. Moreover, by illuminating the C-field region with light from a resonance lamp, it is possible to detect radio-frequency transitions between the hyperfine levels of excited atoms (Senitzki and Rabi 1956). This is a technique which is likely to increase in importance with the further development of high-intensity C.W. tunable dye lasers. However, we turn now to an important practical application of the atomic beam technique: the development of our present standard of time and frequency.

18.7. Cesium beam atomic clock

18.7.1. Time interval standards. The familiar, though expensive, electronic wrist watches are clearly time interval standards since they are read in seconds, minutes, and hours. Yet at the heart of the watch is a crystal-controlled oscillator operating at a frequency of ≈ 1 MHz; so the lucky owner also possesses a personal frequency standard of quite respectable quality. Having made the point that time and frequency standards are basically identical, we recall that for many centuries our methods of determining time intervals have relied on sub-division of a basic standard: the length of the day as measured on earth (see Table 18.4). In the

TABLE 18.4

Definitions of time standards

Era of use	Time system	Definition of the second
-1956	Universal (UT) (earth's rotation on axis)	1/86 400 of the mean solar day
1956-1964	Ephemeris (ET) (earth's orbit about sun)	1/31 556 925·9747 of the tropical year 1900
1964-?	Atomic (AT) (cesium atomic beam clock)	9 192 631 770 periods of cesium hyperfine oscillation in zero magnetic field
Future	(hydrogen maser) ?	1 420 405 751·786 periods of hydrogen hyperfine oscillation in zero magnetic field

1950's, however, the comparison of careful astronomical observations with quartz crystal clocks revealed that there were annual variations of the rate of the earth's rotation on its axis amounting to about 1 part in 10^8. Similar measurements showed that the yearly rotation of the earth about the sun was considerably more stable and in 1956 Ephemeris Time was adopted as the absolute standard. Clearly to obtain Ephemeris Time requires a long series of tedious astronomical observations and is unsatisfactory for laboratory experiments. Therefore a standard is required which does not suffer from the disadvantages of quartz crystal clocks which drift with age and are sensitive to ambient temperature variations.

The precision of the atomic beam magnetic resonance method naturally led to the proposal that it could be used as the basis of an atomic clock. Of the many available elements, cesium was chosen because it combines a hyperfine transition of very high frequency with simplicity of experimental details: an atomic beam can be obtained from a cesium oven at only 200°C, the high atomic mass gives a relatively low thermal velocity of $2{\cdot}5 \times 10^4$ cm s^{-1}, and the beam is easily detected by the surface ionization method.

18.7.2. Cesium beam frequency standard. The only naturally-occurring isotope ^{133}Cs has a nuclear spin of $I = \frac{7}{2}$ and two ground-state hyperfine levels F=4 and F=3 which are separated by an interval of approximately 9192 MHz. When this microwave transition is excited in a beam apparatus in which the r.f. loops are separated by 50 cm, the width of the centre peak of the Ramsey resonance pattern is 325 Hz. The Q-value of the resonance $Q = \nu_{HFS}/\Delta\nu \approx 3 \times 10^7$ is therefore very high. Moreover the precision can be made substantially greater than that indicated by the Q-value alone since with a typical signal-to-noise ratio of 10^4, the centre of the resonance can be located to a small fraction of its width.

A schematic diagram of the cesium beam frequency standard is given in Fig.18.15. As in the rubidium fre-

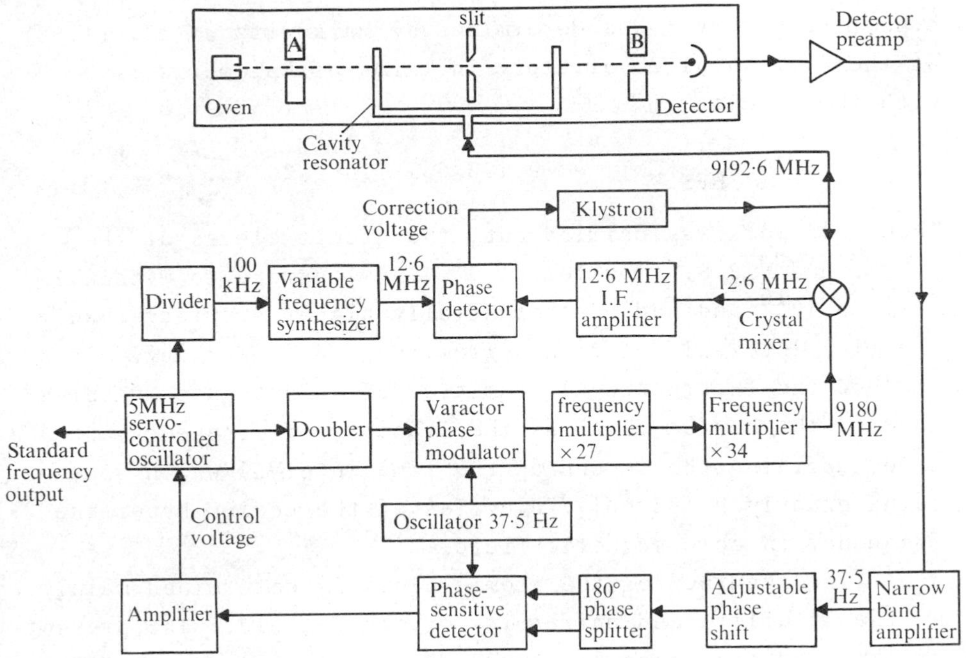

Fig.18.15. Schematic diagram of cesium atomic beam frequency standard.

quency standard discussed in section 18.4, a quartz crystal oscillator is servo-controlled to a sub-harmonic of the frequency of the $(F=4,\ M_F=0) \leftrightarrow (F=3,\ M_F=0)$ hyperfine transition. The frequency of this transition has only a second-order dependence on the strength of the magnetic field in the C-region, as equation (18.29) shows, and for ^{133}Cs we have

$$\nu = \nu_{HFS} + 426\ B^2. \qquad (18.38)$$

A field of approximately 0·050 G is used in the C-region to separate the (0,0) resonance from the other field dependent hyperfine transitions. This field need only be maintained constant to 1 per cent for the frequency to be stable to 1 part in 10^{11}. Following pioneering work by Essen and Parry (1957) at the N.P.L., the frequency of the cesium hyperfine interval was determined by Markowitz *et al.* (1958) at the U.S. Naval Observatory in terms of Ephemeris Time with the result

$$\nu_{HFS}(Cs) = 9\ 192\ 631\ 770 \pm 20\ \text{Hz}.$$

When this work was carried out, the atomic clocks at the N.P.L. and N.B.S. (Mockler *et al.* 1960) agreed to within 1 part in 10^{10} and had a substantially higher accuracy than the time interval determined from the mean solar day. Thus in 1964 the International Committee of Weights and Measures adopted the cesium clock as the standard of frequency and time, defining the second as the time interval which contains exactly 9 192 631 770 cycles of the cesium hyperfine frequency in zero magnetic field.

The accuracy of the atomic clock is determined mainly by the stability and uniformity of the C-field. The present cesium beam frequency standards are stable to ±2 parts in 10^{12} per day and are reproducible to better than 1 part in 10^{11}. However, they may soon be replaced by the hydrogen maser discussed in section 18.8.2.

18.8. Hyperfine structure of atomic hydrogen

<u>18.8.1. Theoretical analysis.</u> Precise theoretical calculations of atomic hyperfine structures are possible only in hydrogenic systems, for example the $1\ ^2S_{1/2}$ ground states of hydrogen, deuterium, and tritium. In the ^2S state the electron wavefunction is spherically symmetric and consequently the internal magnetic field at the nucleus, $\underline{B}_{el}(0)$, arises solely from the spin magnetic moment of the electron. In calculating $\underline{B}_{el}(0)$ we note, however, that the magnetic field at the centre of a uniformly magnetized spherical shell is zero, so that the whole contribution to $\underline{B}_{el}(0)$ comes from the electron magnetization. The magnetization or spin magnetic moment per unit volume is non-vanishing at the nucleus and is given by

$$\underline{M}(0) = \underline{\mu}_s\ |\psi(0)|^2 \qquad (18.39)$$

where $|\psi(0)|^2$ is the electron probability density at the nucleus, r=0. For a uniformly magnetized sphere we have, from classical magnetostatics,

$$\underline{B}_{el}(0) = \frac{2}{3}\mu_0 \underline{M}(0) = \frac{2}{3}\mu_0 \underline{\mu}_s |\psi(0)|^2 . \qquad (18.40)$$

In this field the point nuclear magnetic dipole moment $\underline{\mu}_I$ acquires an additional energy of orientation determined by the interaction

$$\mathcal{H}_1 = -\frac{2}{3}\mu_0 \underline{\mu}_I \cdot \underline{\mu}_s |\psi(0)|^2 = A_J \underline{I}.\underline{J} \qquad (18.41)$$

and the magnetic hyperfine constant for the $^2S_{1/2}$ states of hydrogenic systems is given by

$$A_J = \frac{2}{3}\mu_0 g_I g_s \mu_n \mu_B |\psi(0)|^2 . \qquad (18.42)$$

In atomic hydrogen the nucleus is a single proton with spin $I = \frac{1}{2}$ and the ground-state hyperfine structure consists of the levels F=1 and F=0 separated by the frequency interval $\nu_{HFS} = A_{1/2}/h$. In the 1 $^2S_{1/2}$ state the electron density at the nucleus is given by

$$|\psi(0)|^2 = 1/(\pi a_0^3)$$

where a_0 is the Bohr radius. Thus after substituting for the proton magnetic moment $\mu_I = 2{\cdot}792\ 75\ \mu_n$ and including some small relativistic and reduced mass corrections, the best theoretical value of ν_{HFS} obtained using the value g_s=2 is

$$\nu_{HFS}(H) = 1418{\cdot}900 \pm 0{\cdot}003 \text{ MHz}. \qquad (18.43)$$

This differs considerably from the experimental value obtained by Kusch (1955) using the atomic beam magnetic resonance technique:

$$\nu_{HFS}(H) = 1420{\cdot}405\ 73 \pm 0{\cdot}000\ 05 \text{ MHz}. \qquad (18.44)$$

This discrepancy, first discovered in 1947-8, led to the

suggestion that the electron spin g-factor is greater than $g_s=2$. The inclusion of the anomalous value for the electron magnetic moment given by equation (18.36) greatly improves the agreement between the theoretical and experimental hyperfine intervals. Indeed the agreement in 1956 was better than the accuracy with which the fine-structure constant α was then known and consequently the experimental results were combined with the theoretical analysis to provide an experimental determination of the fine-structure constant. However, these measurements, carried out in the 1950s, have now been superseded by more recent experiments with the hydrogen maser.

18.8.2. The hydrogen maser. The hyperfine structure of atomic hydrogen is shown as a function of the external magnetic field strength in Fig.18.16. When we study this diagram and recall our discussion of the emission and absorption of radiation in Chapter 9, it becomes apparent that

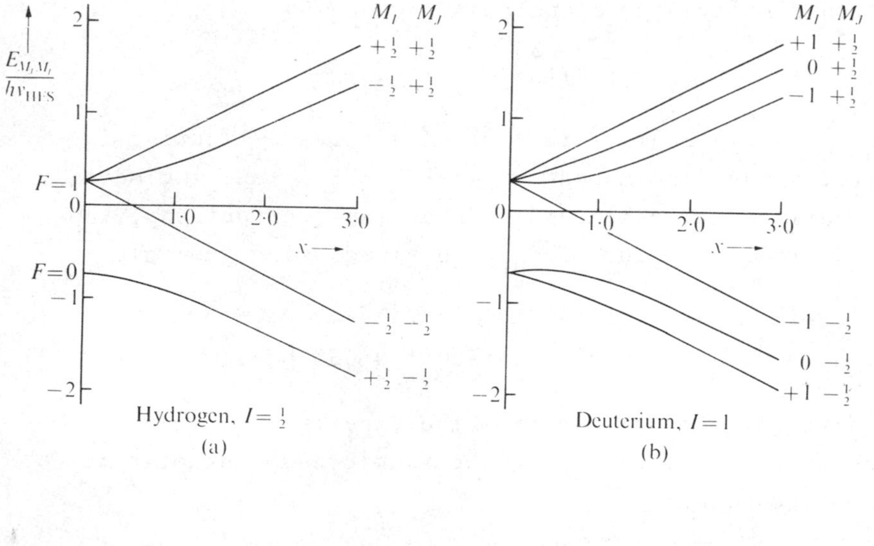

Fig.18.16. The field dependence of the Zeeman sub-levels arising from the hyperfine structure of the $1\,^2S_{1/2}$ ground state for (a) atomic hydrogen, $I = \frac{1}{2}$, and (b) atomic deuterium, I=1.

microwave radiation at the hyperfine frequency could be amplified by stimulated emission if some means could be devised for preparing a sample of hydrogen atoms in which the F=1 level was more highly populated than the F=0 level. Experiments to this end were initiated by Ramsey and his co-workers and successful maser oscillation at the hydrogen hyperfine frequency of approximately 1420 MHz was first reported by Goldenberg *et al.*(1960). A schematic drawing of a typical hydrogen maser is shown in Fig.18.17(a).

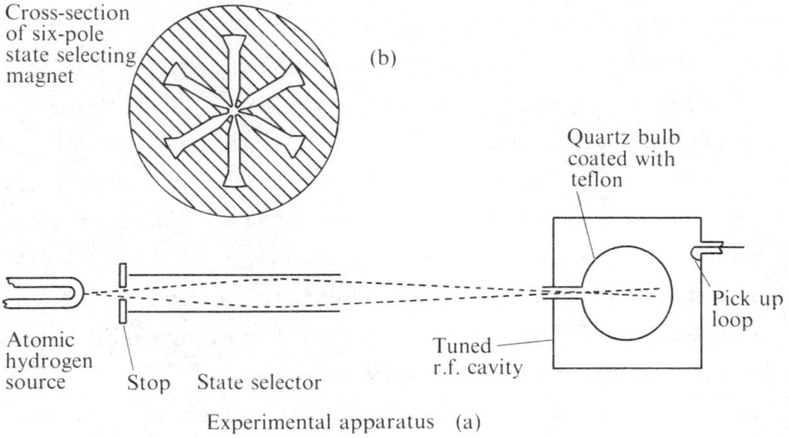

Fig.18.17. Schematic diagram of atomic hydrogen maser. (After Goldenberg *et al.*(1960).).

Molecular hydrogen at 0·2 - 0·02 Torr pressure is dissociated in the source by an r.f. discharge and a beam of atomic hydrogen emerges into the high-vacuum part of the apparatus. The atoms then pass through a six-pole state selecting magnet whose cross-section is shown in Fig.18.17(b). The magnetic field on the axis of the state selector is zero but it increases rapidly off-axis. Thus if an atom enters the state selector in the F=0 state its energy decreases as the atom moves further from the axis, and from equation (18.31) and Fig.18.16 we see that these atoms will be defocussed. On the other hand, those atoms which enter the magnet in the $M_F = +1$ and 0 states belonging to the F=1 level will be spatially focussed on to a small hole in a six inch diameter, Teflon-coated quartz storage bulb. The atoms make random collisions with the walls of the bulb but preserve their orientation, in the absence of microwave radiation, until they eventually effuse out through the entrance aperture.

The storage bulb is surrounded by a microwave cavity tuned to 1420 MHz and a set of Mu-metal shields which reduce the field seen by the atoms to a few milligauss. Owing to the focussing action of the state selector, the bulb is filled predominantly with atoms in the F=1 level and a population inversion is achieved. Consequently the system can be used as a low-noise maser amplifier. However, when the flux of hydrogen atoms into the bulb exceeds about 10^{12} s^{-1}, the cavity losses can be overcome and the device breaks into steady state oscillation at the hydrogen hyperfine frequency. Although the output power of 10^{-12} W is very low compared with laser oscillators in the visible and infrared regions of the spectrum, the frequency and amplitude of oscillation are extremely stable.

The atoms are stored in the bulb for about 1 s and consequently the linewidth, calculated from the uncertainty principle, is narrower by a factor of 10^3 than that achieved in a conventional atomic beam apparatus. This accounts for the great frequency stability of the oscillator: the frequencies of two hydrogen masers have been observed to be stable to 1 part in 10^{13} over a 12-hour period. Unfortunately, however, during the storage time the atom undergoes $\approx 10^5$ collisions with the walls of the bulb. Although the atoms preserve their orientation these collisions cause slight phase changes and result in a shift of the maser oscillation frequency of about 2 parts in 10^{11}. To correct for this systematic error, it is necessary to compare the output frequencies of two masers with bulbs of different sizes and extrapolate the frequency to a bulb of infinite diameter. When this wall shift correction is understood more completely hydrogen masers and saturated absorption stabilized lasers must be considered as possible new frequency standards.

18.8.3. Hyperfine separations in hydrogen and its isotopes. In one of the first applications of the atomic hydrogen maser the hyperfine separations of hydrogen and its isotopes were remeasured to greatly increased accuracy. The results obtained were as follows:

$$\left.\begin{aligned} \nu_{HFS}(H) &= 1420\,405\,751{\cdot}786 \pm 0{\cdot}028 \text{ Hz} \\ \nu_{HFS}(D) &= 327\,384\,352{\cdot}3 \pm 2{\cdot}5 \text{ Hz} \\ \nu_{HFS}(T) &= 1516\,701\,470{\cdot}7938 \pm 0{\cdot}0071 \text{ Hz} \end{aligned}\right\} \quad (18.45)$$

The measurement in the case of hydrogen is based on the international definition of the second in terms of the hyperfine separation in atomic cesium, as discussed in section 18.7, while the results for deuterium and tritium are measured in terms of the hydrogen hyperfine frequency. From equations (18.36) and (18.42) the hyperfine separation for the hydrogen atom is given theoretically by

$$\nu_{HFS}(H) = \frac{16}{3}\left(1 + \frac{m}{M}\right)^{-3} \frac{\mu_p}{\mu_B}\, \alpha^2\, cR_\infty \left\{1 + \frac{\alpha}{2\pi} - 0{\cdot}328\,\frac{\alpha^2}{\pi^2}\right\} \quad (18.46)$$

where μ_p/μ_B is the proton magnetic moment expressed in Bohr magnetons, α is the fine structure constant, R_∞ is the Rydberg constant, and the factor $(1 + m/M)^{-3}$ takes into account the reduced mass of the electron in the hydrogen atom. There are some small additional terms in α^2 and α^3 which have been omitted from equation (18.46). Of the atomic constants appearing in equation (18.46) the least well known is the fine-structure constant α. Consequently from the measured hyperfine frequency given in equation (18.45) a precise value for α may be obtained:

$$\alpha^{-1} = 137{\cdot}036\,6 \pm 0{\cdot}000\,9. \quad (18.47)$$

An analysis of the fundamental constants by Cohen and

Du Mond (1965) showed that this value of α differed significantly from that derived from measurements of the Lamb shift and fine structure in hydrogen. However, subsequent improvements in these fine-structure measurements have confirmed the value of α given in equation (18.47) and the theoretical interpretation of the hyperfine structure of atomic hydrogen.

18.9. Investigation of the hyperfine structure of excited states

18.9.1. Optical double-resonance experiments. So far in this chapter we have concentrated on the measurements of the hyperfine structure of the ground states of hydrogenic and alkali atoms. Owing to the spherical distribution of electron charge in these states, we obtain no information about the nuclear electric quadrupole moment. It is principally for this reason that we now turn to a consideration of the excited levels of atoms. The hyperfine structure of excited states is generally considerably smaller than that of the ground states and consequently conventional high-resolution optical spectroscopy is of limited usefulness. Obviously we must again employ a technique in which the Doppler broadening is negligible or completely eliminated. The optical double-resonance method, which was described in detail in Chapter 16, satisfies this requirement and is the first of the techniques we shall now consider.

A typical experimental arrangement for the investigation of the excited states of the alkali metals is shown in Fig.18.18. The light source is an electrodeless, high-frequency discharge in a vapour of the relevant alkali metal, the pressure of which is controlled by cooling one end of the source by an oil circulation system. By the use of suitable optical filters and a photocell, the intensity of light in the resonance lines is monitored and the signal used to electronically stabilize the fluctuating intensity of the source. Light from the source is polarized

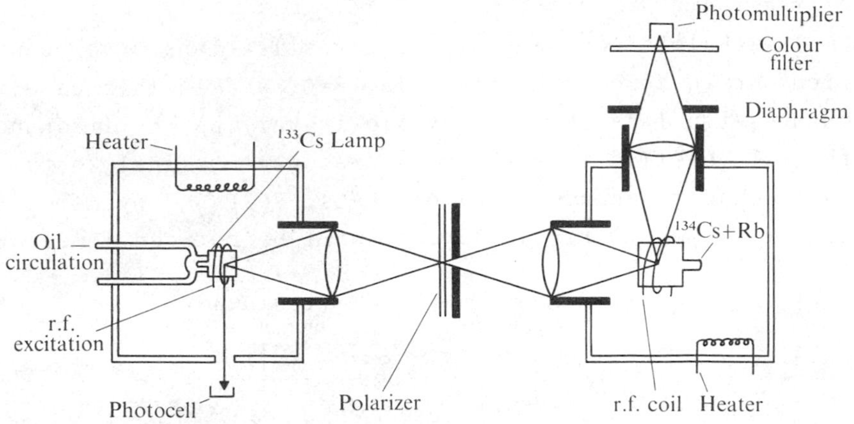

Fig.18.18. Schematic diagram of the apparatus for optical double-resonance experiments in alkali metal vapours. (After zu Putlitz *et al.* (1965).)

and focussed into a resonance cell containing the alkali vapour of interest where the absorption of the polarized resonance radiation creates the non-statistical distribution of population among the hyperfine levels of the excited state which is essential in any resonance experiment.

The fluorescent light emitted by the atoms in a direction at right-angles to the incident beam is collected by a lens and detected by a photomultiplier. Depending on the transitions being studied a polarizer may also be inserted in the detection arm of the apparatus. The resonance cell is located inside the tank coil of an r.f. oscillator and it is this coil which produces the rotating field component $\underline{B}_1$ which stimulates the magnetic dipole transitions between different hyperfine levels. These transitions produce changes in the polarization and spatial distribution of the fluorescent light and enable the magnetic resonance to be monitored. To extract the desired signal from the background of light scattered from the walls of the resonance cell, the r.f. power is modulated at audio frequencies. The a.c. signal from the photomultiplier is amplified by a narrow-band amplifier and detected by a phase-sensitive detector and chart recorder.

In alkali atoms many of the investigations have been carried out in zero magnetic field, $\underline{B}=0$, and in this case the resonance signal is obtained by slowly varying the frequency of the r.f. oscillator. A typical resonance signal, obtained on the second resonance level of ^{133}Cs, $7\ ^2P_{3/2}$, by Althoff (1955), is shown in Fig.18.19. It will be noticed immediate-

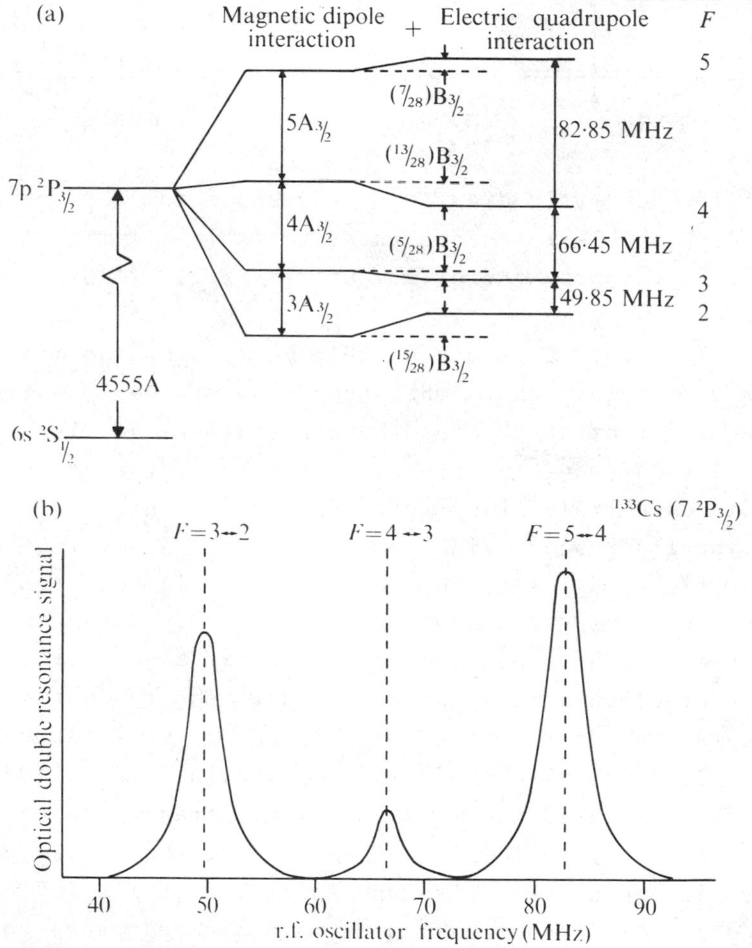

Fig.18.19. (a) Hyperfine structure of the $7\ ^2P_{3/2}$ level of ^{133}Cs, (b) Schematic diagram of zero-field optical double-resonance signals observed on this level. (After Althoff (1955).)

ly that the resonances are considerably broader than those observed in ground-state investigations owing to the short radiative lifetime of the excited atoms. Consequently the results obtained by this technique, some of which are listed in Table 18.5, are considerably less precise than those obtained on ground-state atoms (Problem 18.16). Nevertheless the electric quadrupole moments obtained from an analysis of the hyperfine structure of excited states with $J > \frac{1}{2}$ give valuable information about the shape of the nuclear charge distribution.

In a resonance cell of normal dimensions, $L \approx 4$ cm, the number of atoms in the vapour phase is only $\approx 10^{12}$. Thus it is possible to carry out experiments with very small numbers of atoms provided that chemical reactions with the walls of the cell do not reduce the number of usable atoms. Indeed working resonance cells have been prepared which contain only 1 μg of radioactive cesium isotopes such as ^{131}Cs ($\tau_{1/2} = 9{\cdot}7$ days) and ^{132}Cs ($\tau_{1/2} = 6{\cdot}6$ days). In these cells the loss of cesium atoms to the walls was controlled by adding approximately 1 mg of rubidium to act as a buffer. The rubidium atoms are not, of course, excited by the incident cesium resonance radiation.

In these experiments it is often difficult to hold the strength of the r.f. field constant over the frequency range covered during the recording of the magnetic resonance signal. The r.f. power must be monitored continuously either by a bolometer or by measurement of the oscillator voltage of the r.f. generator in order that the distortions in the observed signals may be corrected. A more serious practical difficulty is the attainment of the necessary r.f. field strengths. In order that an excited atom shall have an appreciable probability of making a magnetic dipole transition during its finite lifetime, $\tau = 1/\Gamma$, the amplitude of the r.f. magnetic field must satisfy the condition $\gamma B_1 \approx \Gamma$. Even for a relatively long lifetime of 10^{-7} s this means that $B_1 \approx 1$ G and it often proves difficult to generate sufficient power to create high-frequency fields

TABLE 18.5

Hyperfine structure of alkali atoms; results mostly from resonance fluorescence techniques†

Isotope	n	$A_J(n\ ^2S_{1/2})$ (MHz)	$A_J(n\ ^2P_{1/2})$ (MHz)	$A_J(n\ ^2P_{3/2})$ (MHz)	$B_J(n\ ^2P_{3/2})$ (MHz)	Q (10^{-24} cm^2)
^{7}Li	2	401·75	46·17(35)	-3·07(13)	-0·18(12)	-0·03(2)
	3	—	13·5(2)	-0·96(13)	—	—
^{23}Na	3	885·82	94·3(2)	18·65(10)	2·82(30)	0·097(11)
	4	202 (3)	—	6·006(30)	0·86(9)	—
^{39}K	4	230·85	28·85(30)	6·09(4)	2·77(10)	+0·059(6)
	5	55·50(60)	8·99(15)	1·972(10)	0·866(15)	—
	6	21·81(18)	—	0·866(8)	0·370(15)	—
^{85}Rb	5	1011·9	120·7(1)	25·029(16)	26·032(70)	+0·298(1)
	6	239·3(1·2)	39·11(3)	8·25(10)	8·16(20)	—
	7	94·00(64)	17·65(2)	3·71(1)	3·65(10)	—

^{87}Rb	5	3417·3	409·1(4·0)	84·852(30)	12·510(57)	+0·141(1)
	6	809·1(5·0)	132·56(3)	27·70(2)	3·947(13)	—
	7	318·1(3·2)	59·92(9)	12·57(1)	1·768(8)	—
	8	158·0(3·0)	—	6·747(14)	0·933(20)	—
^{133}Cs	6	2298·16	292·1(3)	50·31(5)	-0·38(18)	-0·0036(13)
	7	546·3(3·0)	94·5(9)	16·609(5)	-0·15(3)	—
	8	218·9(1·6)	42·97(10)	7·626(5)	-0·090(24)	—
	9	109·5(2·0)	—	4·129(7)	-0·051(25)	—

† Detailed references are given by Rosen and Lindgren (1972) and Happer (1975)

of this magnitude. Fortunately there are two alternative techniques for studying the hyperfine structure of excited states which do not require r.f. fields, namely quantum beats and level-crossing spectroscopy, and these are discussed in the sections which follow.

18.9.2. Quantum beats and hyperfine structure. In sections 15.8 and 16.7 we described experiments in which the resonance fluorescence from excited atoms was observed to be modulated at frequencies in the range 0·5 - 10 MHz, corresponding to the separation of the Zeeman sub-levels of the excited atoms in low magnetic fields. These light beats arise because the atoms were prepared in a coherent superposition of at least two excited states,

$$|\psi(t)\rangle = a \exp\{-i(\omega_1 + \frac{\Gamma}{2})t\}|1\rangle + b \exp\{-i(\omega_2 + \frac{\Gamma}{2})t\}|2\rangle$$

where a and b are expansion coefficients and $\hbar\omega_1$ and $\hbar\omega_2$ are the energies of the two states involved. When atoms in this superposition state decay to a common lower level $|g\rangle$ the intensity of the emitted light is given by

$$I(t) \propto |\langle g|\hat{\underline{e}}.\underline{r}|\psi(t)\rangle|^2$$

$$= \{A + B\cos(\omega_2-\omega_1)t\}\exp(-\Gamma t) \qquad (18.48)$$

and is intensity modulated at the frequency corresponding to the difference in energy between the states $|1\rangle$ and $|2\rangle$. In experiments involving transient excitation the depth of modulation will only be significant when the excitation pulse length Δt satisfies the condition $(\omega_2-\omega_1)\,\Delta t \ll 1$. The experiments we described in section 15.8 were difficult due to the low intensity of the available conventional sources and were limited to the observation of quantum beats corresponding to low-field Zeeman splittings in Cd and Hg by the relatively long pulse of excitation. However, there is in principle no reason why the technique should not be applied to the measurement of the separation between hyperfine levels in zero magnetic field, although of course the fluorescent light will then generally display beats at several frequencies.

(a) Hyperfine quantum beats: optical excitation. The development of high-intensity tunable dye lasers now solves many of the problems associated with the earlier experiments. Of these devices perhaps the most useful for time-resolved studies is the dye system pumped by a pulsed nitrogen laser. The tunable output has a typical pulse duration of 2 ns, which implies a Fourier-limited linewidth of the order of 1 GHz. Thus excited-state hyperfine levels spanning a frequency range of up to 1 GHz may now be coherently populated by optical excitation. The first experiment of this kind was reported by Haroche *et al.*(1973) on the 4555 Å, $6\ ^2S_{1/2} - 7\ ^2P_{3/2}$ transition of ^{133}Cs. The detection apparatus consisted essentially of a fast photomultiplier, sampling oscilloscope, and signal averager. Although the results obtained in this particular experiment were less accurate than the previous measurements of the hyperfine splitting using optical double-resonance, the precision of future experiments could be increased by sampling the oscillatory decay curve at a larger number of points and by averaging the signal over a larger number of excitation pulses. However, in spite of the increased intensity provided by the pulsed laser source, the time resolution of the photomultiplier and the associated signal processing equipment is likely to restrict the hyperfine separations measured by this technique to frequencies $\leq$ 200 MHz. This limitation does not apply in the fast beam experiments which we now consider.

(b) Hyperfine quantum beats: excitation of fast beams. The time-dependent decay of a sample of excited atoms or ions may be converted into a spatial variation of intensity by localized excitation of a beam of fast moving particles. This is the principle of the beam-foil technique which has an attainable time resolution of the order of 10^{-10} s, as described in section 6.1. It is therefore apparent that hyperfine quantum beats will be superimposed on the observed exponential decay curves in a fast beam experiment when the ions of odd isotopes are prepared in a coherent superposition of excited hyperfine levels. In order to achieve a high atomic alignment and to avoid problems associated with radiative cascade, the carbon foil excitation of the normal beam apparatus may be replaced by optical excitation using a C.W. laser as shown in Fig.18.20(a).

In this experiment, performed by Andrä (1975, 1976), the 4545 Å line of an argon ion laser was Doppler tuned to coincide with one or other of the two groups of unresolved hyperfine components of the 4554 Å D_2 resonance line, $6s\ ^2S_{1/2} - 6p\ ^2P_{3/2}$, in $^{137}Ba^+ (I = \frac{3}{2})$ by correct adjustment of the angle of intersection of the laser and ion beams. The laser light was linearly polarized with its electric vector perpendicular to the image plane at the photomultiplier. The time-resolved signal was observed exactly as in a beam-foil experiment by measuring the intensity of light emitted from a well-defined segment of the beam Δl as a function of the distance l downstream from the excitation region.

The quantum beats observed when the laser was tuned to the lower-frequency component of the D_2 resonance line of the barium ion are shown in Fig.18.21(a). These arise from interferences between the states F=3, F=2 and F=1 of the $6\ ^2P_{3/2}$ level in zero magnetic field, as shown in Fig. 18.20(b). The frequencies corresponding to the hyperfine intervals between these pairs of levels may be obtained directly from the Fourier transform of the observed decay signal which is shown in Fig.18.21(b). Similar measurements on the high-frequency component of the D_2 resonance line enable the complete hyperfine structure of the $6\ ^2P_{3/2}$ level to be determined and the results of this preliminary investigation are given in Fig.18.20(b).

It has also been discovered that ion beams excited by the usual interaction with a carbon foil also emit polarized light and may be used for zero-field quantum beat measure-

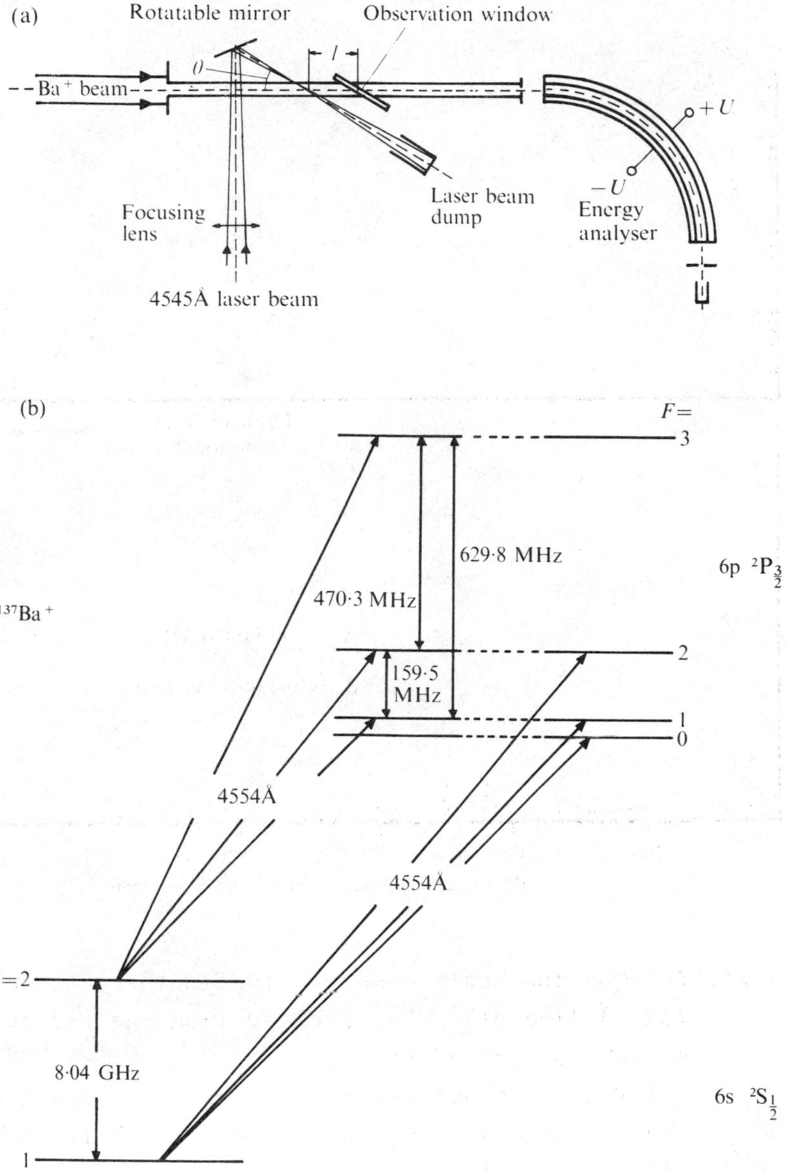

Fig.18.20. (a) Schematic diagram of the apparatus used for laser excitation of fast beams of Ba^+. The observation window is moved along the beam axis relative to the fixed laser excitation region. (b) Level scheme of $^{137}Ba^+$ showing selective excitation from the F=1 and F=2 ground-state hyperfine levels. (After Andrä (1975).)

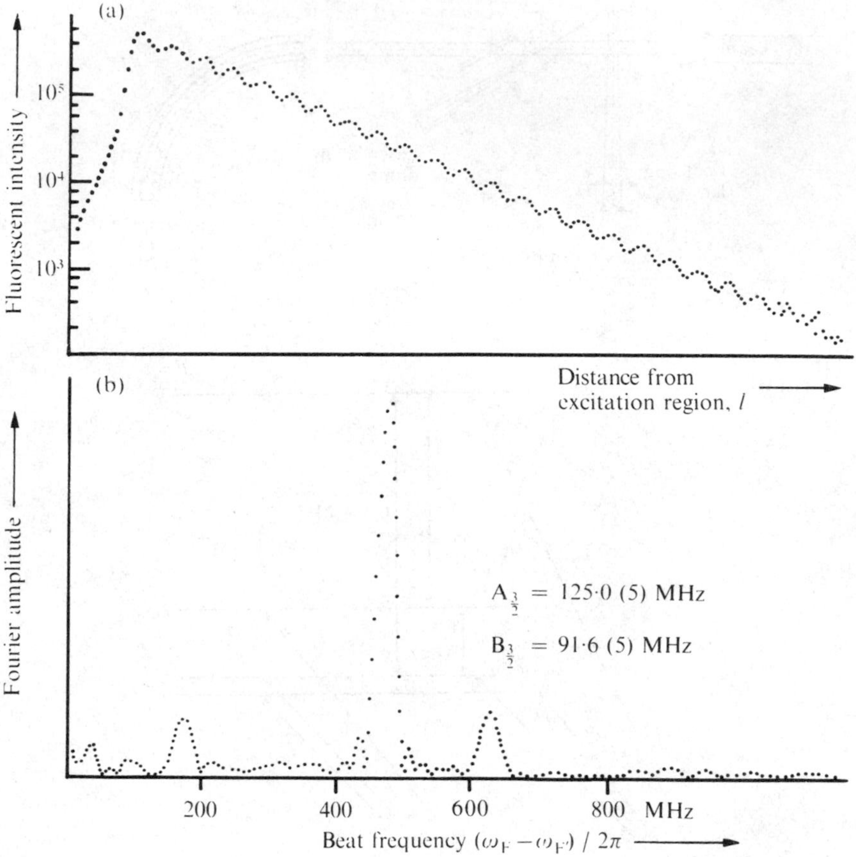

Fig.18.21. (a) Quantum beats observed in fluorescence on the 4554 Å line of $^{137}Ba^+$ excited from the F=2 level of the ion ground state.
(b) Fourier transform of the observed signal and level scheme showing the measured beat frequencies. (After Andrä (1975).)

ments. The accuracy of this technique approaches 1 part in 10^3 and can be used for the direct measurement of hyperfine splittings in the range 30 MHz - 10 GHz. It therefore represents an attractive alternative to the other techniques for studying excited atoms and possesses unique advantages when one is interested in ionized systems.

18.9.3. Level-crossing experiments. In optical double-resonance experiments the excited states of the atom are mixed by the effect of an r.f. magnetic field and this results in changes in the angular distribution and polarization of the fluorescent light. Similar changes in the resonance fluorescence occur in the absence of r.f. fields when two or more energy levels cross at zero magnetic field, as is demonstrated by the Hanle effect experiments discussed in Chapter 15. When the atomic nucleus has a finite spin I, some of the hyperfine Zeeman levels belonging to states with $J > \frac{1}{2}$ intersect not only at zero fields but also in the intermediate-field region where the electronic and nuclear angular momenta are becoming progressively decoupled. An example of this type of high-field level-crossing is shown in Fig.18.22 where the Zeeman splitting of the hyperfine levels belonging to the 5s5p $^3P_1$ state of ^{113}Cd($I = \frac{1}{2}$) are plotted as a function of the strength of the applied field $\underline{B}$. In this case the levels ($F = \frac{1}{2}$, $M_F = -\frac{1}{2}$) and ($F = \frac{3}{2}$, $M_F = \frac{3}{2}$) in the weak field notation intersect at a field of approximately 2055 G and the characteristic Lorentzian- or dispersion-shaped level-crossing signals are observed when the polarization of the 3261 Å ($5\ ^1S_0 - 5\ ^3P_1$) resonance fluorescence is monitored as a function of the applied field.

High-field level-crossing signals were first observed by Colegrove *et al.*(1959) about 34 years after the original discovery of the zero-field crossing signals by Hanle (1925). A theory of the effect was soon worked out by Franken (1961) but again this turned out to be very similar to a much earlier treatment given by Breit (1933). An expression for the intensity of the fluorescent light of a given polarization observed in a well-defined direction may be obtained by a modification of the theoretical Hanle effect signal, equation (15.27). We simply replace the low-field energy level separation $\hbar(m-m')\omega_L$ by the more general expression

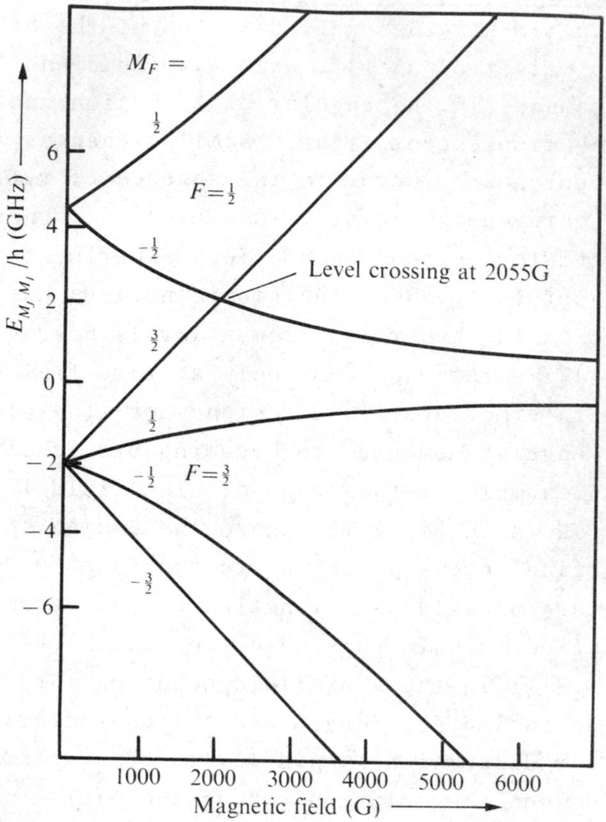

Fig.18.22. Zeeman effect of the hyperfine structure of the (5s5p) $^3P_1$ state of ^{113}Cd showing a high-field level-crossing. (After Thaddeus and Novick (1962).)

$$E_{M_q} - E_{M'_q} = \hbar(\omega_{M_q} - \omega_{M'_q})$$

with the result

$$\frac{dI}{d\Omega} = \frac{U(\omega)}{8\pi\varepsilon_0^2}\left(\frac{\omega}{\hbar c}\right)^3 \sum_{\substack{M_q M'_q \\ \mu\mu'}} \frac{F_{M_q M'_q} G_{M'_q M_q}}{\Gamma + i(\omega_{M_q} - \omega_{M'_q})} \rho_{\mu\mu}. \quad (18.49)$$

Of course, the correct intermediate-field wavefunctions should be used in the evaluation of the electric dipole matrix elements appearing in the excitation and emission

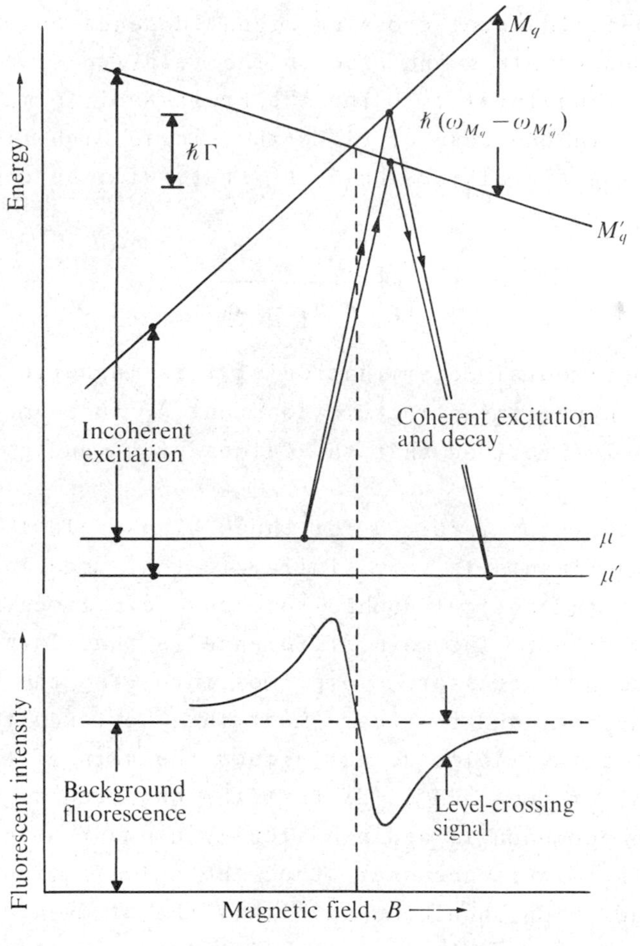

Fig.18.23. Diagram illustrating the origin of level-crossing signals. As the magnetic field is progressively increased the excitation process changes from incoherent to coherent and back again, resulting in a spatial redistribution of the fluorescent light.

matrices, which are now written as $F_{M_q M'_q}$ and $G_{M'_q M_q}$ respectively. We have also assumed that the levels M_q and M'_q both decay at the same rate, Γ, which is valid for substates arising from the same fine-structure level. The width of the high-field level-crossing signal depends on this radiative decay rate Γ and also on the relative slope of the intersecting levels. Using the Breit-Rabi formula, we find that in the case of ^{113}Cd the single high-field level-crossing signal for the 5 $^3P_1$ state will be centred at

$$B = \frac{A_1}{(g_J + g_I \frac{m}{M})\mu_B} . \tag{18.50}$$

Thus an experimental determination of this magnetic field enables the hyperfine structure constant A_1 to be measured very precisely provided that the values of g_J and g_I are accurately known.

The apparatus required for these high-field level-crossing experiments is very similar to that used in the Hanle effect and optical double-resonance experiments previously described. The main difference is that large magnetic fields are necessary, sometimes involving the use of an electromagnet, and that care must then be taken to ensure that stray fields do not affect the gain of the photomultiplier tube and so distort the observed signals. Since the phenomenon is again basically a quantum interference effect it is necessary that the polarization of the incident radiation should be such that the atoms are created in a coherent superposition of the intersecting states and that these states should be monitored in fluorescent radiation of coherent polarization down to a single lower level, as shown in Fig.18.23. This requirement restricts the observable level crossings to those satisfying:

$$\Delta M_q = M_q - M_q' = \pm 2 \quad \text{(using linearly-polarized light)}$$

and

$$\Delta M_q = M_q - M_q' = \pm 1 \quad \text{(using circularly- or elliptically-polarized light)}.$$

Level-crossing experiments on the $^3P_1$ and $^1P_1$ levels of Zn, Cd, and Hg have also been extended to stable and radioactive isotopes having $I > \frac{1}{2}$. In these cases it is obvious that more than one level crossing occurs and from the measured fields the values of the magnetic dipole and electric quadrupole hyperfine interaction constants may be obtained. Some of the results obtained on the group IIB metals by the level-crossing, optical double-resonance, and optical pumping techniques are given in Table 18.6. The level-

TABLE 18.6

Hyperfine structure of excited levels of elements of group IIB measured by resonance fluorescence techniques

Isotope	State	I	A_J (MHz)	B_J (MHz)	μ_I (nuclear magnetons)	Q (10^{-24} cm^2)
^{65}Zn	$4\ ^3P_1$	$\frac{5}{2}$	+535·163(2)†	+2·870(5)	0·7688(2)	-0·024(2)
^{67}Zn	$4\ ^3P_1$	$\frac{5}{2}$	+608·99(5)	-19·37(9)	—	+0·18(2)
^{111}Cd	$5\ ^3P_1$	$\frac{1}{2}$	-4123·81(1)	—	0·591 444(6)	—
^{113}Cd	$5\ ^3P_1$	$\frac{1}{2}$	-4313·86(1)	—	0·590 776(6)	—
^{199}Hg	$6\ ^3P_1$	$\frac{1}{2}$	+14 750·7(50)	—	+0·497 865(6)	—
^{201}Hg	$6\ ^3P_1$	$\frac{3}{2}$	-5454·569(3)	-280·107 (5)	-0·551 344 (9)	+0·50

†The numbers in parentheses indicate the uncertainty in the last digits of each quoted value.

crossing method has clear advantages over the double-resonance method when short-lived excited states with large hyperfine splittings are being investigated, for the production of the intense r.f. fields required to induce transitions between the short-lived states is thereby avoided. Moreover the level-crossing signals are large compared with those observed in most double-resonance experiments since all atoms in the states concerned contribute to the signal whereas in the resonance experiments only a small fraction of the atoms actually undergo r.f. transitions before they decay. However, it should be appreciated that the positions of the level crossings, such as that given by equation (18.50), determine only the ratio of the hyperfine structure constants to the electronic g-factor g_J and that in precise work this must be determined separately by an optical double-resonance experiment in low fields.

18.10. Conclusion

18.10.1. Further applications of the atomic beam and resonance fluorescence techniques. The reader will no doubt have realized that our treatment of the measurements of the hyperfine structure of atoms has perforce been very selective, and for further information about ground-state atoms we recommend the monographs by Ramsey (1956) and Kopfermann (1958). Although no similar treatment exists for excited atoms, excellent reviews of the optical double-resonance and level-crossing techniques have been written by Series (1959), zu Putlitz (1965), and Budick (1967). These describe the techniques in greater detail and also treat their application to the investigation of excited atoms perturbed by external electrostatic fields (Stark effect), a topic which we have not had space to include. A more serious omission is that we have not discussed the application of any of these techniques to the study of the Lamb shift in hydrogenic systems and the fine structure of the light elements H, He, Li, etc. The development of our understanding of the fine structure of hydrogen-like and helium-like atoms is reported in the proceedings of the International Conferences on Atomic Physics which are published under the title *Atomic Physics, Vols. 1-4.*

18.10.2. Future developments in atomic spectroscopy. Most of the precise, Doppler-free investigations of atomic fine and hyperfine structure made prior to 1970 were restricted to measurements on the first few excited levels of neutral atoms and their singly-ionized ions by the low intensities of the available light sources. Following the recent development of tunable dye lasers, step-wise excitation now enables the level-crossing and optical double-resonance techniques to be applied to highly excited states of neutral atoms. Moreover two-photon absorption spectroscopy also gives results with the same kind of precision. Similarly the use of beam-foil and beam laser excitation has enabled the quantum beat technique to be applied to fine- and hyperfine-structure measurements in a wide range of ionized species which were previously inaccessible. This burst of activity in atomic spectroscopy has already significantly increased our knowledge of the Lamb shifts in hydrogenic systems and the hyperfine structure of highly excited states in the alkali metals. Doubtless these techniques will soon be applied to elements in other groups of the periodic table and will stimulate a complementary development of new theoretical calculations. Although some of the experimental techniques we have described will certainly become obsolete in the next few years, the basic concepts will remain valid and we hope that this book will continue to provide a good introduction to the field of advanced atomic physics.

Problems

18.1. The surface of a nucleus of atomic number Z is represented classically by an ellipsoid of revolution:

$$r(\theta) = r_0\{1 + \frac{\alpha}{2}(3\cos^2\theta - 1)\},$$

where θ is the angle between the vector $\underline{r}$ and the axis of symmetry $0z$, and α is a constant deformation parameter. Assuming the nuclear charge density to be constant and equal to that of a uniformly charged sphere of radius r_0, show that to first order in α the nuclear electric quadrupole moment is given by

$$Q \cong 2Z(a^2 - b^2)/5$$

where $a=r(0)$ and $b=r(\pi/2)$ are the semi-axes of the ellipsoid.

18.2. In the spectrum of $^{133}Cs^+$ the transition $6s[1\frac{1}{2}]^o_1 - 6p[2\frac{1}{2}]_2$ at 4953 Å consists of a hyperfine multiplet with lines at positions 0, 0·081, 0·195, 0·337 and 0·513 cm^{-1} above the lowest frequency component of the multiplet. Assuming that the hyperfine structure of the lower term is too small to be resolved, show that the interval rule is obeyed approximately and hence determine the nuclear spin I.

18.3. From equation (18.14) determine the hyperfine structure of levels with total electronic angular momentum $J=\frac{1}{2}$ and $\frac{3}{2}$ for the case $I=\frac{3}{2}$ in terms of the magnetic dipole and electric quadrupole interaction constants.

Assuming, for the sake of simplicity, that the quadrupole interaction constant B_J is identically zero, prove that the centre of gravity of the hyperfine-structure multiplet defined by

$$\bar{E} = \sum_{F=I-J}^{F=I+J} (2F+1)E_F \Big/ \sum_{F=I-J}^{F=I+J} (2F+1)$$

is equal to the energy of the original fine-structure level, E_J.

$$\left(\sum_{m=1}^{m=n} m^2 = \tfrac{1}{6}n(n+1)(2n+1) \quad ; \quad \sum_{m=1}^{m=n} m^3 = \tfrac{1}{4}n^2(n+1)^2.\right)$$

18.4. Show that the scalar product of the electronic and nuclear spin angular momenta may be expressed in the form

$$\underline{I}.\underline{J} = I_zJ_z + \tfrac{1}{2}(I_+J_- + I_-J_+)$$

where

$I_\pm = I_x \pm iI_y$ and $J_\pm = J_x \pm iJ_y$. Hence calculate all the matrix elements of the hyperfine interaction $\langle JIM_JM_I|\mathcal{H}_{HFS}|JIM'_J\ M'_I\rangle$ in the uncoupled representation for the case $I=\frac{1}{2}$, $J=\frac{1}{2}$. The raising and lowering operators have the following properties:

$$I_{\pm}\ |IM_I\rangle\ =\ \{I(I+1)\ -\ M_I(M_I\ \pm\ 1)\}^{1/2}\ \ |I,M_I\pm 1\rangle$$

$$J_{\pm}\ |JM_J\rangle\ =\ \{J(J+1)\ -\ M_J(M_J\ \pm\ 1)\}^{1/2}\ \ |J,M_J\pm 1\rangle .$$

Finally, by solving the secular equation, determine the energies of the hyperfine sub-levels in fields of arbitrary strength. Verify the correctness of the results using the Breit-Rabi formula and Fig.18.16.

18.5. Construct an energy level diagram to show how the Zeeman splitting of the hyperfine levels of a state having $J=\frac{1}{2}$ and $I=\frac{5}{2}$ changes as the external magnetic field is varied from zero to some large value.

18.6. Using the information given in section 18.2.2(a), show that ^{85}Rb and ^{87}Rb have nuclear spins of $I=\frac{5}{2}$ and $\frac{3}{2}$ respectively. On the nuclear shell model the nuclear spin is due to the odd neutron or proton, both with spin $\frac{1}{2}$, moving in an orbit within the nucleus having angular momentum $l\hbar$. This gives an observed nuclear spin of $l\pm\frac{1}{2}$. Apply the general formula for the Landé factor, equation (3.84), to calculate the nuclear magnetic moment, assuming that for a proton $g_l=1$ and $g_s=5\cdot586$, and for a neutron $g_l=0$ and $g_s=-3\cdot826$. Show that the magnetic moment, $\mu_I=g_I\ \mu_n I$, is given by the Schmidt limit formulae:

(a) odd proton $I=l+\frac{1}{2}$ $\qquad \mu_I=\mu_n(I+2\cdot293)$

$\qquad I=l-\frac{1}{2}$ $\qquad \mu_I=\mu_n I(I-1\cdot293)/(I+1)$

(b) odd neutron $I=l+\frac{1}{2}$ $\qquad \mu_I=-1\cdot913\ \mu_n$

$\qquad I=l-\frac{1}{2}$ $\qquad \mu_I=1\cdot913\ \mu_n I/(I+1)$

18.7. For the hyperfine levels $F=I\pm\frac{1}{2}$ of the ground states of alkali atoms in low magnetic fields, verify that the frequencies of the Zeeman magnetic resonance transitions $\Delta F=0$, $\Delta M_F = \pm 1$ are given by equation (18.27). From the frequency separations of these Zeeman resonances, given in section 18.2.2(b), show that the nuclear magnetic moments of ^{85}Rb and ^{87}Rb are 1·351 and 2·748 nuclear magnetons respectively. How well do these values agree with the Schmidt limits derived in Problem 18.6?

18.8. Expand the Breit-Rabi formula up to terms of order x^2 and thus obtain a precise expression for the frequency of the Zeeman magnetic resonance transitions $\Delta F=0$, $\Delta M_F = \pm 1$ in small magnetic fields. For a given hyperfine level F show that the frequencies of the transitions ($M_F \leftrightarrow M_F+1$) and ($M_F+1 \leftrightarrow M_F+2$) are separated by an amount $2\nu_{HFS}x^2/(2I+1)^2$. Given that this frequency separation is about 144 Hz in a field of 10^{-4} T for both rubidium isotopes, show that the zero-field hyperfine separations in ^{85}Rb and ^{87}Rb are approximately 3·03 GHz and 6·82 GHz respectively.

18.9. Construct a diagram showing the low-field Zeeman splittings of the ground-state hyperfine levels of an alkali atom with nuclear spin $I = \frac{3}{2}$. Indicate the permitted magnetic dipole hyperfine transitions $\Delta F = \pm 1$, $\Delta M_F = 0, \pm 1$ and sketch the magnetic resonance spectra observed with both σ and π polarized radio-frequency fields.

By expanding the Breit-Rabi formula in the weak field approximation prove that the frequency of the $\Delta F = \pm 1$, $M_F=0 \leftrightarrow M_F=0$ transition is given by

$$\nu_0 = \nu_{HFS}(1 + x^2/2)$$

and calculate the frequencies of the other transitions to first order in the parameter x.

18.10. In an atomic beam apparatus the components are symmetrically arranged about the collimating slit C with the detector slit placed a distance l' beyond the end of the B-magnet which itself is of length l. By integration of the equation of motion in the direction transverse to the axis of the apparatus, show that in a 'flop-in' arrangement:

(a) those atoms which undergo transitions in the C-field strike a detector on the axis of the apparatus, independent of their velocity

(b) those atoms which do not undergo transitions are deflected by an amount

$$z = \pm \frac{\mu_B}{Mv^2} \cdot \frac{\partial B}{\partial z} l(l+2l').$$

Show that atoms travelling with the most probable velocity $v_\alpha = (2kT/M)^{1/2}$ in an oven at a temperature of 420 K are deflected by 1·17 mm in an apparatus in which l = 8·9 cm, l' = 12·0 cm and $\partial B/\partial z = 50\ \mathrm{T\ m^{-1}}$.

18.11. The r.f. loop in an atomic beam apparatus is 3 cm long. Assuming that the C-field is perfectly homogeneous, show that the observed linewidth of a resonance using ^{39}K atoms emitted from an oven at 420 K should not exceed 15 kHz.

18.12. In an atomic beam magnetic resonance experiment the transitions $\Delta F=0$, $\Delta M_F = \pm 1$ are observed to occur at the frequencies 1·557 MHz and 3·504 MHz for ^{40}K and ^{39}K respectively in the same weak magnetic field. The nuclear spin of ^{39}K is $I = \frac{3}{2}$. Show that the nuclear spin of ^{40}K is $I=4$ and that the applied magnetic field is $5{\cdot}001 \times 10^{-4}$ T.

18.13. The ground state of ^{69}Ga is 4p $^2P_{1/2}$. In an atomic beam experiment the field in the resonance region is set at 6×10^{-4} T and a resonance is observed at 1·400 MHz. Show that the nuclear spin of ^{69}Ga is $I = \frac{3}{2}$.

The low lying metastable state 4p $^2P_{3/2}$ is also

well populated at the oven temperature used and the ratio of the g_J-values was measured by Kusch and Foley (1948) as

$$\frac{g_J(^2P_{3/2})}{g_J(^2P_{1/2})} = 2(1\cdot001\ 72 \pm 0\cdot000\ 06).$$

Hence, using equation (3.84), show that this leads to a value for the electron spin g-factor of

$$g_s = 2(1\cdot001\ 14 \pm 0\cdot000\ 04)$$

and compare this result with the theoretical value given in equation (18.36).

18.14. In atomic beam experiments on ^{39}K and ^{7}Li, both of which have nuclear spin $I = \frac{3}{2}$, Kusch *et al.* (1940) observed that the low field π-polarized hyperfine transitions, shown in Fig.18.14, occurred at the following frequencies:

Transition $(F,M_F) \leftrightarrow (F',M_F')$	^{39}K resonance frequency (MHz)	^{7}Li resonance frequency (MHz)
(2,-2) ↔ (1,-1)	461·64	803·22
(2,-1) ↔ (1, 0) (2, 0) ↔ (1,-1)	461·72	803·45
(2, 1) ↔ (1, 0) (2, 0) ↔ (1, 1)	461·79	803·64
(2, 2) ↔ (1, 1)	461·87	803·84

Show that the zero-field hyperfine separations deduced from these figures are 461·75 MHz and 803·54 MHz for ^{39}K and ^{7}Li respectively and that the magnetic fields employed were 0·05 G and 0·15 G.

18.15. In ^{39}K and ^{7}Li Kusch *et al.* (1940) also observed $\Delta M_J=0$, $\Delta M_I=\pm 1$ transitions at fields which were not quite large enough for the complete Paschen-Back formula, equation (18.37), to be applicable. By expanding the Breit-Rabi formula up to terms of order $\left(\frac{h\nu_{HFS}}{2I+1}\right)^2/g_J\mu_B B$, obtain modified expressions for the frequencies of these transitions. Hence analyse the experimental data given below to redetermine ν_{HFS} and show that the nuclear g-factors are given by $g_I(^{39}K) = -0{\cdot}261$ and $g_I(^{7}Li) = -2{\cdot}18$.

Transition $(M_J,M_I) \leftrightarrow (M_J,M_I')$	B=3950 G ^{39}K resonance frequency (MHz)	B=3060 G ^{7}Li resonance frequency (MHz)
$(-\frac{1}{2}, -\frac{3}{2}) \leftrightarrow (-\frac{1}{2}, -\frac{1}{2})$	119·904	220·72
$(-\frac{1}{2}, \frac{1}{2}) \leftrightarrow (-\frac{1}{2}, -\frac{1}{2})$	117·347	209·96
$(\frac{1}{2}, -\frac{1}{2}) \leftrightarrow (\frac{1}{2}, -\frac{3}{2})$	115·776	199·78
$(-\frac{1}{2}, \frac{3}{2}) \leftrightarrow (-\frac{1}{2}, \frac{1}{2})$	114·961	200·65
$(\frac{1}{2}, \frac{1}{2}) \leftrightarrow (\frac{1}{2}, -\frac{1}{2})$	113·374	190·51
$(\frac{1}{2}, \frac{3}{2}) \leftrightarrow (\frac{1}{2}, \frac{1}{2})$	111·115	182·38

18.16. It may be shown that the magnetic hyperfine interaction constant A_j for the $|nlj\rangle$ excited states of alkali atoms with $l\neq 0$ is given by

$$A_j = \left(\frac{\mu_0}{4\pi}\right)\left(2\,\mu_B\,\frac{\mu_I}{I}\right)\left\langle\frac{1}{r^3}\right\rangle\frac{l(l+1)}{j(j+1)}F_R(j)$$

where r is the radial coordinate of the valence electron and $F_R(j)$ is a relativistic correction factor of order unity.

(a) The fine structure separation $\delta\nu_{FS}$ between the $n\ ^2P_{1/2}$ and $n\ ^2P_{3/2}$ levels is also proportional to $\langle \frac{1}{r^3} \rangle$ (Problem 3.13), and the ratios of the fine-structure separations in the first and second resonance levels of the alkalis have the values:

Atom	Na (n=3)	K (n=4)	Rb (n=5)	Cs (n=6)
$\dfrac{\delta\nu_{FS}(n\ ^2P)}{\delta\nu_{FS}\{(n+1)\ ^2P\}}$	3·13	3·09	3·066	3·061

Use the measurements listed in Table 18.5 to confirm that

$$\frac{A_j(n\ ^2P_{3/2})}{A_j\{(n+1)\ ^2P_{3/2}\}} = \frac{\delta\nu_{FS}(n\ ^2P)}{\delta\nu_{FS}\{(n+1)\ ^2P\}}$$

(b) Show also that

$$\frac{A_j(n\ ^2P_{1/2})}{A_j(n\ ^2P_{3/2})} = 5 \times \frac{F_R(1/2)}{F_R(3/2)} .$$

Use the measurements on ^{39}K and ^{133}Cs given in Table 18.5 to verify this relationship and to estimate the magnitude of the relativistic correction factor $F_R(j)$.

References

Althoff, K. (1955). *Z.Phys.* 141, 33.

Andrä, H.J. (1975). *Atomic Physics 4*, p.635. Plenum Press, New York.

__________ (1976). *Proceedings of the fourth International Beam Foil Spectroscopy Conference.*

Arditi, M. (1960). *Annls. Phys.* 5, 973.

Arditi, M. and Carver, T.R. (1958). *Phys.Rev.* 109, 1012.

________ and ________ (1961). *Phys.Rev.* 124, 800.

Bender, P.L., Beaty, E.C., and Chi, A.R. (1958). *Phys.Rev. Lett.* 1, 311.

Bloom, A.L. and Carr, J.B. (1960). *Phys.Rev.* 119, 1946.

Breit, G. (1933). *Rev.mod.Phys.* 5, 91.

Carpenter, R.J., Beaty, E.C., Bender, P.L., Saito, S., and Stone, R.O. (1960). *I.R.E. Trans.Instrumn.* I-9, 132.

Cohen, E.R. and Du Mond, J.W.M. (1965). *Rev.mod.Phys.* 37, 537.

Colegrove, F.D., Franken, P.A., Lewis, R.R., and Sands, R.H. (1959). *Phys.Rev.Lett.* 3, 420.

Eisinger, J.T., Bederson, B., and Feld, B.T. (1952). *Phys. Rev.* 86, 73.

Essen, L. and Parry, J.V.L. (1957). *Phil.Trans. R.Soc.* A250, 45.

Franken, P.A. (1961). *Phys.Rev.* 121, 508.

Franzen, W. and Emslie, A.G. (1957). *Phys.Rev.* 108, 1453.

Goldenberg, H.D., Kleppner, D., and Ramsey, N.F. (1960). *Phys.Rev.Lett.* 5, 361.

Hanle, W. (1925). *Ergebn.exakt.Naturw.* 4, 214.

Happer, W. (1975). *Atomic Physics 4.* p.651. Plenum Press, New York.

Haroche, S., Paisner, J.A., and Schawlow, A.L. (1973). *Phys. Rev.Lett.* 30, 948.

Hawkins, W.B. (1955). *Phys.Rev.* 98, 478.

Hellwig, H.W. (1975). *Proc.I.E.E.E.* 63, 212.

Kusch, P. (1955). *Phys.Rev.* 100, 1188.

________ and Foley, H.M. (1947). *Phys.Rev.* 72, 1256.

________ and ________ (1948). *Phys.Rev.* 74, 250.

________, Millman, S. and Rabi, I.I. (1940). *Phys.Rev.* 57, 765.

Logan, R.A. and Kusch, P. (1951). *Phys.Rev.* 81, 280.

Markowitz, W., Hall, R.G., Essen, L., and Parry, J.V.L.(1958). *Phys.Rev.Lett.* 1, 105.

Mockler, R.C., Beehler, R.E., and Snider, C.S. (1960). *I.R.E. Trans.Instrumn.* I-9, 120.

Packard, M.E. and Swartz, B.E. (1962). *I.R.E. Trans.Instrumn.* I-11, 215.

Pauli, W. (1924). *Naturwissenschaften* 12, 741.

Penselin, S., Moran, R., Cohen, V.W., and Winkler, G. (1962). *Phys.Rev.* 127, 524.

Pipkin, F.M. and Lambert, R.H. (1962). *Phys.Rev.* 127, 787.

Prodell, A.G. and Kusch, P. (1957). *Phys.Rev.* 106, 87.

Rabi, I.I., Zacharias, J.R., Millman, S., and Kusch, P. (1938). *Phys.Rev.* 53, 318.

Ramsey, N.F. (1949). *Phys.Rev.* 76, 996.

Rosen, A. and Lindgren, I. (1972). *Physica Scripta* 6, 109.

Schlect, R., McColm, D., and Maleh, I. (1962). *Bull.Am.phys. Soc.* 7, 604.

Senitzki, B. and Rabi, I.I. (1956). *Phys.Rev.* 103, 315.

Stacey, D.N. (1966). *Rep.Prog.Phys.* 29, 171.

Thaddeus, P. and Novick, R. (1962). *Phys.Rev.* 126, 1774.

General references and further reading

Our treatment of the theory of hyperfine structure closely follows the account given by

Woodgate, G.K. (1970). *Elementary atomic structure*. McGraw-Hill, London.

More detailed accounts of both the theory and experimental techniques are contained in

Kopfermann, H. (1958). *Nuclear moments*. (Translated by E.E. Schneider). Academic Press, New York.

Kusch, P. and Hughes, V.W. (1959). *Handbuch der Physik*, 37, (ed. S. Flugge,) Springer-Verlag, Berlin.

Ramsey, N.F. (1956). *Molecular beams*, Oxford University Press, London.

Good reviews of the optical double-resonance and level-crossing techniques have been published by

Series, G.W. (1959). *Rep.Prog.Phys.* 22, 280.

zu Putlitz, G. (1965). *Ergebn.exakt.Naturw.* 37, 105.

Budick, B. (1967). *Adv.At.& Mol.Phys.* 3, 73.

The invited papers presented at recent International Conferences on Atomic Physics contain valuable additional information:

Bederson, B., Cohen, V.W., and Pichanick, F.M.J.(eds.) (1969). *Atomic physics 1.* Plenum Press, New York.

Sandars, P.G.H. (ed.) (1971). *Atomic physics 2.* Plenum Press, New York.

Smith, S.J. and Walters, G.K. (eds.) (1973). *Atomic physics 3.* Plenum Press, New York.

zu Putlitz, G., Weber, E.W., and Winnacker, A. (eds.) (1975). *Atomic physics 4.* Plenum Press, New York.

Appendix: Table of Fundamental Constants

Table of Fundamental Constants

Accurate values of some physical constants in SI and cgs units. These values are selected from Table 33.1 of the paper by E.R.Cohen and B.N.Taylor, J.Phys.Chem.Ref.Data, 2, 663, (1973)†

Quantity	Symbol	Value	Units SI	Units cgs††
Elementary charge	e	1·602 189 2(46)	10^{-19} C	10^{-20} emu
		4·803 242(14)		10^{-10} esu
Planck constant	h	6·626 176(36)	10^{-34} J s	10^{-27} erg s
	$\hbar = h/2\pi$	1·054 588 7(57)	10^{-34} J s	10^{-27} erg s
Avogadro constant		6·022 045(31)	10^{23} mol^{-1}	10^{23} mol^{-1}
Electron rest mass	m	9.109 534(47)	10^{-31} kg	10^{-28} g
Proton rest mass	M	1·672 648 5(86)	10^{-27} kg	10^{-24} g
Ratio of proton mass to electron mass	M/m	1836·151 52(70)		
Electron charge to mass ratio	e/m	1·758 804 7(49)	10^{11} C kg^{-1}	10^{7} emu g^{-1}
		5·272 764(15)		10^{17} esu g^{-1}

Rydberg constant, $[\mu_0 c^2/4\pi]^2(me^4/4\pi\hbar^3 c)$	R_∞	1·097 373 177(83)	10^7 m^{-1}	10^5 cm^{-1}
Bohr radius, $[\mu_0 c^2/4\pi]^{-1}(\hbar^2/me^2)=\alpha/4\pi R_\infty$	a_0	5·291 770 6(44)	10^{-11} m	10^{-9} cm
Fine structure constant, $[\mu_0 c^2/4\pi](e^2/\hbar c)$	α	7·297 350 6(60)	10^{-3}	10^{-3}
	α^{-1}	137·036 04(11)		
Speed of light	c	2·997 924 583(12)	10^8 m s^{-1}	10^{10} cm s^{-1}
Classical electron radius, $[\mu_0 c^2/4\pi](e^2/mc^2)=\alpha^3/4\pi R_\infty$	r_0	2·817 938 0(70)	10^{-15} m	10^{-13} cm
Thomson cross-section, $(8/3)\pi r_0^2$	σ_e	0·665 244 8(33)	10^{-28} m^2	10^{-24} cm^2
Free electron g-factor or electron magnetic moment in Bohr magnetons	$g_e/2=\mu_e/\mu_B$	1·001 159 656 7(35)		
Bohr magneton, [c]$(e\hbar/2mc)$	μ_B	9·274 078(36)	10^{-24} J T^{-1}	10^{-21} erg G^{-1}
Electron magnetic moment	μ_e	9·284 832(36)	10^{-24} J T^{-1}	10^{-21} erg G^{-1}
Ratio of electron and proton magnetic moments	μ_e/μ_p	658·210 688 0(66)		
Nuclear magneton, [c]$(e\hbar/2Mc)$	μ_n	5·050 824(20)	10^{-27} J T^{-1}	10^{-24} erg G^{-1}
Boltzmann constant	k	1·380 662(44)	10^{-23} J K^{-1}	10^{-16} erg K^{-1}

Stefan-Boltzmann constant, $\pi^2k^4/60\hbar^3c^2$	σ	5·670 32(71)	10^{-8} W $m^{-2}K^{-4}$	10^{-5}erg $s^{-1}cm^{-2}K^{-4}$
Permeability of free space (by definition)	μ_0	4π= 12·566 370 614 4	10^{-7} H m^{-1}	
Permittivity of free space, $(1/\mu_0c^2)$	ε_0	8·854 187 818(71)	10^{-12} F m^{-1}	
Impedance of free space, (μ_0c)	Z_0	3·767 303 14(15)	10^2 ohm	

† Note that the numbers in parentheses are the one standard-deviation uncertainties in the last digits of the quoted value computed on the basis of internal consistency. In cases where formulae for constants are given (e.g., R_∞), the relations are written as the product of two factors. The second factor, in parentheses, is the expression to be used when all quantities are expressed in cgs units, with the electron charge in electrostatic units. The first factor, in brackets, is to be included only if all quantities are expressed in SI units.

†† In order to avoid separate columns for 'electromagnetic' and 'electrostatic' units, both are given under the single heading 'cgs Units'. When using these units, the elementary charge e in the second column should be understood to be replaced by e_m or e_e, respectively.

<u>Useful relations</u>

$$1 \text{ ev} \cong 1{\cdot}602 \times 10^{-19} \text{ J}$$

$$kT \text{ (at 290K)} \cong 1/40 \text{ eV}.$$

Author Index

Subject Index